Fungal Biology

Fungal biology has an integral role to play in the development of the biotechnology and biomedical sectors. It has become a subject of increasing importance as new fungi and their associated biomolecules are identified. The interaction between fungi and their environment is central to many natural processes that occur in the biosphere. The hosts and habitats of these eukaryotic microorganisms are very diverse; fungi are present in every ecosystem on Earth. The fungal kingdom is equally diverse, consisting of seven different known phyla. Yet detailed knowledge is limited to relatively few species. The relationship between fungi and humans has been characterized by the juxtaposed viewpoints of fungi as infectious agents of much dread and their exploitation as highly versatile systems for a range of economically important biotechnological applications. Understanding the biology of different fungi in diverse ecosystems as well as their interactions with living and non-living is essential to underpin effective and innovative technological developments. This series will provide a detailed compendium of methods and information used to investigate different aspects of mycology, including fungal biology and biochemistry, genetics, phylogenetics, genomics, proteomics, molecular enzymology, and biotechnological applications in a manner that reflects the many recent developments of relevance to researchers and scientists investigating the Kingdom Fungi. Rapid screening techniques based on screening specific regions in the DNA of fungi have been used in species comparison and identification, and are now being extended across fungal phyla. The majorities of fungi are multicellular eukaryotic systems and therefore may be excellent model systems by which to answer fundamental biological questions. A greater understanding of the cell biology of these versatile eukaryotes will underpin efforts to engineer certain fungal species to provide novel cell factories for production of proteins for pharmaceutical applications. Renewed interest in all aspects of the biology and biotechnology of fungi may also enable the development of "one pot" microbial cell factories to meet consumer energy needs in the 21st century. To realize this potential and to truly understand the diversity and biology of these eukaryotes, continued development of scientific tools and techniques is essential. As a professional reference, this series will be very helpful to all people who work with fungi and should be useful both to academic institutions and research teams, as well as to teachers, and graduate and postgraduate students with its information on the continuous developments in fungal biology with the publication of each volume.

More information about this series at http://link.springer.com/series/11224

Amritesh Chandra Shukla
Editor

Applied Mycology

Entrepreneurship with Fungi

Springer

Editor
Amritesh Chandra Shukla
Department of Botany
University of Lucknow
Lucknow, Uttar Pradesh, India

ISSN 2198-7777 ISSN 2198-7785 (electronic)
Fungal Biology
ISBN 978-3-030-90648-1 ISBN 978-3-030-90649-8 (eBook)
https://doi.org/10.1007/978-3-030-90649-8

This Springer imprint is published by the registered company Springer Nature Switzerland AG
The registered company address is: Gewerbestrasse 11, 6330 Cham, Switzerland

This volume is dedicated to my daughters, Baby Aaradhya (08 yrs) and Baby Arunnima (05 yrs), who sacrificed their sport time and helped me during compilation of this work.

Foreword

Fungi is a group of organisms that are an important link in the food chain in all ecosystems with immense potential and myriad of useful bioactive compounds. Fungi feature in a wide range of diverse processes and applications in the modern agriculture, food, and pharmaceutical industry. In the food and drink arena as well as in fermented foods, the role of fungi has been important historically in products such as mushrooms, and yeasts for baking and brewing, respectively. These roles are supplemented by the use of fungal food processing enzymes and additives, and more recently the development of protein-based foodstuffs from fungi. Further, certain mycorrhizal fungi (VAM & AM) may be necessary for seed germination and plant health or may be used as biocontrol agents against weeds and invertebrates; formulations of biofertilizer cum biopesticide can be used as biostimulants and bioprotectants of crops. The successful application of biotechnological processes in agriculture and food using fungi may therefore require the integration of number of scientific disciplines and technologies. These may include subjects as diverse as agriculture, biotechnology, biochemistry, molecular biology, and process engineering. The practical use of newer techniques such as genetic recombination and robotics has revolutionized the modem agricultural biotechnology industry and has created an enormous range of possible applications of fungal products in developing entrepreneurship.

Applied Mycology: Entrepreneurship with Fungi, published by Springer Nature, is a momentous volume, and I strongly deem that it will attract readers working in the field of microbiology, especially in mycology. The current volume has 20 chapters contributed by academicians, scientists, and researchers working in the field of mycology throughout the world. I congratulate the editor for bringing out this volume with amazing commitments from professionals working in the area of fungi and their application for developing entrepreneurship in various sectors including agriculture, food, health, nutraceuticals, pharmaceuticals, and other allied sectors.

President
Microbiologist Society,
Satara, Maharashtra, India

Arvind Deshmukh

Preface

Microorganisms are a part and parcel of our life. They are our friends and can also be our foe. They help in maintaining a balance in our ecosystem. We have tapped their potential for the benefit of society. Hence, the objective of this volume, *Applied Mycology: Entrepreneurship with Fungi*, is to keep the readers informed about the recent developments that took place in the applied mycology and their possible application for developing entrepreneurship.

The volume consists of 20 chapters contributed by authors with vast experience in the field of applied mycology. The main focus of this book is to popularize various methods and protocols related to formulation of different value-added fungal products and their application for developing entrepreneurship. The book chapters cover wide applications of fungi in nutraceutical and baking purposes; fungi in cosmetics; fungi in industrial biotechnology/mycotechnology (such as production of alcohols, organic acids, enzymes, vitamins, polysaccharides, pigments, lipids, glycolipids) and brewing industry; fungi in pharmaceuticals and production of antibiotics; mycometaboiltes from endophytic fungi having potential in pharmaceutical and agricultural industries; fungi in bioremediation/biodegradation; fungal consortium for organic municipal solid waste composting; fungal biopesticides, fungal biofertilizer, plant growth promoter, mycotoxins, and mycorrhizal symbioses; as well as fungi in micro-based bio-refinery for gold nanoparticles production.

Further, advance researches and their applications, as given in the book, will give more ideas to researchers, scientists, policy makers, and professionals working in the area of applied and industrial mycology. Besides, the information will not only be helpful for developing new startup/fungal entrepreneurship and revenue generation for the welfare of society but also be helpful to reduce un-employability at the global level.

Lucknow, India Amritesh Chandra Shukla

Acknowledgments

My sincere thanks are extended to all the academicians, scientists, and researchers, who have contributed chapters and happily agreed to share their work for the *Applied Mycology: Entrepreneurship with Fungi*. I am thankful to the series editors, Dr. Vijai Kumar Gupta and Prof. Maria G. Tuohy, for accepting my proposal and providing their valuable support and cooperation throughout the proposal. I am also thankful to the production team of Springer Nature for all their efforts in publishing the volume on time.

I am highly grateful and thankful to the authorities of the University of Lucknow, especially Professor Alok Kumar Rai, Hon'ble Vice-Chancellor, University of Lucknow, for his endeavor and motivations at all stages of the progress. I am also thankful to my colleagues and research scholars in the Department of Botany, University of Lucknow, for providing their valuable help and support during compilation of this book.

Further, I don't have words to thank my parents, Sri Kailash Chandra Shukla and late Smt. Sandhya Shukla, the makers of my destiny who with all difficulties sent me away from home to get this highly specialized knowledge of 'Science'; without their well wishes, this work would have never been completed.

My acknowledgments are incomplete if I fail to remember the constant encouragement and great sacrifice of my wife Dr. Neetu Shukla, and my daughters (baby Aaradhya Shukla and baby Arunnima Shukla); without her understanding forbearance and help, this work would have been impossible.

I also express my deepest gratitude to my family members, friends, and well-wishers, who rendered their support during this work. Finally, with folded hands and bent knees, I bow before the almighty to thank Him for the continuing graces and blessings He showers on me as I tread through the scientific path of service for humanity.

Lucknow, India — Amritesh Chandra Shukla

Contents

Contributors

Karina Afzal Biocontrol Laboratory, Department of Botany, University of Lucknow, Lucknow, Uttar Pradesh, India

Mamta Bhat Department of Botany, BGSB University, Rajouri, Jammu and Kashmir, India

Moitrayee Banerjee Chakraborty Department of Botany, Cotton University, Guwahati, Assam, India

N. K. Chandrawanshi School of Studies in Biotechnology, Pt. Ravishankar Shukla University, Raipur, Chhattisgarh, India

Subrata Das Department of Life Science and Bioinformatics, Assam University, Silchar, Assam, India

Vasavi Dathar Head, Department of Microbiology, Telangana Social Welfare Residential Degree College for Women, Palamuru University, Mahabubnagar, Telangana State, India

Deepali School of Studies in Biotechnology, Pt. Ravishankar Shukla University, Raipur, Chhattisgarh, India

Ankita Dey Department of Biotechnology, Maulana Abul Kalam Azad University of Technology, Haringhata, West Bengal, India

Sushma Dubey Department of Biotechnology, Kalinga University, Raipur, Chhattisgarh, India

Bandita Dutta Department of Biotechnology, Maulana Abul Kalam Azad University of Technology, Haringhata, West Bengal, India

T. Eaton Gratton Way, Concord, CA, USA

Musarrat Fatima Department of Botany, BGSB University, Rajouri, Jammu and Kashmir, India

K. Geetha Head, Department of Botany, Telangana Social Welfare Residential Degree College for Women, Palamuru University, Mahabubnagar, Telangana State, India

Mayukh Ghosh Multitrophic Interactions and Biocontrol Research Laboratory, Department of Life Sciences, Presidency University, Kolkata, India

Sougata Ghosh Department of Microbiology, School of Science, RK University, Rajkot, Gujarat, India

Saurabh Gupta Department of Biotechnology, Kalinga University, Raipur, Chhattisgarh, India

Mehavesh K. Hameed Department of Chemistry, University of Sharjah, Sharjah, UAE

S. K. Jadhav School of Studies in Biotechnology, Pt. Ravishankar Shukla University, Raipur, Chhattisgarh, India

Akhtari Khatoon Department of Botany, Sri Jayadev College of Education and Technology, Bhubaneswar, Odisha, India

Sibashish Kityania Department of Life Science and Bioinformatics, Assam University, Silchar, Assam, India

Deepali Koreti School of Studies in Biotechnology, Pt. Ravishankar Shukla University, Raipur, Chhattisgarh, India

A. Kosre School of Studies in Biotechnology, Pt. Ravishankar Shukla University, Raipur, Chhattisgarh, India

Ravishankar Krishnamoorthy GM, Technical Services, R&D, K.P. Manish Global Ingredients Pvt. Ltd., Chennai, Tamil Nadu, India

A. Kumar Department of Biotechnology, Sant Gahira Guru Vishwavidyalaya, Sarguja Ambikapur, Chhattisgarh, India

Nitu Kumari Seed Pathology and Mycotoxin Laboratory, University Department of Botany, L. N. Mithila University, Darbhanga, India

Dibyajit Lahiri Department of Biotechnology, University of Engineering & Management, Kolkata, West Bengal, India

M. Mishra E.G.S. Pillay Engineering College, Nagapattinam, Tamil Nadu, India

Renu Mishra Department of Botany and Microbiology, Sri Sathya Sai College for Women, Bhopal, MP, India

Ahmed A. Mohamed Department of Chemistry, University of Sharjah, Sharjah, UAE

Ashirbad Mohapatra Department of Botany, Sri Jayadev College of Education and Technology, Bhubaneswar, Odisha, India

Moupriya Nag Department of Biotechnology, University of Engineering & Management, Kolkata, West Bengal, India

Deepa Nath Department of Botany, Guru Charan College, Silchar, Assam, India

Rajat Nath Department of Life Science and Bioinformatics, Assam University, Silchar, Assam, India

A. A. Ngadin Faculty of Fisheries and Food Science, Universiti Malaysia, Terengganu, Terengganu, Malaysia

Alka Panda Institute of Renewable Energy Technology and Management, Pt Ravishankar Shukla University, Raipur, Chhattisgarh, India

Javad B. M. Parambath Department of Chemistry, University of Sharjah, Sharjah, UAE

Saranga Ranjan Patgiri Department of Botany, Cotton University, Guwahati, Assam, India

Jayanta Kumar Patra Research Institute of Biotechnology and Medical Converged Science, Dongguk University-Seoul, Goyangsi, Republic of Korea

Gajendra Prasad Seed Pathology and Mycotoxin Laboratory, University Department of Botany, L. N. Mithila University, Darbhanga, Bihar, India

Kamal Prasad Absolute Foods, Plant Life Science, Division of Microbiology, Gurugram, Haryana, India

Puja Ray Multitrophic Interactions and Biocontrol Research Laboratory, Department of Life Sciences, Presidency University, Kolkata, India

Rina Rani Ray Department of Microbiology, School of Science, RK University, Rajkot, Gujarat, India

Gokul Shankar Sabesan Department of Microbiology, Faculty of Medicine, Manipal University College, Malacca, Malaysia

Khwaja Salahuddin Department of Botany, M.R.M College, L. N. Mithila University, Darbhanga, India

Mandira Sarma Department of Botany, Cotton University, Guwahati, Assam, India

Reshmi Sasi School of Biotechnology, National Institute of Technology, Calicut, Kerala, India

Kunja Bihari Satapathy Department of Botany, School of Applied Sciences, Centurion University of Technology and Management, Bhubaneswar, Odisha, India

Richa Sharma Department of Botany, Rajiv Gandhi University, Itanagar, Arunachal Pradesh, India

Department of Botany, University of Lucknow, Lucknow, Uttar Pradesh, India

Amritesh Chandra Shukla Biocontrol Laboratory, Department of Botany, University of Lucknow, Lucknow, Uttar Pradesh, India

Manisha Shukla P. G. Department of Plant Science, Avvaiyar Govt. College for Women, Puducherry, Karaikal, India

Jaya Singh Biodiversity Conservation and Rural Biotechnology Centre (BCRBC), Jabalpur, MP, India

D. K. Srivastava Council of Science and Technology, Lucknow, Uttar Pradesh, India

T. V. Suchithra School of Biotechnology, National institute of Technology, Calicut, Kerala, India

E. Taghavi Faculty of Fisheries and Food Science, Universiti Malaysia, Terengganu, Terengganu, Malaysia

Anupam Das Talukdar Department of Life Science and Bioinformatics, Assam University, Silchar, Assam, India

Shraddha Tamrakar Sri Sathya Sai College for Women, Bhopal, MP, India

Sumpam Tangjang Department of Botany, Rajiv Gandhi University, Itanagar, Arunachal Pradesh, India

Hrishikesh Upadhyaya Department of Botany, Cotton University, Guwahati, Assam, India

Sridevi Visvanathan Unit of Biochemistry, Faculty of Medicine, AIMST University, Bedong, Kedah, Malaysia

Serin Zachariah School of Biotechnology, National Institute of Technology, Calicut, Kerala, India

Chapter 1
Fungal Cosmetics: Mushrooms in Beauty Care and the New Age of Natural Cosmetics

Sridevi Visvanathan, Ravishankar Krishnamoorthy, and Gokul Shankar Sabesan

Contents

S. Visvanathan
Unit of Biochemistry, Faculty of Medicine, AIMST University, Bedong, Kedah, Malaysia

R. Krishnamoorthy
GM, Technical Services, R&D, K.P. Manish Global Ingredients Pvt. Ltd., Chennai, India, Tamil Nadu

G. S. Sabesan (✉)
Department of Microbiology, Faculty of Medicine, Manipal University College, Malacca, Malaysia

A. C. Shukla (ed.), *Applied Mycology*, Fungal Biology,
https://doi.org/10.1007/978-3-030-90649-8_1

1.1 Introduction

Fungi are a group of eukaryotic organisms existing in the ecosystem as chemoheterotrophs, predominantly as saprophytes on dead and decaying organic matter. Mushroom is a term usually used to describe the conspicuous 'umbrella-like' shaped fruiting bodies or sporophores of certain fungi. Mushrooms typically are members of the order Agaricales and phylum Basidiomycota (produces basidia or cup-shaped fruiting bodies). Mushrooms are also referred to as 'gill fungi', because they have a characteristic gill-like appearance on the underside of the cap. In general, mushrooms are popular around the globe due to their edible nature and it forms part of the recipe for a variety of delicious food because of the taste and meat-like texture. But it should be remembered that all mushrooms are not edible and there are certain toxic varieties because of the secondary metabolites produced by them. However, the deadly mushroom groups are limited and the toxicity displayed by these species is an inborn mechanism of protection of the basidiocarp from consumption and premature destruction by animals. Nevertheless, the versatility of the edible mushrooms is much and beyond; they are widely exploited for their medicinal values and cosmeceutical properties. Numerous mushrooms are used as functional 'do-good' ingredients in cosmetic formulations. The surging demand for mushroom or fungal cosmetics is on a rise owing to their inherent properties that confer natural beauty benefits in the cosmetics and personal care formulations. The nutrient-rich extracts of various mushrooms can serve as great skin and hair care ingredients in enhancing the quality of beauty care products.

Most of the time, the mushrooms are more easily recognized by their common names than the botanical names. The common names and scientific names of mushrooms are listed in Table 1.1.

The incredible functional benefits of the mushrooms are multifold. Mushrooms are powerhouses of natural bioactive compounds and have been used traditionally

Table 1.1 Popular mushrooms/fungi and their scientific names

S. no.	Common name	Scientific name
1	Snow mushroom or white jelly mushroom or silver ear fungus	*Tremella fuciformis*
2	Reishi mushroom or Lingzhi	*Ganoderma lucidum*
3	Caterpillar fungus	*Ophiocordyceps sinensis*
4	Oyster mushroom	*Pleurotusostreatus*
5	Button mushroom	*Agaricus bisporus*
6	Turkey tail mushroom	*Trametes versicolor/Coriolus versicolor and Polyporus versicolor*
7	Shiitake mushroom	*Lentinula edodes*
8	Chaga mushroom	*Inonotus obliquus*
9	Penny bun or Porcini mushroom	*Boletus edulis*

used for centuries in alternative medicine and beautifying decoctions for skin and hair care benefits (Taofiq et al., 2016).

1.2 Application in Skincare

1.2.1 Anti-microbial Properties

The popular saying goes as 'beauty is skin deep', but the beauty of the skin can be spoiled by a variety of skin infections that are caused by bacteria and fungi. The extracts of mushrooms contain bioactive components that exhibit inhibitory action of several bacteria and fungi that cause skin conditions and therefore find its place in the cosmetic industry.

1.2.1.1 Anti-dermatophytic and Skin Protective

Dermatophytes are a group of keratinophylic fungi which cause superficial mycoses on human beings and animals. The disease caused by dermatophytes is called dermatophytosis (pl dermatophytoses), which affects the superficial keratinized layers of the body, namely, the skin, hair and the nail. Dermatophytes are pathogenic fungi with the ability to colonize the skin and degrade the keratinous layers by virtue of producing proteolytic enzymes such as keratinases and collagenases. Dermatophyte infections are most common among the skin infections and can be easily acquired indirectly by fomite transmission from soil, by sharing of bed linen, clothing and footwears and directly from infected pets or domestic animals. Hot and humid climate, coupled with a habit of tight-fitted clothes like jeans, jeggings, leggings, etc., can further facilitate ward of dermatophyte infection. The exorbitant cost of the conventional antifungal drugs coupled with the rampant development of anti-fungal resistance by the skin fungi justifies the need to look for alternative sources of remedies. The use of mushrooms extracts that have anti-dermatophytic activity in skincare formulations would indirectly help to prevent such infections. Extracts of *Lenzitesquercina, Ganoderma lucidum* and *Rigidoporusulmarius* exhibited a wide range of anti-dermatophytic activity against dermatophytes similar to the commonly used antifungal agents (Ogidi & Oyetayo, 2016). This activity may be attributed to the presence of several compounds such as alkaloids, phenols, steroids, flavonoids, tannins, cardiac glycosides and terpenoids. Further, Wong et al. (2010) had substantiated the anti-fungal activity of certain mushrooms to the activity of glucanase. Glucanase is an enzyme exhibiting anti-fungal and anti-dermatophytic activities. It acts on glycan in the fungal cell wall and thereby weakens the wall by hydrolysing the glycan of the pathogenic fungi.

Further, the fungal extracts of these mushrooms are documented to have a medicinal effect on certain skin conditions such as eczema, rosacea and psoriasis. The extract of chaga mushroom has betulinic acid which has a beneficial role in several skin conditions. Few commercially available medicinal products are also known to contain mushroom extracts with soothing effects on these skin conditions.

1.2.1.2 Anti-acne and Anti-inflammatory Properties

Acne vulgaris is a most common chronic skin disease and a major cosmetic concern for teenagers. The skin condition results because of blockage and/or inflammation of the pilosebaceous units of the skin. The pilosebaceous unit is composed of the hair follicles and their accompanying sebaceous gland. The skin bacterium *Propionibacterium acnes* (currently *Cutibacterium acnes*) plays an important role in acne vulgaris (pimples). Acne can manifest as both inflammatory lesions (pustules and papules) and non-inflammatory lesions (blackheads and whiteheads). The common area affected is mostly the face but can also affect the back and chest. The causative organism produces lipases that break down oily sebum to fatty acids and glycerol which causes inflammation and the sebum starts to accumulate when the gland is blocked, resulting in acne.

Polycephalomycesphaothaiensis, an insect-infecting fungus (entomopathogenic), is known for its anti-acne activity (Sonyot et al., 2020). The anti-bacterial and anti-inflammatory property exhibited by the extracts of this fungus is a promising new avenue for acne treatment. Several mushrooms also exhibit anti-inflammatory and anti-microbial properties which may have a beneficial effect on acne. In addition, mushrooms are also a rich source of vitamin D that, when applied topically, enables and speeds up the healing process.

1.2.2 Emollient Properties

Emollient property is the quality of softening or soothing the skin. Several studies have established that mushrooms such as *Tremella fuciformis* extracts contain certain polysaccharides that are similar to the functional benefits offered by hyaluronic acid. Hyaluronic acid, which is also referred to as hyaluronan, is an important glycosaminoglycan that is widely distributed throughout the epithelial and connective tissues of the body. Since the mushroom extract mimics the function of hyaluronic acid, it serves as an external moisturizer to the body. The application of snow mushroom extracts can provide adequate moisturizing and hydration to our skin and makes it soft and supple, thereby providing a plumping effect making the skin look youthful.

1.2.3 Exfoliation Properties

Exfoliation is a natural process by which the dead skin cells from the outer layer of the skin are removed. It happens at a frequency of around 30 days, but during this nature's process, not all the dead skin cells are shed completely and therefore a part of the dead cells get accumulated on the surface of the skin, making it prone to infections by bacteria. Chemical exfoliators are available, but it can be sometimes harsh on sensitive skin. The extracts of shiitake mushrooms have bioactive components that can work as natural exfoliators.

1.2.4 Skin Lightening and Brightening Properties

Skin colour is of major cosmetic significance and it is perceived that fairness is an indication of beauty for long. In the quest to achieve fairness, the consumers use skin-lightening cosmetics. Most of these cosmetics work by inhibiting melanin, the pigment that is present in the skin and determines the skin colour of man. In humans, melanin pigments are derivatives of the amino acid tyrosine (precursor). The pathway of conversion of tyrosine to melanin involves a key enzyme called tyrosinase, which catalyzes two steps. Most of the skin-lightening agents work by inhibiting the tyrosinase enzyme, thereby preventing the conversion of melanin. Other ingredients like kojic acid are even more special. It can make melanocytes non-dendritic and thereby decreases the melanin content. Shiitake mushroom is a key source of kojic acid and therefore a natural remedy for skin lightening as well as brightening. Shiitake extracts can reduce both melanin production and their transfer; thus, it can limit the melanin content in the newly formed cells and make them appear relatively lighter. It is also helpful to remove blemishes and scars caused by acne.

1.2.5 Anti-aging

The reishi mushroom is a notable skincare component, since it moisturizes and detoxifies the skin while also preventing premature ageing, redness and fine wrinkles. The high content of beta-glucan aids soften and tone skin complexion; furthermore, the free radical–scavenging ability of this mushroom renders age-related problems in skin and protects skin from damages due to pollution and UV rays. Enriched with antioxidants, mushroom extracts can help to reduce oxidative stress caused by free radicals in our body. This free radical–scavenging activity makes it a potent natural do-good ingredient in anti-aging formulations. The benefit would be double-fold if mushrooms were a part of the daily diet besides being applied externally to the skin surface. Mushroom extracts help to ward off signs of aging like fine lines, wrinkles, sunspots, etc. The extract can also keep the skin smooth and firm, toned and young looking.

1.3 Application in Hair Care

1.3.1 Anti-hair Fall

Pleurotusostreatus extract helps to strengthen the root hairs and reduces hair fall. Mushrooms that are rich in iron such as oyster variety when taken orally also benefit by preventing hair fall. Iron is an essential component that boosts circulation of oxygen to the hair follicles; it enriches the hair roots, thereby enabling the hair to grow healthier, longer and faster. Iron deficiency in our body can lead to anaemia. Therefore, iron deficiency indirectly results in loss of vitality in hair, making the hair prone to breakage. Mushrooms are natural iron supplements which help to invigorate the hair roots.

1.3.2 Anti-dandruff

Dandruff is a harmless, chronic condition that usually occurs when the scalp becomes dry or greasy and produces white flakes of dead skin. It is neither contagious nor serious but can be a source of embarrassment with people constantly trying to brush it off their collar or shoulders. Even though dandruff is common, there is no complete cure to this condition till date, although it has been effectively managed. Though dandruff can be related to several predisposing host conditions, it is widely associated with the hyper-proliferation of a commensal lipophilic fungus belonging to the genus, *Malassezia. M. globose, M. furfur* and other *Malassezia* species have different types of lipases, phospholipases and acid sphingomylinases, which help in the breakdown of the sebaceous secretions (oils) in the scalp, thereby providing suitable ecological niche for their rapid multiplication. Therefore, most anti-dandruff (AD) products are designed to exert fungicidal action on *Malassezia* species, the primary target in management of dandruff. However, being a commensal flora, the recurrence of growth of these fungi is inevitable. Hence, the challenges in management should focus on the use of synergistic combination of anti-fungals to minimize resistance development and recurrence while also making the AD product skin friendly.

1.3.3 Hair Moisturizing and Conditioning

Mushroom extracts, besides preventing hair fall, also help to retain hydration and hereby keeps the hair conditioned and shiny.

1.4 Popular Mushrooms with Cosmetic Benefits

1.4.1 Tremella fuciformis

Tremella fuciformis is commonly referred to as a snow mushroom or a white jelly mushroom because of its appearance. The Chinese were the first to discover that this fungus may be used in cosmetics and skincare as well as cuisine. In Korea, Japan and China, the use of mushrooms in skincare has been practised for more than thousands of years. *Tremella fuciformis* has the ability to disseminate and infuse the skin, hydrating it deeply. It also reduces the formation of melanin in the skin, resulting in a brighter complexion. This mushroom is high in vitamin D content that aids in the fading of old acne scars. Oxidative stress is caused by the formation of free radicals in the body, which affects skin structure and promotes premature skin ageing. However, on the other hand, this mushroom has the capacity to combat free radical damage to the skin and may prevent age-related problems. Furthermore, due to its anti-oxidant activity, it can be utilized as an ingredient in eye gel masks to brighten dark under-eye circles and prevent hair loss and is therefore included in shampoos.

1.4.2 Ganoderma lucidum

It is popularly known as Reishi mushroom or Lingzhi, which has been celebrated as a food delicacy for centuries and mostly reserved for royal feasts. In addition to the medicinal value, this mushroom had played a pivotal role in skin and hair care routine. *G. lucidum* has a great potential as a skincare ingredient, especially for 'skin-whitening' products. So far, only a few publications have been published that elucidate the chemical makeup of this fungi and its melanogenesis inhibiting actions. Ganodermanondiol, a bioactive molecule, has been discovered to suppress the activity and expression of cellular tyrosinase, as well as the expression of tyrosinase-related protein-1 (TRP-1), TRP-2, and microphthalmia-associated transcription factor (MITF), lowering melanin formation. Furthermore, ganodermanondiol had an effect on the melanogenesis signalling pathway. As a result of these findings, this mushroom is an excellent choice for use as a skin-lightening agent. (Kim et al., 2016). *G. lucidum* polysaccharide (GLP) is a natural antioxidant with no toxic side effects, which can antagonize UVB-induced fibroblast photo aging (Hu et al., 2019). Based on the recent research, *Ganoderma* extracts that contain *Ganoderma* polysaccharides have been used in promoting skin wound healing, mitigating post-burn infection, and preventing skin flap ischemia-reperfusion injury (Yin et al., 2019).

1.4.3 *Ophiocordyceps sinensis*

Ophiocordyceps sinensis, a genus of ascomycete fungi, parasitize mainly on insects. The presence of polysaccharides, sterols, nucleosides, steroids and several other bioactive compounds in *O. sinensis* makes it useful in the field of alternative medicine for the treatment of several disease conditions. Several research studies have revealed that this mushroom has anti-tyrosinase, anti-elastase and anti-collagenase activity and a photo-protective effect with a sun protection factor up to 25. Therefore, *O. sinensis* can be the best source of cosmetic ingredient for skincare applications. (Cheng et al., 2018)

1.4.4 *Agaricus bisporus*

The button mushroom, *Agaricus bisporus*, has recently acquired attention as a source of active substances such as glucans and proteins, as well as for its antioxidant properties and as a functional ingredient in cream as a replacement for natural antioxidants. Because it is a natural material, this mushroom extract is a suitable substitute for EDTA and BHT in all-natural goods. The levels needed to provide the same effect as EDTA have been assessed, and future use of such extracts in natural cosmetics is encouraged (Tables 1.2, 1.3, and 1.4).

1.5 Formulations

The real challenge in the development of the cosmetic formulation is the right choice of ingredients and the choice of the mushroom/ fungal extract to be used. The choice of the mushroom extract is based on the functional benefits and the planned claims on the product.

Further, in the formulations, the choice of using an oil-soluble or water-soluble extract depends on the type of product. If it is a hair oil formulation, an oil-soluble mushroom extract can be used, whereas for a hair cream, and shampoo formulation, you can use water-soluble extract of mushroom.

In a normal formulation like an anti-dandruff (AD) shampoo or anti-acne cream, one of the major concerns is the possibility of the development of that resistance in commensal flora: For example, (1) resistance in *Malassezia* spp to anti-fungals like zinc pyrithione (ZPTO), octopirox, etc. and (2) resistance in *Cutibacterium acnes* to anti-bacterial ingredients in anti-acne products. This is used in AD products/ anti-acne products due to either cross-resistance or co-resistance mechanisms. Hence, there is a need for the development of eco-friendly and skin-friendly cosmetics and personal care products by lowering the levels of synthetic anti-fungals/anti-bacterials. The use of herbal anti-microbials (fungal/mushroom extracts) in

Table 1.2 Functional/key ingredient in mushrooms with cosmetic applications

Name of the mushroom	Ingredient/active principle	Properties	Applications	Functional benefits	References
Rhizopus species	Lactic acid	Exfoliating agent, Skin lightening and reduce acne eruption	Skin-care cream	Skin moisturizer, impart smoothness and suppleness	Wilkerson and Goldberg (2018)
Candida albicans, Agaricus bisporus, Armillaria tabescens Saccharomyces cerevisiae	Ceramides	Skin-hydrating agents and anti-aging	Skin-care cream, Shampoo, deodorant, restorative facial mask	Prevent moisture loss	Gao et al. (2004) and Anon (2017)
Portabellas and *Criminis*species	L-ergothioneine	Anti-oxidant	Anti-aging creams and lotions	Guard skin from oxidative and DNA damage	Dubost et al. (2006) and Bazela et al. (2014)
*Lentinula edodes, Grifolafondosa, Pholiota nameko*and *Auricularia auricula-judae*	Trehalose	Free radical scavenging	Moisturizing cream, hydrating cleansing gel	Water retention	Hyde et al. (2010)
Inonotus Obliquus	Betulinic acid	Accelerates the regeneration of tissues and skin. Anti-septic helps to heal wounds. Useful in treatment of eczema, psoriasis and dermatitis.	Regenerative creams, anti-septic cream	Prevent skin infections	Megan (2020)
Ganoderma lingzhi	Beta-glucans	Anti-inflammatory, anti-redness, antioxidant, anti-wrinkle benefits.	Anti-aging creams, anti-acne cream and moisturizing cream	Keep the skin youthful, healthy and hydrated Attracts water and helps in synthesis of ceramides, which are the building blocks in the protective layer of the skin	Megan (2020)

(continued)

Table 1.2 (continued)

Name of the mushroom	Ingredient/active principle	Properties	Applications	Functional benefits	References
Lentinula edodes	Kojic acid	Skin brightening	Fairness cream, anti-mark cream	Lighten areas of hyperpigmentation, fade dark spots and scars, improve overall complexion	Megan (2020)
Lentinula edodes	Selenium, vitamin D, and complex B vitamins	Anti-oxidant	Anti-aging cream	Help combat the effects of inflammation	Megan (2020)
Lentinula edodes	L-ergothioneine	Anti-oxidant	Anti-aging cream	Cellular breakdown and encourages a faster cellular renewal process, resulting in healthy collagen and elastin production	Megan (2020)
Tremella fuciformis	Similar to hyaluronic acid	Hydration, Antioxidant	Anti-aging cream	Can retain up to 500 times its weight in water, making it a high-efficacy humectant	Megan (2020)
Trametes Versicolor	Polysaccharopeptides	Anti-tyrosinase and anti-inflammatory	Skin-lightening cream	To help minimize dark spots, as well as can help reduce redness and irritation	Jhan et al. (2016)
Ophiocordyceps sinensis	Ethyl acetate extracts	Anti-elastase activity, Anti-collagenase activity and photo-protective	Anti-aging cream	Can help hydrate dry skin, boost collagen and elastin production, and has even shown some promise as a topical treatment for inflammatory conditions like eczema It helps reduce one of the most important sources of skin aging	Cheng et al. (2018)
Boletus edulis, B. pinicola, B. aereus, B. aestivalis, B. erythropus, and B. appendiculus, can also accumulate considerable amounts of selenium. *Agaricus. bisporus, A. bitorquis, A. campestris, A. cesarea, A. campestris, A. edulis, A. macrosporus, and A. silvaticus.*	Selenium	Anti-dandruff	Anti-dandruff shampoo, cream and oil	Amazing source of selenium, anti-oxidant and anti-dandruff activity	Falandysz (2008)

Table 1.3 Popular skin -care products in the global market with mushroom extract as functional do-good ingredient

Product name	Company name	Name of fungi	Properties	Application	References/ website
SS Smile	Shenzhen Youdi Cosmetics Co., Ltd. and Beijing Tongrentang Skincare Research Institute	*Ophiocordyceps sinensis* (caterpillar mushroom)	Analeptic and beautifying effects, anti-tyrosinase and anti-elastase activity, anti-collagenase and photo-protective effect	Skincare	Cheng et al. (2018) and Ching Li Tor (2020)
Snow mushroom moisturize	The Inkey list	*Tremella fuciformis*	Enhances skin elasticity	Skincare	Devash (2018)
Raw job's tears pore brightening essence	Naruko	*Tremella fuciformis*	Anti-oxidant, moisturizer, humectant	Skincare	https://incidecoder.com/
Virgin skin serum	Code of harmony	*Tremella fuciformis*	For soothe stressed and breakout prone skin, anti-oxidant, moisturizer, humectant	Skincare	https://incidecoder.com/
Virgin skin tightening and brightening serum	Crave skincare	*Tremella fuciformis*	Anti-oxidant, moisturizer, humectant	Skincare	https://incidecoder.com/
Nu-shroom hydrafill serum	Cyberderm	*Tremella fuciformis*	Anti-oxidant, moisturizer, humectant	Skincare	https://incidecoder.com/
Hydrating leave on gel primer	Nuface	*Tremella fuciformis*	Anti-oxidant, moisturizer, humectant	Skincare	https://incidecoder.com/
Hyaluronic concentrate	The chemistry brand	*Tremella fuciformis*	Anti-oxidant, moisturizer, humectant	Skincare	https://incidecoder.com/
Intensive skin boosting toner	Luoki	*Tremella fuciformis*	Exfoliant, anti-oxidant, moisturizer, humectant	Skincare	https://incidecoder.com/
Hydro boost 100% hydrogel mask	Neutrogena	*Tremella fuciformis*	Anti-oxidant, moisturizer, humectant	Skincare	https://incidecoder.com/
Hydro boost the super hydrator hydrogen mask	Neutrogena	*Tremella fuciformis*	Anti-oxidant, moisturizer, humectant	Skincare	https://incidecoder.com/
Hydro boost hydrogel recovery mask hyaluronic acid	Neutrogena	*Tremella fuciformis*	Anti-oxidant, moisturizer, humectant	Skincare	https://incidecoder.com/

(continued)

Table 1.3 (continued)

Product name	Company name	Name of fungi	Properties	Application	References/ website
20% vitamin C serum	Purad'or	*Tremella fuciformis*	Anti-aging, anti-oxidant, moisturizer, humectant	Skincare	https:// incidecoder.com/
Enbioment cleanser	DMK	*Tremella fuciformis*	Skin feels fresh, clean and vibrant	Skincare	https:// incidecoder.com/
Sensitive eye smoother	Bioelements	*Tremella fuciformis*	Shields against irritants and reduce puffiness	Skincare Eye care	https:// incidecoder.com/
Soothing reset mist	Bioelements	*Tremella fuciformis*	Forms a barrier-protective hydrating film to keep skin nourished	Skincare	https:// incidecoder.com/
Ufo H_2O overdose	Foreo	*Tremella fuciformis*	Microfiber facemask instantly drenches dry, thirsty skin in soothing moisture	Skincare	https:// incidecoder.com/
Euphoria face mask	Joanna vargas	*Tremella fuciformis*	Instantly soothes, calms and nourishes the skin	Skincare	https:// incidecoder.com/
Ultra whitening perfect ampoules	Miguhara	*Tremella fuciformis*	Anti-oxidant	Skincare	https:// incidecoder.com/
Volition's Snow mushroom water serum	Sephora	*Tremella fuciformis*	Hydrates and locks in moisture for long-term retention	Skincare	www.sephora. com
Moon juice plump jelly Hydrating serum	Revolve	*Tremella fuciformis*	Instantly quenches the skin, gives it some bounce	Skincare	Devash (2018)
tata harper clarifying moisturizer	Sephora	*Tremella fuciformis*	Moisturizer	Skincare	Devash (2018)
Kypris glow philtra	Credo	*Tremella fuciformis*	Exfoliation and prepare for makeup application	Skincare	Devash (2018)
Mega mushroom relief and Resilience soothing face serum, mask and lotion	Dr. Andrew weil for origins™	*Tremella fuciformis* and *Ganoderma lingzhi*	Soothing	Skincare	www.origins.my

Mega-mushroom skin relief advanced face serum	Dr. Andrew weil for origins™	*Inonotus obliquus*	Revitalizing	Skincare	www.origins.my
Mega-mushroom relief & resilience fortifying emulsion	Dr. Andrew weil for origins™	*Inonotus obliquus*	Revitalizing	Skincare	www.origins.my
Mega-mushroom relief & resilience soothing face mask	Dr. Andrew weil for origins™	*Inonotus obliquus*	Soothing	Skincare	www.origins.my
Mega-mushroom relief & resilience soothing cream	Dr. Andrew weil for origins™	*Inonotus obliquus*	Soothing	Skincare	www.origins.my
Mega-mushroom hydra burst gel lotion	Dr. Andrew weil for origins™	*Inonotus obliquus*	Hydration	Skincare	www.origins.my
Mega-mushroom skin relief face cleanser	Dr. Andrew weil for origins™	*Inonotus Obliquus*	Cleansing	Skincare	www.origins.my
Mega-mushroom skin relief micellar cleanser	Dr. Andrew weil for origins™	*Inonotus obliquus*	Cleansing	Skincare	www.origins.my
Make Prem Chaga concentrate essence	Dr. Andrew weil for origins™	*Inonotus obliquus*	Anti-oxidants, soothing protect the skin from free radical damage	Skincare	Driscoll (2019)
Embellir night cream	Menard	*Ganoderma lingzhi* and *G. sinense*	Eliminate toxins that slow cellular renewal, stimulate cell production and help repair damage	Skincare	www.menard-cosmetic.com
Gempyuri self-aesthetic skin care	DXN	*Ganoderma lucidum*	Multi-massage cleaner Tightening fresh toner Volume energy serum Lifting impact ampoule oil Derma perfection rejuvenation cream DXN Gempyuri mask	Skincare	https://www.dxn2u.com/
Ganozhi E UV defense day cream	DXN	*Ganoderma lucidum*	Protect against UV rays	Photo-protective	https://www.dxn2u.com/

(continued)

Table 1.3 (continued)

Product name	Company name	Name of fungi	Properties	Application	References/ website
Ganozhi Soap	DXN	*Ganoderma lucidum*	Revitalize the skin and delays the aging process	Cleansing	https://www.dxn2u.com/
Ganozhi liquid cleanser	DXN	*Ganoderma lucidum*	Cleanses the skin deep into the pores and leaving it clean and refreshed	Cleansing	https://www.dxn2u.com/
Ganozhimoisturising microemulsion	DXN	*Ganoderma lucidum*	Leaving your skin smooth and radiant all day long	Moisturizer	https://www.dxn2u.com/
Ganozhi toner	DXN	*Ganoderma lucidum*	Minimize skin pores while leaving skin soft and hydrated	Tones skin	https://www.dxn2u.com/
GanozhiHydrasoft toner	DXN	*Ganoderma lucidum*	Cleanses and minimizes pores, penetrates and tones skin from the deepest levels	Tones skin	https://www.dxn2u.com/
Ganozhi nourishing night cream	DXN	*Ganoderma lucidum*	Intensely nourish when the skin naturally renews itself at night	Revitalizes the skin	https://www.dxn2u.com/
Ganozhi™ Lipstick	DXN	*Ganoderma lucidum*	Hydration	Lip care	https://www.dxn2u.com/
AluminEye	Alumier	*Tremella fuciformis*	Improve elasticity and firmness, while diminishing the appearance of fine lines, dark circles and puffiness	Eye cream	Sapphi (2020)
Anti-aging liftaway mud face sheet mask	Starskin	*Tremella fuciformis*	Brightening and rejuvenation	Skincare	Sapphi (2020)
Full of grace-naked facial oil	Lush	*Agaricus bisporus*	Protective and conditioning for skin	Skincare	Sapphi (2020)
Adaptogen deep moisture cream	Youth to the people	*Reishi mushroom*	Hydrate, detoxify and combat premature signs of aging	Skincare	Sapphi (2020)
Time stop collagen ampoule	The plant base	*Hericiumerinaceus*	Anti-oxidants activity, to brighten, soothe and strengthen skin	Skincare	Sapphi (2020)
Evercalm- ultra comforting rescue mask	REN clean skincare	*Albatrellusovinus*	Smooth skin, with even tone and a luminous, healthy complexion	Skincare	Sapphi (2020)

Table 1.4 Popular hair care products in the market with mushroom extract as functional do-good ingredient

Product name	Company name	Name of fungi	Properties	References/website
Ganozhi Plus Shampoo	DXN	Ganoderma	Makes hair silky soft and glossy	https://www.dxn2u.com/
Lingzhi hair strengthening shampoo and conditioner	tianDe	Lingzhi	Powerful anti-oxidant, activating hair growth, protection from UV radiation	https://www.naturalniprodukti.com/
Color me brilliant conditioner	Briogeo	*Trametes versicolor*	Active and natural colour-lock complex that coats colour treated hair, locking in colour and protecting against fade caused by frequent washing	https://briogeohair.com/
Anti-hairloss tonifying shampoo	Anuva cosmetics haircare	Reishi mushroom	Prevents hair loss, acts as anti-oxidant	https://www.amazon.in/Anti-Hair-Loss-Tonifying-Mushroom
Derma solution shampoo and conditioner	Dr. Groot	*Pleurotusostreatus*	strengthens hair elasticity from the roots	https://shopee.com.my/Dr.Groot-Anti-Hair-Loss-Shampoo
Reishi mushroom & Wasabi radiant color protection shampoo	Natural hygiene company limited	Reishi mushroom	Hair care and UV protection	https://www2.thaitrade.com
Ryo Chouibang premium multi-nutri scalp shampoo	Ryo	White tree ear mushroom, Reishi mushroom, Flower mushroom	Scalp care shampoo	Ryo-Chouibang-Premium-Multi-Nutri-Scalp-Shampoo
Organic hair oil-reishi mushroom based	Shazos	Reishi mushroom	Hair oil	https://shazos.com/
Natural nourish snow mushroom deep conditioning mask	AG haircare	*Tremella fuciformis*	Deep conditioning mask	https://www.ulta.com/p/natural-nourish-snow-mushroom-deep-conditioning-mask-pimprod2023923
The key to volume and growth chanterelle mushroom	Madara	Chanterelle mushroom	3 min growth boost scalp treatment	https://www.madaracosmetics.com/en/blog/let-your-hair-grow-like-mushrooms

synergistic combination with synthetic anti-microbial agents with novel delivery methods seems to be a better alternative.

Based on the experimental studies, the authors have shown that the addition of one or more mushroom extracts not only helps to reduce the level of synthetic anti-fungals like zinc pyrithione, climbazole, octopirox in the AD formulations viz shampoo, cream and oil but also make them more efficacious than those popular brands that use stand-alone synthetic anti-fungal ingredients. Similar benefits are obtained in documented skincare products as well.

The complete ready-to-suse data of eight cosmetic formulations are provided for the benefit of entrepreneurs. The data provided includes a complete ingredient list of eight formulations, their lab-scale manufacturing process and test specifications. Each of these formulations is standardized based on intensive research and development. The choice of ingredients is based on established literature and extensive R&D and therefore provides the desired functional benefits. Minor modifications and tweaking to suit the personal likings and preferences on colours, perfumes, etc., can be done and standardized.

Formulation 1

1.1 *Product Name: Amla and Shiitake Hair Oil*

Variant: Coconut Oil

Phase	Ingredients	Qty in %
A	Mineral oil (Light Liquid Paraffin) Cyclopentasiloxane (SF 1202)	63.5911 1.0000
B	Mineral oil (Light Liquid Paraffin) Paraffin (Paraffin Wax) Butylated Hydroxy Toluene (BHT)	10.0000 0.1000 0.0500
C	Mineral oil (Light Liquid Paraffin)	2.0000
D	*Phyllanthus emblica* oil-soluble extract	1.2000
	***Lentinula edodes* extract**	1.0000
E	E1 *Cocos nucifera*(Coconut Oil) Perfume Amla Propylparaben D & C Red No 17, C.I.No 26100 D&C Green No 6, C.I. No 61565 D&C Yellow Np 11, C.I. No 47000	10.0000 1.0000 0.0500 0.0002 0.0017 0.0070
F	Mineral oil (Light Liquid Paraffin)	3.0000
	Total	100.0000

1.2 *Process*

Step 1:

Weigh Mineral oil of Phase A in the main kettle, start stirring. Add SF 1202 into main kettle under mixing.

Step 2:

Weigh mineral Oil of Phase B in the side jacketed kettle, add Paraffin wax and BHT one after another under stirring. Heat to 60 °C under mixing. Stir well to clarity.

Step 3:

Add Phase B into Phase A main kettle under mixing. Rinse side kettle with Phase C mineral oil and add to Phase AB under constant stirring. Stir till phase clear.

Step 4:

Add Amla oil and *Lentinula edodes* soluble extract of Phase D in Phase ABC under constant stirring. Continue stirring for 20 min.

Step 5:

Take Phase E1 coconut oil in side kettle, start stirring.

Take Perfume Phase E2 in separate vessel; add propylparaben and colours one after another under slow mixing. Mix well until Clear. Add E2 into E1 under slow mixing until clear solution.

Rinse this vessel with coconut Oil Phase F and add the rinses to Phase E1 E2. Continue stirring for 20 min.

Step 6:

Add Phase EF into Phase ABCD under continue stirring. Continue stirring for 20 min.

Step 7:

Adjust colour if required.

Step 8:

Sample out for QC Approval.

Step 9:

Discharge bulk through filer after QC Clearance.

1.3 *Analytical Parameters*

S. no.	Test	Specification	Method
01	Appearance	Clear, oily liquid	Visual
02	Colour	yellowish green: To match with master	Visual
03	Perfume	To match with Master	Olfactory
04	Specific gravity at 30 °C	0.84–0.86	Pycnometer
05	Refractive index at 40 °C	1.445–1.455	Refracto meter
06	Acid value max	1.0	Volumetric
07	Peroxide value max	7.5	Volumetric
08	Microbial limit[a]		Micro Manual
	a. Total viable count fu/g	Max.100	
	b. Pathogens	Absent	

[a]Means of the average counts in Nutrient agar and Sabouraud media

Formulation 2

2.1 *Product Name: Antidandruff Hair Oil*

Variant: Hair nourishing and antidandruff

Phase	Ingredients	Qty in %
A	Mineral oil (Light Liquid Paraffin)	91.2361
	Cyclopentasiloxane (SF 1202)	2.0000
	Tocopheryl Acetate (Vitamin E Acetate)	0.0050
B	Perfume amla	0.8000
	Butylated Hydroxy Toluene (BHT)	0.0500
	Piroctone Olamine (Octopirox)	0.0250
	Boletus edulis extract	0.0250
	Propylparaben	0.0500
C	D & C Red No 17, C.I.No 26100	0.0002
	D&C Green No 6, C.I. No 61565	0.0017
	D&C Yellow Np 11, C.I. No 47000	0.0070
D	Sorbitan Oleate (Arlacel 80)	1.6000
E	*Phyllanthus Emblica* Oil Soluble Extract	2.2000
F	Mineral oil (Light Liquid Paraffin)	2.0000
	Total	100.0000

2.2 *Process*

Step 1:

Weigh Mineral oil of Phase A in the main kettle, start stirring. Add SF 1202 and vitamin E Acetate one after another into main kettle under mixing.

Step 2:

Weigh perfume of Phase B in the side vessel with an overhead stirrer. Start stirring, add BHT, Octopirox, *Boletus edulis* extract and Propylparaben. Stir well to clarity.

Then add colours Phase C one after another into Phase B mix under stirring. Continue the stirring under clear solution.

Take Phase D Arlacel 80 into Phase BC under constant stirring. Stir for 30 min.

Then add Amla Extract of Phase E in Phase BCD under constant stirring. Continue stirring for 20 min.

Step 3:

Transfer Phase BCDE mixture under slow stirring in Phase A in the main kettle, continue constant stirring.

Step 4:

Use Phase F for rinsing the vessel of BCDE and add the rinsing to the main kettle of Phase A. Continue stirring Phase APCDEF for 45 min till clarity.

Step 5:

Sample out for QC Approval.

Step 6:

Discharge bulk through filer after QC Clearance.

2.3 *Analytical Parameter*

S. no	Test	Specification	Method
01	Appearance	Clear, Oily Liquid	Visual
02	Colour	Light yellowish green: To match with master	Visual
03	Perfume	To match with Master	Olfactory
04	Specific Gravity at 30 °C	0.824–0.826	Pycnometer
05	Refractive Index at 40 °C	1.445–1.455	Refracto meter
06	Acid Value Max	1.0	Volumetric
07	Peroxide Value Max	7.5	Volumetric
08	Microbial Limit[a]		Micro Manual
	a. Total Viable Count fu/g	Max.100	
	b. Pathogens	Absent	

[a]Means of the average counts in Nutrient agar and Sabouraud media

Formulation 3

3.1 *Product: Hair Cream with Oyster Mushroom*

Variant: Hair Strengthening

Phase	Ingredient	Qty in %
A	A1 Sterile DM water	67.100
	Disodium EDTA	0.050
	Methylparaben	0.230
	Carbomer (Carbopol 940) [To adjust viscosity]	0.500
	A2 Polysorbate	1.000
B	Mineral oil-LLP	13.000
	Stearic acid	2.000
	Emulsifying wax	1.000

(continued)

Phase	Ingredient	Qty in %
	Cetearyl alcohol	3.200
	Dimethicone	0.200
	Propylparaben	0.120
C	Sterile DM water	3.000
	Triethanolamine 97%	1.200
D	Sterile DM water	3.000
	Glycerin	1.000
	Aloe barbadensis Leaf Juice (Aloe Juice-MJAL 41)	0.500
	Pleurotusostreatus **Extract**	**0.500**
E	Perfume Ref.PoloGreen-Paria 174069	0.400
F	Sterile DM water	2.00
	Total	100

3.2 *Process*:

Step 1:

Weigh in sterile DM water from phase A1 in main kettle, add Disodium EDTA and Methylparaben.

Start Stirring. Disperse Carbomer by steady sifting into the vortex. Continue Stirring (with high speed if necessary) for 20 min. Add this to main kettle. Continue stirring. Add Polysorbate of phase A2Heat to 80 °C under constant stirring.

Step 2:

Weigh in all the materials of phase B into a side jacketed vessel. Heat 80 °C under constant gentle stirring. Mix to homogeneity.

Step 3:

Apply vacuum in main kettle. Suck in through muslin cloth of phase B at 90 °C into the main kettle to phase A at 80 °C under full vacuum with slow speed mixing. Mix well at high speed for 5 min and at low speed for 10 min. Cut off steam. Start cooling with cold water.

Step 4:

Suck in TEA solution from [phase C into the kettle to phase AB] at 70 °C. Mix well for 10 min under vacuum. Continue mixing.

Step 5:

Suck in Phase D into the kettle to phase ABC at 55 °C.

Step 6:

At 40 °C, stop mixing and cut off vacuum. Open the kettle, add perfume of phase E into the bulk ABCD. Close the kettle, apply vacuum and mix for 10 min.

Step 7:

Use Sterile DM water of phase F for rinsing of vessels under different steps wherever necessary and suck into the main kettle. Continue mixing for 20 min. Cool to RT.

Step 8:

Stop mixing. Remove vacuum. Sample out the bulk for QC.

Step 9:

Discharge after QC clearance.

Note:

(i) Hot mixing under open vacuum leads to evaporation losses of water. This has to be compensated by extra water of about 1–2% which may be added to the main kettle along with phase A. This is specific to the type of kettle, its capacity, rpm of stirrer during mixing and cycle time.This process requires standardization to ensure that the moisture content of the finished product is not affected.
(ii) Ideally, the mixing has to be carried out under closed vacuum as far as possible to keep the moisture loss to the minimum.
(iii) Cycle time for the batch has to be arrived at on the shop floor over the first 4–5 batches.

3.3 *Analytical parameters*

S. no	Test	Specification	Method
01	Appearance	Opaque, homogenous cream	Visual
02	Colour	White: to match master	Visual
03	Perfume	To match master	Olfactory
04	Viscosity, cps, Brookfield T#6,0.3rpm,at 30C	1.2–3.2 Lakhs	Viscometer
05	pH	6.50–7.50	pH meter
06	Bulk density at 30C,g/cc	0.970 ± 0.010	Density Cup
07	Total fatty matter %B by weight	19.0 ± 1.0	Extraction
08	Loss on drying at 105 C % by weight	77.5 ± 1.0	Oven Method
09	Rancidity	Absent	Extraction & Visual
10	Microbial Test[a] 1. Total count, cfu/gm 2. Pathogens	<100 Absent	Micro manual

[a]Mean of two average numbers on Nutrient agar and Sabouraud Agar

Formulation 4

4.1 *Product: Reishi Mushroom Hair Cream*

Variant: Hair Fall Control

Hair cream–hair fall control		
Phase	Raw materials	%Qty
A	A1 Sterile DM water	36.100
	Disodium	0.050
	Methyl Paraben	0.250
	A2 Sterile DM water	26.000
	Carbomer (Carbopol 940)	0.400
	A3 Sterile DM water	2.000
	Polyvinylpyrrolidone (PVP K30)	0.300
	A4 Polysorbate(Tween 20)	1.500
B	Mineral oil (Light Liquid Paraffin)	12.500
	Glycerylmonostreate-SE	3.000
	Emulsifying wax (Monecol-EW)	1.000
	White bees wax	1.000
	Olive oil	1.000
	Fatty alcohol ethoxylate (Noigen 375)	0.800
	Cetyl alcohol	0.600
	Sorbitan Mono Oleate(Arlacel80)	0.500
	Phenoxyethanol	0.400
	Octymethoxycinnamate	0.200
	Propylparaben	0.150
C	Sterile DM water	2.000
	Triethanolamine 98%	0.550
D	Cyclopentasiloxane(SF1202)	1.000
	Cyclopentasiloxane & Dimethiconol(DC 1502)	1.000
	Amino silicone fluid (DC 2 8566)	0.300
E	Sterile DM water	2.000
	Glycerine	2.000
	Aloe juice	0.500
	***Ganoderma lingzhi* extract**	0.200
F	Perfume	0.600
	Wheat germ oil	0.100
G	Sterile DM water	2.000
	Total	1000.000

4.2 *Process*

Step 1:

Weigh in sterile DM water from phase A1 in main kettle, add Disodium EDTA and Methylparaben.

Start Stirring.

Disperse Carbomer by steady sifting into the vortex. Continue Stirring (with high speed if necessary) for 20 min. Add this to main kettle. Continue stirring.

Prepare Phase A2 and A3 and add one after another into Main Kettle Phase A1 under slow mixing.

Add Polysorbate of phase A4 to main kettle. Heat to 80 °C under constant stirring.

Step 2:

Weigh in all the materials of phase B into a side jacketed vessel. Heat 80 °C under constant gentle stirring. Mix to homogeneity

Step 3:

Apply vacuum in main kettle. Suck in through muslin cloth of phase B at 90 °C into the main kettle to phase A at 80 °C under full vacuum with slow speed mixing. Mix well at high speed for 5 min and at low speed for 10 min. Cut off steam. Start cooling with cold water

Step 4:

Suck in TEA solution from phase C into the kettle to phase AB at 70 °C. Mix well for 10 min under vacuum. Continue mixing.

Step 5:

Prepare Phase D, Suck in Phase D into the kettle to phase ABC at 55 °C. Continue mixing.

Step 6:

Prepare Phase E, Suck in Phase E into the kettle to phase ABCD at 50 °C. Continue mixing.

Step 7:

At 40 °C, stop mixing and cut off vacuum. Open the kettle; add perfume of phase F into the bulk ABCDE. Close the kettle, apply vacuum and mix for 10 min.

Step 8:

Use Sterile DM water of phase F for rinsing of vessels under different steps wherever necessary and suck into the main kettle. Continue mixing for 20 min. Cool to RT.

Note:

(i) Hot mixing under open vacuum leads to evaporation losses of water. This has to be compensated by extra water of about 1–2%, which may be added to the main kettle along with phase A. This is specific to the type of kettle, its capacity, rpm of stirrer during mixing and cycle time. This process requires standardization to ensure that the moisture content of the finished product is not affected.
(ii) Ideally the mixing has to be carried out under closed vacuum as far as possible to keep the moisture loss to the minimum
(iii) Cycle time for the batch has to be arrived at on the shop floor over the first 4–5 batches.

4.3 *Analytical parameters*

S. no	Test	Specification	Method
01	Appearance	Opaque, homogenous cream	Visual
02	colour	White: to match master	Visual
03	Perfume	To match master	Olfactory
04	Viscosity, cps, Brookfield T#6,0.3rpm,at 30C	2–4 Lakhs	Viscometer
05	pH	6.50–7.50	pH meter
06	Bulk density at 30C,g/cc	0.970 ± 0.010	Density Cup
07	Total fatty matter %B by weight	23.0 ± 1.0	Extraction
08	Loss on drying at 105 C % by weight	75.0 ± 1.0	Oven Method
09	Rancidity	Absent	Extraction and Visual
10	Microbial Test[a] 1. Total count, cfu/gm 2. Pathogens	<100 Absent	Micro manual

[a]Mean of two average numbers on Nutrient agar and Sabouraud Agar

Formulation 5

5.1 *Product: Shampoo with Snow Mushroom*

Variant: Herbal Conditioner

Phase	Raw materials	%
A	DM water	35.72
	Polyquaternium 10	0.30
B	SLES 28%	27.00
	DSLS	4.00

(continued)

Phase	Raw materials	%
	Dimethiconal	2.00
	Flocare 7	1.00
	Glycerine	1.00
	Kathon CG	0.05
	Amla extract	0.25
	Henna extract	0.25
	Hibiscus extract	0.25
	Tremella fuciformis extract	0.25
C	DM Water	5.00
	EDTA DS	0.05
	SLES	9.00
	CMEA	1.50
	EGDS	0.05
D	CAPB	3.00
	CDEA	1.50
E	DM Water	2.00
	D panthanol	0.10
F	Perfume	0.50
G	Citric acid 25%	0.20
H	Sodium Chloride 25%	1.00
I	Tartrazine 1%	0.50
	Caramnel 10%	1.40
	B. Blue 0.1%	0.13
J	DM Water	2.00
		100.00

5.2 *Process*

Check the kettle and clean previous residue before starting the batch. Confirm the cleaning of Kettle.

Water should be heated at 90 °C and cooled.

Step 1

Transfer DM Water from Phase A into main kettle and start stirring and disperse Polyquaternium 10 into water under slow stirring. Avoid lump formulation during dispersion of Polyquaternium.

Step 2

Transfer slowly all ingredients from Phase B into main kettle Phase A one after another by slow stirring. Continue stirring (avoid foaming during mixing).

Step 3

Add Phase C one after another inside jacketed oil kettle and heat to 70 °C under slow mixing. Once bulk reaches temperature 70 °C, stop heating. Cool room temperature. Transfer Phase C mixture in main kettle bulk Phase ABC under slow mixing. Continue the mixing for 15 min.

Step 4

Add Phase D materials one after another into main kettle Phase ABC by slow mixing. Mix continuously at slow speed-mixing for 15 min.

Step 5

Prepare Phase E & F and add one phase after another phase into bulk main kettle Phase ABCD r slow stirring. Continue the mixing.

Step 6

Adjust pH of the bulk with Citric acid 25% solution. Add Phase G to phase ABCDEF (adjust pH to 6.0 – 7.0). Continue mixing for 5 min.

Step 7

Add sodium chloride 25% solution of phase H into Phase ABCDEFG in small instalments under constant mixing to adjust the viscosity 3000–4000 cps. Continue mixing for 10 min.

Step 8

Add colour Phase I into bulk Phase ABCDEFGH under slow constant mixing.

Step 9

Use Phase J DM water for rinsing of vessels under different steps wherever necessary and add leftover into the bulk.

Step 10

Sample out for QC approval

Step 11

Adjust pH, adjust viscosity if required

Step 12

Discharge after QC clearance

Note:

(i) Adjusted for SLES active content according to the active content in the supply. The short fall/excess quantity has to be adjusted with DM water of Phase j.
(ii) Thoroughly mix well the Zinc pyrithione 48% solution drum/pack before weighing for proper homogeneity of the material.
(iii) Adjust the shortfall/excess quantity of citric acid solution of phase G with DM water of Phase J.
(iv) Adjust the shortfall/excess quantity of Sodium chloride solution of phase H with DM water of Phase J.

(v) Stirring has to be slow and uniform so that development of foam during process would be very normal.
(vi) Cycle time for the batch has to be arrived at on the shop floor over the 4–5 batches.

5.3 *Analytical Parameters*

S.no	Test	Specification	Method
01	Appearance	Pearly, light red coloured viscose liquid	Visual
02	Colour	Match with master	Visual
03	Perfume	Match with master	Olfactory
04	pH of product	6.0–7.0	pH Meter
05	Viscosity in cps LV#3, 12 rpm	3000–4000	Brookfield Viscometer
06	Active content as SLES (Mol Wt 384), % by weight	12–14	Volumetric
07	Sodium chloride content % by weight max	2.0	Volumetric
08	Microbial test[a]		Micro Manual
	1. Total count cfu/g	<100	
	2. Pathogens	Absent	

Note:

(i) Goods should be transferred to dispatch only after QC and microbial clearance.

Formulation 6

6.1 *Product: Shampoo with Button Mushroom*

Variant: Anti-dandruff

Phase	Raw materials	%
A	DM water	29.92
	ETD 2020	0.20
	TEA	0.30
B	SLES 28%	36.00
	DSLS	4.00
	ZPTO (Zinc Pyrithione 48%)	1.00
	***Agaricus bisporus* extract**	1.00
	Flocare 7	1.00
	Propylene glycol	1.00
	Kathon CG	0.05
C	DM water	5.00
	Zinc chloride	0.05
	SLES 28%	9.00
	CMEA	1.00

(continued)

Phase	Raw materials	%
	EGDS	0.05
D	CAPB	3.00
	CDEA	1.00
E	DM water	2.00
	D panthanol	0.10
F	Perfume	0.80
	Menthol	0.20
G	Citric acid 25%	0.20
H	Sodium chloride 25%	1.00
I	B. Blue 0.1%	0.13
J	DM water	2.00
		100.00

6.2 *Process*

Check the kettle and clean previous residue before starting the batch. Confirm the cleaning of Kettle.

Water should be heated to 90 °C and cooled.

Step 1

Transfer DM Water from Phase A into main kettle and start stirring and disperse carbomer into water under slow stirring. Avoid lump formulation during dispersion of carbomer. Neutralize carbomer using TEA from Phase A under missing.

Step 2

Transfer slowly all ingredients from Phase B into main kettle Phase A one after another by slow stirring. Continue stirring (Avoid foaming during mixing).

Step 3

Add Phase C one after another inside jacketed oil kettle and heat to 70 °C under slow mixing. Once bulk reaches temperature 70 °C, stop heating. Cool room temperature.

Step3

Add Phase D materials one after another into main kettle Phase AB by slow mixing. Mix continuously at slow speed mixing for 15 min.

Step 4

Prepare Phases E and F. Add one phase after another into the bulk main kettle Phase ABCD r slow stirring. Continue the mixing.

Step 5

Adjust pH of the bulk with citric acid 25% solution of Phase G to phase ABCDEF to 6.0–7.0. mix 5 min.

Step 6

Add sodium chloride 25% solution of phase H into Phase ABCDEFG in small instalments under constant mixing to adjust the viscosity 3000–4000 cps. Continue mixing for 10 min.

Step 7

Add colour Phase I into bulk Phase ABCDEFGH under slow constant mixing.

Step 8

Use Phase J DM water for rinsing of vessels under different steps wherever necessary and add leftover into the bulk.

Step 9

Sample out for QC approval

Step 10

Adjust pH, adjust viscosity if required

Step 11

Discharge after QC Clearance

Note:

(i) Adjusted for SLES active content according to the active content in the supply. The short fall/excess quantity has to be adjusted with DM water of Phase j.
(ii) Thoroughly mix well the Zinc pyrithione 48% solution drum/pack before weighing for proper homogeneity of the material.
(iii) Adjust the shortfall/excess quantity of citric acid solution of phase G with DM water of Phase J.
(iv) Adjust the shortfall/excess quantity of Sodium chloride solution of phase H with DM water of Phase J.
(v) Stirring has to be slow and uniform so that development of foam during process would be very normal.
(vi) Cycle time for the batch has to be arrived at on the shop floor over the 4–5 batches.

6.3 *Analytical Parameters*

S. no	Test	Specification	Method
01	Appearance	Pearly, light blue coloured viscous liquid	Visual
02	Colour	Match with master	Visual
03	Perfume	Match with master	Olfactory
04	pH of product	6.0–7.0	pH Meter
05	Viscosity in cps LV#3, 12 rpm	3000–4000	Brookfield Viscometer

(continued)

S. no	Test	Specification	Method
06	Active content as SLES (Mol Wt 384), % by weight	12–14	Volumetric
07	Sodium chloride content % by weight max	2.0	Volumetric
08	Microbial Test[a]		Micro Manual
	1. Total count cfu/g	<100	
	2. Pathogens	Absent	

Note:

(i) Goods should be transferred to dispatch only after QC and microbial clearance.

Formulation 7

7.1 *Product* Face Scrub Cream with Snow Mushroom

Variant Walnut and Snow Mushroom

Phase	Ingredients	%
A	DM water	63.000
	EDTA DS	0.050
	SLS powder	0.200
	Allantoin	0.100
B	Stearic acid	16.000
	LLP	2.000
	IPM	2.000
	EGDS	1.000
	Cetyl alcohol	1.000
	Phenoxy ethanol	0.400
	Silicone DC 200	0.500
C	DM water	3.000
	KOH	0.500
D	Walnut scrub	2.000
E	Glycerine	1.000
	Cucumber powder	1.000
	Tremella fuciformis extract	1.000
F	DM Water	2.000
	Niacinamide	0.500
	Kathon CG	0.050
G	Vitamin E acetate	0.100
	Vitamin A palmitate	0.050
	Perfume	0.500
H	DM water	2.150
		100.000

7.2 *Process*

Check the Kettle and clean previous residue before starting the batch. Confirm the cleaning of Kettle.

Water should heat 90 °C and cool.

Step 1

Transfer DM Water from Phase A into main kettle and start heat at 80 °C. Once water reaches 80 °C, add one after another of remaining phase A material in main kettle under slow stirring.

Step 2

Transfer all ingredients from Phase B into side jacketed oil kettle and heat to 80 °C under slow mixing. Once bulk reaches temperature 80 °C, stop heating.

Step 3

Apply vacuum in main kettle. Suck In, through muslin cloth of Phase B oil Phase at 80 °C into the main kettle to Phase A at 80 °C under full vacuum with under slowing speed mixing. Mix continuously at high-speed mixing for 5 min and at low speed for 10 min at 80 °C. Cut off stream and start cooling.

Step 4

Suck in KOH solution from Phase C into the kettle to phase AB at 70 °C. Mix well for 10 min under vacuum. Continue mixing.

Step 5

Stop stirring and remove the vacuum, start slow stirring and add slowly phase D into Phase AB at 60 °C under mixing. Continue mixing for 10 min and confirm even spreading of scrub in bulk. Continue mixing,

Step 6

Prepare Phase E & F and suck Phase E &F one after another into bulk main kettle Phase ABCD at 55 °C under slow stirring. Continue the mixing under vacuum.

Step 7

Prepare Phase G, and suck Phase G into bulk main kettle Phase ABCDEF at 40 °C under slow mixing. Continue mixing under vacuum 10 min.

Phase 8

Use DM water Phase H for rinsing of vessels under different steps wherever necessary and suck into main kettle. Continue mixing for 20 min. Cool to RT.

Step 9

Stop mixing, remove vacuum. Sample out the bulk for QC approval.

Step 10

Discharge after QC Clearance.

Note:

(i) Hot mixing under vacuum leads to evaporation losses of water. This has to be compensated by extra water of about 1–2% which may be added to the main kettle along with phase A. This is specific to the type of kettle, its capacity, and rpm of stirrer during mixing and cycle time. The requires to standardize vis-à-vis the moisture content of the finished product over the first 4–5 batches.
(ii) Ideally, the mixing has to be carried out under closed vacuum as far as possible to keep the moisture loss to the minimum.
(iii) Cycle time for the batch has to be arrived at on the shop floor over the 4–5 batches.

7.3 *Analytical Parameters*

S. no	Test	Specification	Method
01	Appearance	Opaque	Visual
02	Colour	Light brownish white Match with Master	Visual
03	Perfumes	Match with master sample	olfactory
04	Viscosity cps, Brookfield T#4, 0.3 rpm, at 30 °C	2–4 lakhs	viscometer
05	pH	6.50–7.50	pH Meter
06	Bulk Density at 30 °C	0.960–0.980	Density Cup
07	Total Fatty Mater % by weight	14–18	Extraction method
08	Loss on drying at 105 °C by weight	23–26	oven
09	Rancidity	absent	Olfactory
10	Microbial Test[a]		Micro Manual
	1. Total Count cfu/g	<100	
	2. Pathogens	Absent	

[a]Means of the two average numbers on Nutrient and Sabouraud Agar

Formulation 8

8.1 *Product: Hand and Body Lotion with snow mushroom*

Variant: Moisturizer

Phase	Ingredients	%
A	DM Water	74.0500
	EDTA	0.0500
	Carbomer 940	0.2500
	Allantoin	0.1000
	Methylparaben	0.2500
	Polsysorbate 20	0.5000
B	LLP	10.0000

(continued)

Phase	Ingredients	%
	Stearic Acid	2.0000
	Cetostearyl alcohol	2.0000
	Emulsifying Wax	1.0000
	Isopropyl Myristate	2.0000
	DC 200	0.5000
	Almond Oil	0.2000
	Propyl paraben	0.1500
	phenoxy ethanol	0.4000
C	DM Water	1.0000
	TEA	0.7000
D	Glycerin	1.0000
	***Tremella fuciformis* extract**	1.0000
	Sodium Lactate 60%	0.2000
E	Perfume	0.5000
	Vitamin E Acetate	0.1000
	Vitamin A Palmitate	0.0500
F	DM Water	2.000
	Total	100.000

8.2 *Process*

Check the Kettle cleaning and pervious residue before starting the batch. Confirm the cleaning of Kettle.

Water should heat 90 °C and cool.

Step 1

Transfer DM Water form Phase A into main kettle and start stirring, dissolve EDTA, and disperse slowly carbomer in the water, start to heat at 80 °C. Once bulk reaches 80 °C, add one after another of remaining phase A material in main kettle under slow stirring.

Step 2

Transfer all ingredients from Phase B into side jacketed oil kettle and heat to 80 °C under slow mixing. Once bulk reaches temperature 80 °C, stop heating.

Step 3

Apply vacuum in main kettle. Suck in through the muslin cloth. Add Phase B oil at 80 °C into the main kettle to Phase A at 80 °C under full vacuum and slowing speed mixing. Mix continues high speed mixing for 5 min and at low speed for 10 min at 80 °C. cut off stream and start cooling.

Step 4

Suck in TEA solution from Phase C into the kettle to phase AB at 70 °C. Mix well for 10 min under vacuum. Continue mixing.

Step 5

Prepare Phase D and add into the main kettle Phase ABC at 60 °C under mixing. Continue mixing for 10 min. Continue mixing.

Step 6

Prepare Phase E and suck Phase E into bulk main kettle Phase ABCD at 40 °C under slow stirring. Continue the mixing under vacuum.

Step 7

Use DM water Phase F for rinsing of vessels under different steps wherever necessary and suck into main kettle. Continue mixing for 20 min. Cool to RT.

Step 8

Stop mixing, remove vacuum. Sample out the bulk for QC approval.

Step 9

Discharge after QC Clearance.

Note:

(i) Hot mixing under vacuum leads to evaporation losses of water. This has to be compensated by extra water of about 1–2% which may be added to the main kettle along with phase A. This is to specific to the type of kettle, it is capacity, rpm of stirrer during mixing and cycle time

   (a) The requires to standardize vis-à-vis the moisture content of the finished product over the first 4–5 batches.

(ii) Ideally, the mixing has to be carried out under closed vacuum as far as possible to keep the moisture loss to the minimum.

(iii) Cycle time for the batch has to be arrived at on the shop floor over the 4–5 batches.

8.3 *Analytical Parameters*

S.No	Test	Specification	Method
01	Appearance	Opaque lotion	Visual
02	Colour	Light brownish white Match with Master	Visual
03	Perfumes	Match with master sample	olfactory

(continued)

S.No	Test	Specification	Method
04	Viscosity cps, Brookfield 63 0.3 rpm, at 30 °C	10,000–20,000 cps	viscometer
05	pH	5.50–7.50	pH Meter
06	Bulk density at 30 °C	0.960–0.980	Density Cup
07	Total fatty matter % by weight	2 4	Extraction method
08	Loss on drying at 105 °C by weight	23–26	oven
09	Rancidity	absent	Olfactory
10	Microbial test		Micro Manual
	3. Total Count cfu/g	<100	
	4. Pathogens	Absent	

1.6 Conclusion

Mushrooms as a rich source of natural bioactive compounds are finding growing applications in nutraceutical, cosmetical and nutritional industries. Several mushrooms and their extracts had already found their places as functional do-good ingredients in various skin-care, lip care and hair care products. Besides cosmetics, mushrooms are now being exploited for their role in personal care products. Several global companies have patented the use of mushrooms/fungi in their formulations with functional claims such as anti-aging, anti-oxidant, anti-wrinkle, anti-aging, skin lightening, moisturizing, anti-acne and anti-dandruff effects. The scope for entrepreneurs to exploit the global opportunity to utilize the appropriate mushroom in their commercial personal care products is expanding. With the immense hidden potential and the resource of a variety of bio-active compounds, there are still several species of fungi/ mushrooms which remain untapped in the industry. In future, more fungi and mushrooms would be studied for their efficacy, safety and utility in cosmetic products. A new world of unexplored fungi/ wild mushrooms would also be discovered, identified, taxonomically classified, verified and cultivated in the future which would boost the cosmetic industry. The future research should be strengthened with cutting-edge technology coupled with the growing omics-driven knowledge on mushrooms. The use of mushroom extracts with multi-functions in the skin and hair care industry would be a trend of the future glamour-driven cosmetic industry.

References

Bazela, K., Solyga-Zurek, A., Debowska, R., Rogiewicz, K., Bartnik, E., & Eris, I. (2014). l-ergothioneine protects skin cells against UV-INDUCED damage – A preliminary study. *Cosmetics, 1*, 51–60.

Cheng, W.-Y., Wei, X.-Q., Siu, K.-C., Song, A.-X., & Wu, J.-Y. (2018). Cosmetic and skincare benefits of cultivated mycelia from the chinese caterpillar mushroom, *Ophiocordyceps sinensis* (Ascomycetes). *International Journal of Medicinal Mushrooms, 20*, 623–636.

Devash, M. (2018). *Why snow mushroom is the trendiest new ingredient in skin care.* Available from https://www.allure.com/story/snow-mushroom-skin-care-benefits

Driscoll, K. (2019). *Make Prem Chaga concentrate essence review*. Available from https://iamkellydriscoll.com/2019/09/09/make-prem-chaga-concentrate-essence-review/

Dubost, N. J., Beelman, R. B., Peterson, D., & Royse, D. J. (2006). Identification and quantification of ergothioneine in cultivated mushrooms by liquid chromatography-mass spectroscopy. *International Journal of Medicinal Mushrooms, 8*, 215–222.

Falandysz, J. (2008). Selenium in Edible Mushrooms. Journal of environmental science and health. Part C, *Environmental carcinogenesis & ecotoxicology reviews 26*, 256–99.

Gao, J.-M., Zhang, A.-L., Chen, H., & Liu, J.-K. (2004). Molecular species of ceramides from the ascomycete truffle *Tuber indicum*. *Chemistry and Physics of Lipids, 131*, 205–213.

Hu, S., Huang, J., Pei, S., Ouyang, Y., Ding, Y., Jiang, L., Lu, J., Kang, L., Huang, L., Xiang, H., et al. (2019). *Ganoderma lucidum* polysaccharide inhibits UVB-induced melanogenesis by antagonizing cAMP/PKA and ROS/MAPK signaling pathways. *Journal of Cellular Physiology, 234*, 7330–7340.

Hyde, K. D., Bahkali, A. H., & Moslem, M. A. (2010). Fungi – An unusual source for cosmetics. *Fungal Diversity, 43*, 1–9.

Jhan, M.-H., Yeh, C.-H., Tsai, C.-C., Kao, C.-T., Chang, C.-K., & Hsieh, C.-W. (2016). Enhancing the antioxidant ability of *Trametes versicolor* polysaccharopeptides by an enzymatic hydrolysis process. *Molecules, 21*.

Kim, J.-W., Kim, H.-I., Kim, J.-H., Kwon, O.-C., Son, E.-S., Lee, C.-S., & Park, Y.-J. (2016). Effects of ganodermanondiol, a new melanogenesis inhibitor from the medicinal mushroom *Ganoderma lucidum*. *International Journal of Molecular Sciences, 17*.

Lewis, N. *Mushrooms in personal care*. https://www.cosmeticsandtoiletries.com/formulating/function/repair/11644231.html

Megan McIntyre. (2020). This is your skin on shrooms. https://aremesfermentis.com/blog/this-is-your-skin-on-shrooms/

Ogidi, O. C., & Oyetayo, V. O. (2016). Phytochemical property and assessment of antidermatophytic activity of some selected wild macro fungi against pathogenic dermatophytes. *Mycology, 7*(1), 9–14.

Sapphi Littleton. (2020). Spike your skincare with these 'magic' mushroom products. Available from: https://www.dazeddigital.com/head/article/47739/1/skincare-mushroom-products-alumier-lush-moon-juice-ren

Sonyot, W., Lamlertthon, S., Luangsa-Ard, J. J., et al. (2020). In vitro antibacterial and anti-inflammatory effects of novel insect fungus *Polycephalomyces phaothaiensis* extract and its constituents against *Propionibacterium* acnes. *Antibiotics (Basel), 9*(5), 274.

Taofiq, O., Heleno, S. A., Calhelha, R. C., Alves, M. J., Barros, L., Barreiro, M. F., González-Paramás, A. M., & Ferreira, I. C. F. R. (2016). Development of mushroom-based cosmeceutical formulations with anti-inflammatory, anti-tyrosinase, antioxidant, and antibacterial properties. *Molecules, 21*, 1372.

Tor, C. L., (2020). *The magic of mushrooms: Fungi's medicinal value gain attention as cosmetics go green*. By BeautyTech.jp. Available from https://medium.com/beautytech-jp/the-magic-of-mushrooms-fungis-medicinal-value-gain-attention-as-cosmetics-go-green-42c87447de33

Wilkerson, E. C., Goldberg, D. J., & Cosmet. Dermatol, J. (2018). Poly-L-lactic acid for the Improvement of photodamage and rhytids of the décolletage. *Journal of Cosmetic Dermatology, 17*(4), 606–610.

Wong, J. H., Ng, T. B., Cheung, R. C. F., Ye, X. J., Wang, H. X., Lam, S. K., et al. (2010). Proteins with antifungal properties and other medicinal applications from plants and mushrooms. *Applied Microbiology and Biotechnology, 87*, 1221–1235.

Yin, Z., Yang, B., & Ren, H. (2019). Preventive and therapeutic effect of *Ganoderma* (*Lingzhi*) on skin diseases and care. In Z. Lin & B. Yang (Eds.), *Ganoderma and health: Pharmacology and clinical application* (pp. 311–321). *Springer*.

Zhang, Z. Y. Production of lactic acid from renewable materials by *Rhizopus* fungi. *Science Direct*.

Chapter 2
Fungal Endophytes: A Potential Source of Low-Cost Entrepreneurship

Richa Sharma, Amritesh Chandra Shukla, and Sumpam Tangjang

Contents

R. Sharma
Department of Botany, Rajiv Gandhi University, Itanagar, Arunachal Pradesh, India

Department of Botany, University of Lucknow, Lucknow, Uttar Pradesh, India

A. C. Shukla (✉)
Department of Botany, University of Lucknow, Lucknow, Uttar Pradesh, India

S. Tangjang
Department of Botany, Rajiv Gandhi University, Itanagar, Arunachal Pradesh, India

A. C. Shukla (ed.), *Applied Mycology*, Fungal Biology,
https://doi.org/10.1007/978-3-030-90649-8_2

2.1 Introduction

De Barry first time coined the term "endophyte". (De Barry, 1866) and to detect fungi, which live intracellular as well as intracellular in plant tissues without giving any symptom to the plant (Jeewon et al., 2017). While the means of endophytes is "in the plant" (in Greek word *endon*= within, *phyton* = plant) and endophytes emerged mainly from the rhizosphere along with phyllosphere (Ryan et al., 2008). Further, the definition of fungal endophyte as fungal endophytes are colonizers of the alive internal tissues of their host plant (Rollinger & Langenheim, 2018). Endophytic fungi were reported in almost all the plant species viz. algae, mosses, ferns, mainly in angiosperms, as well as gymnosperms from various areas of the world (Doilom et al., 2017).

Several current studies showed the ubiquity of these microbes, and approximately 1 million species of fungal endophytes have been reported from plants (Dreyfuss & Chapela, 1994) and even from lichens (Li et al., 2007a). This prediction is based on a rational assumption that every individual plant species can be occupied by one or more fungal species (Huang et al., 2007).

2.1.1 Classification of Fungal Endophytes

Fungal endophytes generally belong to Ascomycetes, Basidiomycetes, Oomycetes, or Zygomycetes and anamorphic fungi (Arnold, 2007; Sieber, 2007). In addition, there are also various ways of assemblage of fungal endophytes. Transmission way in specific and between the fungal endophytes of the Clavicipitaceae known as clavicipitaceous endophytes while rest are known as non-clavicipitaceous endophytes (White, 1988). Numerous studies have recognized that different fungal endophytes interactions align in the environmental conditions and depend on the host conditions (Faeth et al., 2006).

2.1.2 Reproduction Way of Fungal Endophytes

Reproductive ways of fungal endophytes are normally used synonymously to move to the host along with the population of host plants. Fungal endophytes may produce spores and promote horizontal transmission. Otherwise, hyphae of fungal endophyte may grow into host seeds and transmit to offspring of infested plants, which is generally termed as vertical transmission. However, it is essential to identify those endophytic fungi that may produce mitotic asexual or meiotic sexual spores. While sexual reproduction needs to the production of sexual spores and is therefore always horizontal. Further, asexual reproduction of endophytic fungi is probably through vertical transmission through the host seeds and horizontal transmission via spores or maybe hyphae (Tadych et al., 2014). Literature reveals that endophytic fungi associated with grasses have concentrated with two associated genera such as *Neotyphodium* and *Epichloë*. Both of them arise systemically and are transmitted vertical way from maternal plants to progeny. However, *Epichloe* fungal endophytes are capable to transmit sexually via spores (Schard et al., 2004). But *Neotyphodium* fungal endophytes are assumed to be harshly transmitted vertically and, hence, considered cached in the host plant (Eaton et al., 2011). On other hand, *Neotyphodium* produces asexual conidia on living plants and on growth culture media (di Menna et al., 2012). Moreover, in woody plants, fungal endophytes have also been observed in seeds and acorns (Petrini et al., 1992), but vertical transmission fungal endophytes in woody plant are probably unusual (Saikkonen et al., 2004). Current facts indicate the horizontal transmission in usually common in grass (Tadych et al., 2014).

2.1.3 Fungal Endophytes of Tissue, Organ, and Host Specificity

Fungal endophytes have been screened from roughly entire aerial tissues; there are very few studies on separately from tissue, organ, and host specificity. Literature reveals that fungal endophytes show remarkable host specificity, comparable to another group, such as smut, rust, and mycorrhizal fungi (Fisher & Petrini, 1990). Organ specificity of fungal endophytes have been reported from wheat plant, Piceaabies white fir Norway and spruce (Sieber, 1985, 1988, 1989). The organ specificity of fungal endophytes occurred due to adaptation to specific microecological and physiological situations (microcosm) found in a particular organ (Petrini et al., 1992). Moreover, previous studies also established that certain level of tissue specificity in fungal endophytes (Wang & Guo, 2007). The host specificity of fungal endophytes from *Fagus sylvatica* and *Pinus sylvestris* and both were growing at the same place, as reported by Petrini and Fisher (Petrini & Fisher, 1988).

2.1.4 Interactions Between Fungal Endophytes and Host Plants

Endophytic fungi can develop a range of forms of interactions with their host plants covering from beneficent mutualism to hidden and dormant pathogens transient through symbiotic microorganisms (Schulz & Boyle, 2005). The asymptomatic colonization of fungal endophytes inside the host-plant tissues might be transient throughout the host plants. The steadiness of the position of communication or the exchange from a type of interaction to a different depends on factors: for example, their nutritional status, developmental stages, genetic character of the two partners, and different environmental factors (Schulz & Boyle, 2006). However, the development phases of the pathogens of their occurrence within the plants are not observable till the symptoms of the disease come out. This is common to separate identified pathogens as fungal endophytes (Bacon & Hinton, 1996). Asymptomatic colonization may correspond to a prolonged incubation period or a recess infection, which is a style of tolerance of the host plant to meticulous micro-organisms. Different species of *Colletotrichum*, *Alternaria alternata*are, *Phomopsis* spp., and *Fusarium* sp., regularly reported in hidden fungal endophytes infections of various host plants (Sinclair & Cerkauskas, 1996). Latent-infecting endophytic fungi may be accountable for the significant deprivation in pre- and postharvest of economically essential crops (Hartman et al., 1986). Fungal endophytes represent thus a continuum of different interactions (Schulz & Boyle, 2005). Like pathogens, fungal endophytes possess necessary tools such as phytotoxic metabolites and exoenzymes to cross plant barriers and to grow within plants (Castro & Fontes, 2005). However, in variation of pathogens, the phenotypic plasticity of fungal endophytes permits them to set up asymptomatically within host-plant tissue, intercellularly or intracellularly with additional options such as hidden pathogens, mutualism, and saprophytism, local or also systemic colonization (Schulz & Boyle, 2005).

2.1.5 Plant Protection by Fungal Endophytes

Endophytic fungi are known to protect the host plant in numerous ways, such as producing toxic alkaloids in grasses to prevent herbivores, increasing the tolerance in hot springs to grow, and providing defense against pests in plants (Zhang et al., 2006). Endophytic fungi give to everything with existing pathogenic microorganisms in the host plant. Increasing evidences suggested that fungal endophytes interconnect with the pathogenic microorganisms in diverse ways in different hosts' plants and, resultantly, changes in nutrient balance in support of endophytic fungi or support the defense mechanism of plants, may suppress the growth of the pathogens may be due to altered physiology conditions (Busby et al., 2016). Many fungal endophytes species generate antifungal, antibacterial, antiviral, and antibiotics novel compounds (Istifadah & Mcgee, 2006) and also provide protection against pathogenic microorganisms with compact harshness (Zabalgogeazcoa, 2008). Colonization of fungal endophytes inside the plants enhance the protection against plant nematodes. Fungal endophytes control the functioning of consequently plant's survival, pathogen system, diversity, and conservation (Busby et al., 2016).

Fungal endophytes also protect different herbivores and pests. Roughly thousands of insect pathogenic fungi belong to the class Basidiomycetes and Chytridiomycetes are identified as fungal endophytes; they are closely related with grass fungal endophytes: for example, Epichloë and Claviceps (Moonjely et al., 2016). Many different mechanisms of fungal endophytes are capable to suggest for this struggle and inducing resistance to diseases. The mechanisms of fungal endophyte encourage decreasing the fitness of plants by increasing their tolerance to abiotic and biotic stress, resistance is also connected with the nutritional status of the host plants (Aguilar & Barea, 1996). Endophytic fungi *Cryptosporiopsis cf. quercina*and *Colletotrichum* sp. are found capable against phytopathogens such as *Gaeumannomycesgraminis, Pyriculariaoryzae, Rhizoctonia cerealis,* and *Phytophthora capsici* (Lu et al., 2000).

Current study suggests that plants infected with fungal endophytes have distinguishable benefits against biotic and abiotic stresses, other nonendophytic counterparts (Bae et al., 2008). Many valuable features have regularly been reported improved competitiveness (Hill et al., 1991), improved tolerance to stressful factors such as microbial infections, heavy metal presence (Reiter et al., 2002), low pH and high salinity (Waller et al., 2005), drought acclimation (Cheplick et al., 2000), enhanced resistance to insect and pests, and herbivores (Akello et al., 2008).

2.2 Fungal Endophytes Producer of Secondary Metabolites

Endophytic fungi are among the generally major groups of eukaryotic microorganisms that are well known for producing numerous new metabolites, which are directly used as drugs or function as lead structures for synthetic modifications (Chin et al., 2006; Mitchell et al., 2008). Endophytes guard their hosts from infectious agents and adverse conditions by secreting bioactive secondary metabolites (Nongalleima et al., 2013). Endophytic fungi benefit the infected plants, for example, it increases the resistance to herbivore grazing through the production of different alkaloids (Owen & Hundley, 2004), improves the growth and competitive capability. They also influence the nutrients uptake and endurance of these plants (Jia et al., 2016), temperature and drought tolerance, plant phenotypic traits, leaf chemistry, propensity for vegetative reproduction and tolerance of heavy metals in soils (Redman et al., 2002). In addition, endophytic fungi have the ability to protect host plants from diseases caused by pathogen microorganisms either through direct effects, for example, by producing antibiotics and lytic enzymes, or indirect effects, for example, by enhancing the plant defense (Ye et al., 2017). Endophytic fungi are able of producing bioactive compounds, which are alike to their host plants, and are able in the conservation of world's deteriorating biodiversity (Mane et al., 2017).

Endophytic fungi producing toxic substances to protect the host plant from pets, insects, and herbivores. Host beneficial endophytes such as *Neotyphodium* and *Epichloë*, which provide the drought tolerance, antiherbivore protection, as well as improved nutrient uptake to the host plants (Schard et al., 2004). Some species of *Stagonospora*, *Piriformosporaindica*, and *Acremoniumstrictum* show parallel functions of defense and growth promotion (Hol et al., 2007). Moreover, endophytic fungi have developed mechanisms to live inside the host plant by shielding themselves beside every chemical and physical weapon of the plants: for example, *Fusarium solani* customized its poisomerase-biding site by means of alterations in amino acids to escape from damaging effects of camptothecin (Hesterkamp, 2017). Endophytic fungi are gaining importance due to their role in plant growth stimulation, defense against biotic and abiotic stresses, plant growth hormones, higher seed yield, and pests via modulation of growth hormone signaling (Vandenkoornhuyse et al., 2015). Wild relatives of *Aegilops sharonensis* and *Triticum dicoccoides* (wheat) have several valuable endophytic fungi of diverse taxonomic groups, which are not present in cultivated modern-day *T. Aestivum* (Ofek-Lalzar et al., 2016).

Studying endophytic fungal metabolites found that a correlation of habitat and entire biological activity of fungal metabolites (Schulz et al., 2002). Several significant medicines are derived from plants such as taxol, camptothecin, quinine, vincristine, and vinblastine (Ramawat et al., 2009), whereas more than eight thousand five hundred bioactive compounds or secondary metabolites are identified by fungal endophyte origin (Goyal et al., 2017). Fungal endophytes may produce different chemicals as overstated by standard example, like gibberellin production by *Fusarium oxysporum* causing agent of foolish seedling disease in rice. Fungal endophytes produce a plethora of bioactive metabolites of distinctive structure, including alkaloids, phenolic acids, benzopyranones, quinones, flavonoids, steroids, xanthones, terpenoids, and tetralones (Tan & Zou, 2001). At the present, it has been found that *Cladosporium cladosporioides* and *Metarhizium anisopliae* fungal endophyte can be produced taxol (El-Maali et al., 2018). Accumulation of a fungal endophyte *Taxomyces andreanae* from *Taxus baccata* to produce taxol biosynthesis fueled discover for fungal endophyte associated with promise bioactive compounds and their derivatives (Nicoletti & Fiorentino, 2015). Fungal endophyte strains have been found to be rich sources of numerous novel and valuable bioactive compounds like antioxidants, pigments, anticancer compound, antiviral, antibacterial, antifungal, anti-inflammatory, antitumor, immuno-suppressants, antimalarial, insecticidal, immunomodulatory, antitubercular activities and antimicrobials like azadirachtin A, B, cytochalasin N, citrinal B, diosgenin, germacrane-type ginkgolide-B, gliotoxin, huperzine A, penicillide derivatives, a-pyrone analogues, piperine, taxolcamptothecin, podophyllotoxin, and biocontrol agents (Nassimi & Taheri, 2017). Many researchers have proven that fungal endophytes from medicinal plants are a possible source of novel natural products, especially valuable resources for pharmaceutically and industrially important compounds that can be used in the treatment of diverse life-threatening diseases, as well as different industrial applications (Kaul et al., 2013; Bengtsson-Palme, 2018). Table 2.1 shows the diversity of fungal endophytes and a large quantity of different secondary metabolite or bioactive compounds or production by various fungal endophytes reported from different host plants universally.

2.3 Extracellular Enzyme Activities from Fungal Endophytes

Generally, fungal endophytes have the ability to utilize diverse organic compounds (carbon), which enables them in degradation of structural components such as proteins, lipids, cellulose, hemicelluloses, lignin, glucose, keratin, pectin, and

Table 2.1 Production of secondary metabolites by fungal endophytes

Fungal endophytes	Secondary metabolites	References
Phoma sp.	(3*S*,4*S*)-3,8-dihydroxy-6-methoxy-3,4,5-trimethylisochroman-1-one	Kim et al. (2019)
Aspergillus niger	Trypacidin A, Methylsulochrin.	Isabelle et al. (2019)
Chaetomium, Aspergillus, Alternaria, Penicillium and*Charobacter*	favonoids, terpenoids	Yuan et al. (2019)
Alternaria sp. *Fusarium* sp. and mycelia sterilia	bioactive secondary metabolites	Prabha et al. (2018)
Aspergillusflocculus	Ergosterol peroxide, ergosterol and campesterol	Tawfike et al. (2019)
Trichoderma atroviride	Atrichodermone A, B, C	Zhou et al. (2017)
Fusarium oxysporum	4-hydroxybenzoic acid	Bogner et al. (2017)
Alternaria sp. Samif01	Altenuisol	Tian et al. (2017)
Chaetomium globosum	Chaetoglobosin A	Dissanayake et al. (2016)
Pestalotiopsisneglecta	Phenols, flavonoids	Sharma et al. (2016)
Hypocrea virens	Gliotoxin	Ratnaweera et al. (2015)
Diaporthephaseolorum,	Baccatin III	Li et al. (2015)
*Trichoderma stromaticum*and *Rhizoctonia solani*	Ethanolic extract and Solanioic acid	Ratnaweera et al. (2015)
Xylaria sp.	Helvolic acid	Ratnaweera et al. (2014)
Stemphylium sp.	Infectopyrones A and B	Zhou et al. (2014)
Taxomycesandreanae	Taxol (paclitaxel)	Kusari et al. (2014)
Eupenicillium sp.	Eupenicinicols	Li et al. (2014)
Aspergillus versicolor	8-O-methylversicolorin	Dou et al. (2014)
Myxotrichum sp.	Myxotrichin A, Myxotrichin D	Yuan et al. (2013)
Chaetomium elatum	Xanthoquinodin	Chen et al. (2013)
Fusarium oxysporum	Vincristine	Kumar and Kaushik (2013)
Ulocladium sp.	Ophiobolin P	Wang et al. (2013)
Phomopsis longicolla	Dicerandrol C (24)	Erbert et al. (2012)
Penicillium vinaceum	Quinazoline alkaloid	Erbert et al. (2012)
Phaeosphaeria sp.	Phaeosphaerin A	Li et al. (2012)
Acremonium persicinium	Cordyheptapeptides C–E	Chen et al. (2012)
Eupenicillium parvum	Azadirachtin, Azadirachtin A	Kusari and Spiteller (2012))
Preussiaafricana	Preussochrome C	Zhang et al. (2012)
Phoma sp.	b-sitosterol	Wang et al. (2012)

(continued)

Table 2.1 (continued)

Fungal endophytes	Secondary metabolites	References
Muscodorsutura	Isocaryophyllene	Kudalkar et al. (2012)
Chaetomium globosum	Gliotoxin	Li et al. (2011)
Phomopsis sp.	Cytochalasin N	Fu et al. (2011)
Colletotrichum sp.	Methanol	Arivudainambi et al. (2011)
Coniochaeta sp.	Coniothiepinol A	Wang et al. (2012)
Coniochaeta sp.	Conioxepinol B	Wang et al. (2010)
Alternaria sp.	Altenusin	Kjer et al. (2009)
Nigrospora sp. YB-141	Solanapyrone C	Wu et al. (2009)
Phomapinodella	Phomodione (43)	Hoffman et al. (2008)
Penicillium sp.	Penicillenols A1 and B1	Lin et al. (2008)
Ampelomyces sp.	Altersolanol A	Alves et al. (2008)
Pestalotiopsis sp.	Ambuic acid derivative	Ding et al. (2008)
Geotrichum sp. AL4	1,3-oxazinane	Li et al. (2007a)
Cytonaema sp.	Cytonic acid B	Li, Zhou, et al. (2007b)
Aspergillus niger	Naptha-y-pyrone	Zhang and Qi-Yong (2007)
Alternaria sp.	Podophyllotoxin	Eyberger et al. (2006)
Phialocephalafortinii	Podophyllotoxin	Eyberger et al. (2006)
Penicillium janthinellum	Polyketide citrinin	Marinho et al. (2005)
Entrophospora	Campothecin	Puri et al. (2005)
Dothiorella sp.	Cytosporone B	Xu et al. (2005)
Phomopsis sp.	Phomol	Weber et al. (2004)
Fusarium subglutinans	Periconicins	Kim et al. (2004)
Pestalotiopsismicrospora	Pestacin and isopestacin	Harper et al. (2003)
P. microspora	Isopestacin and pestacin	Strobel et al. (2002)
Xylaria sp.	Cytochalasins	Wagenaar et al. (2000)
Cytonaema sp.	Cytonic acid A	Guo et al. (2000)

oligosaccharides (Kudanga & Mwenje, 2005). Microorganisms signify a viable substitute source of enzymes, as they may be cultured in large quantities within short time frames by fermentation, they are biochemically diverse, and are agreeable to genetic manipulation (Anbu et al., 2017). Literature reveals that fungal endophytes produce numerous extracellular enzymes, such as amylases, cellulases, protease, lipases, and laccases. The association of such fungal endophytes with the host plant initiates the release of extracellular hydrolysis (Leo et al., 2016). They hydrolysis the food substances and protect against pathogens (Desire et al., 2014). Moreover, endophytic fungal enzymes are used in the beverage, food, confectionary, pulp and paper, textile and leather industries to shorten the processing of unprocessed materials and pharmaceutical uses (Raju et al., 2015). Further, endophytic fungi were isolated from several medicinal plants, traditionally used in the local communities of the Western Ghats, India, the analysis exposed cellulase, amylase, and pectinase enzymes (D'Souza and Hiremath 2015). Similarly, the qualitative and quantitative analyses of the fungal endophyte have also been reported from some medicinal plants, for the production of extracellular enzymes, including amylase, cellulase, lipase, and protease (Patil et al. 2015a). Fungal endophytes are recognized for accumulating diverse extracellular enzymes such as cellulases, amylase, esterase, protease, lipases, pectinases, laccase, and xylanases which play a major role in defending themselves from the protection reply of the host plants or in up taking the food from the soil (Bezerra et al., 2012; Suto et al., 2002). The several industries that devour microbial enzymes together with biomaterials, cellulose, paper, detergents, leather, cosmetics, food, pharmaceuticals, agriculture, and textiles industries (Yadav, 2015). Table 2.2 shows the production of extracellular enzymes by different endophytic fungi reported from various host plants worldwide.

Several fungal endophytes have also been investigated in the production of enzymes. *Penicillium*, *Aspergillus*, *Trichoderma*, and *Humicola* fungal endophytes were identified to produce cellulase enzyme (Sukumaran et al., 2005). *Sclerocystis*, *Nigrospora, Microsphaeropsis*, *Cephalosporium*, and *Phomopsis* fungal endophytes were recognized to produce cellulase enzyme (Peng & Chen, 2007). *Pestalotiopsis aurantiogriseum, P. guepini, P. glandicola, Fusarium lateritium, Xylaria* sp., *Acremonium terricola, A. japonicas, N. sphaerica,* and *Cladosporium cladosporioides* fungal endophytes were identified to produce pectinases, cellulases, proteases, and xylanase enzyme (Bezerra et al., 2012). *Fusarium oxysporum* fungal endophytes have been identified to produce cellulases enzyme (Onofre et al., 2013). *Chaetomium* sp., *Bisporus* sp., *Cladosporium* sp., *Colletotrichum* sp., *Fusarium* sp., *Rhizoctonia* sp., *Aspergillus* sp., and *Curvularia* sp., fungal endophytes, have been identified to produce proteases, cellulases, amylases, lipases, and activity (Patil et al., 2015a). *Pringsheimia smilacis, Ulocladium* sp., *Hormonema* sp., *Neofusicoccum austral*, and *Neofusicoccum luteum* fungal endophytes hasve been identified to produce cellulases, Glucosidase, and phosphatises enzyme (Khan et al., 2016). *Neofusicoccum austral, Ulocladium* sp., *Neofusicoccum luteum, Pringsheimia smilacis*, and *Hormonema* sp. fungal endophytes have been identified to produce Laccase enzyme (Fillat et al., 2016). *Colletotrichum* sp., *Nigrospora sphaerica, Penicillium funiuclas, Fusarium solani, Macrophomina phaseolina* and *Trichoderma viride, Pochonia chlamydosporia*, and *Aspergillus* sp. fungal endophytes have been identified to produce Amylase, cellulose, and lipase enzyme (Wagenaar et al., 2000; Escudero et al., 2016). *Alternaria tenuissima, P. rubens, Curvularia kusanoi, Humicola*sp, *P. pinophilum, Phoma* sp., and *Fusarium* sp. fungal endophytes have been identified to produce amylase, cellulase, cellulase

Table 2.2 Extracellular enzyme production by fungal endophytes

Name of fungal endophytes	Enzymes produced	References
Fusarium sp., *Humicola*sp, *Phoma* sp.	Cellulase	Liu et al. (2019)
Alternaria tenuissima, Penicillium pinophilum and *P. rubens*	Cellulase tyrosinase and amylase	Yasser et al. (2019)
Curvularia sp.	Lipase	Liu et al. (2019)
Thielavia arenaria, Aureobasidium pullulans, Phoma sp., *Preussia* sp., *Sordariomycetes* sp., *Fusarium proliferatum, Preussia* sp., *Chaetomium* sp., *Penicillium citrinum, Phomamedicaginis, Aureobasidium* sp., *A. pullulans, Dothideomycetes* sp., *Penicillium citrinum,*	Glucosidase, phosphatases and cellulases	Khan et al. (2016)
Neofusicoccum luteum, Ulocladium sp., *Pringsheimiasmilacis, Neofusicoccum austral, Hormonema* sp.,	Laccase	Fillat et al. (2016)
Nigrosporasphaerica, Macrophominaphaseolina, Colletotrichum sp., *and Fusarium solani*	Cellulase, amylase, and protease	Ayob and Simarani (2016)
*Trichoderma viride*and *Penicillium funiuclas*	Amylase, lipase, cellulose, protease,	Chathurdevi and Gowrie (2016)
Pochoniachlamydosporia	Protease	Escudero et al. (2016)
Aspergillus sp.	Amylase	Jurynelliz et al. (2016)
Aureobasiduaceae, Eurotiales, Chaelomiaceae, SporomiaceaeIncertaesadis, Nectriaceae,	Glucosidases, celluloses, and phosphatases	Khan et al. (2016)
Aspergillus terreus	l-asparaginase	Kalyanasundaram et al. (2015)
Fusarium, Phoma, Colletrotrichum, Penicillium	l-Asparaginase	Chow and Ting (2015)
Trichoderma piluliferum, Nodulisporium, Cochliobolus lunatus, C. lunatus, C. australiensis, Gibberellabaccata, G. fujikuroi, Myrmecridiumschulzeri, Phomaputaminum, Penicillium commune, Acremoniumcurvulum, Aspergillus niger, A. Ochraceus, A. ochraceus, A. chartarum, Myrothecium verrucaria, Pithomycesatro-olivaceus and *P. glabrum*	Xylanase, lipase, cellulase, and protease	Bezerra et al. (2015)
Alternaria alternate, Penicillium chrysogenum, Sterile hyphae	Gelatinase, cellulase, amylase, pectinase, xylanase, and tyrosinase	Fouda et al. (2015)

(continued)

Table 2.2 (continued)

Name of fungal endophytes	Enzymes produced	References
Colletotrichum sp., *Fuzarium* sp., *Rhizoctonia* sp., *Cladosporium* sp., *Chaetomium* sp., *Curvularia* sp., *Aspergillus* sp., *Cladosporium* sp., *Biosporus* sp.,	Cellulase, lipase, amylase, and protease	Patil et al. (2015b)
Colletotrichum gloeosporioides	Protease, chitinase, and amylase	Rabha et al. (2014)
Aspegillusniger, A. Fumigatus, A. flavus, Chaetomium sp., *Nigrosporasphaerica, Fusarium* sp., *F. Chlamydosporum, Phoma* sp., *Discosia* sp., *Colletotrichum* sp., *C. falcatum, , Pestalotiopsis disseminate, Talaromycesemersonii, Alternaria* sp., *Basidiomycetes* sp., *Cylindrocephalum* sp., *Mycelia streilia* sp., *Mycelia sterilia* sp., *Myrothecium* sp., *Coniothyrium* sp., *Fusicoccum* sp., *Xylaria* sp., *Pyllosticta* sp., *Aspergillus* sp., *Paecilomycesvariotii, Fusarium oxysporum, F.chlamydosporum, F. solani,,* Myceliasterilia sp., *Penicillium* sp., *P. longicolla, Myrothecium* sp. , *Acremonium implicatum,*	Amylase, cellulase, laccase lipase protease, and pectinase	Sunitha et al. (2013)
Nigrosporasphaerica, Drechsleraawaiiensis, Lasiodiplodiatheobromae, ,Cladosporiumcladosporioides, Curvulariabrachyspira, C. verruciformis, Colletotrichum carssipes, C. gloeosporioides, C. falcatum, Phyllosticta sp.,	Amylase, protease cellulase, laccase, and lipase	Amirita et al. (2012)
Discosia sp.	Amylase	Hegde et al. (2011)
Preussia minima and *Alternaria* sp.	Amylase	Zhang et al. (2010)
Monotospora sp.	Laccase	Weihua and Hongzhang (2008)
Melanconiumapiocarpum	Laccase, amylase, and cellulase	Guo et al. (2008)
Alternaria sp, *Pestalotiopsis* sp., *Acremonium* sp. And *Fusarium* sp.	Amylase, cellulase, protease, and lipase	Maria et al. (2005)
Phomopsis sp. and *Colletotrichum* sp.	Cellulase, Protease, xylanase, and mannanase	Moy et al. (2002)

tyrosinase, and laccase enzymes (Vázquez et al., 2018; Yasser et al., 2019). *Acremonium implicatum*, *Phyllosticta* sp., *Cylindrocephalum* sp., *C. falcatum*, *C. gleosporoides*, *C. truncatum*, *C. brachyspora*, *C. vermiformis*, *Alternaria* sp., *A.* calidoustus, *A. niger*, A. fumigates, *Cercospora kikuchii, Chaetomium* sp., *Fuzarium* sp., *Drechslera hawaiiensis.*, *Drechslera* sp., *Mycelia streilia* sp., *F. oxysporum*, *Isaria* sp., *Pestalotiopsis, Penicillium* sp., *P. marneffei*, sp., *Phoma* sp., *P. viridicatum*, *Phomopsis longicolla*, *Catharanthus roseus*to, *Xylaria* sp.,

Calophyllumino phyllum, Rhizoctonia sp., *Chaetomium* sp., *Trichophyton tonsurans, Cladosporium* sp., *Biosporus* sp., *Curvularia* sp., and *Microsporum gypseum* fungal endophytes have been identified to produce lipase enzyme (Sunitha et al., 2013; Fareed et al., 2017). *Fusarium* sp., *F. Proliferatum, Xylaria* sp., *Phomopsisliquidambari, Chaetomium* sp., *C. globosum, C. gloeosporioides, Podosporaanserine, Neofusicoccumaustrale, N. luteum, Botryosphaeria* sp., *T. harzianum, Botryosphaeria rhodina, B.obtuse, B. ribis, B. dothidea, Monotospora* sp., and *Hormonema* sp. Fungal endophytes have been identified to produce lipase enzyme (Sara et al., 2016; El-Ghonemy et al., 2017).

2.4 How to Isolate Fungal Endophytes

The samples thus collected were washed gently in running tap water to remove the soil and debris. Further, isolation of fungal endophytes was investigated, using the method of Suryanarayanan (Suryanarayan et al., 1998). A total of 1350 plant bits of leaves, stems, and roots cut into small pieces of 0.5 cm^2; stems and roots samples were cut into 0.5–1.0 cm. The pieces were then surface-sterilized by dipping them serially in 70% ethanol for 5 seconds and in 4% NaOCl for 90 seconds respectively and finally rinsed in sterile distilled water for 10 seconds. Now, one hundred and fifty segments of the samples were randomly selected and plated on Potato Dextrose Agar (PDA) medium (supplemented with 150 mg/L chloramphenicol) contained in Petri dishes (7.5 cm diam). Petri dishes were incubated (12 h dark: 12 h light cycle) for 25 days at 28 °C, to observe the growth of endophytes (Suryanarayanan, 1992). The fungal hyphal tips, which come out from the sample segments, were isolated and then subcultured on PDA plates without using antibiotics. The pure cultures were maintained on PDA slants and identified morphologically as well as used as stock culture for further experimental studies (Fig. 2.1).

2.4.1 Morphological Observations of the Isolated Fungal Endophyte

The various sporulating structures of fungal endophytes, such as conidia, condiogenous cells, pycnidia (Coelomocytes), ascospores, asci, and ascocarp (Ascomycetes), conidia, and conidiophores (Hyphomycetes), were prepared semipermanent slides of various stages of fungal spores and studied beneath the microscope. The photographs of the mounted specimens were recorded properly. The fungal endophytes thus characterized were positioned in suitable genera, species, and strains of fungi, with standard monographs and taxonomic keys including the morphology of the fungal colony or hyphae, characteristics of the spores, and reproductive structures (Bhat, 2010).

Sample collection from of the healthy plant

↓

Cleaning and washing with running tap water

↓

Samples categorization in to leaves stems & roots, and chopping into small pieces 0.5 cm^2, 0.5-1.0 cm respectively

↓

1 st Surface sterilization of chopped samples with 70% ethnol for 30 seconds

↓

2 nd Surface sterilization of chopped samples with 4% NaOCl for 90 seconds

↓

Washing with sterilized distil water, 3 times

↓

Placement of the samples in petriplate containing potato dextrose agar media

↓

Incubate at room temperature (25 ± 2°C), for 5-7 days

↓

Isolation of the emerging fungi and sub-culturing to obtain a pure culture of endophytic fungi

Fig. 2.1 Flow chart procedural steps adopted for isolation of fungal endophytes

2.5 Extracellular Enzyme Analyses from Endophytic Fungi

The production of enzymes by the fungal endophyte was qualitatively screened by growing the fungal mycelium on the solid media having dissolved substrates. All the enzyme activities were carried out at pH 6.0 with slight modifications (Fig. 2.2).

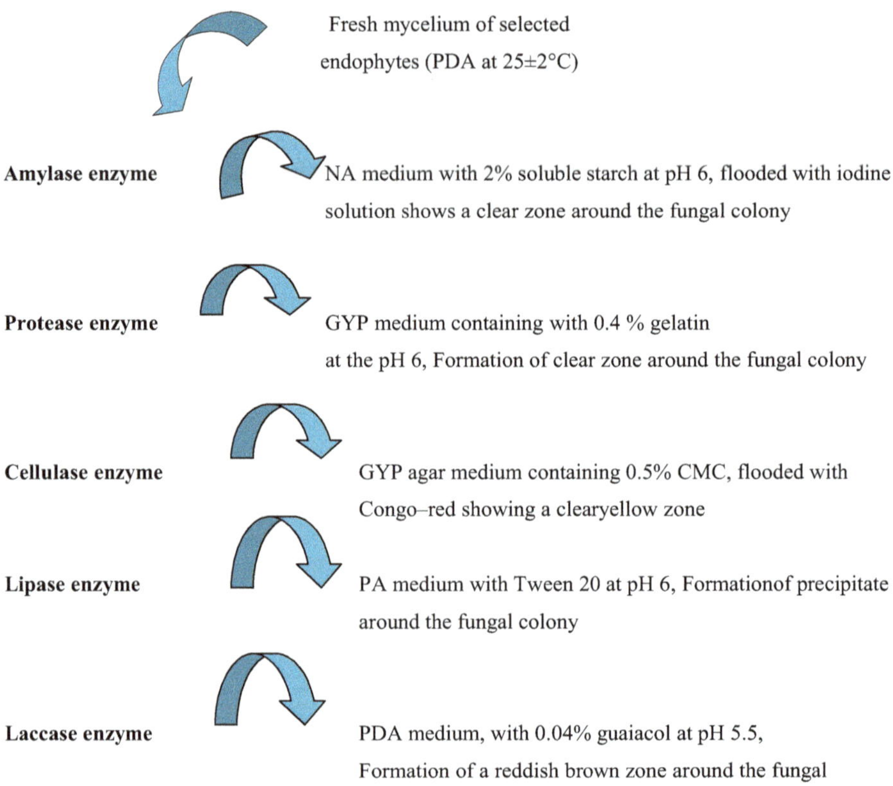

Fig. 2.2 Ray diagram of various extracellular enzyme screening from fungal endophytes

2.5.1 *Amylolytic Activity*

Amylolytic activity was assessed by growing the isolated fungal endophyte on nutrient agar media with 2% soluble starch at pH 6. After the incubation period of 5–6 days, the plates were applied with 1% iodine along with 2% potassium iodide. The appearance of clear zone surrounding the colony was considered positive for the amylolytic activity (Senthilmurugan et al., 2013).

2.5.2 *Proteolytic Activity*

Proteolytic activity was screened by growing the fungal isolates on Glucose Yeast Extract Peptone Agar medium (GYP) containing 0.4 % gelatine (pH 6). After 5–6 days of incubation, plates were flooded with saturated aqueous ammonium sulfate, which resulted in a precipitate. This makes the agar hazy and clear zone over all fungal colony indicating proteolytic activity (Prabavathy & Nachiyar, 2012; Sunitha et al., 2013).

2.5.3 *Cellulolytic Activity*

The cellulolytic activity of the isolated fungal endophyte was determined by growing it on Glucose yeast extract peptone agar (GYP agar) medium having 0.5% carboxy-methylcellulose (CMC). After incubation (5–6 days) periods, plates were filled with 0.2% aqueous congo red solution and doomed with 1M NaCl for 30 minutes. Which formed yellow areas over all fungal colony indicating cellulose activity (Sunitha et al., 2013; Prabavathy & Nachiyar, 2013).

2.5.4 *Lipase Activity*

Lipase activity of the isolated fungal endophyte was determined by growing it on Peptone Agar Medium along with sterilized Tween 20 (2 ml). After the incubation period, a clear zone around overall the fungal colony specifies lipase activity (Senthilmurugan et al., 2013; Prabavathy & Nachiyar, 2013).

2.5.5 *Laccase Activity*

In case of laccase activity, fungal endophyte was cultured in sterilized potato dextrose agar (PDA) medium, supplemented with 0.04% guaiacol (Hi-Media), and adjusted the pH 5.5. These Petri plates were incubated at 28–30 °C for 72 h and then screened for the formation of reddish-brown zones around the fungal colonies indicating the laccase activity (Kalra et al., 2013).

2.6 Fungal Endophytes as a New Source of Entrepreneur

An endophytic fungus is the main source of extracellular enzymes. The association of such fungal endophytes with the host plant initiates the release of extracellular hydrolysis (Leo et al., 2016). Moreover, enzymes are biocatalysts that are involved in catalysis reactions without needing acute conditions, such as high pressures or corrosive environments, temperatures, all of which are often required in chemical processes. Hence, enzymes are used to catalyze reactions in production processes of numerous sectors including environmental bioremediation, industrial bioconversion (biocatalyst), agricultural sectors, pharmaceuticals, and also biotransformations of several compounds such as flavonoids, alkaloids, and steroids (Wohlgemuth, 2010; Choi et al., 2015). Moreover, there are numerous sources of enzymes including plants, animals, bacteria, fungi, and protists. Microbial enzymes have usually been used because of their easier isolation in high amounts, stability at various extreme conditions, their cocompounds, which are also more controllable low-cost production and less injurious. Microbial enzymes concealed into the media are highly dependable for industrial processes and applications. Microbes isolated from diverse sources, still among species and strains of the same genus, may produce altering levels of enzymes of various properties.

Endophytic fungal enzymes have paying attention for numerous applications, because endophytic fungi secrete large amounts of enzymes into the culture medium and also grow on low-cost materials (Anitha & Palanivelu, 2013). Several fungal enzymes are available commercially including lipases, amylases, cellulases, xylanases, phytases, and proteases (Srilakshmi et al., 2015). The encouraging environmental impact of the manufacture processes is of universally interest and the use of enzymatic reactions as an alternative to organic solvents or chemical reactions is highly valued. Figure 2.3 and Table 2.3 show examples of significant endophytic fungal enzymes and the enzyme sources that are used in several applications; however, only a few fungal strains meet the criteria for industrial production. Generally, applications of enzymes in the food industry have focused on hydrolytic reactions

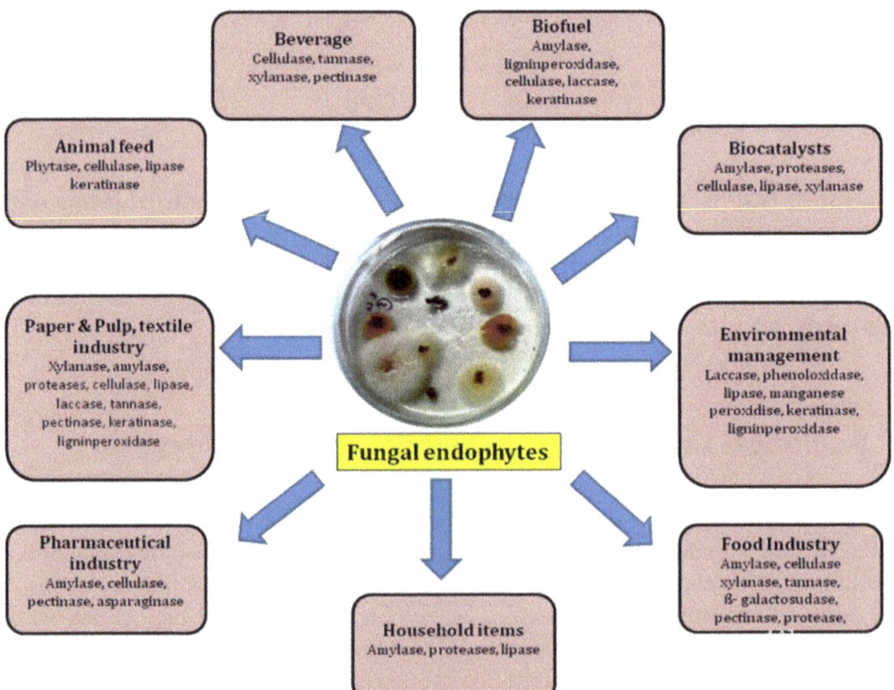

Fig. 2.3 Applications of fungal endophyte origin enzymes

Table 2.3 Examples of enzyme sources and applications of fungi

Enzymes	Applications	Fungal sources	References
Tannases	Food processing; removal of phenolic substances, juices and wines clarification and stabilize the beverage, reduction of hydrolysable tannin in poultry feeds, production of gallic acid from gallotannins	*Aspergillus aculeatus, A. Awamori, A. Caespitosus, A. Niger, A. versicolor Penicillium charlesii, P. crustosum, P. variable, P. restrictum*	Ahmed and Rahman (2014), Bagga et al. (2015), Batra and Saxena (2005) and Srivastava and Kar (2009)
Amylases	Hydrolysis of starch in starch processing industry; food and dairy industry; textile industry; pulp and paper industry; detergent industry; pharmaceutical industry; animal feed industry	*Aspergillus fumigatus, A.Niger, Cylindrocephalum* sp., *Lentinula edodes, Penicillium citrinum, P. fellutanum, Rhizopus stolonifer*	Ko et al. (2005), Sahoo et al. (2014) and Saleem and Ebrahim (2014)
Keratinases	Biomass conversion into biofuels; hydrolysis of keratinous wastes such as feather, hair, and horn; eliminating horny epithelial cells adhering to textile fibers; reducing the environmental pollution	*Aspergillus oryzae, A. Parasiticus, Doratomyces microspores, Paecilomycesmarquandii*	Anitha and Palanivelu (2013), Farag and Hassan (2004), Friedrich et al. (2005), Gradisar et al. (2000) and Veselá and Friedrich (2009).
Cellulases	Animal feed industry; pulp and paper industry; detergent industry; food processing; juices and wines clarification; textile industry; biomass conversion into biofuels	*Aspergillus niger, Lentinula edodes, Trichoderma Longibrachiatum, Volvarielladiplasia*	Pachauri et al. (2017), Wang and Hsu (2006) and Puntambekar (1995)
Lignin peroxidase	Lignin degradation; biomass conversion into biofuels	*Aspergillus sclerotiorum Cladosporium Cladosporioides, Mucor racemosus, Phanerochaete Chrysosporium, Sparassis latifolia*	Chandrasekaran et al. (2014), Brugger et al. (2004) and Wen et al. (2009)
Laccase	Biopulping biobleaching deinking in pulp and paper industry; lignin degradation; pharmaceutical industry; removal of phenolic substances and stabilize the beverage; biomass conversion into biofuels	*Agaricus subrufescens (as "blazei"), Coniophoraputeana, Ganoderma sp., Omphalotusolearius, Phanerochaetefloridensis, Pleurotusostreatus*	Arora et al. (2002), Lee et al. (2004), Ullrich et al. (2005) and Songulashvili et al. (2007)

Table 2.3 (continued)

Enzymes	Applications	Fungal sources	References
Xylanases	Pulp and paper industry; animal feed industry; bread-making; juice and wine industries; xylitol production; prebiotics production; food processing; textile industry; juices and wines clarification; biomass conversion into biofuels	*Aspergillus foetidus, A. Niger, Talaromyces amestolkiae*	Chapla et al. (2012), de Alencar et al. (2013) and Nieto-Domínguez et al. (2017)
Lipases	Degradation of fat in wastewater treatment; animal feed industry; pulp and paper industry; detergent industry; food processing; leather processing; textile industry; pharmaceutical industry	*Aspergillus* sp., *Curvularia* sp., *Fusarium solani,, F. Verticillioides, Penicillium* sp., *P. restrictum, P.wortmanii, Rhizopus oligosporus, Trichoderma* sp., *T. atroviride, T. harzianum, Mucor* sp.	Facchini et al. (2015), El-Ghonemy (2017), Marques et al. (2014), Nwuche and Ogbonna (2011), Ul-Haq et al. (2002) and Maia et al. (2001)
Manganese peroxidase	Lignin degradation; biomass conversion into biofuels	*Lentinula edodes, Phlebiaradiate, Omphalotusolearius, Phellinus robustus*	Arora et al. (2002) and Songulashvili et al. (2007)
Pectinases	Juices and wines clarification; textile industry *Aspergillus japonicus*	*Aspergillus oryzae, Penicillium viridicatum, P.chrysogenum, Thermoascusaurantiacus*	Songulashvili et al. (2007), Banu et al. (2010), Biz et al. (2016) and Martins et al. (2002)
Phytases	Feed supplement in diets in animal feed industry; improvement of soil fertilization and nutrient uptake by plants; reduction in the excretion of phosphorus in manure; reducing phosphate pollution in soil and water	*Aspergillus* sp., *A.niger, Mucor* sp., *Rhizopus oligosporus, Rhizomucor pusillus*	Bei et al. (2009), Casey and Walsh (2004), Chadha et al. (2004) and Saxena et al. (2005)
Proteases	Detergent industry; food processing; pharmaceutical industry; leather processing; textile industry	*Aspergillus* sp., *Fomitellafraxinea, Humicola* sp., *Mucor* sp., *Penicillium* sp., *Pleurotuscitrinopileatus, Rhizopus* sp., *Thermoascus* sp., *Thermomyces* sp.	Cui et al. (2007) and Lee et al. (2006)

(Choi et al., 2015). Glycoside β-galactosidase and hydrolases are involved in the production of a dietary material composed of nonstarch polysaccharides, prebiotics, and oligosaccharides, such as galacto-oligosaccharides, inulin fructo-oligosaccharides, breast milk oligosaccharides, and lactulose that selectively encourage the growth of beneficial intestinal microorganisms in humans (Torres et al., 2010). Pectinases, xylanases, and cellulase are extensively used for the clarification of juices and wines.

Amylases are used to hydrolyze starch in the starch liquefaction process and convert starch into glucose syrups (Souza & Magalhaes, 2010). Further, starch is used as the strengthening agent in the desizing process to prevent the breaking of the warp thread used within the weaving process in the textile industry. Amylases are used for desizing starch in the textile industry, because they provide products that do not degrade the fibers. Moreover, α-amylases are used in the pulp and paper industry instead of starch for coated paper and textiles. The quality of textiles and paper coated with desizing starch is secluded against mechanical harm during processing and finishing. The recycling of waste paper is an eco-friendly trend in the paper industry. Further, amylases, proteases, and lipases enzymes are used for dishwashing detergents. However, amylases are the second other types of enzymes used in the formulation of enzymatic detergent, and 90% of all liquid detergents contain amylases (Souza & Magalhaes, 2010).

Tannases are used to decrease tannin levels in food products such as tea, fruit juices, wines, and beer (Yao et al., 2014). Moreover, these enzymes are used to hydrolyze gallic acid esters and produce gallic acid, which are used as a substrate for the synthesis of food preservatives (Yao et al., 2014; Dhiman et al., 2017). Tannase enzymes are also used for high-grade leather tannin preparation in the leather industry (Dhiman et al., 2017) and this enzymatic process is eco-friendly and also helps to manage waste resulting from production processes.

However, the enzymatic reactions of pectinases, cellulases, xylanases, and lipases aid in the elimination of contaminated ink (Bhat, 2000). Cellulases and proteases are used in the polishing step for clear dyeing to get better color and surface brightness, and oppose textile wrinkling. Lignin peroxidases, xylanases, manganese peroxidases, and laccases are all used to improve the quality of the pulp by removing hemicelluloses and lignin, which are typical impurities in the pulp and paper industries (Choi et al., 2015).

Lipases are hydrolytic enzymes capable of cleaving the ester bond of triacylglycerol as well as catalyze ester synthesis in vitro by shifting the equilibrium of the reaction (Mohamed et al., 2011). These enzymes have a broad range of industrial applications in the production and processing of detergents, oils, fats, dairy products, therapeutic agents, and biodiesel (Gog et al., 2012). Ligninolytic enzyme has a variety of isoforms, of which manganese-dependent lignin peroxidise, peroxidase, and laccase are the three main classes. These enzymes are directly concerned with the degradation of diverse xenobiotic compounds (polycyclic aromatic hydrocarbons) and lignin in their normal lignocellulosic substrates (Pointing, 2001). White-rot fungi are capable of degrading the lignin and regarded as the most capable producers, which secrete ligninolytic enzymes in nature. Laccases (oxygen oxido-reductases and benzenediol) are glycosylated polyphenol oxidases, which degrade lignin powerfully (Kunamneni et al., 2008). These enzymes have significant applications in detoxification of phenolic pollutants, animal biotechnology, pulp and paper industry, and biotransformation (Kunamneni et al., 2008). Protease enzymes are generally used in leather processing for the dehairing and debating of skins and hides instead of chemical processes that have negative implications for the environment. While these proteases enzymes (aspartic protease, protease, cysteine, serine, metalloprotease, and protease) are a significant class of enzymes that catalyze hydrolysis of protein (Oliveira et al., 2014). These enzymes have important applications in food processing industries, leather industries, detergent industry, pharmaceuticals, and biotechnology (Joo et al., 2003).

Xylanases, β-glucanases, and Cellulases have used been in cereal-based feed for monogastric animals for example poultry and pigs, which are not capable to fully degrade and utilize plant-based feeds containing huge amounts of hemicellulose and cellulose (Kirk et al., 2002). However, adding phytases to the feed improves utilization of normal phosphorus bound in phytic acid in cereal-based feed for monogastrics. Usually, 80–90% of whole plant phosphorus is bound to phytic acid, which makes it complicated to monogastrics to utilize. Thus, accessible inorganic phosphates have to be supplemented to the feed in order to reach the necessary concentration. Phytases can reduce the need for supplementation of inorganic phosphorus in the feed (Kirk et al., 2002). However, cellulases are usually used along with the supplement of pectinases, hemicellulases, ligninases, and associated enzymes (Adav & Sze, 2014).

Generally, those fungal endophytes belong to the basidiomycetes, which can enzymatically attack the polymers in the complex-structured lignocellulose. The white rot fungi involve the nutrient recycling of wood required for new plant growth as well as evolutionary impact on diversification of plants (Kües, 2015). However, fungal endophytes are documented as one of the excellent lipase sources among microorganisms (Facchini et al., 2015). The most significant step to recovering the biological degradation of fatty wastewater is pretreatment procedures in order to hydrolyze and dissolve fats, which are able to accelerate the process by declining the fat adsorption to the surface of the anaerobic sludge. Whereas not limiting the transport of the soluble substrate to the biomass (Facchini et al., 2015; Valladao et al., 2009). Moreover, phenoloxidases, laccases, peroxidises, and dioxygenases enzymes are good examples for wastewater treatment (Duran & Esposito, 2000). However, effluents from slaughterhouses have high concentrations of biodegradable organic matter, mainly of which consists of proteins and lipids with low degradability. Subsequently, before effluents are free into the environment, it needs to reduce fat oil and protein from these wastewaters.

Probable applications of enzyme treatments are reduction of the organic matter, which contributes to a cleaner effluent (Valladao et al., 2011). Further, white rot fungi make lignin-degrading enzymes such as, lignin peroxidises, manganese peroxidases and laccases used in biotechnology for degradation of broad-spectrum intractable organic pollutants, bioremediation of polycyclic aromatic hydrocarbons (PAHs) as well as chlorinated hydrocarbons in the environment (Gao et al., 2010).

Biotechnological applications for practical use have restraining factors, as they have need of large amounts of enzymes. Hence, the productions of low-cost and readily available enzymes possessing satisfactory operating characteristics are demanding. After all, each industrial application can need specific properties of the biocatalysts. Further, researchers still face challenges in searching new enzymes that could result in new applications through better isolation, study of enzyme constancy at severe conditions, and purification procedures. The selection of enzymes for the more stable and stringent conditions is significant for industrial processes and applications. Hence, the search for new microorganisms that have the required properties is a continuous process. Besides, the production and expression of recombinant enzymes through protein engineering technology, it is also required to find enzymes of required characteristics in specific host cell Industrial. Enzymes are usually produced in bioreactors which contain up to over two lakhs liters and are consequently purified in the industrial downstream processes.

Thus, the fungal endophytes that are able to use for such processes have to be speedy growing but not be pathogens or mycotoxins producers. The use of heterologous appearance, recombinant DNA technology, gene cloning to get better production yield of enzymes and their activity can avoid the difficulties related to the production of large quantities.

2.7 International Companies Manufacturing Enzyme from Fungal Endophytes

Fungal endophyte origin enzymes are progressively replacing conformist chemical catalysts in numerous industrial processes. Enzymes have many advantages over chemical catalysts, as well as the capability to function under comparatively gentle conditions of pH, pressure, and temperature. Amylase enzyme was first produced via Solid State Fermentation (SSF) by *Aspergillus oryzae* on wheat bran or moist rice. Jokichi Takamine initially developed this process and got patent in USA in 1884. Moreover, the major production of fungal enzymes was approved out by inundated fermentation technology in the 1940s. Current days, there are numerous international companies having wager in producing industrial enzymes from endophytic fungi. In India, Biocon India Ltd. is a major mass enzyme producer company, but it is not a major at the international level. Table 2.4 is showing the top fourteen companies at the international level.

Table 2.4 Major bulk enzyme-producing companies

S. no.	International companies
1	Amano Pharmaceutical Co., Japan
2	Meito Sankyo Co., Japan
3	Enzyme Development Corp., USA
4	Nagase Biochemicals Ltd., Japan
5	Solavy Enzymes gmbh, Germany
6	Biocatalysis Ltd., Wales
7	Rhone-Poulenc, England
8	Novo Nordisk, Denmark
9	Rohm gmbh, Germany
10	Danisco Cultar, Finland
11	Sankyo Co., Japan
12	Yakult Biochemical Co., Japan
13	Shin-Nihon Chemical Co., Japan
14	DSM-GIST, Netherlands

2.8 Biotechnological Potential of Fungal Endophytes

Approximately four thousand secondary bioactive compounds or metabolites of fungal origin were reported as biologically active (Dreyfuss & Chapela, 1994). Endophytic fungi are now established as an unused resource for producing significant natural products contributing potential for pharmaceuticals, agriculture, food, and other industrial exploitation (Suryanarayanan et al., 2009). Many novel natural bioactive compounds or secondary metabolic are now known to have immunosuppressive properties, antidiabetic, antioxidants antiviral, anticancer, and insecticidal, etc. (Strobel & Daisy, 2003).

Nowdays broad variety of bioactivities of unstable organic compounds produced by endophytic fungi, *Muscodor albus* Worapong, Strobel & W.M. Hess and *M. roseus*-Worapong, Strobel & W.M. Hess was established in controlling the plant pathogenic fungi (Stinson et al., 2003) and nematodes (Riga et al., 2008). Moreover, *M. albus* against various plants and human-pathogenic fungi and bacteria as an antibiotic (Strobel et al., 2001). In current times *Coryne* sp. (Strain-NRRL 50072) elected as anamorphic form of an *Ascocoryne* sp, was established to produce hydrocarbon derivative (Strobel et al., 2010). Further, *M. albus* strain GBA was newly isolated from *Ginkgo biloba* collected in Newport, RI, USA, which inhibited and killed sure test microbes by production of a mixture of volatile compounds (Banerjee et al., 2010).

However, several endophytic fungi were reported as taxol producers, viz. *Stegoleriumkukenani, Alternaria* sp., *Pestalotia* sp., *Pithomyces* sp., *Fusarium* sp., *Tubercularia* sp., *Monochaetia* sp., Bat. & Peres, *Pestalotiopsis guepini* (Desm.), *Pestalotiopsis microspora* (Speg.) Steyaert, *Sporormia minima* Auersw., *Periconia* sp., *Trichothecium* sp. and *Seimatoantleriumnepalense* Bashyal *et al.* which were reported from yew as well as other plants (Strobel et al. 20001; Wang et al., 2000). In India, attempts were finished to isolate taxol among potent cytotoxic action from two fungal endophytes, viz. *Bartaliniarobillardoides* Tassi and *Colletotrichum gloeosporioides* isolated from *Aegle marmelos* and *Justicia gendarussa* Burm F respectively (Gangadevi & Muthumary, 2009).

Fungal endophytes belong to the genus *Pestalotiopsis* have been accepted as the ordinary endophytic fungi from rainforests (Nag Raj, 1993). *Pestalotiopsis* species were produced numerous bioactive compounds and secondary metabolites (Li et al., 2001). In the similar way, *Pestalotiopsis terminaliae* G.P. Agarwal & Hasija from *Terminalia arjuna* plant and *P. Pauciseta* (Sacc.) Y. X. Chen from *Cardiospermum helicacabum* Linn. fungal endophytes were produced taxol (Gangadevi & Muthumary, 2009).

Camptothecin was isolated from different plants such as *Nothapodytesfoetida, Camptotheca acuminata* Decais, and *Ervatomiahyneana Ophiorrhizamungos* (Rehman et al., 2009). Further, Camptothecin screened from an unidentified fungal endophyte from the inner bark of plant *Nothapodytesfoetida* and *N. nimmoniana* in the Western coast of India (Chapla et al., 2012). However, Camptothecin screened from *Entrophosporainfrequens* (I.R. Hall) R.N. Ames & R.W. Schneid fungal endophyte isolated from the inner bark of *N. foetida* in Jammu and Mahabaleshwar regions of India (Amna et al., 2006).

2.9 Conclusion

Fungal endophytes are the group of microorganisms, which dwell in the inner tissues of the plant parts (viz., roots, stems, leaves, bark, blossoms, and so forth) in symbiotic manner without bringing on any harmful impacts on the plants. Endophytes assume a significant part in assisting plants with battling against biotic and abiotic stresses and improve plant development. Fungal endophytes, obtained from medicinally important plants, are considered as a choice to deliver the bioactive constituents that could be used for the management of various human diseases. Taxol is an anticancer bioactive metabolites that has been extracted from various genera of fungal endophytes. Similarly, fungal endophytes show their efficacy to fight against several pathogens which contain drug resistance activity.

Thus, its wider applicability ranges from agricultural to pharmaceutical sectors could also be explored as potential agent for developing entrepreneurship.

References

Adav, S. S., & Sze, S. K. (2014). *Trichoderma* secretome: An overview. In V. K. Gupta, M. Schmoll, A. Herrera-Estrella, R. S. Upadhyay, I. Druzhinina, & M. G. Tuohy (Eds.), *Biotechnology and biology of Trichoderma* (pp. 103–114). Elsevier.

Aguilar, A. C., & Barea, J. M. (1996). Arbuscularmycorrhizasand biological control of soil-borne plant pathogens – An overview of the mechanisms involved. *Mycorrhiza, 6*, 457–464.

Ahmed, M. E., & Rahman, A. (2014). Detection of the perfect condition to produce the tannase from Aspergillus niger at different medium. *Journal of University of Babylon for Pure and Ap plied Sciences, 22*, 1363–1371.

Akello, J., Dubois, T., Coyne, D., & Kyamanywa, S. (2008). Effect of endophytic Beauveria bassiana on populations of the banana weevil, Cosmopolites sordidus, and their damage in tissue cultured banana plants. *Entomologia Experimentalis et Applicata, 129*(2), 157–165.
Alves, A., Crous, P. W., Correia, A., & Phillips, A. J. L. (2008). Morphological and molecular data reveal cryptic speciation in Lasiodiplodiatheobromae. *Fungal Diversity, 28*, 1–13.
Amirita, A., Sindhu, P., Swetha, J., Vasanthi, N. S., & Kannan, K. P. (2012). Enumeration of endophytic fungi from medicinal plants and screening of extracellular enzymes. *World Journal of Science and Technology, 2*, 13–19.
Amna, T., Puri, S. C., Verma, V., Sharma, J. P., Khajuria, R. K., Musarrat, J., Spiteller, M., & Qazi, G. N. (2006). Bioreactor studies on the endophytic fungus *Entrophosporainfrequens*for the production of an anticancer alkaloid camptothecin. *Canadian Journal of Microbiology, 52*, 189–196.
Anbu, P., Gopinath, S. C. B., Chaulagain, B. P., & Lakshmipriya, T. (2017). Microbial enzymes and their applications in industries and medicine. *BioMed Research International, 2017*, 3. Article ID 2195808. https://doi.org/10.1155/2017/2195808.
Anitha, T. S., & Palanivelu, P. (2013). Purification and characterization of an extracellular keratinolytic protease from a new isolate of *Aspergillus parasiticus*. *Protein Expression and Purification, 88*, 214–220.
Arivudainambi, U. E., Anand, T. D., Shanmugaiah, V., Karunakaran, C., & Rajendran, A. (2011). Novel bioactive metabolites producing endophytic fungus *Colletotrichum gloeosporioides*against multidrug-resistant *Staphylococcus aureus*. *FEMS Immunology and Medical Microbiology, 61*, 340–345.
Arnold, A. E. (2007). Understanding the diversity of foliar endophytic fungi: Progress, challenges, and frontiers. *Fungal Biology Reviews, 21*, 51–66.
Arora, D. S., Chander, M., & Gill, P. K. (2002). Involvement of lignin peroxidase manganese peroxidase and laccase in degradation and selective ligninolysis of wheat straw. *International Bio deterioration and Biodegradation, 15*, 115–120.
Ayob, F. W., & Simarani, K. (2016). Endophytic filamentous fungi from a *Catharanthus roseus:* Identification and its hydrolytic enzymes. *Saudi Pharmaceutical Journal, 24*(3), 273–278.
Bacon, C. W., & Hinton, D. M. (1996). Symptomless endophytic colonisation of maize by *Fusarium moniliforme*. *Canadian Journal of Botany, 75*, 1195–1202.
Bae, H., Kim, S., Sicher, R. C., Jr., Kim, M. S., Strem, M. D., & Bailey, B. A. (2008). The beneficial endophyte, *Trichoderma hamatum*, delays the onset of drought stress in *Theobroma cacao*. *Biological Control, 46*, 24–35.
Bagga, J., Pramanik, S. K., & Pandey, V. (2015). Production and purification of tannase from Aspergillus aculeatus using plant derived raw tannin. *International Journal of Scientific Engin eering and Technology, 4*, 50–55.
Banerjee, D., Strobel, G., Geary, B., Sears, J., Ezra, D., Liarzid, O., & Coombs, J. (2010) *Muscodor albus* strain GBA, an endophytic fungus of *Ginkgo biloba* from United States of America, produces volatile antimicrobials. *Mycology, 1*, 17–186
Banu, A. R., Devi, M. K., Gnanaprabhal, G. R., Pradeep, B. V., et al. (2010). Production and characterization of pectinase enzyme from Penicillium chrysogenum. *Indian Journal of Science and Technology, 3*, 377–381.
Batra, A., & Saxena, R. K. (2005). Potential tannase producers from the genera Aspergillus and Penicillium. *Process Biochemistry, 40*, 1553–1557.
Bei, J., Chen, Z., Fu, J., Jiang, Z., et al. (2009). Structure-based fragment shuffling of two fungal phytases for combination of desirable properties. *Journal of Biotechnology, 139*, 186–193.
Bengtsson-Palme, J. (2018). The diversity of uncharacterized antibiotic resistance genes can be predicted from known gene variants-but not always. *Microbiome, 6*, 125.
Bezerra, J., Santos, M., Svedese, V., Lima, D., Fernandes, M., Paiva, L., & Souza-Motta, C. (2012). Richness of endophytic fungi isolated from *Opuntia ficus-indica* Mill.(Cactaceae) and preliminary screening for enzyme production. *World Journal of Microbiology and Biotechnology, 28*, 1989–1995.

Bezerra, J. D., Nascimento, C. C., Barbosa, R. D., da Silva, D. C., Svedese, V. M., Silva-Nogueira, E. B., Gomes, B. S., Paiva, L. M., & Souza-Motta, C. M. (2015). Endophytic fungi from medicinal plant *Bauhiniaforficata*: Diversity and biotechnological potential. *Brazilian Journal of Microbiology, 46*, 49–57.
Bhat, M. K. (2000). Cellulases and related enzymes in biotechnology. *Biotechnology Advances, 18*, 355–383.
Bhat, J. (2010). *Fascinating microfungi (Hyphomycetes) of Western Ghats* (p. 221). Broadway Book Centre, . ISBN:978-3-642-23341-8.
Biz, A., Finkler, A. T. J., Pitol, L. O., Medina, B. S., et al. (2016). Production of pectinases by solid-state fermentation of a mixture of citrus waste and sugarcane bagasse in a pilot-scale packed-bed bioreactor. *Biochemical Engineering Journal, 111*, 54–62.
Bogner, C. W., Kamdem, R. S., Sichtermann, G., Matthäus, C., Hölscher, D., Popp, J., Proksch, P., Grundler, F. M., & Schouten, A. (2017). Bioactive secondary metabolites with multiple activities from a fungal endophyte. *Microbial Biotechnology, 10*, 175–188.
Brugger, R., Nunes, C. S., Hug, D., Vogel, K., et al. (2004). Characteristics of fungal phytases from Aspergillus fumigatus and Sartorya fumigate. *Applied Microbiology and Biotechnology, 63*, 383–389.
Busby, P. E., Ridout, M., & Newcombe, G. (2016). Fungal endophytes: Modifiers of plant disease. *Plant Molecular Biology, 90*, 645–655.
Casey, A., & Walsh, G. (2004). Identification and characterization of a phytase of potential commercial interest. *Journal of Biotechnology, 110*, 313–322.
Castro, M. S., & Fontes, W. (2005). Plant defense and antimicrobial peptides. *Protein and Peptide Letters, 12*, 11–16.
Chadha, B. S., Harmeet, G., Mandeep, M., Saini, H. S., & Singh, N. (2004). Phytase production by the thermophilic fungus Rhizomucorpusillus. *World Journal of Microbiology & Biotechnology, 20*, 105–109.
Chandrasekaran, G., Choi, S. K., Lee, Y. C., Kim, G. J., & Shin, H. J. (2014). Oxidative biodegradation of single-walled carbon nanotubes by partially purified lignin peroxidase from Sparassis latifolia mushroom. *Journal of Industrial and Engineering Chemistry, 20*, 3367–3374.
Chapla, D., Pandit, P., & Shah, A. (2012). Production of xylooligosaccharides from corncob xylan by fungal xylanase and their utilization by probiotics. *Bioresource Technology, 115*, 215–221.
Chathurdevi, G., & Gowrie, S. U. (2016). Endophytic fungi isolated from medicinal plant – A source of potential bioactive metabolites. *International Journal of Current Pharmaceutical Research, 8*, 50–56.
Chen, Z., Song, Y., Chen, Y., Huang, H., Zhang, W., & Ju, J. (2012). Cyclic heptapeptides, cordyheptapeptides C–E, from the marine-derived fungus *Acremonium persicinum* SCSIO 115 and their cytotoxic activities. *Applied Microbiology and Biotechnology, 75*, 1215–1219.
Chen, G. D., Chen, Y., Gao, H., Shen, L. Q., Wu, Y., Li, X. X., Li, Y., Guo, L. D., Cen, Y. Z., & Yao, X.-S. (2013). Xanthoquinodins from the endolichenic fungal strain *Chaetomium elatum*. *Journal of Natural Products, 76*, 702–709.
Cheplick, G. P., Perera, A., & Koulouris, K. (2000). Effect of drought stress on the growth of *Lolium perenne*genotypes with and without fungal endophytes. *Functional Ecology, 14*, 657–667.
Chin, Y. W., Balunas, M. J., Chai, H. B., & Kinghorn, A. D. (2006). Drug discovery from natural sources. *Advances in Applied & Pharmaceuticals Sciences Journal, 8*, 239–253.
Choi, J. M., Han, S. S., & Kim, H. S. (2015). Industrial applications of enzyme biocatalysis: Current status and future aspects. *Biotechnology Advances, 33*, 1443–1454.
Chow, Y., & Ting, A. S. (2015). Endophytic L-asparaginase producing fungi from plants associated with anticancer properties. *Journal for Artistic Research, 6*, 869–876.
Cui, L., Liu, Q. H., Wang, H. X., & Ng, T. B. (2007). An alkaline protease from fresh fruiting bodies of the edible mushroom Pleurotuscitrinopileatus. *Applied Microbiology and Biotechnology, 75*, 81–85.

D'Souza, M. A., & Hiremath, K. G. (2015). Isolation and bioassay screening of medicinal plant endophytes from Western Ghats forests, Goa, India. *International Journal of Advanced Research in Biological Sciences, 2*, 176–190.
de Alencar, G. N. C. A., Sorgatto, M., Peixoto-Nogueira, S. C., Betini, J. H. A., et al. (2013). Bioprocess and biotechnology: effect of xylanase from Aspergillus niger and Aspergillus flavus on pulp biobleaching and enzyme production using agroindustrial residues as substrate. *Springerplus, 2*, 380.
De Barry, A. (1866). *Morphologie und Physiologie der Pilze, Flechten, und Myxomyceten* (Vol. II). Hofmeister's Handbook of Physiological Botany.
Desire, M. H., Bernard, F., Forsah, M. R., Assang, C. T., et al. (2014). Enzymes and qualitative phytochemical screening of endophytic fungi isolated from Lantana camara Linn.leaves. *Journal of Applied Biology and Biotechnology, 2*, 1–6.
Dhiman, S., Mukherjee, G., Kumar, A., Mukherjee, P., et al. (2017). Fungal tannase: Recent advances and industrial applications. In T. Satyanarayana, S. Deshmukh, & B. Johri (Eds.), *Developments in fungal biology and applied mycology* (pp. 295–313). Springer.
di Menna, M., Finch, S. C., Popay, A., & Smith, B. L. (2012). A review of the Neotyphodiumlolii/Lolium perenne symbiosis and its associated effects on animal and plant health, with particular emphasis on ryegrass staggers. *New Zealand Veterinary Journal, 60*(6), 315–328.
Ding, G., Li, Y., Fu, S., Liu, S., Wei, J., & Che, Y. (2008). Ambuic acid and torreyanic acid derivatives from the endolichenic fungus *Pestalotiopsis* sp. *Journal of Natural Products, 72*, 182–186.
Dissanayake, R. K., Ratnaweera, P. B., Williams, D. E., Wijayarathne, C. D., Wijesundera, R. L., Andersen, R. J., & de Silva, E. D. (2016). Antimicrobial activities of endophytic fungi of the Sri Lankan aquatic plant *Nymphaea nouchali* and chaetoglobosin A and C, produced by the endophytic fungus *Chaetomium globosum*. *Mycology, 7*, 1–8.
Doilom, M., Manawasinghe, I. S., Jeewon, R., Jayawardena, R. S., Tibpromma, S., Hongsanan, S., Meepol, W., Lumyong, S., Jones, E. B. G., & Hyde, K. D. (2017). Can ITS sequence data identify fungal endophytes from cultures? A case study from *Rhizophora apiculata*. *Mycosphere, 8*, 1869–1892.
Dou, Y., Wang, X., Jiang, D., Wang, H., Jiao, Y., Lou, H., & Wang, X. (2014). Metabolites from *Aspergillus versicolor,* an endolichenic fungus from the lichen *Lobariaretigera*. *Drug Discoveries & Therapeutics, 8*, 84–88.
Dreyfuss, M. M., & Chapela, I. H. (1994). Potential of fungi in the discovery of novel, low molecular weight pharmaceuticals. In V. P. Gulb (Ed.), *The discovery of the natural products with therapeutic potential* (pp. 49–80). Butterworth-Heinemann.
Duran, N., & Esposito, E. (2000). Potential applications of oxidative enzymes and phenoloxidase-like compounds in wastewater and soil treatment: A review. *Applied Catalysis B, 28*, 83–99.
Eaton, C. J., Cox, M. P., & Scott, B. (2011). What triggers grass endophytes to switch from mutualism to pathogenism? *Plant Science, 180*(2), 190–195.
El-Ghonemy, A. (2017). Extracellular alkaline lipase from a novel fungus Curvularia sp. DHE 5: Optimisation of physicochemical parameters, partial purifcation and characterisation. *Food Technology and Biotechnology, 55*(2), 206–217.
El-Ghonemy, D. H., El-Gamal, M. S., Tantawy, A. E., & Ali, T. H. (2017). Extracellular alkaline lipase from a novel fungus Curvularia sp. DHE 5: Optimisation of physicochemical parameters partial purification and characterization. *Food Technology and Biotechnology, 55*, 206–217.
El-Maali, N., Mohrram, A., El-Kashef, H., & Gamal, K. (2018). Novel resources of Taxol from endophytic and entomopathogenic fungi: Isolation, characterization and LC-Triple mass spectrometric quantification. *Talanta, 190*, 466–474.
Erbert, C., Lopes, A. A., Yokoya, N. S., Furtado, N. A. J. C., Conti, R., Pupo, M. T., et al. (2012). Antibacterial compound from the endophytic fungus *Phomopsis longicolla*isolated from the tropical red seaweed *Bostrychia radicans*. *Botanica Marina, 55*, 435–440.
Escudero, N., Ferreira, S. R., Lopez-Moya, F., Naranjo-Ortiz, M. A., Marin-Ortiz, A. I., Thornton, C. R., & Lopez-Llorca, L. V. (2016). Chitosan enhances parasitism of Meloidogyne javanica eggs by the nematophagous fungus *Pochoniachlamydosporia*. *Fungal Biology, 120*, 572–585.

Eyberger, A. L., Dondapati, R., & Porter, J. R. (2006). Endophyte fungal isolates from Podophyllum peltatum produces podophyllotoxin. *Journal of Natural Products, 69*(8), 1121–1124.

Facchini, F. D. A., Vici, A. C., Pereira, M. G., Jorge, J. A., & Polizeli, T. M. (2015). Enhanced lipase production of *Fusarium verticillioides* by using response surface methodology and wastewater pretreatment application. *Journal of Biochemical Technology, 6*, 996–1002.

Faeth, S. H., Gardner, D. R., Hayes, C. J., Jani, A., Wittlinger, S. K., & Jones, T. A. (2006). Temporal and spatial variation in alkaloid levels in Achnatherumrobustum, a native grass infected with the endophyte Neotyphodium. *Journal of Chemical Ecology, 32*(2), 307–324.

Farag, A. M., & Hassan, M. A. (2004). Purification, characterization and immobilization of a keratinase from *Aspergillus oryzae*. *Enzyme and Microbial Technology, 34*, 85–93.

Fareed, S., Jadoon, U. N., Ullah, I., Ayub, M., Jadoon, M. U. R., Bibi, Z., Waqas, M., & Nisa, S. (2017). Isolation and biological evaluation of endophytic fungus from *Ziziphus nummularia*. *Journal of Entomology and Zoology Studies, 5*(3), 32–38.

Fillat, Ú., Martín-Sampedro, R., Macaya-Sanz, D., Martín, J. A., Ibarra, D., Martínez, M. J., & Eugenio, M. E. (2016). Screening of eucalyptus wood endophytes for laccase activity. *Process Biochemistry, 51*, 589–598.

Fisher, P. J., & Petrini, O. (1990). A comparative study of fungal endophytes in xylem and bark of *Alnus* species in England and Switzerland. *Mycological Research, 94*, 313–319.

Fouda, A. H., Hassan, S. E., Eid, A. M., & Ewais, E. E. (2015). Biotechnological applications of fungal endophytes associated with medicinal plant *Asclepiassinaica* (Bioss.). *Annals of Agricultural Science, 60*, 95–104.

Friedrich, J., Gradisˇar, H., Vrecl, M., & Pogacˇnik, A. (2005). In vitro degradation of porcine skin epidermis by a fungal keratinase of Doratomycesmicrosporus. *Enzyme and Microbial Technolog, 36*, 455–460.

Fu, J., Zhou, Y., Li, H. F., Ye, Y. H., & Guo, J. H. (2011). Antifungal metabolites from *Phomopsis* sp. By254, an endophytic fungus in *Gossypium hirsutum*. *Asian Journal of Research in Microbiology, 5*, 1231–1236.

Gangadevi, V., & Muthumary, J. (2009). Taxol production by *Pestalotiopsisterminaliae,* an endophytic fungus of *Terminalia arjuna* (arjun tree). *Biotechnology and Applied Biochemistry, 52*, 9–15.

Gao, D., Du, L., Yang, J., Wu, W. M., & Liang, H. (2010). A critical review of the application of white rot fungus to environmental pollution control. *Critical Reviews in Biotechnology, 30*, 70–77.

Gog, A., Roman, M., Tos, M., Paizs, C., & Irimie, F. D. (2012). Biodiesel production using enzymatic transesterification-current state and perspectives. *Renewable Energy, 39*, 10–16.

Goyal, S., Ramawat, K. G., & Mérillon, J. M. (2017). Different shades of fungal metabolites: An Overview. In J. M. Mérillon & K. G. Ramawat (Eds.), *Fungal metabolites*. Springer.

Gradisar, H., Kern, S., & Friedrich, J. (2000). Keratinase of Doratomycesmicrosporus. *Applied Microbiology and Biotechnology, 53*, 196–200.

Guo, B., Dai, J.-R., Ng, S., Huang, Y., Leong, C., Ong, W., & Carté, B. K. (2000). Cytonic acids A and B: Novel tridepside inhibitors of hCMV protease from the endophytic fungus *Cytonaema*species. *Journal of Natural Products, 63*, 602–604.

Guo, L. D., Huang, G. R., & Wang, Y. (2008). Seasonal and tissue age influences on endophytic fungi of *Pinus tabulaeformis*(*Pinaceae*) in the Dongling Mountains, Beijing. *International Journal of Plant Biology, 50*, 997–1003.

Harper, J. K., Arif, A. M., Ford, E. J., Strobel, G. A., Porco, J. A., Tomer, D. P., Oneill, K. L., Heider, E. M., & Grant, D. M. (2003). Pestacin: A 1, 3-dihydro isobenzofuran from *Pestalotiopsismicrospora*possessing antioxidant and antimycotic activities. *Tetrahedron, 59*, 2471–2476.

Hartman, G. L., Manandhar, J. B., & Sinclair, J. B. (1986). Incidence of *Colletotrichum* spp. on soybeans and weeds in Illinois and pathogenicity of *Colletotrichum truncatum*. *Plant Disease, 70*, 780–782.

Hegde, S. V., Ramesha, A., & Srinvas, C. (2011). Optimization of amylase production from an endophytic fungi Discosia sp. isolated from *Calophylluminophyllum. International Journal of Agricultural Technology, 7*, 805–813.
Hesterkamp, T. (2017). Antibiotics clinical development and pipeline. In How to overcome the antibiotic crisis—Facts, challenges, technologies and future perspective. In M. Stadler & P. Dersch (Eds.), *Current topics in microbiology and immunology* (Vol. 398, pp. 447–474).
Hill, N. S., Belesky, D. P., & Stringer, W. C. (1991). Competitiveness of tall fescue as influenced by *Acremonium coenophialum. Crop Science, 31*, 185–190.
Hoffman, A. M., Mayer, S. G., Strobel, G. A., Hess, W. M., Sovocool, G. W., Grange, A. H., Harper, J. K., Arif, A. M., Grant, D. M., & Kelley-Swift, E. G. (2008). Purification, identification and activity of phomodione, a furandione from an endophytic *Phoma* species. *Phytochemistry, 69*(1049), 1056.
Hol, W. H. G., La, P. E., De Moens, M., & Cook, R. (2007). Interaction between a fungal endophyte and root herbivores of Ammophila arenaria. *Basic and Applied Ecology, 8*, 500–509.
Huang, J. W., Shih, H. D., Huang, H. C., & Chung, W. C. (2007). Effects of nutrients on production of fungichromin by Streptomyces padanus PMS-702 and efficacy of control of Phytophthora infestans. *Canadian Journal of Plant Pathology, 29*(3), 261–267.
Isabelle, K. M., Celine, N., Augustin, N., Jean, B. J., Paul-Keilah, L., Pierre, E., Valere, T. F., & Jules-Roger, K. (2019). Antimicrobial activities of two secondary metabolites isolated from Aspergillus niger, endophytic fungus harbouring stems of Acanthus montanus. *Biological Sciences and Pharmaceutical Research, 7*(1), 7–15.
Istifadah, N., & Mcgee, P. A. (2006). Endophytic *Chaetomium globosum* reduces development of tan spot in wheat caused by Pyrenophoratritici-repentis. *Australasian Plant Pathology, 35*, 411–418.
Jeewon, R., Wanasinghe, D. N., Rampadaruth, S., Puchooa, D., Zhou, L.-G., Liu, A.-R., & Wang, H.-K. (2017). Nomenclatural and identification pitfalls of endophytic mycota based on DNA sequence analyses of ribosomal and protein genes phylogenetic markers: A taxonomic dead end? *Mycosphere, 8*(10), 1802–1817.
Jia, M., Chen, L., Xin, H., Zheng, C., Rahman, K., Han, T., & Qin, L. (2016). A friendly relationship between endophytic fungi and medicinal plants: A systematic review. *Frontiers in Microbiology, 7*, 906.
Joo, H. S., Kumar, C. G., Park, G. C., Paik, S. R., & Chang, C. S. (2003). Oxidant and SDS-stable alkaline protease from *Bacillus clausii* I-52: Production and some properties. *Journal of Applied Microbiology, 95*, 267–272.
Jurynelliz, R-V., David, P., Gloria, O., Carlos, R., Jesus-Bonilla, D., & Walleska. (2016). Enzymatic and bacterial activity of fungal strains isolated from *Alpiniazerumbet. Abstracts of papers, 251st ACS national meeting & exposition*, San Diego, CA, United States, March 13–17, CHED-1130.
Kalra, K., Chauhan, R., Shavez, M., & Sachdeva, S. (2013). Isolation of laccase producing Trichoderma spp. and effect of pH and temperature on its activity. *International Journal of ChemTech Research, 5*(5), 2229–2235.
Kalyanasundaram, I., Jayaprabha, N., & Srinivasan, M. (2015). Antimicrobial activity of endophytic fungi isolated and identified from salt marsh plant in Vellar Estuary. *Journal of Microbiology and Antimicrobials, 7*(2), 13–20.
Kaul, S., Ahmed, M., Zargar, K., Sharma, P., & Dhar, M. K. (2013). Prospecting endophytic fungal assemblage of digitalis lanata Ehrh, (foxglove) as a novel source of digoxin: A cardiac glycoside. *Biotechnology, 3*, 335–340.
Khan, A. L., Al-Harrasi, A., Al-Rawahi, A., Al-Farsi, Z., Al-Mamari, A., Waqas, M., Asaf, S., Elyassi, A., Mabood, F., Shin, J. H., & Lee, I. J. (2016). Endophytic fungi from frankincense tree improves host growth and produces extracellular enzymes and indole acetic acid. *PLoS One, 11*(6), e0158207.
Kim, S., Shin, D.-S., Lee, T., & Oh, K.-B. (2004). Periconicins, two new fusicoccane diterpenes produced by an endophytic fungus *Periconia* sp. with antibacterial activity. *Journal of Natural Products, 67*, 448–450.

Kim, J. W., Choi, H. G., Song, J. H., et al. (2019). Bioactive secondary metabolites from an endophytic fungus *Phoma* sp. PF2 derived from *Artemisia princeps* Pamp. *Journal of Antibiotics, 72*, 174–177.

Kirk, O., Borchert, T. V., & Fuglsang, C. C. (2002). Industrial enzyme applications. *Current Opinion in Biotechnology, 13*, 345–351.

Kjer, J., Wray, V., Edrada-Ebel, R., Ebel, R., Pretsch, A., Lin, W., & Proksch, P. (2009). Xanalteric acids I and II and related phenolic compounds from an endophytic *Alternaria* sp. isolated from the mangrove plant *Sonneratia alba*. *Journal of Natural Products, 72*, 2053–2057.

Ko, H. G., Park, S. H., Kim, S. H., Park, H. G., & Park, W. M. (2005). Detection and recovery of hydrolytic enzymes from spent compost of four mushroom species. *Folia Microbiologica, 50*, 103–106.

Kudalkar, P., Strobel, G., Riyaz-Ul-Hassan, S., Geary, B., & Sears, J. (2012) *Muscodorsutura*, a novel endophytic fungus with volatile antibiotic activities. *Mycoscience, 53*, 319–325.

Kudanga, T., & Mwenje, E. (2005). Extracellular cellulase production by tropical isolates of *Aureobasidium pullulans*. *Canadian Journal of Microbiology, 51*, 773–776.

Kües, U. (2015). Fungal enzymes for environmental management. *Current Opinion in Biotechnology, 33*, 268–278.

Kumar, S., & Kaushik, N. (2013). Endophytic fungi isolated from oil-seed crop *Jatropha curcas* produces oil and exhibit antifungal activity. *PLoS One, 8*, e56202.

Kunamneni, A., Camarero, S., García-Burgos, C., Plou, F. J., Ballesteros, A., & Alcalde, M. (2008). Engineering and applications of fungal laccases for organic synthesis. *Microbial Cell Factories, 7*(1), 32.

Kusari, S., & Spiteller, M. (2012). Metabolomics of endophytic fungi producing associated plant secondary metabolites: Progress, challenges and opportunities. In U. Roessner (Ed.), *Metabolomics* (pp. 241–266). InTech.

Kusari, S., Singh, S., & Jayabaskaran, C. (2014). Rethinking production of Taxol®(paclitaxel) using endophyte biotechnology. *Trends in Biotechnology, 32*, 304–311.

Lee, K. H., Wi, S. G., Singh, A. P., & Kim, Y. S. (2004). Micromorphological characteristics of decayed wood and laccase produced by the brown-rot fungus Coniophoraputeana. *Journal of Wood Science, 50*, 281–284.

Lee, J. S., Baik, H. S., & Park, S. S. (2006). Purification and characterization of two novel fibrinolytic proteases from mushroom Fomitellafraxinea. *Journal of Microbiology and Biotechnology, 16*, 264–271.

Leo, V. V., Passari, A. K., Joshi, J. B., Mishra, V. K., Uthandi, S., Ramesh, N., Gupta, V. K., Saikia, R., Sonawane, V. C., & Singh, B. P. (2016). A novel triculture system (CC3) for simultaneous enzyme production and hydrolysis of common grasses through submerged fermentation. *Frontiers in Microbiology, 7*, 447.

Li, J. Y., Harpe, J. K., Grant, D. M., Tombe, B. O., Bashyal, W. M., & Hess, & Strobel, G.A. (2001). Ambuic acid, a highly functionalized cyclohexane with antifungal activity from *Pestalotiopsis* sp. and Monochaetia sp. *Phytochemistry, 56*, 463–468.

Li, G. H., Yu, Z. F., Li, X., Wang, X. B., Zheng, L. J., & Zhang, K. Q. (2007a). Nematicidal metabolites produced by the endophytic fungus *Geotrichum* sp. AL4. *Chemistry & Biodiversity, 4*, 1520–1524.

Li, W. C., Zhou, J., Guo, S. Y., & Guo, L. D. (2007b). Endophytic fungi associated with lichens in Baihua mountain of Beijing, China. *Fungal Diversity, 25*, 69–80.

Li, H. Q., Li, X. J., Wang, Y. L., Zhang, Q., Zhang, A. L., Gao, J. M., & Zhang, X. C. (2011). Antifungal metabolites from *Chaetomium globosum*, an endophytic fungus in *Ginkgo biloba*. *Biochemical Systematics and Ecology, 39*, 876–879.

Li, G., Wang, H., Zhu, R., Sun, L., Wang, L., Li, M., Li, Y., Liu, Y., Zhao, Z., & Lou, H. (2012). Phaeosphaerins A–F, cytotoxic perylenequinones from an endolichenic fungus, *Phaeosphaeria* sp. *Journal of Natural Products, 75*, 142–147.

Li, G., Kusari, S., Lamshöft, M., Schüffler, A., Laatsch, H., & Spiteller, M. (2014). Antibacterial secondary metabolites from an endophytic fungus, *Eupenicillium* sp. LG41. *Journal of Natural Products, 77*, 2335–2341.

Li, Y., Yang, J., Zhou, X., Zhao, W., & Jian, Z. (2015). Isolation and identification of a 10-deacetyl baccatin-III- producing endophyte from *Taxus wallichiana*. *Applied Biochemistry and Biotechnology, 175*, 2224–2231.

Lin, Z.-J., Lu, Z.-Y., Zhu, T.-J., Fang, Y.-C., Gu, Q.-Q., & Zhu, W.-M. (2008). Penicillenols from *Penicillium* sp. GQ-7, an endophytic fungus associated with *Aegicerascorniculatum*. *Chemical & Pharmaceutical Bulletin, 56*, 217–221.

Liu, Z. Z., Chen, Y., Lian, B., Zhang, Z., Zhao, Y. Y., Ji, Z. H., Lv, Y. N., & Li, H. G. (2019). Comparative study on population ecological distribution and extracellular enzyme activities of endophytic fungi in *Artemisia annua*. *Journal of Biological Sciences and Medicine, 7*, 94–105.

Lu, H., Xou, W. X., Meng, J. C., Hu, J., & Tan, R. X. (2000). New bioactive metabolites produced by Colletotrichum sp., an endophytic fungus in Artemisia annua. *Plant Science, 151*, 67–73.

Maia, M. M. D., Heasley, A., Camargo de Morais, M. M., Melo, E. H. M., et al. (2001). Effect of culture conditions on lipase production by Fusarium solani in batch fermentation. *Bioresource Technology, 76*, 23–27.

Mane, R. S., Shinde, M. B., Wagh, P. R., & Malkar, H. M. (2017). Isolation of endophytic microorganisms as a source of novel secondary metabolite producers against Tuberculosis. *International Journal for Science and Advance Research in Technology, 3*, 1267–1269.

Maria, G. L., Sridhar, K. R., & Raviraja, N. S. (2005). Antimicrobial and enzyme activity of mangrove endophytic fungi of southwest coast of India. *Journal of Agricultural Technology, 1*, 67–80.

Marinho, A. M., Rodrigues-Filho, E., Moitinho, M. L. R., & Santos, L. S. (2005). Biologically active polyketides produced by *Penicillium janthinellum*isolated as an endophytic fungus from fruits of *Melia azedarach*. *Journal of the Brazilian Chemical Society, 16*, 280–283.

Marques, T. A., Baldo, C., Borsato, D., Buzato, J. B., & Celligo, M. A. P. C. (2014). Production and partial characterization of a thermostable alkaline and organic solvent tolerant lipase from *Trichoderma atroviride* 676. *International Journal of Scientific and Technology Research, 3*, 77–83.

Martins, E. S., Silva, D., Da, S., & R., & Gomes, E. (2002). Solid state production of thermostable pectinases from thermophilic Thermoascusaurantiacus. *Process Biochemistry, 37*, 949–954.

Mitchell, A.M., Strobel, G.A., Hess, W.M., Vargas, P.N. & Ezra, D. (2008) *Muscodorcrispans*, a novel endophyte from *Anansananassoides*in the Bolivian Amazon. *Fungal Diversity, 31*, 37–43.

Mohamed, S. A., Abdel-Mageed, H. M., Tayel, S. A., El-Nabrawi, M. A., & Fahmy, A. S. (2011). Characterization of *Mucor racemosus*lipase with potential application for the treatment of cellulite. *Process Biochemistry, 46*, 642–648.

Moonjely, S., Barelli, L., & Bidochka, M. K. (2016). Insect pathogenic fungi as endophytes. *Advances in Genetics, 94*, 107–135.

Moy, M., Li, H. M., Sullivan, R., White, J. F., Jr., & Belanger, F. C. (2002). Endophytic fungal β-1, 6-glucanase expression in the infected host grass. *Plant Physiology, 130*, 1298–1308.

Nag Raj, T. R. (1993). *Coelomycetous anamorphs with appendage bearing conidia* (p. 1101). Edwards Brothers Publishing Co.

Nassimi, Z., & Taheri, P. (2017). Endophytic fungus *Piriformospora indica* induced systemic resistance against rice sheath blight via affecting hydrogen peroxide & antioxidants. *Biocontrol Science and Technology, 27*, 1–16.

Nicoletti, R., & Fiorentino, A. (2015). Plant bioactive metabolites and drugs produced by endophytic fungi of spermatophyta. *Agriculture, 5*, 918–970.

Nieto-Domínguez, M., de Eugenio, L. I., York-Durán, M. J., Rodríguez-Colinas, B., et al. (2017). Prebiotic effect of xylooligosaccharides produced from birchwood xylan by a novel fungal GH11 xylanase. *Food Chemistry, 232*, 105–113.

Nongalleima, K. H., Dey, A., Deb, L., Singh, C., Biseshwori, T., Devi, H. S., & Devi, S. I. (2013). Endophytic fungus isolated from *Zingiber zerumbet* (L.) Sm. inhibits free radicals and cyclooxygenase activity. *International Journal of PharmTech Research, 5*, 301–307.
Nwuche, C. O., & Ogbonna, J. C. (2011). Isolation of lipase producing fungi from palm oil mill effluent (POME) dump sites at Nsukka. *Brazilian Archives of Biology and Technology, 54*, 113–116.
Ofek-Lalzar, M., Gur, Y., ben-Moshe, S., Sharon, O., Kosman, E., Mochli, E., & Sharon, A. (2016). Diversity of fungal endophytes in recent and ancient wheat ancestors *Triticum dicoccoides* and *Aegilops sharonensis*. *FEMS Microbiology Ecology, 92*(10), fiw152.
Oliveira, A. C. D., Fernandes, M. L., & Mariano, A. B. (2014). Production and characterization of an extracellular lipase from *Candida guilliermondii*. *Brazilian Journal of Microbiology, 45*(4), 1503–1511.
Onofre, S. B., Mattiello, S. P., da Silva, G. C., Groth, D., & Malagi, I. (2013). Production of cellulases by the endophytic fungus *Fusarium oxysporum*. *Journal of Microbiology, 3*, 131–134.
Owen, N. L., & Hundley, N. (2004). Endophytes – The chemical synthesizers inside plants. *Science Progress, 87*, 79–99.
Pachauri, P., Aranganathan, V., More, S., Sullia, S. B., & Deshmukh, S. (2017). Purification and characterization of cellulase from a novel isolate of *Trichoderma longibrachiatum*. *Biofuels, 1*, 7.
Patil, M. G., Pagare, J., Patil, S. N., & Sidhu, A. K. (2015a). Extracellular enzymatic activities of endophytic fungi isolated from various medicinal plants. *International Journal of Current Microbiology and Applied Sciences, 4*, 1035–1042.
Patil, M. P., Patil, R. H., & Maheshwari, V. L. (2015b). Biological activities and identification of bioactive metabolite from endophytic Aspergillus flavus L7 isolated from *Aeglemarmelos*. *Current Microbiology, 71*, 39–48.
Peng, X. W., & Chen, H. Z. (2007). Microbial oil accumulation and cellulase secretion of the endophytic fungi from oleaginous plants. *Annals of Microbiology, 57*(2) 239–242.
Petrini, O., & Fisher, P. J. (1988). A comparative study of fungal endophytes in xylem and whole stem of *Pinus sylvestris* and *Fagus sylvatica*. *Transactions of the British Mycological Society, 91*, 233–238.
Petrini, O., Sieber, T. N., Toti, L., & Viret, O. (1992). Ecology, metabolite production and substrate utilization in endophytic fungi. *Natural Toxins, 1*, 185–196.
Pointing, S. B. (2001). Feasibility of bioremediation by white-rot fungi. *Applied Microbiology and Biotechnology, 57*, 20–33.
Prabha, P., Govindan, K., Suganya, K., & Murugan, M. (2018). Bioactive potential of secondary metabolites derived from medicinal plant endophytes. *Egyptian Journal of Basic and Applied Sciences, 5*(4), 303–312.
Prabavathy, D., & Nachiyar, V. (2012). Screening for extracellular enzymes and production of cellulase by an endophytic *Aspergillus* sp, Using Cauliflower Stalk as substrate. *International Journal of Applied Biomedical Engineering, 6*(2).
Prabavathy, D., & Nachiyar, V. (2013). Screening the Bioactivity of ethyl acetate extract of endophytic *Phoma* sp. isolated from *Vitex negundo*. *International conference on chemical and environmental engineering,* pp. 15–16.
Puntambekar, U. S. (1995). Cellulase production by the edible mushroom Volvarielladiplasia. *World Journal of Microbiology and Biotechnology, 11*, 695.
Puri, S. C., Verma, V., Amna, T., Qazi, G. N., & Spiteller, M. (2005). An endophytic fungus from *Nothapodytesfoetida*that produces Camptothecin. *Journal of Natural Products, 68*, 1717–1719.
Rabha, A. J., Naglot, A., Sharma, G. D., Gogoi, H. K., & Veer, V. (2014). In vitro evaluation of antagonism of endophytic *Colletotrichumgloeosporioides* against potent fungal pathogens of *Camelliasinensis*. *Indian Journal of Microbiology, 54*, 302–309.
Raju, D. C., Thomas, S. M., & Thomas, S. E. (2015). Screening for extracellular enzyme production in endophytic fungi isolation from *Calophylluminophyllum* L leaves. *Journal of Chemical and Pharmaceutical Research, 7*, 900–904.

Ramawat, K. G., Dass, S., & Mathur, M. (2009). The chemical diversity of bioactive molecules and therapeutic potential of medicinal plants. In K. G. Ramawat (Ed.), *Herbal drugs: Ethnomedicine to modern medicine* (pp. 7–32). Springer.

Ratnaweera, P. B., Williams, D. E., de Silva, E. D., Wijesundera, R. L., Dalisay, D. S., & Andersen, R. J. (2014). Helvolic acid, an antibacterial nortriterpenoid from a fungal endophyte, *Xylaria* sp. of orchid *Anoectochilussetaceus*endemic to Sri Lanka. *Mycology, 5*, 23–28.

Ratnaweera, P. B., Williams, D. E., Patrick, B. O., de Silva, E. D., & Andersen, R. J. (2015). Solanioic acid, an antibacterial degraded steroid produced in culture by the fungus *Rhizoctonia solani*isolated from tubers of the medicinal plant *Cyperus rotundus*. *Organic Letters, 17*, 2074–2077.

Redman, R. S., Sheehan, K. B., Stout, R. G., Rodrigues, R. J., & Henson, J. M. (2002). Thermotolerance conferred to plant host and fungal endophyte during mutualistic symbiosis. *Science, 298*, 1581.

Rehman, S., Shawl, A. S., Sultana, S., Kour, A., Riyaz-ul-hassan, S., Ghulam, N., & Q. (2009). In vitro cytotoxicity of an endophytic fungus isolated from *Nothapodytesfoetida*. *Annales de Microbiologie, 59*, 157–161.

Reiter, B., Pfeifer, U., Schwab, H., & Sessitsch, A. (2002). Response of endophytic bacterial communities in potato plants to infection with *Erwinia carotovora* sub sp. *atroseptica*. *Applied and Environmental Microbiology, 68*, 2261–2268.

Riga, E., Lacey, L. A., & Guerra, N. (2008). Muscodor albus, a potential biocontrol agent against plant-parasitic nematodes of economically important vegetable crops in Washington State, USA. *Biological Control, 45*, 380–385.

Rollinger, J. L., & Langenheim, J. H. (2018). Geographic survey of fungal endophyte community composition in leaves of coastal redwood. *Mycologia, 85*(2), 149–156.

Ryan, R. P., Germaine, K., Franks, A., Ryan, D. J., & Dowling, D. N., (2008). Bacterial endophytes: recent developments and applications. *FEMS Microbiology Letters, 278*, 1–9.

Sahoo, K., Dhal, N. K., & Das, R. (2014). Production of amylase enzyme from mangrove fungal isolates. *African Journal of Biotechnology, 13*, 4338–4346.

Saikkonen, K., Wäli, P., Helander, M., & Faeth, S. H. (2004). Evolution of endophyte-plant symbioses. *Trends in Plant Science, 9*, 275–280.

Saleem, A., & Ebrahim, M. K. H. (2014). Production of amylase by fungi isolated from legume seeds collected in AlmadinahAlmunawwarah Saudi Arabia. *Journal of Taibah University for Science, 8*, 90–97.

Sara, B., Noreddine, K. C., & Jacqueline, D. (2016). Production of laccase without inducer by *Chaetomium* species isolated from Chettaba forest situated in the East of Algeria. *African Journal of Biotechnology, 15*, 207–213.

Saxena, R. K., Malhotra, B., & Batra, A. (2005). Commercial importance of some fungal enzymes. In D. K. Arora (Ed.), *Handbook of fungal biotechnology* (2nd ed., pp. 287–298). Marcel Dekker.

Schard, C. L., Leuchtmann, A., & Spiering, M. J. (2004). Symbioses of grasses with seed borne fungal endophytes. *Annual Review of Plant Biology, 55*, 315–340.

Schulz, B., & Boyle, C. (2005). The endophytic continuum. *Mycological Research, 109*, 661–686.

Schulz, B., & Boyle, C. (2006). What are endophytes? In B. Schulz, C. Boyle, & T. N. Sieber (Eds.), *Microbial root endophytes. Soil biology* (Vol. 9, pp. 1–13). Springer.

Schulz, B., Boyle, C., Draeger, S., Rommert, A. K., & Krohn, K. (2002). Endophytic fungi: A source of novel biologically active secondary metabolites. *Mycological Research, 9*, 996–1004.

Senthilmurugan, V. I. J. I. G., Sekar, R., Kuru, S., & Balamurugan, S. (2013). Phytochemical Screening, Enzyme and Antibacterial activity analysis if Endophytic Fungi Botrytis sp. Isolated From *Ficus Benghalensis*(L.). *International Journal of Pharma and Bio Sciences, 2*(4), 264–273.

Sharma, D., Pramanik, A., & Agrawal, P. K. (2016). Evaluation of bioactive secondary metabolites from endophytic fungus *Pestalotiopsis neglecta*BAB-5510 isolated from leaves of *Cupressus torulosa* D. Don. *3Biotech, 6*(2), 210.

Sieber, T. N. (1985) *EndophytischePilze von Winterweizen (Triticum aestivum L.). Ein Vergleich Zwischen Weizenausgebeiztem und solchemausungebeiztemSaatgut*. Ph.D. thesis. Zurich.

Sieber, T. N. (1988). EndophytischePilze in Nadeln von gesunden und geschädigten Fichten [*Piceaabies* (L.) Karst.]. *European Journal of Plant Pathology, 18*, 321–342.

Sieber, T. N. (1989). Endophytic fungi in twigs of healthy and diseased Norway spruce and white fir. *Mycological Research, 92*, 322–326.

Sieber, T. N. (2007). Endophytic fungi in forest trees: Are they mutualists? *Fungal Biology Reviews, 21*, 75–89.

Sinclair, J. B., & Cerkauskas, R. F. (1996). Latent infection vs. endophytic colonization by fungi. In S. C. Redlin & L. M. Carris (Eds.), *Endophytic fungi in grasses and woody plants: Systematics, ecology, and evolution* (pp. 3–29). APS Press.

Songulashvili, G., Elisashvili, V., Wasser, S. P., Nevo, E., & Hadar, Y. (2007). Basidiomycete's laccase and manganese peroxidase activity in submerged fermentation of food industry wastes. *Enzyme and Microbial Technology, 41*, 57–61.

Souza, P. M., & Magalhaes, P. O. (2010). Application of microbial a-amylase in industry—A review. *Brazilian Journal of Microbiology, 41*, 850–861.

Srilakshmi, J., Madhavi, J., Lavanya, S., & Ammani, K. (2015). Commercial potential of fungal protease: Past, present and future prospects. *International Journal of Pharmaceutical, Chemic al and Biological Sciences, 2*, 218–234.

Srivastava, A., & Kar, R. (2009). Characterization and application of tannase produced by Aspergillus niger ITCC 651407 on pomegranate rind. *Brazilian Journal of Microbiology, 40*, 782–789.

Stinson, M., Ezra, D., Hess, W. M., Sears, J., & Strobel, G. A. (2003). An endophytic *Gliocladium* sp. of *Eucryphiacordifolia* producing selective volatile antimicrobial compounds. *Plant Science, 165*, 913–922.

Strobel, G., & Daisy, B. (2003). Bioprospecting for microbial endophytes and their natural products. *Microbiology and Molecular Biology Reviews, 67*, 491–502.

Strobel, G. A., Dirksie, E., & Sears, & Markworth, C. (2001). Volatile antimicrobials from *Muscodor albus*, a novel endophytic fungus. *Microbiology, 147*, 2943–2950.

Strobel, G., Ford, E., Worapong, J., Harper, J. K., Arif, A. M., Grant, D. M., Fung, P. C., & Chau, R. M. W. (2002). Isopestacin, an isobenzofuranone from *Pestalotiopsismicrospora*, possessing antifungal and antioxidant activities. *Phytochemistry, 60*, 179–183.

Strobel, G., Tomsheck, A., Geary, B., Spakowicz, D., Strobel, S., Mattner, S., & Mann, R. (2010). Endophyte Strain NRRL 50072 producing volatile organics is a species of Ascocoryne. *Mycology, 1*, 187–194.

Sukumaran, R. K., Singhania, R. R., & Pandey, A. (2005). Microbial cellulases-production, applications and challenges. *Journal of Scientific and Industrial Research, 64*(11), 832–844.

Sunitha, V. H., Devi, D. N., & Srinivas, C. (2013). Extracellular enzymatic activity of endophytic fungal strains isolated from medicinal plants. *World Journal of Agricultural Sciences, 9*, 1–9.

Suryanarayan, T. S., Kumarsan, V., & Jonson, J. A. (1998). Foliar fungal endophytes from two species of the mangrove Rhizophora. *Canadian Journal of Microbiology, 44*, 1003–1006.

Suryanarayanan, T. S. (1992). Light-incubation: a neglected procedure in mycology. *Mycologist, 6*, 144.

Suryanarayanan, T. S., Thirunavukkarasu, N., Govindarajulu, M. B., Sasse, F., Jansen, R., & Murali, T. S. (2009). Fungal endophytes and bioprospecting: An appeal for a concerted effort. *Fungal Biology, 23*(1-2), 9–19.

Suto, M., Takebayashi, M., Saito, K., Tanaka, M., Yokota, A., & Tomita, F. (2002). Endophytes as producers of xylanase. *Journal of Bioscience and Bioengineering, 93*, 88–90.

Tadych, M., Bergen, M. S., & White, J. F., Jr. (2014). Epichloe spp. associated with grasses: New insights on life cycles, dissemination and evolution. *Mycologia, 106*(2), 181–201.

Tan, R. X., & Zou, W. X. (2001). Endophytes: A rich source of functional metabolites. *Natural Product Reports, 18*, 448–449.

Tawfike, A. F., Romli, M., Clements, C., Abbott, G., Young, L., Schumacher, M., Edrada, E., & R. (2019). Isolation of anticancer and anti-trypanosome secondary metabolites from the endophytic fungus *Aspergillusflocculus* via bioactivity guided isolation and MS based metabolomics. *Journal of Chromatography B, 1106–1107*, 71–83.

Tian, J., Fu, L., Zhang, Z., Dong, X., Xu, D., Mao, Z., Liu, Y., Lai, D., & Zhou, L. (2017). Dibenzo-α-pyrones from the endophytic fungus *Alternaria* sp. Samif01: Isolation, structure elucidation, and their antibacterial and antioxidant activities. *Natural Product Research, 31*, 387–396.

Torres, D. P. M., Goncalves, M. P. F., Teixeir, J. A., & Rodrigues, L. R. (2010). Galacto-oligosaccharides: Production properties applications and significance as prebiotics. *Comprehensive Reviews in Food Science and Food Safety, 9*, 438–454.

Ul-Haq, I., Idrees, S., & Rajoka, M. I. (2002). Production of lipases by Rhizopus oligosporus by solid-state fermentation. *Process Biochemistry, 37*, 637–641.

Ullrich, R., Huong, L. M., Dung, N. L., & Hofrichter, M. (2005). Laccase from the medicinal mushroom Agaricus blazei: Production purification and characterization. *Applied Microbiology and Biotechnology, 67*, 357–363.

Valladao, A. B. G., Sartore, P. E., Freire, D. M. G., & Cammarota, M. C. (2009). Evaluation of different pre-hydrolysis times and enzyme pool concentrations on the biodegradability of poultry slaughterhouse wastewater with a high fat content. *Water Science and Technology, 60*(1), 243–249.

Valladao, A. B. G., Torres, A. G., Freire, D. M., & Cammarota, M. C. (2011). Profiles of fatty acids and triacylglycerols and their influence on the anaerobic biodegradability of effluents from poultry slaughterhouse. *Bioresource Technology, 102*, 7043–7050.

Vandenkoornhuyse, P., Quaiser, A., Duhamel, M., Le, V., & A., & Dufresne, A. (2015). The importance of the microbiome of the plant holobiont. *The New Phytologist, 206*(4), 1196–1206.

Vázquez, M. A., Cabrera, E. C. V., Aceves, M. A., & Mallol, J. L. F. (2018). Cellulolytic and ligninolytic potential of new strains of fungi for the conversion of fbrous substrates. *Biotechnology Research and Innovation, 3*(1), 177–186.

Veselá, M., & Friedrich, J. (2009). Amino acid and soluble protein cocktail from waste keratin hydrolysed by a fungal keratinase of Paecilomycesmarquandii. *Biotechnology and Bioprocess Engineering, 14*, 84–90.

Wagenaar, M. M., Corwin, J., Strobel, G., & Clardy, J. (2000). Three new cytochalasins produced by an endophytic fungus in the genus *Rhinocladiella*. *Journal of Natural Products, 63*, 1692–1695.

Waller, F., Achatz, B., Baltruschat, H., Fodor, J., Becker, K., Fischer, M., Heier, T., Hückelhoven, R., Neumann, C., Von Wettstein, D., Franken, P., & Kogel, K. H. (2005). The endophytic fungus *Piriformsporaindica* reprograms barley to salt-stress tolerance, disease resistance, and higher yield. *Proceedings of the National Academy of Sciences USA, 102*, 13386–13391.

Wang, Y., & Guo, L. D. (2007). A comparative study of endophytic fungi in needles, bark, and xylem of *Pinus tabulaeformis*. *Canadian Journal of Botany, 85*, 911–917.

Wang, H. T., & Hsu, J. T. (2006). Usage of enzyme substrate to protect the activities of cellulase protease and α-amylase in simulations of monogastric animal and avian sequential total tract digestion. *Asian-Australasian Journal of Animal Sciences, 19*, 1164–1173.

Wang, J., Li, G., Lu, H., Zheng, Z., Huang, Y., & Su, W. (2000). Taxol from *Tubercularia* sp. strain TF5, an endophytic fungus of *Taxusmairei*. *FEMS Microbiology Letters, 193*, 249–253.

Wang, Y., Zheng, Z., Liu, S., Zhang, H., Li, E., Guo, L., & Che, Y. (2010). Oxepinochromenones, furochromenone, and their putative precursors from the endolichenic fungus *Coniochaeta* sp. *Journal of Natural Products, 73*, 920–924.

Wang, L. W., Xu, B. G., Wang, J. Y., Su, Z. Z., Lin, F.-C., Zhang, C. L., & Kubicek, C. P. (2012). Bioactive metabolites from *Phoma*species, an endophytic fungus from the Chinese medicinal plant *Arisaemaerubescens*. *Applied Microbiology and Biotechnology, 93*, 1231–1239.

Wang, Q. X., Bao, L., Yang, X. L., Liu, D. L., Guo, H., Dai, H. Q., Song, F. H., Zhang, L. X., Guo, L. D., & Li, S. J. (2013). Ophiobolins P–T, five new cytotoxic and antibacterial sesterterpenes from the endolichenic fungus *Ulocladium* sp. *Fitoterapia, 90*, 220–227.

Weber, D., Sterner, O., Anke, T., Gorzalczancy, S., Martino, V., & Acevedo, C. (2004). Phomol, a new anti-inflammatory metabolite from an endophyte of the medicinal plant *Erythrina crista-galli*. *The Journal of Antibiotics, 57*, 559–563.
Weihua, Q., & Hongzhang, C. (2008). An alkali-stable enzyme with laccase activity from entophytic fungus and the enzymatic modification of alkali lignin. *BioresourTechnol, 99*, 5480–5484.
Wen, X., Jia, Y., & Li, J. (2009). Degradation of tetracycline and oxytetracycline by crude lignin peroxidase prepared from Phanerochaetechrysosporium—A white rot fungus. *Chemosphere, 75*, 1003–1007.
White, J. F. (1988). Endophyte-host associations in forage grasses. A proposal concerning origin and evolution. *Mycologia, 80*(4), 442–446.
Wohlgemuth, R. (2010). Biocatalysis-key to sustainable industrial chemistry. *Current Opinion in Biotechnology, 21*, 713–724.
Wu, S. H., Chen, Y. W., Shao, S. C., Wang, L. D., Yu, Y., Li, Z. Y., Yang, L. Y., Li, S. L., & Huang, R. (2009). Two new solanapyrone analogues from the endophytic fungus *Nigrospora* sp. YB-141 of *Azadirachta indica*. *Chemistry & Biodiversity, 6*, 79–85.
Xu, Q. Y., Huang, Y. J., Zheng, Z. H., & Song, S. Y. (2005). Purification, elucidation and activities study of cytosporone. *Journal of Xiamen University Natural Science, 44*, 425–428.
Yadav, A. N. (2015). *Bacterial diversity of cold deserts and mining of genes for low temperature tolerance*. Ph.D. thesis, IARI, New Delhi/BIT, Ranchi. p. 234.
Yao, J., Guo, G. S., Ren, G. H., & Liu, Y. H. (2014). Production, characterization and applications of tannase. *Journal of Molecular Catalysis B, 101*, 137–147.
Yasser, M. M., Mousa, A. S. M., Marzouk Marym, A., & Tagyan, A. I. (2019). Molecular identification, extracellular enzyme production and antimicrobial activity of endophytic fungi isolated from solanum tuberosum L. in Egypt. *Biosciences Biotechnology Research Asia, 16*(1), 135–142.
Ye, Z., Pan, Y., Zhang, Y., Cui, H., Jin, G., McHardy, A. C., Fan, L., & Yu, X. (2017). Comparative whole-genome analysis reveals artificial selection effects on *Ustilago esculenta* genome. *DNA Research, 24*, 635–648.
Yuan, C., Wang, H. Y., Wu, C. S., Jiao, Y., Li, M., Wang, Y. Y., Wang, S. Q., Zhao, Z. T., & Lou, H. X. (2013). Austdiol, fulvic acid and citromycetin derivatives from an endolichenic fungus, *Myxotrichum* sp. *Phytochemistry Letters, 6*, 662–666.
Yuan, Z., Tian, Y., He, F., & Zhou, H. (2019). Endophytes from *Ginkgobiloba* and their secondary metabolites. *Chinese Medicine, 14*(1), 14–51.
Zabalgogeazcoa, I. (2008). Fungal endophytes and their interections with plant pathogens. *Spanish Journal of Agricultural Research, 6*, 138–146.
Zhang, S. S.-Q. O. M., & Qi-Yong, Z.-D. T. (2007). Isolation and characterization of endophytic microorganisms in glycyrrhiza inflata Bat. from Xinjiang [J]. *Microbiology, 5*, 14.
Zhang, H. W., Song, Y. C., & Tan, R. X. (2006). Biology and chemistry of endophytes. *Natural Product Reports, 23*, 753–771.
Zhang, J. Y., Tao, L. Y., Liang, Y. J., Chen, L. M., Mi, Y. J., & Zheng, L. S. (2010). Anthracenedione derivatives as anticancer agents isolated from secondary metabolites of the mangrove endophytic fungi. *Marine Drugs, 8*, 1469–1481.
Zhang, F., Li, L., Niu, S., Si, Y., Guo, L., Jiang, X., & Che, Y. (2012). A thiopyranchromenone and other chromone derivatives from an endolichenic fungus, *Preussia*africana. *Journal of Natural Products, 75*, 230–237.
Zhou, Y. G., Gong, S., Zhang, J., Wang, L., & Yu, X. (2014). A new species of the genus *Trematosphaeria*from China. *Mycological Progress, 13*, 33–43.
Zhou, P., Wu, Z., Tan, D., Yang, J., Zhou, Q., Zeng, F., Zhang, M., Bie, Q., Chen, C., & Xue, Y. (2017). Atrichodermones A–C, three new secondary metabolites from the solid culture of an fungal strain, *Trichoderma atroviride*. *Fitoterapia, 123*, 18–22.

Chapter 3
The Development of White-Rot Fungi as a Mycoremediation Product

A. A. Ngadin, E. Taghavi, and T. Eaton

Contents

A. A. Ngadin (✉) · E. Taghavi
Faculty of Fisheries and Food Science, Universiti Malaysia Terengganu, Terengganu, Malaysia
e-mail: andrew@umt.edu.my

T. Eaton
Gratton Way, Concord, CA, USA

A. C. Shukla (ed.), *Applied Mycology*, Fungal Biology,
https://doi.org/10.1007/978-3-030-90649-8_3

3.1 Introduction

3.1.1 Habitat's Contamination

A variety of environmental problems have affected economies and policies around the world (Tsui, 2020). There are many environmental problems related to community behavior, such as producing habitat pollution, health risks and problems, government organizations, the private sector, and industry.

Habitat contamination is one of the environmental issues that all human beings are responsible to solve immediately. Water pollution will lead to the disappearance of biodiversity and aquatic ecosystems. In other words, the aquatic organisms will lose their home. Many reported articles include from Global Water Pollution and Human Health, stating pollution of freshwater resources, such as lakes, rivers, and groundwater, has emphasized chemical pollution. This includes both organic and inorganic pollutants affecting the level of water pollution

A major challenge in determining environmental risks for heavy metals and metalloids is associated with their contrast activity under different redox conditions. Oxidation or reduction reactions, complexes, absorption, and precipitation are key processes that determine their transport and bioavailability. If there is oxygen and in a state of reduction, the solubility of most metal elements is very different. Redox-sensitive metals, iron, and manganese being the most abundant, forming finely dispersed oxide particles that highly absorb heavy metals and metals under toxic conditions.

Interestingly, when oxygen has been depleted, these oxide particles undergo a reductive decomposition solution and release a toxic load. While microorganisms also function to control the precipitation and breakdown of reactive particles in the atmosphere. Most aquatic organisms including microbes are very sensitive to any changes in their environment. They respond to and adapt to this new environment, such as polluted conditions, in many ways. Fewer responses may include reduced reproductive capacity and also the removal of certain enzymatic systems necessary for conventional metabolism. Potential fungi were determined to contribute to the trophic dynamics of freshwater environments. These organisms not only modulate aquatic productivity through an intermediary level of occupying in the food chain, but also indicate the environment status in a definite period. Furthermore, their range has grown in importance in recent years as a result of their specific species' ability to detect any decline in water quality as a result of contamination and eutrophication (Bassem, 2020).

Soil pollution also may affect habitat contamination when the presence of toxic chemicals like herbicides, pesticides, and fungicides in soil, which has caused a high risk to the ecosystem. Various compounds often enter the soil from the atmosphere, such as precipitation water, as well as from surface water sources and shallow groundwater flowing through the soil, as well as from wind movement or other forms of soil disturbances. When the amounts of soil contaminants exceed natural levels, pollution is generated (Osama et al., 2020).

The agricultural industry is very important for food production, but inefficient agricultural practices will continue to contribute to severe soil pollution. For example, copper has been widely used as a fungicide and its concentration can increase soil pollution to a higher level based on how this chemical is used by farmers. These chemicals are also used in animal feed and released into the atmosphere as fertilizer products that are spread through grasslands as well as other agricultural land. In addition, Cadmium is used as a fertilizer and a highly toxic metal in agricultural land. All these pollutants if not handled properly will then produce a lot of metals and natural pollutants from these organic materials such as sewage sludge, dirt, compost, and biowaste (European Environment Agency, 2020).

3.1.2 Health Risks and Issues

Polluted environmental problems can result in serious health risks for humans. The factors of clean air, unpolluted drinking water, nutritious food, and safe living all depend on its environment being well managed by the community. The increase in human population often causes unhealthy environmental problems and affects the health of the population as well as increased medical costs. This will result in a short human life span if the quality of the environment is affected.

Resnik and Portier (2015) stated in the article Environment, Ethics, and Human Health that although the environment can sustain human life, it can also be a source of disease if the environment is mismanaged. Human mortality rates can be high if basic medical needs are inadequate. The risk of cancer, heart disease, asthma, and various other diseases are all due to polluted environmental factors. Solid waste, toxic materials, and food contaminants, or hazardous are factors that can affect the quality of human health such as used vehicle batteries will be dumped on the ground and hydrogen gas can leak from batteries; potassium hydroxide, a caustic agent that can cause respiratory, eye, and skin irritations. If battery acid makes contact with your skin, it can create a skin reaction. Chemical burns can be the result. Unlike thermal burns caused by fire or heat, burns caused by batteries can quickly dissolve your skin. In addition, various infectious illnesses can be caused by contaminated drinking water, inadequate sanitation, and poor hygiene.

Exposure to agricultural chemicals is one of the human challenges in maintaining health and the environment. For example, pesticides play a critical role in improving agricultural yields while also posing health and environmental hazards. Alternatives to the use of pesticides result in the deterioration of human health. Avoiding the use of all pesticides will have a major impact on agricultural production, resulting in food shortages and higher food prices. To increase food production as well as minimize the use of harmful chemicals, public health authorities have chosen to control the use of pesticides and encourage innovation projects based on the idea of integrity in science and technology. These efforts can produce products and reduce human health risks. One of the serious health risks is cancer and this is due to the agricultural sector such as chemicals, combustion smoke, contaminated drinking water, chemicals found in food, and poor air quality.

3.1.3 Industry Sectors

Private industry needs to have higher standards to prevent any kind of contamination. Most of industrial pollution can be directly linked worldwide. There are a number of forms of industrial pollution that can impact air quality, and it can enter the soil, causing widespread environmental problems. An article on environmental impacts of industrialization highlights four primary impact points: air, water, soil, and habitat (Fig. 3.1).

The most serious problem of the industrial sector is air pollution, caused by smoke or burning of fossil fuels but water pollution is also a serious issue in factory areas especially those built near natural water sources. These contaminants may be solid, liquid, or gaseous, and all have the potential to contaminate local water sources. Industrial sites can also result in soil pollution. Although lead contamination is the most common, heavy metals and other toxic chemicals can also get into the soil and contaminate agricultural crops. In addition, contaminated soil is also not suitable for use if it contains contaminated substances that are harmful to the health of organisms. The destruction of these habitats can disrupt local ecosystems and can result in the extinction of plants and animals if the species is unable to migrate or adapt to its new environment (Folk, 2018).

The harmful impacts of pharmaceutical manufacturing on the environment are well-known. However, because it is not fully managed, its toxic effects on animals

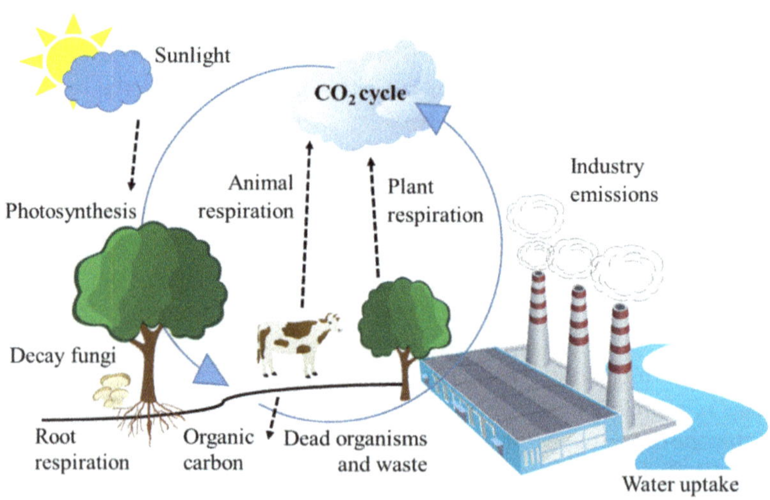

Fig. 3.1 The carbon cycles in natural ecosystem

and humans continue to linger until it is difficult to control. Pharmaceutical wastes are frequently released into the atmosphere. Throwing antibiotics into the environment poses a high risk of the level of resistance of various natural pathogens and pushes their strength for resistance to be treated if infected by humans or animals. It is clear that this global problem is often discussed in terms of pharmaceutical contamination. This problem will have a more significant impact especially on those living near water-processing plants and food supplies that have been contaminated with pharmaceutical products.

Animals, particularly fish that dwell in contaminated water, are negatively affected by pharmaceutical product contamination in the environment. Many rivers in Europe, for example, are home to intersex male fish that show feminine sexual traits, including female reproductive anatomy, according to a research published in the scientific journal *Nature* in 2009. Males can also create vitellogenin, a protein present in eggs that can be stimulated by hormones (Nawrat, 2018).

3.1.4 Communities

Environmental problems are the result of human decisions and activities that harm the environment. Individuals who act irresponsibly will make choices and participate in actions that affect the climate, such as dumping garbage, clearing forest areas for agriculture, roads, and factory development. Many companies, government agencies, and other organizations make decisions and participate in activities that have a significant negative impact on the environment. Individuals and organizations may be fully aware that their actions have damaged the environment, or they may be insensitive without thinking about the possible environmental consequences of their actions.

Humans can influence environmental problems such as the release of chemicals in soil and rivers. These include drain cleaners, laundry powder, paints, and pesticides, all things to avoid at any cost. It has been reported that arts and crafts materials as well as lawn care items can harm our environment and pollute the environment (Lindsey, 2017). The health risks from applying pesticides on crops or animals, drinking dirty water, inhaling polluted air, and being exposed to animal hormones and genetically modified organisms are all present in the lives of every family and farmworker today. Farmers are often affected by risks arising from disturbances to the environment, such as chemical pollution of water and air, sediments, and harmful odors. However, it gave rise to the idea to help them through innovative products that can reduce the low level of health of farmers such as the use of products that can absorb these chemicals. Environmental challenges to land production are more long-term and indirect, but they are still real threats to the viability of farms and farming (Ikerd, 1999).

3.2 The Role of White-Rot Fungi

White-rot fungi are a heterogeneous group that have an extraordinary function in breakdown components of plant cell wall and play an important role in global carbon cycling in forest ecosystems. Interestingly about 90% of all wood-degrading basidiomycetes species are white-rot type (Eriksson et al., 1990; Hatakka & Hammel, 2010). This group of WRF is the most efficient lignin-degrading organisms to convert it into carbon dioxide and water (Fig. 3.2). The WRF are often found on wood trees of angiosperm species and can cause nonselective or simultaneous damage or decay. Some basidiomycetes species can degrade lignin and leave the lighter-colored cellulose, or some of them degrade both lignin and cellulose in tree wood tree cells and this is called selective WRF (Eriksson et al., 1990).

Several previous studies have reported that the WRF are promising bioremediation agents because they have a high tolerance to toxic environment and the ability to transform a complex variety of aromatic pollutants and reduce their concentration in soil or water. In addition, they can also adapt to a wide range of temperatures, pH, humidity, aeration, and biotic factors (Nur et al., 2019). As a result, WRF have the ability to generate enzymes such as lignin peroxidase, manganese peroxidase, and laccase in different abiotic and biotic factors in order to be used in a wide range of mycoremediation or bioenergy applications (Cohen et al., 2002). Fig. 3.3

White-rot fungi potential is an economic resource, since its value may be exploited as a product in commercial biotechnology applications (Table 3.1). Mycoremediation technology, which has the ability to reduce various pollutants in

Fig. 3.2 The wild oyster mushroom (white-rot fungus), *Pleurotus ostreatus*, is grows on wood, and biodegradation processes can be seen based on discoloration and mechanical changes in their natural habitat of forest

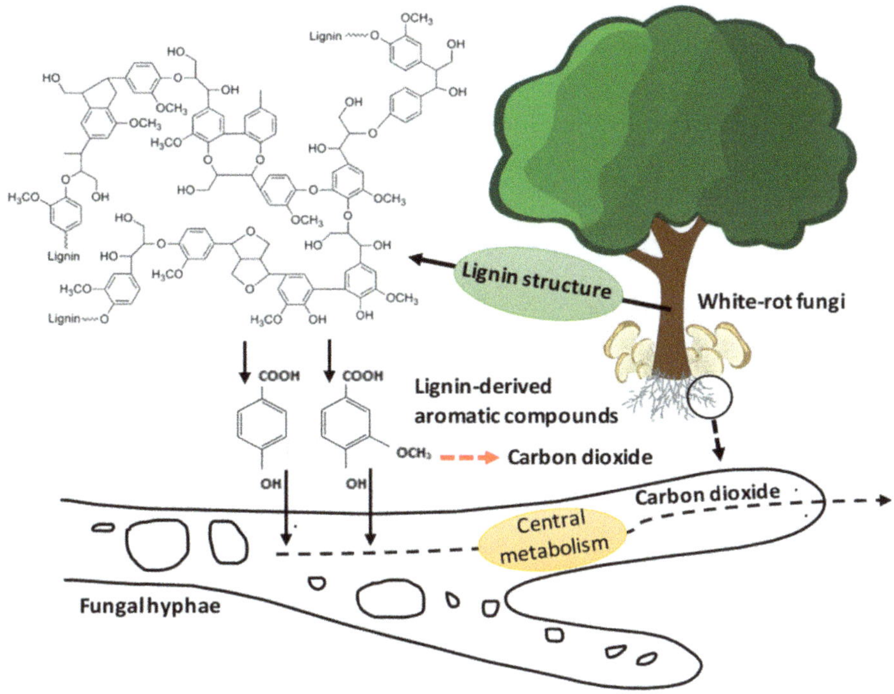

Fig. 3.3 The complex process of lignin-degrading fungi of white-rot fungi and convert it into carbon dioxide and water. (Modified Image from Davinia Salvachúa Rodríguez, National Renewable Energy Laboratory)

soils and sediments, including the dominant conventional explosives, DDT (dichlorodiphenyltrichloroethane), polynuclear aromatic hydrocarbons (PAH), and polychlorine biphenyl PCBs, can be applied to the entrepreneurial aspect of the WRF production process as a product.

3.2.1 Biodegradation Process by White-Rot Fungi

In general, WRF are responsible for the biodegradation of complex biomass, including lignin and a wide variety of environmental pollutants. There are several methods for reducing environmental pollutants and obviously, the WRF technology is different than other well-known methods of bioremediation such as bacteria. The advantage that WRF than bacterial systems are that they do not require preconditioning against certain pollutants. On the other hand, bacteria must be pre-exposed to pollutants to allow the enzymes to be induced and degrade the pollutants. The concentration of pollutants must also be present in significant amounts; otherwise, induction of enzyme synthesis will not occur by bacteria. In addition, the limited nutrients will

Table 3.1 Role of fungi in biotechnology processes of pollutants

Organism (fungus)	Chemicals/pollutants	References
Agaricus bisporus, *Lactarius piperatus*	Cadmium (II) ions	Nagy et al. (2014)
Allescheriella sp., *Stachybotrys* sp., *Phlebia* sp.	Soil contamination	Mancera-López et al. (2008)
Coriolus versicolor	Polycyclic aromatic hydrocarbons	Jang et al. (2009)
Fomes fasciatus	Copper (II)	Sutherland and Venkobachar (2013)
Flammulina velutipes	Copper	Luo et al. (2013)
Ganoderma sp.	Textile dyes Yellow, Blue and Red Procion	Barreto et al. (2011)
Jely sp., *Schizophyllum commune* and *Polyporous* sp.	Malachite green	(Tsujiyama et al., 2013)
Lentinula edodes	2,4-dichlorophenol Eucalyptus waste Barley straw, Vineyard pruning and Wheat straw	Brienzo et al. (2007), Gaitán-Hernández et al. (2006), and Lechner and Papinutti (2006)
Lentinula tigrinus	Wheat straw	Yogita et al. (2011)
Pleurotus citrinopileatus	Handmade paper and cardboard industrial waste	Kulshreshtha et al. (2013)
Pleurotus eous and *Lentinus connotus*	Paddy straw, sorghum stalk, and banana pseudostem	Rani et al. (2008)
Pleurotus florida	Handmade paper and cardboard industrial waste	Kulshreshtha et al. (2010)
Pleurotus ostreatus	Oxo-Biodegradable plastic Cadmium Extract from the sawdust	da Luz et al. (2013), Tay et al. (2011), and Akinyele et al. (2012)
Pleurotus pulmonarius	Radioactive cellulosic-based waste Crude oil	Eskander et al. (2012) and Olusola and Anslem (2010)
Pleurotus sajor-caju	Heavy metal Zn	Jibran (2011)
	Indigo	Kamida et al. (2005)
Pleurotus tuber-regium	Heavy metals Nigerian trees; *Terminalia superba*, *Mansoniaaltissima*, *Holopteliagrandis* and *Miliciaexcelsa* Cotton waste, sawdust of *Khaya ivorensis* and rice straw	Oyetayo et al. (2012), Jonathan et al. (2008), and Kuforiji and Fasidi (2008)
Pleurotus platypus, *Agaricus bisporus*, *Calocybe indica*	Copper, Zinc, Iron, Cadmium, Lead, Nickle	Prasad (2013)
Volvariella volvacea	Banana leaves (*Musa sapientum lina*) Agro-industrial residues such as cassava, sugar beet pulp, wheat bran, and apple pomace	Belewu and Belewu (2005) and Akinyele et al. (2011)

also react with white rot enzymes to perform the degradation process. Fungi continue to develop several survival mechanisms in limited and extreme conditions. These include glutathione s-transferases (GSTs) as an intracellular enzyme are also involved in stress responses and defense systems such as *Phanerochaete chrysosporium* (Morel et al., 2009).

The processes by WRF usually can be recognized when they are degrading the wood structures or any undesirable change in the properties of a wood material based on three stages: biodeterioration, biofragmentation, and assimilation (Luyt & Malik, 2019). They used wood as their source of carbon or nutrients, which will benefit other organisms when the particles of degradation process are distributed in soil. The biodeterioration process occurs on the polymer surface or inside the polymer, and it is also called a surface-level degradation that usually changes the mechanical, physical, and chemical properties of the wood material. This surface degradation occurrs rapidly when it is exposed to abiotic factors (nonbiological agents) such as sunlight, temperature, and chemicals in the environment. The influence of abiotic factors allows the degradation process due to the weak wood material structure changes (mechanical), and potential WRF performance in a wide range of exposure conditions (Luyt & Malik, 2019; Lucas et al., 2008). Although biodeterioration is the first stage, it is possible that the process of biodegradation occurred parallelly with biofragmentation (Müller, 2005). The early stages of biodeterioration are often accompanied by a discoloration of the wood, which can be difficult to recognize until the late stages of biodegradation and are easily recognized by their changes in color and chemical properties.

The biofragmentation of complex polymers in the wood cell wall is a process in which the bonds in the polymer are cleaved, producing oligomers and monomers (Luyt & Malik, 2019). White-rot fungi carry out aerobic digestion, which uses oxygen to break down these polymeric materials and it produces carbon dioxide, water, some types of residues, and new biomass (Van Der Zee, 2011). Most of WRF are potentially used due to their aerobic digestive capabilities in waste management systems and as a local renewable energy source (Klinkner, 2014).

Meanwhile, the process of assimilation occurs from biofragmentation and the product is then integrated into microbial cells. The WRF used the mycelium to carry these products into the cell by membrane carriers. These biofragmentation products will enter the catabolic pathway, leading to the production of adenosine triphosphate (ATP) or elements of cell structure (Lucas et al., 2008).

The results of this biodegradation process can give ideas to researchers to produce potential products for entrepreneurs who want to overcome environmental problems. The ability of WRF to use wood material (substrate) as a food source with physical changes in the substrate is an important indicator. Also, potential in the mycoremediation of chemical pollutants when proven by changes in the chemical properties of wood material. This will provide an opportunity to produce many products related to mycoremediation to solve the problem of polluted environment.

3.2.2 Mycoremediation of Environmental Pollutants

Various types of environmental pollution have been produced in public areas including chemicals from various sectors such as agriculture, industry, and pharmacy. The mycoremediation process uses enzymes secreted by hyphae or mycelium, often producing a mass of fibers to whiten the substrate or wood at high levels of biodegradation. The process of developing this product can be combined with the use of potential WRF with engineering technology to clean up environmental pollution. The use of bead hydrogels is a technology that needs to be developed with a combination of mycelial fungi that have potential in mycoremediation. This technology was able to maintain the structure of the hydrogel and does not significantly reduce the quality of propagule within seven months. (Kuek et al., 1992).

The ability to use wood material (substrate) as a food source by WRF by producing physical changes to the substrate is an important idea for producing new products.

3.2.3 Capsule Technology of WRF Product

Hydrogel products constitute a group of polymeric materials, the hydrophilic structure of which renders them capable of holding large amounts of water in their three-dimensional networks. In another definition, hydrogel is a polymeric material that exhibits the ability to swell and retain a large amount of water in its structure (Ahmed, 2015), but will not dissolve in water (Ahmed, 2015; Oyen, 2014). Due to their cross-linked polymer networks, hydrogels exhibit elastic solid properties with deformation and softness (Liu et al., 2020; Hong et al., 2008).

Hydrogels can be classified into various groups based on their: physical structure (amorphous, semicrystalline, hydrogen bonded, or supramolecular); electric charge (ionic (charged or neutral); crosslink (physically or chemically cross-linked); responses to external effects (stimulus-sensitive and insensitive ones); and origin (synthetic and natural) (Deligkaris et al., 2010).

Extensive employment of hydrogel products in a number of industrial and environmental areas of application is considered to be of prime importance (Ahmed, 2015). Hydrogels have been widely used in applications that closely interact with biological organisms, such as tissue engineering, drug delivery, and biological research (Liu et al., 2020). Moreover, the unique properties of hydrogels, such as superior softness, wetness, responsiveness, biocompatibility, and bioactivity (Liu et al., 2020), indeed suggest the possibility of their crucial functions in devices and machines, such as sensors, actuators, coatings, optics, electronics, and water harvesters (Liu et al., 2020; Gerlach & Arndt, 2010).

The hydrogels for agriculture can be used as a water-repellent granule product as it is capable of swelling from its original size many times when in contact with water. It has been widely proposed for use in agriculture with the aim of improving root conditions as well as supplying water to crops, by improving the water-retaining

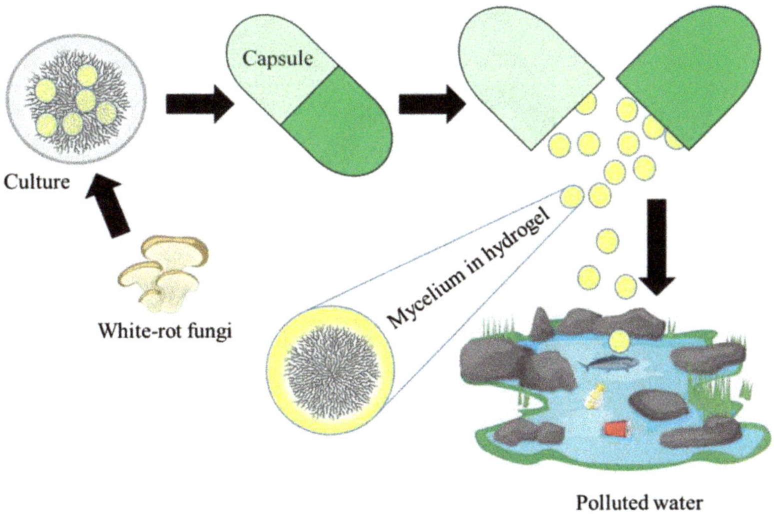

Fig. 3.4 Inoculated hydrogel by mycelium as a model product for decontamination of contaminated habitats

properties of growing media (soil or substrate without soil). Moreover, low price, high durability, and stability in swelling environment and during storage, gradual biodegradability without formation of toxic species characters encouraged scientists to apply hydrogel system in agriculture (Neethu et al., 2018).

According to the above definition, one of the emerging fields for both scientists and entrepreneurs is inoculation of hydrogels by mycelium in capsule format. This can solve the problem of pollution in high-risk areas such as contaminated wastewater pollution areas, which are close to pharmacies, food industry factories, and others. The type of hydrogel capsule inoculated with mycelium is able to retain the mycelium for a long-time during storage and control its release (i.e., once exposed to the wastewater) (Fig. 3.4).

3.2.4 Cleaning and Removal Contaminations Product

White-rot fungi have the ability to degrade the contaminated liquid, soil, sediments and demostrated their ability to survive in extreme environment conditions (Stella et al., 2017). Tišma et al., 2010 investigated the possibility of degradation of the high polyphenolic compounds, dyes, and other xenobiotics. Also, WRF have a great range of different enzymes such as hydrolytic enzymes (cellulase, pectinase, xylanase) (Tišma et al., 2010; Teerapatsakul et al., 2007). Most of these enzymes are industrially important and have great potential in the processes of bioremediation, biodegradation, biopulping, degradation, and detoxification of recalcitrant substances (Tišma et al., 2010) Fig. 3.5.

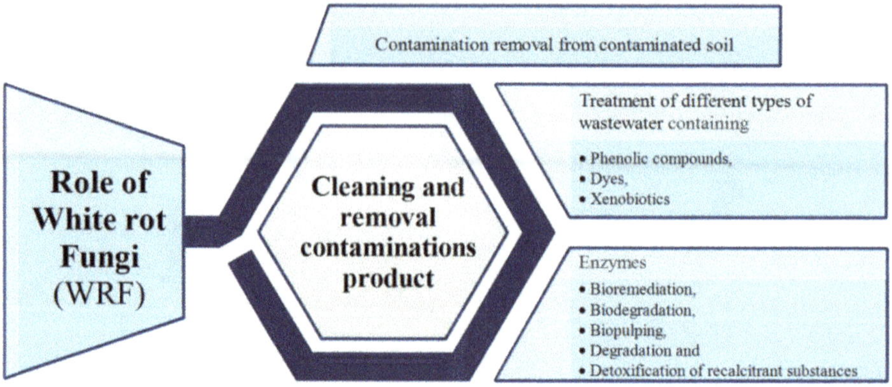

Fig. 3.5 Summary of WRF role in the removal of contamination

A previous study by Stella et al. (2017) examined the ability of *Pleurotus ostreatus* and *Irpex lacteus*, as WRF, to degrade polychlorinated biphenyls (PCBs) in actual contaminated soils with varying chemical characteristics and autochthonous microflora. *Pleurotus ostreatus* had the best results, with PCB removal rates of 18.5, 41.3, and 50.5 percent from the bulk soil, top (surface), and rhizosphere after 12 weeks of treatment, respectively. They concluded that the extensive PCB degradation performance in a relatively short period of time and the exceptional competitiveness of *P. ostreatus* make its application in the remediation of soils contaminated with PCBs potentially transferred to a larger scale (Stella et al., 2017).

Polluted waters with phenol are formed from the production of olive oil in Mediterranean region (Gianfreda et al., 2006). Olive mill wastewater is highly toxic effluent obtained from the extraction process by the olive oil industry. Numerous studies were done on using WRF in the treatment of phenols and related compounds (Laconi et al., 2007; Justino et al., 2009, 2010; D'Annibale et al., 2006; Lu et al., 2009; Bollag et al., 2003; d'Acunzo et al., 2002). For instance, Justino and coworkers in 2009 and 2010 reported the efficiency of three different approaches on phenols removal from olive mill wastewater (Justino et al., 2009, 2010). They oxidized different types of phenols that were extracted from Olive mill wastewater via biological (i.e., treatment with two fungal species *Trametes versicolor* and *Pleurotus sajor*), enzymatic, and chemical treatment. They found that chemical treatment by photo-Fenton was more efficient than biological or enzymatic treatments.

3.3 Development of Potential WRF Product to Entrepreneurs

3.3.1 Natural Environment and Municipal

The municipal management organizations might be interested based on WRF product, such as mycobead to land and water pollution issues. This product can be used as a bioremediation to chemicals. It is clear that polluted chemicals that dissolved in water cannot be treated by physical and chemical methods but rather by biological treatment. The efforts to produce WRF products in interaction with contaminants through bio-mediated and engineering techniques are required. Research and development of these products can benefit countries that produce pollutants, improve environmental policies and manage with appropriate procedures, future regulations, address bioremediation targets, identify pollutant availability, and potential threats to natural ecosystems and human health. This is considered an excellent effort for the country to protect us from pollution in natural systems and the level of threat to human health caused by various human activities.

3.3.2 Agriculture Industry

Isolation of potential species from the soil can benefit the agricultural industry and increase yields. This is through technology and products produced such as hydrogels, which can absorb chemicals such as herbicides or pesticides using the mycelial structure of the fungi. These chemicals can be converted to carbon dioxide and water by fungi. Toxins that are considered waste or used excessively can become pollutants in the soil and these will be absorbed by agricultural crop products and pose a risk to its users.

This innovative mycobead product can also be used to store potential fungi for at least 7 months without losing capacity as propagules (Kuek et al., 1992). A number of these hydrogel products are also used by fungi that can function in symbiosis with plants.

3.3.3 Food Industry

As mentioned earlier in the previous section of cleaning and removal contaminations product, WRF can help to treat the contaminated soil, water, and sediments. It is suggested to the entrepreneurs to build the greenhouse and to use the WRF near to the contaminated wastewater. This strategy can fulfill two goals: (a) application of contaminated wastewater and (b) producing WRF for human consumption. In other words, eliminate the contaminated wastewater while producing a new product (Fig. 3.6).

3.3.4 Pharmaceutical Industry

Waste generated by the pharmaceutical industry has harmful implications for the environment and public health if disposed of untreated (Shah & Shah, 2020). Most of these wastes are released into the environment at low concentrations through routine pharmaceutical use, damage and expiration. The treatment performed by

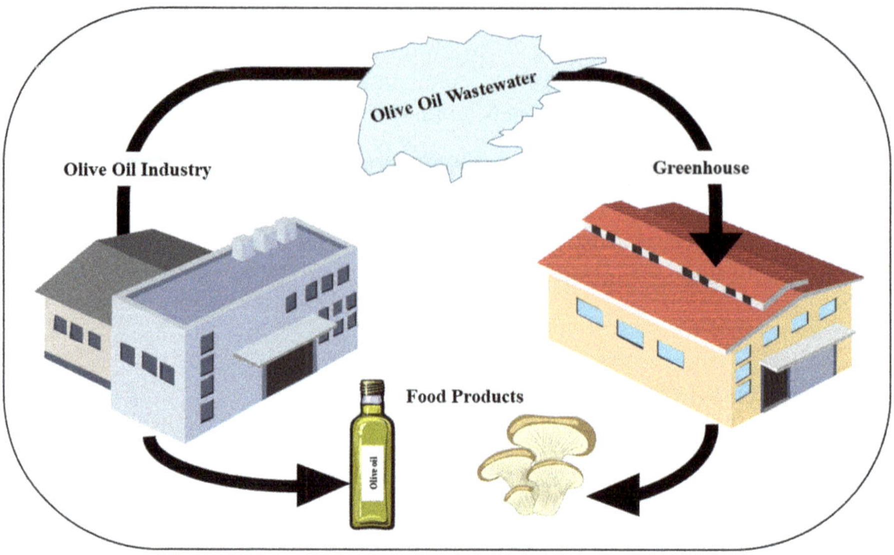

Fig. 3.6 Development of potential WRF product by entrepreneurs in the food industry

WRF can offer an environmentally friendly platform for the removal of pharmaceuticals and personal care products (PPCP) from wastewater as PPCP may have adverse effects on aquatic organisms as well as humans. This will be of great interest to the water industry. It has been reported that intact cell WRF or their extracellular lignin-converting enzyme (LME) can degrade PPCP efficiently. Treatment of PPCP mediated by WRF depends on several factors including the physicochemical properties of PPCP (e.g., hydrophobic and chemical structure) and wastewater matrix (e.g., pH, temperature, and dissolved constituents), the type of WRF species, and specific extracellular enzymes (Asif et al., 2017).

3.4 Economic Value of Mycoremediation Product in the Market

3.4.1 Benefit to Researchers and Entrepreneurs

Bioentrepreneurship is an activity that has a scientific element in the field of entrepreneurship. This is the most recent knowledge shift being discussed by academics toward industry. Biotechnology entrepreneurship is greatly aided by experienced academic research. Bio-entrepreneurship is all the activities required to set up a business that develops, manufactures, and markets biotechnology products to solve research problems. We can see that bio-entrepreneurship is more advanced now, where the biotechnology sector is producing real-world results that enable technological advancement and integration with other fields. Also, it contributes to environmental conservation, disease research, new biomedical tools, product refining using biosynthetic processing, biofabrics and biofuels, and more (Afandi, 2018).

A biotechnology company is usually started by researchers to produce a potential product. This will provide and open up many business and employment opportunities. Through the manufacture of new hydrogel bead products, bioentrepreneurs will have a lot of assessment data to improve the quality and effectiveness of the product for the future. Products will also generate profits for entrepreneurs and product creators. In this way, technology not only benefits the environment but also society. These include ideas, the basics of biological science and materials engineering technology can be improved.

3.4.2 Income of Country

The modern transition of health, food, agriculture, and bio-based industries is referred to as bioeconomy. Pharmacy, biofuels, farming, enzymes, atmosphere and climate, bio-based polymers, biorefining, and well-being and nutrition are all examples of biotechnology assets (Fig. 3.7).

The invention products of white-rot fungi that able to be marketed and used to clean environment issue are very crucial to help the environmental management

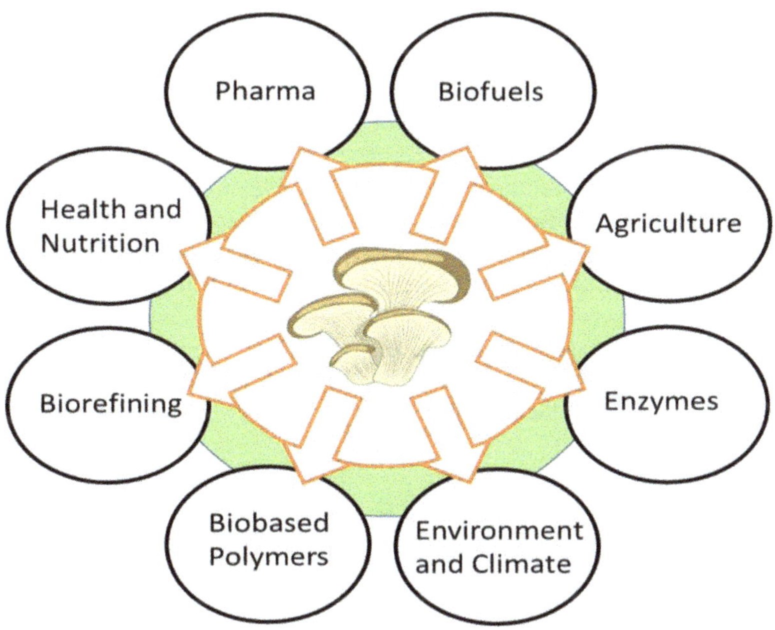

Fig. 3.7 The branch of biotechnology in the field of mushrooms

system. Products used to control or recover toxic waste are beneficial to society and the economy. In the twenty-first century, strategies to clean up toxic wastes and pollutants using affordable and efficient low costs are underway. Hydrogel bead products with capsules can actually provide a high opportunity to the income of the country, because now many environmental areas are polluted. The future of produced products will help all scientists to explore more opportunities to produce the best products. This will also benefit entrepreneurs as they will be able to market the product to the rest of the world. This will generate the income of all industries involved, including increasing economic income.

3.5 Conclusions

The environmental issues are the result of human decisions and their inappropriate activities that harm the environment implies within the communities. White-rot fungi, which are the most efficient lignin-degrading organisms to convert pollutants into carbon dioxide and water, are promising bioremediation agents. They have a high tolerance to toxic environment and the ability to transform a complex variety of aromatic pollutants and reduce them concentration in soil or water. In general, WRF are responsible for the biodegradation of complex biomass, including lignin and a wide variety of environmental pollutants. The degradation of surface structure occurred rapidly when it is exposed to abiotic factors. The integrated development of the use of hydrogel bead and mycelium fungi are potentially in mycoremediation and this technology does not significantly reduce the quality of propagule within seven months. This biodegradation process can be used by researchers to manufacture products that will benefit some entrepreneurs to solve problems related to environmental pollution. It will provide an opportunity for more mycoremediation-related products to be produced with the properties of WRF. Interestingly, hydrogel capsules inoculated with mycelium can maintain the mycelium for a long time during storage. The effort to produce WRF product in interaction with contaminants through the most appropriate bioremediation techniques and other relevant techniques are needed. These are considered very good efforts to protect us from contaminants in our environment caused by various human activities. Therefore, the WRF process can regenerate energy and save the environment. It is suggested to the entrepreneur in the food industry area to use the WRF product near the contaminated wastewater of the food and pharmaceutical factory to refine the wastewater.

References

Afandi, M. H. (2018). *What is bioentrepreneurship*. https://nuclineers.com/whats-bioentrepreneurship/. Accessed 2021.

Ahmed, E. M. (2015). Hydrogel: Preparation, characterization, and applications: A review. *Journal of Advanced Research, 6*(2), 105–121.

Akinyele, B. J., Olaniyi, O. O., & Arotupin, D. J. (2011). Bioconversion of selected agricultural wastes and associated enzymes by *Volvariella volvacea*: An edible mushroom. *Research Journal of Microbiology, 6*(1), 63–70.
Akinyele, J. B., Fakoya, S., & Adetuyi, C. F. (2012). Anti-growth factors associated with *Pleurotus ostreatus* in a submerged liquid fermentation. *Malaysian Journal of Microbiology, 8*(3), 135–140.
Asif, M. B., Hai, F. I., Singh, L., Price, W. E., & Nghiem, L. D. (2017). Degradation of pharmaceuticals and personal care products by white-rot fungi – A critical review. *Topical Collection on Water Pollution. Current Pollution Report, 3*(2), 88–103.
Barreto, W. J., Bernardino, N. D. E., & Afonso, R. (2011). Biodegradation of dyes in textile wastewater by fungus Ganoderma sp: A kinetic study. *Quím Nova, 34*(4), 568–572.
Bassem, S. M. (2020). Water pollution and aquatic biodiversity – Review Article. *Biodiversity International Journal, 4*(1), 10–16.
Belewu, M. A., & Belewu, K. Y. (2005). Cultivation of mushroom (*Volvariella volvacea*) on banana leaves. *African Journal of Biotechnology, 4*(12), 1401–1403.
Bollag, J. M., Chu, H. L., Rao, M., & Gianfreda, L. (2003). Enzymatic oxidative transformation of chlorophenol mixtures. *Journal of Environmental Quality, 32*(1), 63–69.
Brienzo, M., Silva, E. M., & Milagres, A. M. F. (2007). Degradation of eucalypt waste components by *Lentinula edodes* strains detected by chemical and near-infrared spectroscopy methods. *Applied Biochemistry and Biotechnology, 141*(1), 37–50.
Cohen, R., Persky, L., & Hadar, Y. (2002). Biotechnological applications and potential of wood-degrading mushrooms of the genus Pleurotus. *Applied Microbiology and Biotechnology, 58*(5), 582–594.
d'Acunzo, F., Galli, C., & Masci, B. (2002). Oxidation of phenols by laccase and laccase-mediator systems: Solubility and steric issues. *European Journal of Biochemistry, 269*(21), 5330–5335.
D'Annibale, A., Quaratino, D., Federici, F., & Fenice, M. (2006). Effect of agitation and aeration on the reduction of pollutant load of olive mill wastewater by the white-rot fungus *Panus tigrinus*. *Biochemical Engineering Journal, 29*(3), 243–249.
da Luz, J. M. R., Paes, S. A., Nunes, M. D., & da Silva, M. d C. S., & Kasuya, M. C. M. (2013). Degradation of oxo-biodegradable plastic by *Pleurotus ostreatus*. *PLoS One, 8*(8), e69386.
Deligkaris, K., Tadele, T. S., Olthuis, W., & Van den Berg, A. (2010). Hydrogel-based devices for biomedical applications. *Sensors and Actuators B: Chemical, 147*(2), 765–774.
Eriksson, K.-E. L., Blanchette, R. A., & Ander, P. (1990). Microbial and enzymatic degradation of wood and wood components. *Springer Science & Business Media, 1*, 1–399.
Eskander, S. B., Abd El-Aziz, S. M., El-Sayaad, H., & Saleh, H. M. (2012). Cementation of bioproducts generated from biodegradation of radioactive cellulosic-based waste simulates by mushroom. *International Scholarly Research Network, 2012*, 1–6.
European Environment Agency. (2020). *Land and soil pollution – Widespread, harmful and growing, October 15, 2020*. https://www.eea.europa.eu/signals/signals-2020/articles/land-and-soil-pollution
Folk, E. (2018). *Environmental impacts of industrialization*. https://www.ecomena.org/environmental-impacts-of-industrialization. Accessed 2021.
Gaitán-Hernández, R., Esqueda, M., Gutiérrez, A., Sánchez, A., Beltrán-García, M., & Mata, G. (2006). Bioconversion of agrowastes by *Lentinula edodes*: The high potential of viticulture residues. *Applied Microbiology and Biotechnology Springer Berlin Heidelberg, 71*(4), 432–439.
Gerlach, G., & Arndt, K.-F. (2010). Hydrogel sensors and actuators: Engineering and technology. *Springer Science & Business Media, XI*, 1–272.
Gianfreda, L., Iamarino, G., Scelza, R., & Rao, M. A. (2006). Oxidative catalysts for the transformation of phenolic pollutants: A brief review. *Biocatalysis and Biotransformation, 24*(3), 177–187.
Hatakka, A., & Hammel, K. E. (2010). Fungal biodegradation of lignocelluloses. In M. Hofrichter (Ed.), *Industrial applications* (2nd ed., pp. 319–340). Springer.

Hong, W., Zhao, X., Zhou, J., & Suo, Z. (2008). A theory of coupled diffusion and large deformation in polymeric gels. *Journal of the Mechanics and Physics of Solids, 56*(5), 1779–1793.

Ikerd, J. E. (1999). *Environmental risks facing farmers*. Tri-state conference for risk management education, Pocono Manor, Pennsylvania, May 5–6, 1999.

Jang, K.-Y., Cho, S.-M., Seok, S.-J., Kong, W.-S., Kim, G.-H., & Sung, J.-M. (2009). Screening of biodegradable function of indigenous ligno-degrading mushroom using dyes. *Mycobiology, 37*(1), 53–61.

Jibran, A. (2011). *Pleurotus sajor-caju* protein: A potential biosorptive agent. *Advances in BioTechnology, 11*(6), 25–27.

Jonathan, S., Fasidi, I., Ajayi, A., & Adegeye, O. (2008). Biodegradation of Nigerian wood wastes by *Pleurotus tuber-regium* (Fries) *Singer*. *Bioresource Technology, 99*(4), 807–811.

Justino, C. I., Duarte, K., Loureiro, F., Pereira, R., Antunes, S. C., Marques, S. M., Gonçalves, F., Rocha-Santos, T. A. P., & Freitas, A. C. (2009). Toxicity and organic content characterization of olive oil mill wastewater undergoing a sequential treatment with fungi and photo-Fenton oxidation. *Journal of Hazardous Materials, 172*(2–3), 1560–1572.

Justino, C., Marques, A. G., Duarte, K. R., Duarte, A. C., Pereira, R., Rocha-Santos, T., & Freitas, A. C. (2010). Degradation of phenols in olive oil mill wastewater by biological, enzymatic, and photo-Fenton oxidation. *Environmental Science and Pollution Research International, 17*(3), 650–656.

Kamida, H. M., Durrant, L. R., Monteiro, R. T. R., & Armas, E. D. d. (2005). Biodegradação de efluente têxtil por *Pleurotus sajor-caju*. *Chemical Nova, 28*(4), 629–632.

Klinkner, B. A. (2014). Anaerobic digestion as a renewable energy source and waste management technology: What must be done for this technology to realize success in the United States. *UMass Law Review, 9*, 68.

Kuek, C., Tommerup, I. C., & Malajczuk, N. (1992). Hydrogel bead inocula for the production of ectomycorrhizal eucalypts for plantations. *Mycological Research, 96*(4), 273–277.

Kuforiji, O. O., & Fasidi, I. O. (2008). Enzyme activities of *Pleurotus tuber-regium* (Fries) Singer, cultivated on selected agricultural wastes. *Bioresource Technology, 99*(10), 4275–4278.

Kulshreshtha, S., Mathur, N., Bhatnagar, P., & Jain, B. (2010). Bioremediation of industrial waste through mushroom cultivation. *Journal of Environmental Biology, 31*(4), 441–444.

Kulshreshtha, S., Mathur, N., Bhatnagar, P., & Kulshreshtha, S. (2013). Cultivation of *Pleurotus citrinopileatus* on handmade paper and cardboard industrial wastes. *Industrial Crops and Products, 41*, 340–346.

Laconi, S., Molle, G., Cabiddu, A., & Pompei, R. (2007). Bioremediation of olive oil mill wastewater and production of microbial biomass. *Biodegradation, 18*(5), 559–566.

Lechner, B. E., & Papinutti, V. L. (2006). Production of lignocellulosic enzymes during growth and fruiting of the edible fungus *Lentinus tigrinus* on wheat straw. *Process Biochemistry, 41*(3), 594–598.

Lindsey, M. (2017). *Community environmental concerns*. https://swehsc.pharmacy.arizona.edu/outreach/community-environmental-concerns. Accessed 2021.

Liu, X., Liu, J., Lin, S., & Zhao, X. (2020). Hydrogel machines. *Materials Today, 36*, 102–124.

Lu, Y., Yan, L., Wang, Y., Zhou, S., Fu, J., & Zhang, J. (2009). Biodegradation of phenolic compounds from coking wastewater by immobilized white rot fungus *Phanerochaete chrysosporium*. *Journal of Hazardous Materials, 165*(1–3), 1091–1097.

Lucas, N., Bienaime, C., Belloy, C., Queneudec, M., Silvestre, F., & Nava-Saucedo, J.-E. (2008). Polymer biodegradation: Mechanisms and estimation techniques – A review. *Chemosphere, 73*(4), 429–442.

Luo, D., Xie, Y.-F., Tan, Z.-L., & Li, X.-D. (2013). Removal of Cu^{2+} ions from aqueous solution by the abandoned mushroom compost of *Flammulina velutipes*. *Journal of Environmental Biology, 34*(2 suppl), 359.

Luyt, A. S., & Malik, S. S. (2019). Can biodegradable plastics solve plastic solid waste accumulation? In *Plastics to energy* (pp. 403–423) Elsevier.

Mancera-López, M. E., Esparza-García, F., Chávez-Gómez, B., Rodríguez-Vázquez, R., Saucedo-Castaneda, G., & Barrera-Cortés, J. (2008). Bioremediation of an aged hydrocarbon-contaminated soil by a combined system of biostimulation–bioaugmentation with filamentous fungi. *International Biodeterioration and Biodegradation, 61*(2), 151–160.

Morel, M., Ngadin, A. A., Droux, M., Jacquot, J.-P., & Gelhaye, E. (2009). The fungal glutathione S-transferase system. Evidence of new classes in the wood-degrading basidiomycete *Phanerochaete chrysosporium. Cellular and Molecular Life Sciences, 66*(23), 3711–3725.

Müller, R.-J. (2005). Biodegradability of polymers: Regulations and methods for testing. In I.S.A (Ed.), *General aspects and special applications*. Wiley-VCH Verlag.

Nagy, B., Măicăneanu, A., Indolean, C., Mânzatu, C., Silaghi-Dumitrescu, L., & Majdik, C. (2014). Comparative study of Cd (II) biosorption on cultivated *Agaricus bisporus* and wild *Lactarius piperatus* based biocomposites. *Linear and nonlinear equilibrium modelling and kinetics. Journal of the Taiwan Institute of Chemical Engineers, 45*(3), 921–929.

Nawrat, A. (2018). *Pharma and the environment: Pollution continues despite public pressure.* https://www.pharmaceutical-technology.com/features/pharma-and-the-environment-pollution-trend. Accessed 2021.

Neethu, T., Dubey, P., & Kaswala, A. (2018). Prospects and applications of hydrogel technology in agriculture. *International Journal of Current Microbiology and Applied Sciences, 7*(5), 3155–3162.

Nur, D. R., Chuen, N. L., & Ngadin, A. A. (2019). First record of in vitro growth evaluation of wild mushroom, *Schizophyllum commune* from Pulau Kapas in Malaysia. *Asian Journal of Agriculture and Biology, 7*(4), 602–609.

Olusola, S., & Anslem, E. (2010). Bioremediation of a crude oil polluted soil with *Pleurotus pulmonarius* and *Glomus mosseae* using Amaranthus hybridus as a test plant. *Journal of Bioremediation & Biodegradation, 1*, 113.

Osama, R. S., Namat, M. H., Fares F. F., Hwedi E., Farag M. E. O., & Elena M., (2020). *Soil pollution*. Virtual Conference on Environment and Health (VCEH) at: Agricultural University of Iceland. October 4, 2020.

Oyen, M. L. (2014). Mechanical characterisation of hydrogel materials. *International Materials Review, 59*(1), 44–59.

Oyetayo, V. O., Adebayo, A. O., & Ibileye, A. (2012). Assessment of the biosorption potential of heavy metals by *Pleurotus tuber-regium. International Journal of Biological Research, 2*(2), 293–297.

Prasad, Y. (2013). Biosorption of Cu, Zn, Fe, Cd, Pb and Ni by non-treated biomass of some edible mushrooms. *Materials Science, 4*(2), 190–195.

Rani, P., Kalyani, N., & Prathiba, K. (2008). Evaluation of lignocellulosic wastes for production of edible mushrooms. *Applied Biochemistry and Biotechnology, 151*(2), 151–159.

Resnik, D. B., & Portier, C. (2015). *Environment, ethics, and human health.* https://www.thehastingscenter.org/briefingbook/environmental-health/. Accessed Apr 2021.

Shah, A., & Shah, M., (2020). Characterisation and bioremediation of wastewater: A review exploring bioremediation as a sustainable technique for pharmaceutical wastewater. *Groundwater for Sustainable Development, 100383*.

Stella, T., Covino, S., Čvančarová, M., Filipová, A., Petruccioli, M., D'Annibale, A., & Cajthaml, T. (2017). Bioremediation of long-term PCB-contaminated soil by white-rot fungi. *Journal of Hazardous Materials, 324*, 701–710.

Sutherland, C., & Venkobachar, C. (2013). Equilibrium modeling of Cu (II) biosorption onto untreated and treated forest macro-fungus *Fomes fasciatus. International Journal of Plant, Animal and Environmental Sciences, 3*(1), 193–203.

Tay, C. C., Liew, H. H., Yin, C.-Y., Abdul-Talib, S., Surif, S., Suhaimi, A. A., & Yong, S. K. (2011). Biosorption of cadmium ions using *Pleurotus ostreatus*: Growth kinetics, isotherm study and biosorption mechanism. *Korean Journal of Chemical Engineering, 28*(3), 825–830.

Teerapatsakul, C., Parra, R., Bucke, C., & Chitradon, L. (2007). Improvement of laccase production from *Ganoderma* sp. KU-Alk4 by medium engineering. *World Journal of Microbiology and Biotechnology, 23*(11), 1519–1527.

Tišma, M., Zelić, B., & Vasić-Rački, Đ. (2010). White-rot fungi in phenols, dyes and other xenobiotics treatment – A brief review. *Croatian Journal of Food Science and Technology, 2*(2), 34–47.

Tsui, J. (2020). *Five biggest environmental issues affecting the U.S.* https://eponline.com/articles/2020/02/24/five-biggest-environmental-issues-affecting-the-us.aspx. Accessed 24 Feb 2020.

Tsujiyama, S., Muraoka, T., & Takada, N. (2013). Biodegradation of 2, 4-dichlorophenol by shiitake mushroom (*Lentinula edodes*) using vanillin as an activator. *Biotechnology Letters, 35*(7), 1079–1083.

Van Der Zee, M. (2011). Analytical methods for monitoring biodegradation processes of environmentally degradable polymers. In A. S. A. Lendlein (Ed.), *Handbook of biodegradable polymers: Isolation, synthesis, characterization and applications* (pp. 263–281). Wiley-VCH Verlag.

Yogita, R., Simanta, S., Aparna, S., & Kamlesh, S. (2011). Biodegradation of malachite green by wild mushroom of Chhatisgrah. *Journal of Experimental Sciences, 2*(10), 69–72.

Chapter 4
Production, Partial Optimization, and Characterization of Keratinase Enzyme by Fungal Species Isolated from Soil of Bhopal

Renu Mishra and Shraddha Tamrakar

Contents

R. Mishra (✉)
Department of Botany & Microbiology Sri Sathya Sai College for Women, Bhopal, MP, India

S. Tamrakar
Sri Sathya Sai College for Women, Bhopal, MP, India

A. C. Shukla (ed.), *Applied Mycology*, Fungal Biology,
https://doi.org/10.1007/978-3-030-90649-8_4

4.1 Introduction

Soil encourages a large number of microorganisms and is one of the highest complicated microbial habitats that permit bacteria, fungi, and other micro-organisms to nurse their entire life. Soil is the shelter for several fungi, which are not noticed and explored in the environment (Pahare et al., 2018).

Keratinophilic fungi are a little, all around characterized and essential gathering of growths that colonize diverse keratinous substrates, and generate them to fragments of low atomic weight. These fungi are available on the earth with variable dispersion designs, which rely upon components, for example, human or animals (Kumar et al., 2013).

Keratinophilic means keratin-loving micro-organism. Keratinophilic fungi are natural colonizers of keratinous substrates. Few of them are keratinolytic and play a massive ecological character in the decay of insoluble fibrous proteins, keratins (Kushwaha & Guarro, 2000). Keratin has a place with the exceptionally heterogeneous group of fibrillar proteins of conservative structure. Countless extensions of disulfide bridges in keratins structure altogether impact its properties, specifically, high mechanical and substance obstruction. Keratin is highly stable and water insoluble. Common proteases like pepsin and papain are not capable of degrading keratin (Staroń et al., 2011).

Keratinase is a specific class of extracellular proteolytic inducible chemical with the ability of debasing insoluble keratin substrates. It is vital for hydrolyzing hair, plume, and collagen in the sewage framework amid wastewater treatment.

Human hair and nails are always in interaction with pollutants in the environment and waste exposed to fungal and bacterial contamination (Alghamdi et al., 2018). Recent studies have become an important clinical condition, which require attention to public health, because some of them are seriously harmful to human health. Health authorities must build up their health information campaigns, which cover not only serious disease prevention and treatment but also environmental hygiene.

4.1.1 Keratinases: Applications

Keratinolytic micro-organisms and keratinases may be utilized for multiple applications in a myriad of industries, like the detergent industry, leather production, food and feed industries, medical and pharmaceutical industries, waste management industries, etc. Compared to traditional microbial proteases, production of keratinases can be achieved at lower costs as the organisms are able to utilize the abundant keratinous waste as nitrogen and carbon source. Since elevated production cost is one of the main issues in the industrial production of enzymes, this methodology of utilization of abundant quantities of keratinous waste will make the process comparatively more economical while also increasing its application portfolio.

4.1.1.1 Detergent Industry

Proteases are widely utilized in the detergent industry as a safe alternative replacing harmful chemicals such as caustic soda. In addition to being used in washing detergents, keratinases may also be used as an additive in lens cleaning solutions, drain cleaning solutions, etc. (Nasaimento & Martin, 2006).

4.1.1.2 Leather Industry

The leather industry involves several steps of leather treatment with chemicals, such as sodium sulfide, sodium hydroxide, and lime. These chemicals are potent water pollutants along with solid wastes emerging due to these treatments (Sivakumar et al., 2012). Processing of leather with enzymes such as proteolytic, lipolytic, and glycolytic enzymes has emerged as a safe alternative (Bihari et al., 2010).

4.1.1.3 Food and Feed Industry

The National Research Development Corporation (NRDC), India, has recently utilized bifunctional chimeric keratinase from *Bacillus* sp. for the purpose of debittering amino acids in the processing of soy sauce. It was observed that this increased the taste of tea, and feather meal preparation along with other applications (www.nrdcindia.com). Usually, enzymes are useful in feed industries because of their part in the breakdown of the antinutritional elements in feed ingredients. In addition to that, this keratinase treatment improves the bioavailability of biomolecules in animal feed. It does so by assisting endogenous proteolytic enzymes of feeding animals to unwind the rigid structure of feed. However, often, this is restricted due to the high production costs, and therefore, low-cost keratinases can prove to be a useful alternative as a feed additive. Recently, many commercial enzymes, such as Versazyme, and Valkerase, etc., have been used as applications in animal feed preparations.

4.1.1.4 Fertilizer for Organic Farming

Composting of keratinous wastes using keratinolytic micro-organisms leads to gradual release of nitrogen fertilizers. This makes it a viable application in organic farming. This methodology has become one of the safest and most cost-effective technologies to resolve keratinous waste materials.

4.1.1.5 Biofuel Production

Poultry waste can be used as a viable source of biofuel. Keratinolytic microorganisms and their keratinases can be employed in the utilization of these wastes for generation of natural gas (Puhl et al., 2009), methane gas fuel pellets (Gushterova et al., 2005), and biohydrogen (Saha et al., 2013).

4.1.1.6 Cosmetic and Pharmaceutical Applications

Non-collagenolytic keratinases have emerged as promising biocatalysts in the pharmaceutical as well as cosmetic industries. They have been illustrated as a useful ingredient in depilatory compositions for hair shaving formulations and skin lightening agents (Yang, 2012). Certain crude keratinases have also been demonstrated to elevate hair characteristics like weight, flexibility, brightness, softness, and strength; this illustrates that these can be applied in the creation of hair care products (Cao et al., 2012). Additionally, keratinases have shown the ability to degrade thickened layers of dead skin (hyperkeratosis) found around the toes and fingers, thereby promising to be an alternative to the conventional salicylic acid used for the treatment of such predilections (Gupta & Ramnani, 2006). In the same fashion, keratinases can be capable of peeling skin to remove acne, which is caused by the blockage of the sebaceous gland by keratins (Selvam &Vishnupriya, 2012).

4.2 Materials and Methods

4.2.1 Collection of Soil Samples

Soil samples of keratin containing waste were collected in polythene bags from the different sampling sites, brought to the laboratory, and analyzed.

4.2.2 Isolation of Keratinophilic Fungi

For the isolation of fungi, hair baiting technique was used (Vanbreuseghem, 1952).

4.2.3 Morphological Microscopic and Molecular Identification of Keratinophilic Fungi

Firstly, the fungi were indentified on the basis of morphological characteristics of the colony and microscopic examination. Molecular identification of isolated microorganisms was confirmed by the sequence of ITSI, 508 s, and ITS2 region of their DNA sequences.

4.2.4 Optimization of Culture Condition for Mass Production of Keratinase Enzyme

To optimize the culture conditions for mass production of keratinase enzyme by isolated soil microorganisms, a basal medium containing insoluble keratin was prepared with various pH (6, 7, 8, and 9) incubated on different temperatures (40, 45, 50, 55, and 60 °C) for various incubation periods (12, 15, 18, and 21 days) (Table 4.1).

4.2.5 Enzyme Production

The keratinase enzyme production was carried out in the basal medium by using a shaking flask. To begin the process, 5 mg of azokeratin was added in the basal medium: 2.5 ml of the all isolated fungal inoculums was added in 50 ml of medium and cultured on a rotary shaking incubator at 150 rpm and 37 °C for 72 h. After incubation, fermented broth was centrifuged at 5000 rpm for 20 min at 4 °C. The cell-free supernatant was collected and used for the assay of keratinase activity (Sivakumar et al., 2012).

Table 4.1 Composition of basal medium for keratinase production

S. no.	Component	Gm/L
1	Meat extract	1
2	Peptone	1
3	NaCl	0.5
4	Keratin	5

4.2.6 Keratinase Assay

Keratinase activity was determined by taking 20 ml of 0.1 M Tris buffer (pH 8) containing 0.1% keratin and 40 μl of enzyme solution and was incubated for 30 minutes at 55 °C. The reaction was stopped with 500 μl of 0.1 mol-1 trichloroacetic acid (TCA) in 0.1 mol-1 Tris buffer, pH 8. The amount of protein present was assayed by the Lowry method using bovine serum albumin (BSA) as standard. The color developed was read at 660 nm (Jin et al., 2017).

4.3 Results

4.3.1 Morphological Characterization

Morphological characterization was performed and further confirmed by the National Center for Fungal Taxonomy (NCFT), New Delhi. All positive keratinophilic fungi were identified by colony morphology and lactophenol cotton blue microscopic staining method (Table 4.2).

Table 4.2 Morphological and microscopical characterization of isolated keratinophilic fungi

S. no.	Sample ID	Species (morphological basis)	Color of mycelia	Pigmentation	Spore arrangement	Sporulation
1	S1	*Fusarium solani*	White cottony	Pink	Long branched	Sexual
2	S2	*Curvularialunata*	Blackish	Greenish brown	Branched septa	Sexual
3	S3	*Chrysosporium keratinophilum*	Cream	Pale	Long lateral branched	Asexual
4	S4	*Acremonium kiliense*	Cream	Colorless	Branched	Sexual/ asexual
5	S5	*Aspergillus Niger*	Black	Pale	Septet	Sexual
6	S6	*Acremonium restrictum*	White orange	Not appear	Septet	Sexual/ asexual
7	S7	*Absidia californica*	Grayish brown	Colorless	Branched	Sexual
8	S8	*Absidia repens*	White	Pale gray	Branched	Sexual
9	S9	*Penicillium chrysogenum*	White	Colorless	Highly branched	Asexual
10	S10	*Gibberella pulicaris*	Bluish	Light pink	Long branched	Sexual
11	S11	*Aspergillus flavus*	Greenish	White	Branched	Sexual/

4.3.2 Molecular Identification of Fungi

The genomic DNA was isolated from selected keratinophilic fungi using the CTAB method. The DNA was visualized on agarose gel and quantified for an accurate amount of DNA isolated. We have observed an average of 2.86 mg/μl genomic DNA from all major keratinophilic fungi.

The gene-specific primers were designed using coding sequence available at National Centre for Biological Information (NCBI). The primers were designed based on in silico PCR validations. Primers are supplied in lyophilized form and dissolved in recommended volume of sterile water or TE buffer result in 100 μM concentration.

The 100 μM concentration primers are used for master stock and aliquoted into 10 μM for PCR reactions. The targeted genes were amplified using thermal cycler and PCR reaction mixture. Different primers are required for identification of different set of genes in PCR Programme.

The primer may also vary in its melting temperature, and hence Tm values were pre-calculated and PCR program was designed based on Tm value for efficient annealing. The Tag Polymerase, a thermostable DNA polymerase, had a chain elongation capacity, 500 bases per second, and hence PCR program was designed based on gene length. Here, we have run several individual reactions and PCR trials for gene amplification and amplified gene was visualized on agarose gel with a standard DNA ladder. The primer used in the study was 550 bp in length.

The amplification products were sequenced for species identification. The DNA sequences of isolated strains were recited using basic local alignment search tool (BLAST) for species identification. The BLAST search showed that the sequence data of the isolated strain S1 shared 99% similarity with *Fusarium solani* and S2 shared 99% similarity with *Curvularia lunata*. The three isolated strains, identified as *Gibberella pulicaris, Chrysoporium keratinophilum* and *Penicillium chrysogenum* were S10, S3 and S9. DNA sequences of S5 and S 11 showed 90% matching with *Aspergillus niger* and *A. flavus*. Similarly, S7 and TS8 fungi found close similarity with *Absidia* species. Amplification of each primer pair was tested on a panel of strains representing *Acremonium kiliense and A. restrictum*, along with the genomic DNA of the isolate from which the library was generated.

All isolated strains were submitted in the Gen-Bank databases for the purpose of phylogenetic study. The gene bank accession numbers for fungi from S1 to S11 are shown in (Fig. 4.1).

4.3.3 Phylogenetic Analysis

In the phylogenetic tree (Fig. 4.1), all major keratinophilics were reported to be of a distant origin and do not have any close resemblance. The phylogenetic tree was constructed by the neighbor-joining method and Jaccard coefficient. There are three

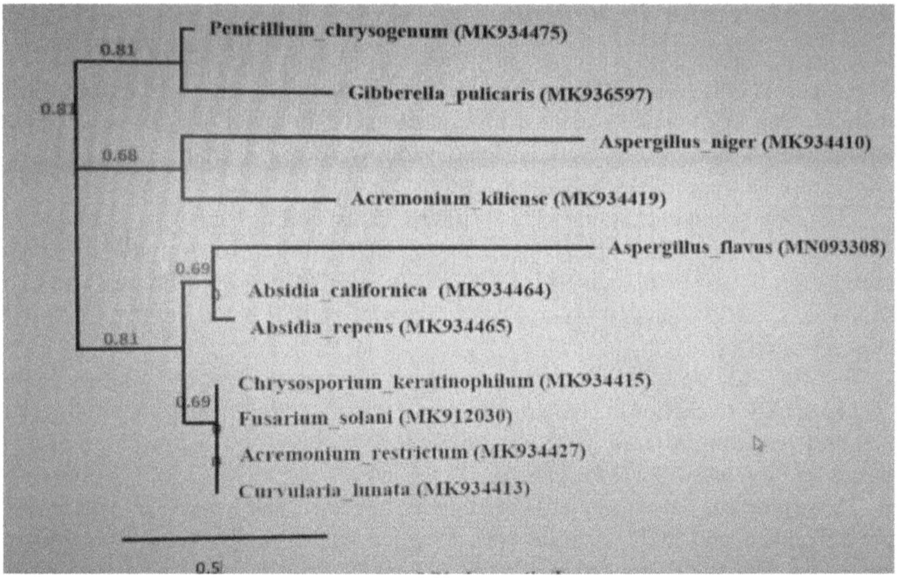

Fig. 4.1 Phylogenetic tree of the major keratinophilic fungi

ancestors of isolates found in this study. Group one has two species, that is, *Penicillum chrysogenum and Gibberella pulicaris. Aspergillus niger* and *Acremonium* are in another group, whereas the third has two subgroups, one including *Aspergillus flavus, Absidia californica*, and *Absida repens.* The other subgroup includes *Chrysosporium keratinophilum, Fusarium solani, Acremonium* restrictum, and *Curvularialuntana* (Fig. 4.1).

4.3.4 Optimization of Culture Media for Keratinase Enzyme Production

4.3.4.1 Effect of Fermentation on Keratinase Enzyme

Maximum keratinase production was recorded in the shake culture method compared to static culture. Table 4.3 and Fig. 4.2 show the results of keratinase production by both techniques. *Curvularialunata* produced a maximum 43.5 U/ml keratinase while minimum 17.5 U/ml was observed in the culture medium containing *Aspergillus flavus* during the shake culture method. The highest amount (38.6 U/ml) of keratinase was produced by *Fusarium solani* and minimum 14.6 U/ml produced by *Gibberellapulicaris* during static culture technique. Data represent the mean and ± standard deviation for different methods. Although Table 4.3 and Fig. 4.2 show varied keratinase production in the two different methods, this

Table 4.3 Optimization of culture media for keratinase production by shake and static culture sample ID

Sample ID	Species	Enzyme (U/ml) in shake culture	Enzyme (U/ml) static culture
S1	*Fusarium solani*	41.6	38.6
S2	*Curvularialunata*	43.5	28.7
S3	*Chrysosporiumkeratinophilum*	34.1	30.2
S4	*Acremonium kiliense*	40.1	36.2
S5	*Aspergillus Niger*	20.5	18.4
S6	*Acremonium restrictum*	22.1	19.6
S7	*Absidia californica*	19.6	16.9
S8	*Absidia repens*	42.3	38.4
S9	*Penicillium chrysogenum*	39.1	34.5
S10	*Gibberellapulicaris*	19.8	14.6
S11	*Aspergillus flavus*	17.5	15.4
Mean ± SD		**30.9 ± 10.8**	**26.5 ± 9.6**

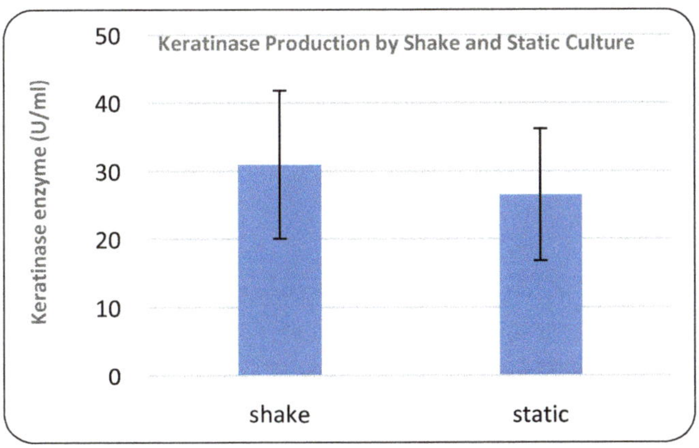

Fig. 4.2 Keratinase production by shake and static culture (error bars show 1 st dev)

difference, however, is not statistically significant (*p* is 0.32 at t-test static value) and t-test is calculated using "Microsoft Excel Version 2019."

4.3.4.2 Effect of Temperature on Keratinase Enzyme Production

Different temperatures, such as 40 °C, 45 °C, 50 °C, 55 °C, and 60 °C, were tested to obtain the optimum temperature for the maximum production of keratinase by all isolates. Out of the temperatures tested as above, the maximum production of keratinase was observed to be at 50 °C (Table 4.4 and Fig. 4.3). Maximum keratinase (43.5 U/ml) at optimum temperature 50 °C was produced by *Penicillium chrysogenum,* while minimum keratinase (5.9 U/ml) was recorded at 40 °C temperature

Table 4.4 Effect of temperature on keratinase enzyme production U/ml

Sample ID	Species Temp (°C)	40 °C	45 °C	50 °C	55 °C	60 °C
S1	*Fusarium solani*	8.6	12.9	19.4	11.8	7.9
S2	*Curvularialunata*	10.3	14.6	22.3	15.9	9.4
S3	*Chrysosporiumkeratinophilum*	9.9	14.9	23.5	14.9	8.6
S4	*Acremonium kiliense*	12.4	24.8	32.1	34.2	14.6
S5	*Aspergillus Niger*	13.3	28.5	35.1	28.4	14.2
S6	*Acremonium restrictum*	11.4	29.1	40.1	28.9	18.4
S7	*Absidia californica*	14.6	28.7	40.3	27.5	16.2
S8	*Absidia repens*	14.9	21	41.7	30.1	11.4
S9	*Penicillium chrysogenum*	13.8	24.9	43.5	30	14.8
S10	*Gibberellapulicaris*	6.2	10.3	19.8	18.6	11
S11	*Aspergillus flavus*	5.9	8.8	17.5	16.2	10.3
Mean ± SD		**11.0 ± 3.1**	**19.8 ± 7.7**	**30.5 ± 10.1**	**23.3 ± 7.8**	**12.4 ± 3.4**

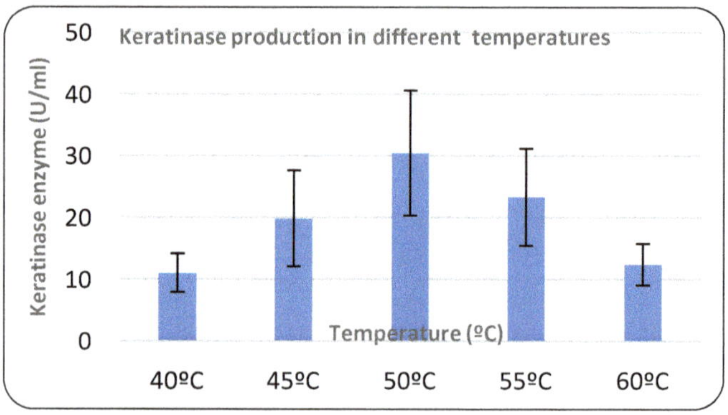

Fig. 4.3 Keratinase production in different temperature

produced by *Aspergillus flavus*. Proteolytic activities of most organisms are seen to be at 30–37 °C, whereas some keratinolytic microbes show keratin-degrading capacities at higher temperatures. A nonparametric Kruskal-Wallis test applied on the data (the data fulfill the requirements for the test) shows that there was a statistically significant difference in the keratinase activity under different temperatures studied ($p < 0.0001$ for the test static $H = 31.08$).

4.3.4.3 Effect of pH on Keratinase Enzyme Activity

The medium adjusted on different pH values, (such as pH 6, 7, 8, and 9) was assayed to obtain the optimum pH for the maximum production of keratinase by all isolated keratinophilic microbes. From Table 4.5, it is clear that *Penicillium chrysogenum* is found to be the maximum enzyme-producing microorganism, which produced

Table 4.5 Effect of pH on Keratinase Enzyme Activity of Keratinophilic Fungi (U/ml enzyme produced)

Sample ID	Species pH	6	7	8	9
S1	*Fusarium solani*	0.620	1.010	1.910	0.960
S2	*Curvularialunata*	0.920	1.920	2.110	1.640
S3	*Chrysosporiumkeratinophilum*	0.940	1.840	2.210	1.680
S4	*Acremonium kiliense*	1.160	2.980	3.970	1.890
S5	*Aspergillus Niger*	1.210	3.010	3.940	1.760
S6	*Acremonium restrictum*	0.960	2.670	3.210	1.540
S7	*Absidia californica*	1.280	3.240	3.860	2.140
S8	*Absidia repens*	1.190	2.970	3.390	1.790
S9	*Penicillium chrysogenum*	1.210	3.910	4.010	2.050
S10	*Gibberellapulicaris*	0.840	1.230	2.010	1.130
S11	*Aspergillus flavus*	0.720	1.120	1.840	1.050
Mean ± SD		**1.0 ± 0.2**	**2.3 ± 0.9**	**2.9 ± 0.9**	**1.6 ± 0.4**

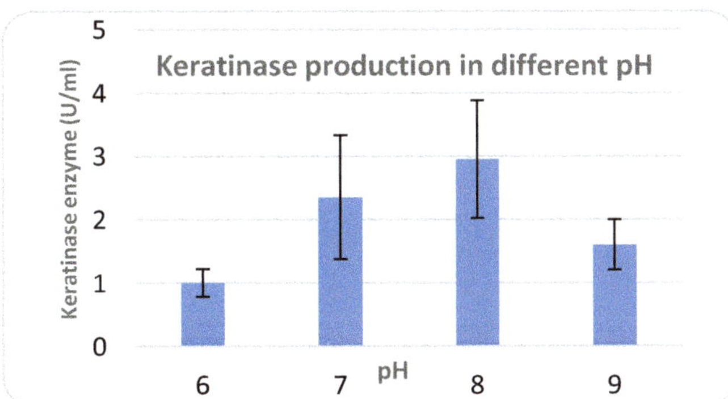

Fig. 4.4 Keratinase Production in different pH

4.01 U/ml keratinase at 8pH at 50 °C optimum temperature. pH 6 reduced the proteolytic activity of all isolates. The lowest enzyme (0.62 U/ml) production was found at pH 6 by *Fusarium solani*. There was a statistically significant difference in the keratinase activity under different pH values studied ($p = 0.00001$ for the test static $H = 29.25$) (Fig. 4.4) (Bewick et al., 2004).

4.3.4.4 Effect of Incubation Time

The effect of incubation period on keratinase production from keratinophilic fungi was studied for the incubation period from 12 days to 21 days as shown in Table 4.6 and Fig. 4.5. It was observed that the maximum enzyme production was attained between 15–18 days of incubation period. Incubation beyond the optimum time showed a rapid decline in the enzyme yield, as compared to the maximum (39.8 U/ml)

Table 4.6 Effect of incubation period on keratinase production

Sample ID	Harvesting time species	12 days /U/ml	15 days /U/ml	18 days /U/ml	21 days/U/ml
S1	*Fusarium solani*	19.8	39.8	38.4	30.6
S2	*Curvularialunata*	18.6	24.3	27.6	26.4
S3	*Chrysosporiumkeratinophilum*	17.2	20.1	29.8	27.4
S4	*Acremonium kiliense*	18.7	35.6	34.6	30.2
S5	*Aspergilliusniger*	11.2	21.6	21.9	14.6
S6	*Acremonium restrictum*	11.6	15.6	16.8	12.4
S7	*Absidia californica*	10.3	16.1	17.8	12.3
S8	*Absidia repens*	19.7	28.4	29.8	16.8
S9	*Penicillium chrysogenum*	10.9	35.4	34.2	24.4
TS10	*Gibberellapulicaris*	8.9	18.7	19.6	17.6
TS11	*Aspergillus flavus*	7.9	16.4	17.6	15.4
Mean ± SD		**13.5 ± 4.5**	**24.7 ± 8.7**	**26.2 ± 7.8**	**20.7 ± 7.1**

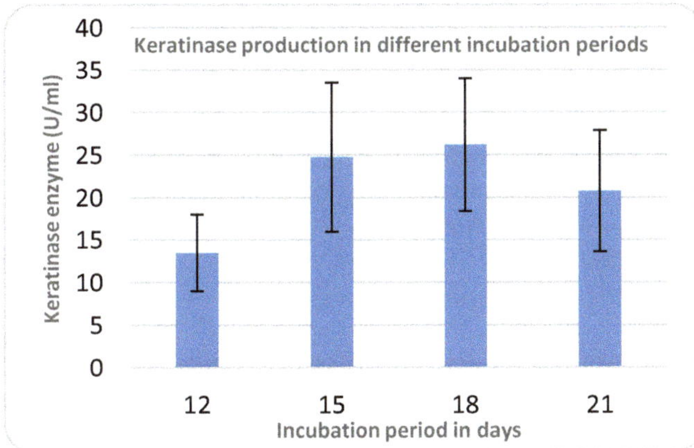

Fig. 4.5 Keratinase production in different periods

at 15 hrs. *Fusarium solani* produced maximum (39.8 U/ml) keratinase enzyme at the incubation of 15 days and minimum (7.9 U/ml) was produced by *Aspergillus flavus* after 12 days of incubation. The Kruskal-Wallis test showed that there was a statistically significant difference between the overall keratinase production by the fungi under the different incubation periods studied ($p = 0.004$ at H statistic 13.15).

4.4 Discussion

The molecular identification was carried out by DNA bar-coding using the ITS region sequencing. The DNA sequences were compared to those in the databases using NCBI-BLAST. Eleven species were identified using DNA bar-coding with an identical range between 97–99%. It is also proposed that DNA region sequence is one of the most important tools for the identification of the fungal species isolated from environmental sources. Hence, it has been widely used to detect the soil fungal community, and as an improvement of the classical identifications (Alsohaili & Bani-Hasan, 2018).

Pakshir et al. (2013) used the molecular technique for the first time in Iran to identify keratinophilic fungi. In the current investigation, 11 species of eight genera were identified by molecular marker–based sequence analysis (Fig. 4.1).

In the present examination, *Absidia californica* and *Absidia repens* were also found majorly in the soil sample of dumping sites. These fungi also have a great keratin-solubilizing potential. The current findings are in agreement with previous reports about *Absidia*as keratinophilic fungi. Morphological criteria and molecular marker for clear distinction of *Absidia* was used. Different species of clinically significant zygomycetes including 2 species of *Rhizopus*, 3 species of *Mucor*, 2 species of *Cunninghamella,* as well as 1 strain each of *Rhizomucorpusillus* and *Absidiacorymbifera* were identified by molecular techniques (Iwen et al., 2011).

Majority of reports on Keratinase production are under submerged shaking/ static conditions (Cai et al., 2011). It is difficult to compare the production condition for Keratinase due to a variety of organisms and the methods of cultivation.

In comparison to the present findings, other studies reported that maximum enzyme activity was obtained within 96 h (150 U/ml and 90 U/ml for *Chrysosporium*and *Microsporum*respectively) of cultivation at pH 7.0, 30 °C. The optimal conditions for the keratinolytic activity of both enzymes were found to be at pH 9.0 and temperature 50 °C; however, the enzymes showed stability over a broad range of pH between 7.0 and 10.0 and temperature 30 °C–50 °C (Kanchana, 2013).

Similar studies were conducted by different scientists. Three factors, temperature, pH, and manitol, were used for keratinase optimization (Shankar et al. 2014). Similar work for keratinase production employing response surface methodology was performed (Ramnani & Gupta, 2004; Hashemet al., 2018). Impact of different substrates, temperature, pH and protease inhibitors, lessening operators, and metal particle supplements in the generation medium on enzyme creation was studied (Yadav et al., 2011).

Previous reports show that the activity of keratinolytic enzyme produced by *Aspergillus flavus* effectively degraded feather substrate (Mini et al., 2015). Extremely high and low pH values showed complete loss of activity for most enzymes. Alkaline pH possibly supported keratin degradation as higher pH modified cysteine residues, making it accessible for enzyme action. Keratin degradation took incubation time for enzyme production from 24 h to several days. This is probably involved in the complex procedure of keratinolysis of these microbes (Kumawat et al., 2013). The harvesting time of 72 h was found to be ideal for maximum enzyme production.

These results showed similarity with our findings that keratinase enzyme production was maximum at 8 pH, between 15 and 18 days of incubation at 50 °C. Although the mean enzyme activity is maximum at 18 days of incubation period (Table 4.6), fungi such as *Fusarium solani* and *Penicillium chrysogenum* did show better activity after 15 days of incubation. Also, some fungi showed good activity after 21 days of incubation. For more precise optimization, enzyme activity could be observed on consecutive days between 15 to 21 days of incubation.

4.5 Conclusion

The findings concluded that total eleven keratinophilic fungi were isolated from the soil of Bhopal. All these fungi showed good keratinase production in shake culture method at 8 pH, 50 °C temperature, between 15–18 days of incubation period. These fungal isolates will be useful in eco-friendly and environment cleanup process of keratinic wastes. They may also be used for large-scale production of keratinase for industrial purposes. This research is suggestive of potential optimization scheme and its scope alongside the development of optimal medium compositions and culture conditions, for the production of keratinolytic enzyme that can be useful for developing entrepreneurship.

References

Alghamdi, S., Alotaibi, A., Haya, A., Al-SubaiMunira, Z., & Alwakeel Suaad, S. (2018). Isolation and identification of microbial and fungal flora from female hair samples in Riyadh Saudi Arabia. *International Journal of Environment, Agriculture and Biotechnology, 3*(11–5), 2456–1878.

Alsohaili, S. A., & Bani-Hasan, B. M. (2018). Morphological and molecular identification of fungi isolated from different environmental sources in the Northern Eastern Desert of Jordan. *Jordan Journal of Biological Sciences, 11*(3), 329–337.

Bewick, V., & Chakl, & Bali, J. (2004). Statistics review 10: Further non parametric methods. *Critical Care, 8*, 196–199.

Bihari, Z., Videki, D., Mihalik, E., et al. (2010). Degradation of native feathers by a novel keratinase-producing, thermophilic isolate, *Brevibacillus thermoruber* T1E. *Zeitschrift für Naturforschung. Section C, 65*, 134–140.

Cai, S. B., Huang, Z. H., Zhang, X. Q., Cao, Z. J., Zhou, M. H., et al. (2011). Identification of a keratinase-producing bacterial strain and enzymatic study for its improvement on shrink resistance and tensile strength of wool- and polyester-blended fabric. *Applied Biochemistry and Biotechnology, 163*(1), 112–126.

Cao, Z. J., Lu, D., Luo, L. S., Deng, Y. X., Bian, Y. G., et al. (2012). Composition analysis and application of degradation products of whole feathers through a large scale of fermentation. *Environmental Science and Pollution Research, 19*(7), 2690–2696.

Gupta, R., & Ramnani, P. (2006). Microbial keratinases and their prospective applications: An overview. *Applied Microbiology and Biotechnology, 70*, 21–33.

Gushterova, A., Vasileva-Tonkova, E., Dimova, E., et al. (2005). Keratinase production by newly isolated Antarctic actinomycete strains. *World Journal of Microbiology and Biotechnology, 21*, 831–834.

Hashem, A. M., Abdel-Fattah, A. M., Ismail, S. A., El-Gamal, M. S., Esawy, M. A., & Emran, M. A. (2018). Optimization, characterization and thermodynamic studies on *B.* 137 *licheniformis* ALW1 keratinase. *Egyptian Journal of Chemistry, 61*(4), 591–607.

Iwen, P. C., Thapa, I., & Bastola, D. (2011). Review of methods for the identification of zygomycetes with an emphasis on advances in molecular diagnostics. *Laboratory Medicine, 42*(5), 260–266.

Jin, H. S., Park, S. Y., Kin, K., Leu, Y.-J., Nam, G.-W., Kang, N. J., et al. (2017). Development of a Keratinase activity assay using recombinant chicken feather keratin substrates. *PLoS One, 12*(2), e0172712.

Kanchana, R. (2013). Utilization of biodegradable keratin containing wastes by enzymatic treatment. *International Journal of Pharmacological and Biological Sciences, 4*(1B), 117–126.

Kumar, R., Mishra, R., Maurya, S., & Sahu, H. B. (2013). Isolation and identification of keratinophilic fungi from garbage waste soils of Jharkhand region of India. *European Journal of Experimental Biology, 3*(3), 600–604.

Kumawat, T. K., Sharma, V., Seth, R., & Sharma, A. (2013). Diversity of keratin degrading fungal flora in industrial area of Jaipur and keratinolytic potential of *Trichophyton mentagrophytes* and *Microsporumcanis*. *International Journal of Biotechnology and Bioengineering Research, 4*(4), 359–364.

Kushwaha, R. K. S., & Guarro, J. (Eds.). (2000). *Biology of dermatophytes and other keratinophilic fungi*. RevistaIbero americana de Micología.

Mini, K. D., George, S. M., & Mathew, J. (2015). Screening and selection of fungus for keratinase production by solid state fermentation and optimization of conditions of SSF and formulation of low-cost medium for the production of keratinase by *Aspergillus flavus* S125. *International Journal of Current Microbiology and Applied Sciences, 4*(9), 535–548.

MS Excel (dentopreed by Microsoft).

Nasaimento, W. C. A., & Martin, M. L. L. (2006). Studies on the stability of Protease from *Bacillus sp.* & its Compatibility with commercial detergent. *Bazilian Journal of Microbiology, 37*, 307–311.

Pahare, S., Shukla, K., & Shukla, R. V. (2018). Keratinophilic fungi from warm, moist, cattle – house of Bilaspur Central – India. *Journal of Microbiology & Experimentation, 6*(2), 46–48.

Pakshir, K., Ghiasi, M. R., Zomorodian, K., & Gharavi, A. R. (2013). Isolation and molecular identification of keratinophilic fungi from public parks soil in Shiraz, Iran. *BioMed Research International, 619576.*

Puhl, A. A., Selinger, L. B., McAllister, T. A., et al. (2009). Actinomadurakeratinilytica sp. nov., a keratin degrading actinobacterium isolated from bovine manure compost. *International Journal of Systematic and Evolutionary Microbiology, 59*, 828–834.

Ramnani, P., & Gupta, R. (2004). Optimization of medium composition for keratinase production on feather by *Bacillus licheniformis* RG1 using statistical methods involving response surface methodology. *Biotechnology and Applied Biochemistry, 40*(20), 191–196.

Saha, S., Dhanasekaran, D., Shanmugapriya, S., et al. (2013). *Nocardiopsis sp.* SD5: A potent feather degrading rare actinobacterium isolated from feather waste in Tamil Nadu, India. *Journal of Basic Microbiology, 53*, 608–616.

Selvam, K., & Vishnupriya, B. (2012). Biochemical and molecular characterization of microbial keratinase and its remarkable applications. *International Journal of Pharmaceutical and Biological Archive, 3*(2), 267–275.

Shankar, T., Thangamathi, P., Sathiya, C., & Sivakumar, T. (2014). Statistical optimization of keratinase production by *Bacillus thuringiensis*. *Journal of Global Biosciences, 3*(2), 477–483.

Sivakumar, T., Shankar, T., Vijayabaskar, P., & Ramasubramanian, V. (2012). Optimization for keratinase enzyme production using *bacillus thuringiensis* TS2. *Academic Journal of Plant Sciences, 5*(3), 102–109.

Staroń, P., Banach, M., & Kowalski, Z. (2011). Keratin – Origins, properties, application. *Chem, 65*(10), 1019–1026.

Vanbreuseghem, R. (1952). Technique biologique pour I' isolement des dermatophytes dusol. *Annales de la Société Belge de MédecineTropicale, 32*, 173–178.

www.nrdcindia.com

Yadav, A. K., Vardhan, S., Yandigeri, M. S., Srivastava, A. K., & Arora, D. K. (2011). Optimization of keratin degrading enzyme from thermophilic strain of *Streptomyces sclerotialus*. *Research Journal of Microbiology, 6*, 693–705.

Yang, Y. (2012). *Skin-whitening and freckle-dispelling essence and preparation method thereof.* Patent: CN102614104.

Chapter 5
Potential Impression of Arbuscular Mycorrhizal Fungi on Agricultural Growth, Productivity, and Environment Toward Global Sustainable Development for Green Technology

Kamal Prasad

Contents

K. Prasad (✉)
Absolute Foods, Plant Life Science, Division of Microbiology, Gurugram, Haryana, India

A. C. Shukla (ed.), *Applied Mycology*, Fungal Biology,
https://doi.org/10.1007/978-3-030-90649-8_5

5.1 Introduction

The term *Green Technology* is an umbrella that designates the utilization of technology and science to make products that are additional eco-friendly. The chief goal of green technology is to shield the surroundings and repair past damage to the environment. Sustainable agriculture could be an important role in today's biosphere, because it offers the potential to fulfill our agricultural needs. A perfect agricultural system is sustainable, maintains and improves human health, surroundings edges, and produces sufficient food for the increasing current world population. Microbes play a crucial role in the rhizosphere and greatly have an effect on the availability of obtainable soil nutrients to plants and thereby modify the standard and quality of root exudates (Barea, 2000). The root exudates are rich sources of carbon, and therefore, the carbon fluxes are crucial determinants of rhizosphere operation. The discharge of root exudates and decaying plant material offer sources of carbon compound for the heterotrophic soil biota as either growth substrates or structural material for root-associated microbial flora (Werner, 1998). Numerous soil organisms act with one another to beat the limitation. Positive effects of arbuscular mycorrhizal fungal biofertilizer with and without combination with species of genus *Rhizobium, Azotobacter, Pseudomonas, Fraturia, Azospirillum* inoculants are reported for various plant species (Prasad et al., 2019). These microbes are usable in several post-harvest agricultural processes, in manure production or perhaps in control of pests/ diseases. Extraction of pricey minerals from poor or leftover mines or production of raw materials for production of chemicals of huge utility or perhaps production of fuel gases are all modern achievements within the field of microorganisms. Presumably, thousands of different technologies could also be waiting behind the screen for their release in future agricultural development programmers for green technology. AM fungal biotechnology and its application are comparatively safe to be utilized in the field for all sorts of plants worldwide. Mycorrhiza are capable to provide micronutrients for effective plants growth, development and quality produce.

Food and nutrition security has long been one of the best challenges for humanity and given existing population and climate change eventualities is a progressively difficult task in current scenario. A number of the newest estimates predict the requirement to extend agricultural productivity by more than 70% by the year 2050, and therefore, the focus shifts progressively to the role of soil biodiversity in normally (Bender et al., 2016) and significantly AM fungi (Thirkell et al., 2017), in an exceedingly achieving in a sustainable methodology. Symbiosis could be a biological phenomenon involving dynamic changes within the genome metabolism and signaling network (Kawaguchi & Minamisawa, 2010). Most of the legumes possess two varieties of microorganism symbionts: mycorrhizal fungi and nitrogen-fixing bacteria, thereby establishing triple association, capable of supplying nitrogen and phosphorous contents to the plants (Prasad et al., 2019; Gautam & Prasad, 2001; Prasad, 2006a, b; Prasad, 2013; Prasad and Meghavanshi (2005). Each of the mycorrhizal fungi and *Rhizobium* act as microbial biofertilizer and have the distinctive ability to convert nutritionally vital components from unavailable to available form through organic processes (Prasad et al., 2019; Vessey, 2003; Prasad, 2000, 2015, 2017, 2020; Harwani et al., 2009; Prasad & Pandey, 2012). The standard for mycorrhizal biofertilizers necessitates an in-depth examination of microorganism features, efficacy, consistency, as well as precautions and limitations, not only in the lab and in the factory, but also in the field. The AM fungal biotechnology is incredibly important, 100% natural, low cost, eco-friendly organic inputs, and has supplementary ability to the chemical fertilizers. The most effective quality AM fungal biofertilizers assist in providing all of the nutrients required by plants as well as maintaining the soil's standard in a natural environment. Mycorrhizal fungi have been used over the past fifty years to varied strategies for food safety; quality, environmental protection, and more practical treatment of agricultural crops give a most spectacular record of achievement. In recent years, however, mycorrhizal biotechnologies have been successfully applied to a variety of agricultural and environmental challenges. AM fungi are terribly helpful in eliminating issues related to the utilization of chemical fertilizers and pesticides; they are currently wide applied in natural farming and organic agriculture (Prasad et al., 2019; Prasad, 2015, 2017; Parr & Hornick, 1994). Soil microbiologists and microbial ecologists have well-tried to differentiate soil microbes as useful or harmful consistent with their functions. Microbes have the ability to effect soil quality, plant health, growth, and productivity. The present manuscript has been critically mentioned numerous aspects of mycorrhizae inputs application in agriculture and their economic importance for sustainable development for green technology.

5.2 Positive and Active Microbes

The mismanagement and extreme use of chemical fertilizers and pesticides have typically adversely affected the soil ecology, atmosphere and created food safety, and human and animal health issues. Consequently, there has been a growing interest in natural and organic agriculture by customers, microbiologists, and ecologists as potential alternatives to chemical-based mostly conventional agriculture. Agricultural systems that adapt to the principles of natural ecosystems are currently receiving an excellent deal of attention in each developed and developing country. A variety of articles has been published that deal with several aspects of natural and organic farming systems. Alternative and sustainable agriculture, soil quality, integrated pest and nutrient management, and even helpful microorganisms are being explored by the agricultural research analysis institution (Prasad et al., 2019; National Academy of Sciences (1989) Alternative Agriculture; Parr & Hornick, 1992; Prasad et al., 2005a, b; Meghavanshi et al., 2008). The use of single and consortium cultures of helpful microbes as soil inoculants relies on the principles of natural ecosystems that are sustained by their constituents; that is, by the standard quality and quantity of their inhabitants and specific ecological parameters, that is the bigger the variety and range of the inhabitants, the upper the order of their interaction and also a lot of stable the ecosystem. The consortium culture approach is just an endeavor to use these principles to agricultural soils, and to shift the biological equilibrium in favor of increased plant growth, productivity, and disease protection (Prasad et al., 2019; Higa, 1994).

5.3 Arbuscular Mycorrhizal Fungi (AM Fungi)

An AM fungus is functionally an integral part of the plant community. AM fungi and therefore the plant community are interdependent. The term mycorrhiza was coined a century past to designate a mutual symbiotic association of plant roots and fungi. There are numerous forms of mycorrhizal fungi. Orchids and ericaceous plants have their own particular brand whole; however, those fungi that colonize *in vivo* and *in vitro* conditions are renowned as arbuscular mycorrhizal fungi. AM fungi are colonizing the root system of most native land plants. The AM fungal hyphae extend from root surfaces like little conduits out into the soil, greatly enhancing the absorptive surface area of root systems. These fungous colonize plants roots from a vital link between the plant and its soil environment, and work as an effective biological tool in making a healthy ecosystem (Prasad et al., 2005, 2019; Gautam & Prasad, 2001; Prasad, 1993, 2006a, b, 2010, 2011a, b, 2013, 2015, 2017; Prasad & Pandey, 2012; Prasad & Deploey, 1999; Prasad & Warke, 2018a, b). Approximately more than 90% of land plant species form association with AM fungi (Prasad, 2017). These plant species represent a broad variety of agricultural crops, agro forestry and horticultural crops, vegetables, oilseeds, grains crops, millets, and decorative and forage crops (Gautam & Prasad, 2001; Prasad, 2015, 2017; Trappe, 1987). The roots of most plant species from symbiotic association with specialized fungi (Genus *Glomus, Gigaspora, Acaulospora, Enterophospora* and *Sclerocystis*) are referred to as arbuscular mycorrhizal fungi (AMF). An AM plant is that the amplest kind of mycorrhizal fungi delineated as universal plant symbioses. Researches on mycorrhizal fungi conducted throughout the previous few decades envisaged their prevalence during a giant variety of hosts, completely different habitats and variability in quality and quantity (Gautam & Prasad, 2001; Prasad, 2015, 2017; Curl & Truelove, 1986; Deacon, 2019; Prasad & Kaushik, 2004). Prasad and Kaushik (2004) describe ecology, physiology, biochemistry, and taxonomy of mycorrhizal fungi. Previous few decades mycorrhizae fungi frequently used in several trees, agricultural crops, and decorative plants worldwide. AM fungi increase approximately 20–50% yield and enhance uptake of macro- and micronutrients, especially P, K, Zn, S, Fe, mg, Mn, Cu, boron, molybdenum, and water. Even it survived in harsh conditions in soil ecosystems. AMF is sometimes inoculated to seedlings, coated on seeds moreover as soil application. AM fungi also are applied as barren wasteland development, industrial contaminated and polluted land moreover as plant diseases and pest management (Prasad & Rajak, 2002; Johnson et al., 1997).

5.3.1 Development of Mycorrhizal Fungal Infection

The growths of AM fungal colonization within the host has been demonstrated in Figs. 5.1 and 5.2, and therefore the following steps: (1) The preinfection resting reproductive spore or hyphae can initiate infection during a host cell (Fig. 5.2). Germination of reproductive spore is primarily ruled by physical and chemical factors of soil such as temperature, CO_2, moisture, pH, presence of phosphate, etc. (Gautam & Prasad, 2001; Prasad, 1993; Smith & Read, 1997). Throughout the development of germ tubes, no divisions of nuclei occur, rather get disturbed in newly fashioned germinating structures. Once germ tubes encounter an appropriate living host, additional development of infection proceeds, whereas in absence of appropriate host, the infection potential of germinated fungal propagule is lost. (2) The AM fungal hyphae penetrate the root between epidermal cells and create an entry (Gautam & Prasad, 2001; Prasad, 1993; Li et al., 2006). The appressorium is the first structure formed within the host cells, and it signifies the start of a successful infection. After this, the autotrophic growth of fungus ceases. (3) The AM fungal hyphae grow inter- and intracellular and develop infection segment spread either by the expansion of hyphae within the root (internal spread) and/or by the formation of secondary entry points adjacent to the primary (Gautam & Prasad, 2001; Prasad, 1993; Li et al., 2006). Within the deeper root cortex, the living intracellular hyphae are positioned paralleled to the long axis of the root system. Shortly when hyphae penetration, arbuscules formation can be initiated in cortical region. Arbuscules are powerfully branched hyphae, and its formation is accompanied by stretching of host cell wall and invagination of plasma membrane. Most intensive affiliation between fungus and plants are maintained through arbuscular and it is believed to be the particular site of phosphate transfer (Prasad et al., 2019; Gautam & Prasad, 2001; Prasad, 1993; Adesemoye et al., 2009). The arbuscules have distinct wall morphology that is improving up to ten times thinner than extraradical hyphae and fibrillar nature of wall be usually lost and becomes amorphous (Prasad, 2007). When

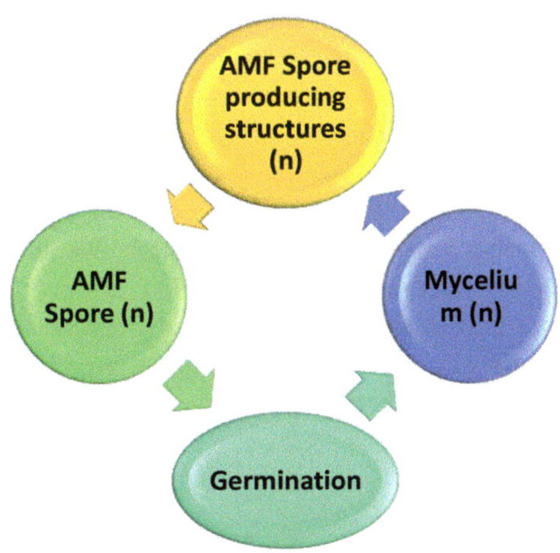

Fig. 5.1 Sketching diagram of asexual life cycle of arbuscular mycorrhizal fungi

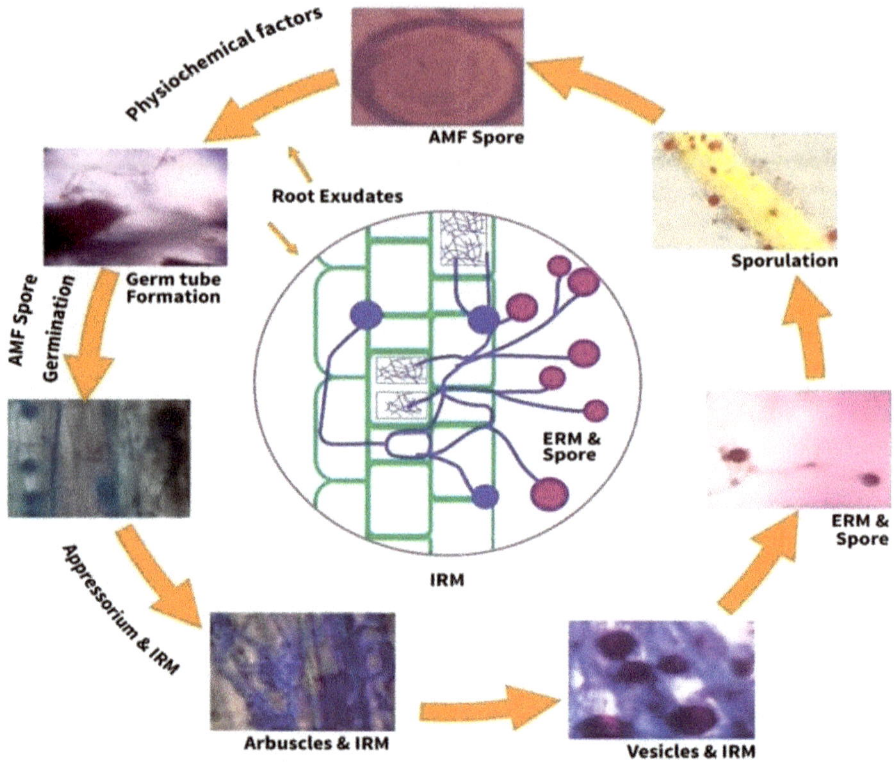

Fig. 5.2 Structural characterization and development of AM Fungal infection within the host cell and their life cycle in soil ecosystem

colonization in root tissue, apical or intercalary swelling of hyphae known as vesicle containing lipid. They have relatively thinner wall than spore. In conjunction with the interior spread of infection, the fungus ramifies the rhizospheric soil and generally extends as over eight cm from the root (Gautam & Prasad, 2001; Prasad, 1993). (4) AMF forms resting spores to external hyphae known as arbuscular mycorrhizal fungal spores. Spore formation can begin as early as 2-4 weeks when primary infection or can take up to three–six months. Short-lived fine hyphae were primary concerned in nutrient acquisition and ne'er colonized roots, whereas a cluster of thicker hyphae were exclusively concerned in colonizing roots (Gautam & Prasad, 2001; Prasad, 1993). As the host plant grows, new roots are produced that become successively infected by runner hyphae or new developed spores. Arbuscules and vesicles are unceasingly fashioned and degraded. Spore germination and spore formation occur at a similar time. Thus, at any stage, numerous phases of infection can be observed (Prasad, 1993).

5.3.2 Characteristics of Arbuscular Mycorrhizal Fungi

The vesicles and arbuscules formation within the root cortex are most vital characteristics of mycorrhizal fungi (Fig. 5.2). The root colonization is directly linked to external AM fungal mycelium that spreads and ramifies within the soil. The appressorium of AM fungal-infected plants forms on the host root surface and penetrates the epidermal and exodermal cells, with hyphal branches passing into the root's middle and inner cortex. So, they sometimes form coil-like structures. The intercellular distributive hyphae grow parallel to the root axis. The lateral branches then penetrate root cells leading to the formation of arbuscules (haustoria-like structure) that develop further by repeated dichotomous branching among the host cells to yield a cluster of fine filaments. Within the next stage (after 2–4 days), the arbuscules disintegrate (digestion stage) to produce a granular mass. At the tip or within the middle of distributive hyphae swellings develop inner or intracellular to create vesicles. AM fungal species of *Gigaspora* and *Scutellospora*are don't seem to be forming vesicle, they solely form arbuscules; thus, presently vesicular arbuscular mycorrhizal fungi (VAMF) are known as arbuscular mycorrhizal fungi (AM fungi). AM fungi colonization is nearly no modification in external morphology of the root. The size and shape of the vesicles are storage and reproductive structure. With relevancy to transfer of nutrients, it had been thought to require place through arbuscule digestion. AM fungi uninterrupted in soil as spores or propagules for sizable amount over a long time. Spores can germinate within the soil with appropriate host plants. Completely different species of AM fungi endophyte have different temperature limits and optimum for germination. Mycelial growth forms interconnections between interconnected plants. It survives in dry soil environment for a long time.

5.4 Positive Impact of AM Fungi on Agricultural Crops and Environments

5.4.1 AM Fungi and Plant Growth

Mycorrhizae play positive role in phosphate (P) uptake by almost all terrestrial plants. The matter for P_2O_5 nutrition comes from heterogeneity of soils; low-quality ends increase in native depletions. In most soils, massive a part of P is held in mineral lattices and so isn't at once exchangeable with ions in soil solution. The huge synthesis of cytoplasm is proportional to extend in plasmalemma area in arbuscular region. AM fungi are capable for convert insoluble P to soluble form. Roots of AM fungi–mediated plants uptake again and again additional P from soil. Presently it's been created clear that replenishment of supply of P from distant soil (beyond the eight cm distance from roots) to root surfaces being low, the mycorrhizal hyphae, having organic connection with the root, extend to these distance and absorb macro and micro nutrients (Prasad et al., 2005a, b; Prasad, 1993). This AM fungi–mediated phosphate transport turn up in three phases specifically uptake by fungal hyphae in soil, movement to hyphae within root cortex, and release in host plant. Typically, the fungus uptake P from soil as orthophosphate ions from the labile pool and its translocation to root is believed to take place mainly by cytoplasm streaming of polyphosphate granules within the vacuoles of the AM fungi. Presently AM fungi are being thought of as a system of various importance in hemisphere. Their effects in inflicting shifts in chlorophyll pigment and phytohormone gradients, resistance to water stress, and attenuate escape of electrolytes form injured or pathologic cells are gaining importance and drawing attention of scientists (Prasad et al., 2019; Gautam & Prasad, 2001; Prasad, 2011a; Prasad & Warke, 2018a, b; Prasad & Rajak, 2002). Scientists have determined that growth of varied crops once infected with completely different species of AM fungi was increased, thereby improving crops growth and productivity. Rotations of crops influence the stability of mycorrhizal flora in soil ecosystems. The importance of AM fungi as a tool for improving the expansion and productivity in various groups of plants was recognized solely once the work of Baylis (Baylis, 1959), and Mosse (Mosse, 1962). However, throughout last five decades, plenty of documents have been gathered concerning the taxonomy, ecology, physiology, and anatomy of AM fungi and their relation with their hosts particularly with relation to uptake of water, enhance root absorption capability, uptake of macro- and micronutrients, growth hormone production, antibiotic secretion, barren wasteland management, biotic and abiotic management, environmental protection, and management of diseases and pests (Fig. 5.3), (Prasad et al., 2019; Prasad & Kaushik, 2004; Prasad & Rajak, 2000, 2001, 2002; Rewari & Tilak, 1988; Warke et al., 2018; Bitterlich et al., 2019; Mathur et al., 2019). Trees are being cultivated in order to use them in a variety of plant production methods (Prasad et al., 2019; Prasad, 1993; Prasad & Warke 2018a; Rahimzadeh & Pirzad, 2019; Ji et al., 2019; Ghorchiani et al., 2018; Mayer et al., 2019).

5.4.2 AM Fungi and Fertilization

The most widely appreciated contribution of AM fungi to plant performance is their ability to extend nutrient uptake, notably of P (Prasad, 2002a, b; Smith & Smith, 2011). Harnessing the nutrient provides by AM fungi, the amount of applied fertilizer, and therefore the energy linked to its production can be reduced. A significant issue in optimizing economical fertilization is reducing the amount of nutrients lost to the system via leaching. AM fungi decrease nutrients leaching not solely increasing the nutrient interception zone due to the development of a mycorrhizosphere; however, additionally increased nutrients uptake increased soil structure and fostering of the microbial community with associated nutrient immobilization (Cavagnaro et al., 2015). Kohl and van der Heijden (Kohl & van der Heijden, 2016) confirmed that different AM fungal species differ in their ability to decrease nutrient leaching. AM fungi increase nutrients leaching in extremely fertile agro ecosystems due to

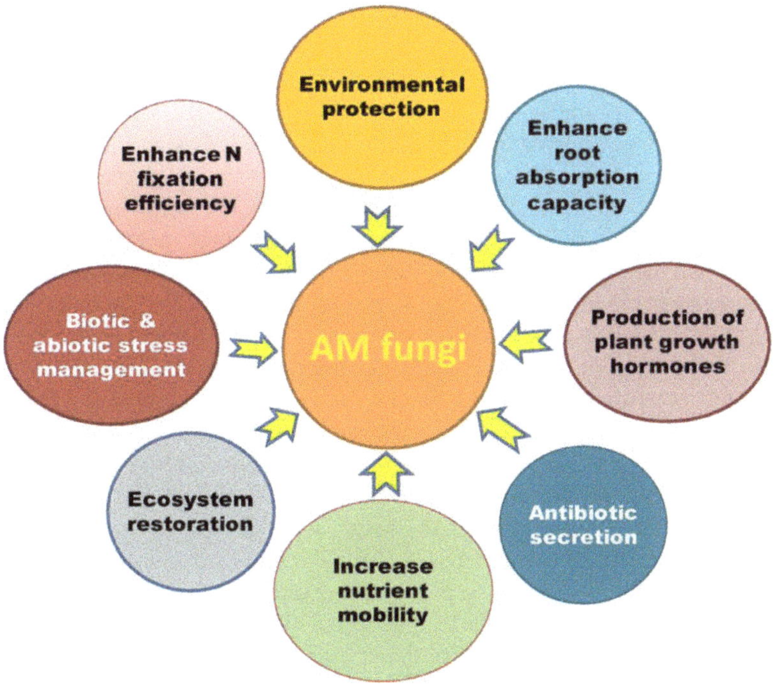

Fig. 5.3 A diagrammatic representation of arbuscular mycorrhizal fungi functions to regulate various processes in the ecosystem and plant growth promotion and protections

larger soil nutrient content. (van der Heijden, 2010). AM fungi are shown to stabilize community productivity across gradients of nutrient availableness and to reduce plant tissue nutrient content variability on such gradients in grassland (Yang et al., 2016). They transferable to agricultural systems, these effects would be crucial in achieving food and nutrition security notably in regions wherever access to fertilizers could be restricted or irregular. Furthermore, extending the available soil nutrient pool to deeper soil might improve resistance, for example, maintaining plant growth under dry situations, where minerals in the soil may be there but are positional untouchable for the roots. Altogether, with the endlessly increasing costs of chemical fertilizers and their foretold deficiency in a very close to future, creating the foremost out of those resources is that the solely approach forward and soil and AM fungi could prove necessary in agriculture industry.

5.4.2.1 AM Fungi and Nitrogen Absorption and Translocation

Nitrogen (N) may be a premier growth-limiting issue for higher plants and agricultural crops. Varied studies have processed that AM fungi have the power to absorb and transfer N to the proximate plants as well as host plants (Prasad, 1993; Hodge

& Storer, 2015; Battini et al., 2017; Turrini et al., 2018). Nitrogen applied in agricultural fields is lost via leaching or in type of gaseous emissions. Leachate N happens principally in form of dissolved nitrate (NO3−), a particularly mobile type of N in soil. AM fungi promote soil aggregation (Zhu et al., 2012; Leifheit et al., 2014; Nouri et al., 2015) by improving soil structure and thus increasing soil water-holding capability. Moreover, AM fungi take up nitrogen preferentially within the type of ammonia (NH4+), reducing the pool of N out there for nitrification and consequently reducing the mobility of nitrogen. In soil, AM fungi might intercept N that migrated down the profile and immobilize it or deliver it to the plant, so avoiding N losses. Moreover, the proportion of NH4+ to different N sources can increase in soil (Prasad, 2007; Meghavanshi et al., 2010; Kautz et al., 2013), increasing the potential role of AM fungi in mobilizing and delivering N to the plant, reassuring access to an antecedently untouchable pool and reducing the requirement for N fertilization. Zhang et al. (2018) have verified AM fungi–mediated plants increased allocation of shoot biomass to panicles and grains through increased N and P redistribution to panicles significantly below low fertilizer levels. Translocation of N into seeds is increased from heading to maturity. AM fungal interdependency turnsa out in-depth underground extraradical hyphae starting from the roots increase to the encompassing rhizosphere, thereby serving to in improving the uptake of nutrients specifically nitrogen (Battini et al., 2017). The interaction of salinity stress and AM fungi considerably affects the concentrations of P and N and also the N: P magnitude relation in plant shoots (Wang et al., 2018). Recently, it has been reported that native AM fungi treatments produce significant alterations within the N contents of crop plants (Turrini et al., 2018). It has been widely accepted that fungi have the power to require substantial quantity of N from dead and decomposed material that later can increase their fitness to grow and keep alive. With the exception of this, massive biomass and increased N needs for AM fungi render them the most stakeholder of global N pool that's equivalent in scale to fine roots. Thus, they play a crucial role within the N cycle (Hodge & Fitter, 2010). The AM fungi extraradical hyphae can absorb and assimilate inorganic N (Jin et al., 2005). Many studies have shown that approximately 20–75% of the overall N uptake of AM plants is transferred by the AM fungi to their hosts (Tanaka & Yano, 2005; Govindarajulu et al., 2005; Ahanger et al., 2014; Hameed et al., 2014; Hashem et al., 2018). Increased N in AM fungal–mediated plants evidently results in higher chlorophyll pigments, as chlorophyll molecules can effectively trap N (Prasad, 1993; de Andrade et al., 2015). AM fungal immunization improves C and N accumulation and N assimilation below ambient and elevated Co_2 concentrations (Zhu et al., 2012).

5.4.2.2 AM Fungi and Phosphorus Translocation

Phosphorus (P) is typically assumed low quality in soils. P action is important, and most effort has been spent on avoiding P loss and P-mediated eutrophication via wearing. However, excessive manuring, existence of advantageous pathways, or a sandy soil texture can cause vital P action (Djodjic et al., 2004; Schoumans, 2015), with its associated economic and environmental consequences. The role of AM fungi in P uptake has been extensively researched (Prasad, 2000; Smith & Smith, 2011), which they are going to cut back the need of significant manuring attributable to exaggerated economical P uptake. AM fungi increase water-holding capability, reducing the danger of action. AM fungi might intercept P that has migrated down the profile and deliver to the plant. Uptake of organic P from soil, mainly via roots but to boot with direct injection of organic matter, can keep inaccessible to the plant attributable to attenuated decomposition and mineralization rates. Similarly, in another study, it was recorded that AM fungi in soil might contribute further to plant P nutrition than soil AM fungi, below serious P fertilization (Wang et al., 2017). Consequently, soil AM fungi have potential to be of nice association inside the rejection of P loss, notably in sandy soils or once the soil is P saturated. Phosphorus is usually assumed low quality in soils. P leaching is importance and most effort has been spent on avoiding P loss and P mediated eutrophication via soil erosion. However, excessive manuring, existence of discriminatory pathways, or a sandy soil texture can lead to significant Pleaching (84–85), with its associated economic and environmental consequences. The role of AM fungi in P uptake has been extensively researched (Prasad, 2000; Smith & Smith, 2011), and that they can scale back the necessity of heavy manuring due to increased economical P uptake. AM fungi are increase water holding capability, reducing the chance of leaching. AM fungi may intercept P that has migrated down the profile and deliver to the plant. Uptake of organic P from soil, principally via roots however conjointly with direct injection of organic matter, can remain inaccessible to the plant due to decreased decomposition and mineralization rates. Wang et al. (2017) found some proof that AM fungi in soil would possibly contribute a lot of to plant P nutrition than soil AM fungi, under heavy P fertilization. As a result, soil fungi have the potential to play an important role in the rejection of P loss, particularly in sandy soils or when the soil is P saturated.

5.4.2.3 AM Fungi and Potassium Nutrition

The role of potassium remains poorly investigated in mycorrhizal analysis. Plant potassium nutrition is improved by mycorrhization in crops, particularly beneath K^+ limiting conditions as found in forest ecosystems. Moreover, this improvement may act on abiotic stress tolerance, P equilibrium maintenance, or exclusion of soil contaminants. Moreover, different reports have shown increased activity of a K^+ transporter within the mycorrhizal roots of *Lotus japonicus* (Guether et al., 2009; Berruti et al., 2016).

5.4.2.4 AM Fungi and Other Mineral Nutrition

AM fungal immigration is stimulating nutrients uptake in overall agricultural crops. AM fungi symbiotic association is to transfer organic carbon within the kind of lipids and sugars (Jiang et al., 2017; Luginbuehl et al., 2017). AM fungal vaccination in plant can enhance the concentration of assorted macro- and micronutrients considerably, that ends up in exaggerated photosynthate production and therefore exaggerated biomass accumulation (Prasad, 2017, 2020; Mitra et al., 2019). AM fungi have the flexibility to extend the uptake of inorganic nutrients in nearly all plants. Except for macronutrients, AM fungal colonisation has been done in order to increase the provision of micronutrients like zinc and copper (Smith & Read, 1997). AM fungi develop in a living extension with root system and dramatically increasing access to get essential nutrients such as N, P, K, Ca, Zn, Mn, Mg, Cu, B, S, and Mo from the soil surroundings. Consequently reciprocally, the host plant feeds the sugar and different organic substance that the AM fungi need to survive and grow. Therefore, AMF give nutritional support to the crop plants even beneath inappropriate conditions within the root cells. AM fungus produce arbuscules, which aid in the interchange of inorganic minerals and thus carbon and phosphorus compounds, giving host plants a significant boost in vigour (Prasad, 2017, 2020; Li et al., 2016). Increased photosynthetic activities and different leaf functions are directly associated with higher growth frequency of AM fungi colonization that is directly linked to the uptake of nutrients and promote the alteration of crop yield. AM fungi interdependency absolutely exaggerated the concentrations of N, P, and Fe in *Pelargonium graveolens* L. beneath drought stress (Amiri et al., 2017). Improved levels of P, Ca, and K were reported in AM fungal associated *Euonymus japonica* beneath salinity stress condition (Gomez-Bellot et al., 2015).vIt's supposed that AM fungi increase the uptake of virtually all essential macro and micro nutrients and contrarily decrease the uptake of Na and Cl, resulting in growth stimulation (Evelin et al., 2012). The extraradical hyphae can effectively improve nutrient uptake, thereby improving plant growth and development (Lehmann & Rillig, 2015). Numerous scientists have conferred the role of AM fungi in uptake of soil essential nutrients and effectively promote the expansion of host plants (Prasad, 2020; Smith et al., 2011). Improvement of plant nutrition and maintenance of Ca_2+ and Na+ ratio are the significant dynamic attributes that facilitate improving useful aspects of AMF colonization on overall plant performance (Evelin et al., 2012; Abdel Latef & Miransari, 2014). Increased growth and levels of protein, Fe, and Zn were recorded in AM fungal–treated chickpea (Pellegrino & Bedini, 2014). Asrar et al. (2012) reported that the required fungal association increased the contents of macronutrients such as N, P, K, Ca, and Mg of *Antirrhinum majus* beneath drought condition. AM fungi additionally well tried to be effective in limiting the high accumulation of Na, Mn, Mg, and Fe in roots (Bati et al., 2015). Many studies unconcealed that AM fungous (*Glomus mosseae* and *Rhizophagus irregularis*) exhibited improved heavy metal translocation within the shoot (Zaefarian et al., 2013; Ali et al., 2015). Micronutrients such as Zn and cu being diffusion restricted in soils are absorbed by plants with the assistance of mycorrhizal hyphae.

5.5 AM Fungi and Greenhouse Gas Emission

Modern agriculture is responsible for around 12% of global anthropogenic greenhouse gas emissions (Linquist et al., 2012). Most of the emissions are related to chemical fertilizer production and therefore the use of significant heavy machinery; however, most of them occur within the type of direct emissions from the field. AM fungi is having potential for reducing emissions from chemical fertiliser (Prasad, 2017). The role of AM fungi in reducing the discharge of carbon dioxide and nitrous oxide greenhouse gasses related to agriculture.

5.6 AM Fungi and Carbon Farming

The traditional, terribly stable carbon found in soil is problematic. Stable soil carbon is also promptly decomposed once contemporary carbon is added. AM fungi have the potential to counteract this development due to their function in soil structure and within the capture of nutrients (Prasad, 2017). AM fungi contribute in the plant to soil organic matter by making a sink demand for plant carbon and distributing to hyphal biomass. The role of AM fungi is essentially unnoticed in terrestrial carbon cycling and global climate change models despite their larger involvement in net primary productivity augmentation and any accumulation of this extra photosynthetic fastened carbon within the soil. However, this buffering mechanism against elevated carbon dioxide condition to sequester additional carbon by AM fungal interaction.

5.7 AM Fungi and Soil Management

Soil is also restricted by physical properties, just like the existence of a troublesome plow pan that stops root growth. The advantages of deep tillage and various soil tillage management options are also debatable and contextdependent; yet, assuming the presence of a plough pan, they are very frequent. Yields are also significantly increased once deep ploughing (Schneider et al., 2017). The existence of AM fungal mediated plant phylotypes and their inability to survive soil mixture events, however, entails precaution and conjointly the overall turning of any methodology that inverts the profile (Sosa-Hernandez et al., 2018). Intensive tillage has been referred to as a major issue reducing AM plant abundance and variety in agriculture (Kabir, 2005). Recently, Sale et al. (2015) has examined the outcomes of reduced and traditional tillage, down to 40cm within the profile exploitation reproductive structure based on community analysis. Their results make sure the expected shifts in reproductive structure abundance and variety in soil but those effects weren't necessary in deeper layers, despite a shift in community composition. The absence of AM fungi reproductive structure abundance shifts do not basically imply an absence of impact on hyphal abundance or settlement rates, but changes in soil community composition highlight that tillage can have an impact on AM fungi in deeper layers, with unknown consequences for his usefulness. AM fungal immunisation has managed the soil structure and performance.

5.8 AM Fungi and Heavy Metals

AM fungi extensively support plant establishment in soils contaminated with heavy metals, due to their potential to strengthen defense of the AM fungal–mediated plants to promote growth and development. Heavy metals could build up in food crops, fruits, vegetables, and soils, posing a health risk to the common people (Liu et al., 2013; Yousaf et al., 2016). AM fungi association with wheat completely multiplied nutrient uptake aluminum stress (Aguilera et al., 2014). Plants well-grown on soils enriched with Cd and Zn exhibit significant suppression in shoot and root growth, leaf chlorosis, and even death (Moghadam, 2016). There are several reports within the literature on uncovering the AM fungi elicited effects on the build-up of metals in plants (Souza et al., 2012). Heavy metals are immobilized within the fungal hyphae of internal and external origin (Ouziad et al., 2005) that have the power to repair heavy metals within the cell wall and store them within the vacuole or could chelate with other substances within the cytoplasm (Punamiya et al., 2010) and therefore reduce the metal toxicity within the plants. The strong impacts of AM fungus on plant development and growth under harsh conditions have commonly reported (Kanwal et al., 2015; Miransari, 2017). It is additionally believed that increased growth or chelation within the rhizospheric soil can cause metal dilution in plant tissues (Kapoor et al., 2013; Audet, 2014). AM fungi reportedly bind Cd and Zn within the cell membrane of mantle hyphae and cortical cells, thereby limiting their uptake and leading to improved growth, yield, and nutrient status (Andrade & Silveira, 2008; Garg & Chandel, 2012) AM fungal mycelia having a high cation exchange capability and absorption of metals (Takacs & Voros, 2003). AM fungi are believed to manage the uptake and accumulation of some key inorganic nutrients and increased uptake of Si has been reportable in mycorrhiza-inoculated plants in soybean and *Zea mays* (Clark & Zeto, 2000). Hammer et al. (2011) additionally recorded significant uptake of Si in spores and hyphae of *Rhizophagus irregularis* and its transfer to the host roots. It is pertinent that low Cd mobility and toxicity can even be addressed with AM fungi by increasing soil pH (Shen et al., 2006), restoring Cd within the additional radical mycelium (Janouskova & Pavlikova, 2010), and binding Cd to glomalin, a glycoprotein. AM fungi are very effective in lowering the level of Cd in the vacuoles and cell wall, leading to Cd detoxification (Li et al., 2016). Wang et al. (2012) determined that AM fungal–mediated alfalfa (*Medicago sativa* L.) improved Cd tolerance because of the modification of chemical form of Cd in numerous plant tissues. Various processes that occur through the AM fungi are immobilization/restriction of metal compounds, precipitation of polyphosphate granules within the soil, adsorption to fungal cell wall chitin and heavy metal chelation within the fungus.

5.9 AM Fungi and Environmental Temperature

Soil temperature increases, plant community reactions are also captivated with AM fungal interactions for sustainable yield and production (Bunn et al., 2009). Heat stress has a significant impact on plant growth and development, causing loss of plant vigour, seed germination suppression, scrubby growth rate, and lower biomass production, wilting and burning of leaves and generative organs, abscission and senescence of leaves, injury additionally as discoloration of fruit, cell death, reduction in yield (Wahid et al., 2007; Hasanuzzaman et al., 2013), and increased oxidative stress.

Generally, AM fungal–immunized plants show higher growth below heat stress compared to non-AM-fungal–inoculated ones (Gavito et al., 2005). Maya and Matsubara (Maya & Matsubara, 2013) have reported the association of AM fungi with plant growth and development, resulting in positive changes in growth below the conditions of height temperature (Fig. 5.4).

AM fungi can increase plant tolerance to cold stress (Liu et al., 2013; Birhane et al., 2012; Chen et al., 2013). Furthermore, according to the majority of findings, plants inoculated with AM fungus at low temperatures grow faster, whereas plants not inoculated with AM fungi grow slowly (Liu et al., 2013; Chen et al., 2013; Zhu et al., 2010a; Abdel Latef & Chaoxing, 2011a). AM fungi support plants in

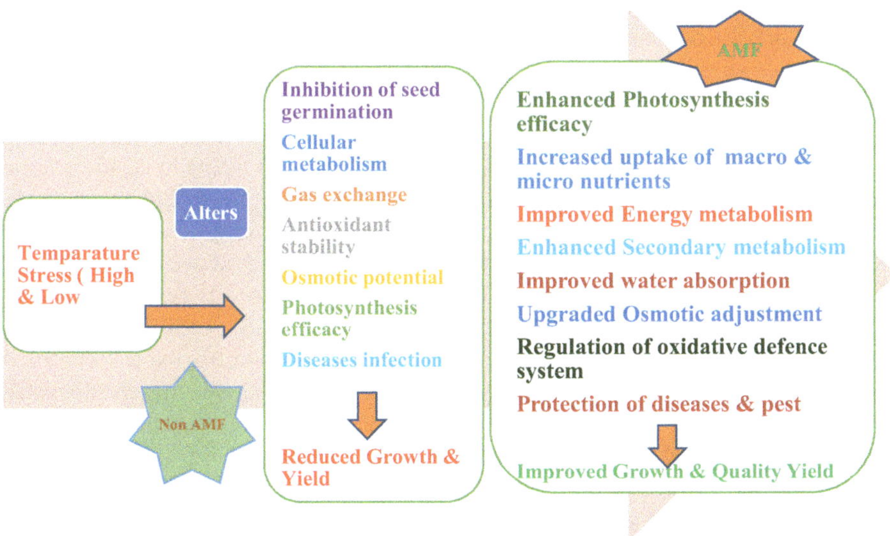

Fig. 5.4 AM Fungi mediation alleviates temperature stress in plants

combating cold stress and eventually improve plant development (Birhane et al., 2012; Gamalero et al., 2009). Moreover, AM fungi can also retain moisture within the host plant (Zhu et al., 2010), increase plant secondary metabolites resulting in strengthen plant immune system, and increase protein content for supporting the plants to combat cold stress conditions (Abdel Latef & Chaoxing, 2011a). AM fungal symbiotic relationship improves water and plant relationships and can increase gas exchange potential and osmotic adjustment (Zhu et al., 2012). AM fungi improve the synthesis of chlorophyll, resulting in a major improvement within the concentrations of varied metabolites in plants in cold stress conditions (Zhu et al., 2010; Abdel Latef & Chaoxing, 2011a). The role of AM fungi throughout cold stress has conjointly been according to change protein content in tomato and different vegetables (Abdel Latef & Chaoxing, 2011a).

5.10 AM Fungi and Abiotic Stresses

It is widely accepted that AM fungi may improve various stresses or combinations of stresses condition such as drought, salinity, temperature, nutrients, and heavy metals. Exposure of plants to a combination of drought and salinity causes an increased production of reactive oxygen species, which might be extremely injurious to plants (Bauddh & Singh, 2012). Detoxification of reactive oxygen species is finished by the enzymes that included commonly superoxide dismutase, catalase, peroxidase, and glutathione enzyme (Ahanger & Agarwal, 2017). AM fungi are improving plant growth and yield beneath stress (Abdel Latef & Chaoxing, 2011; Abdel Latef & Chaoxing, 2011; Abdel Latef, 2011; Abdel Latef & Chaoxing, 2014). AM fungal interdependency protects plants against a range of abiotic stresses exploitation varied processes such as improved photosynthetic rate, uptake and accumulation of mineral nutrients, accumulation of osmoprotectants, regulation of antioxidant enzyme activity, and alter within the rhizosphere ecosystem (Barzana et al., 2015; Calvo-Polanco et al., 2016; Yin et al., 2016). Numerous studies have shown improved nutritional status of AM fungal–mediated plants beneath osmotic stress conditions (Lehmann & Rillig, 2015; Auge et al., 2014; Lehmann et al., 2014), ensuing from deficit irrigation or salinity. Similarities among the tolerance mechanisms could occur in response to AM fungal–mediated combined stress diversifications. AM fungal colonization alterations in growth regulator profile, mineral uptake and assimilation, accumulation of compatible osmolytes and secondary metabolites, and upregulation of antioxidant system can be the common mechanisms induced throughout different stresses. However, specific mechanisms like compartmentation and sequestration of toxic ions, production of phytochelatins, and macromolecule (protein) expression are specific and exhibit a significant amendment with stress type and also the AM fungal concerned. Changes in root characteristics improve the osmatic stress tolerance to extended levels (Evelin et al., 2009). The AM fungi could elevate the nutraceutical quality of crops and could be of extended scientific agronomic importance for the production and management of various potential crops.

5.11 AM Fungi: A New Dimension for Green Technology

Universal microbiologists take into account that the number of soil microbes can be improved by applying organic amendments to the soil. The majority of soil bacteria are heterotrophic, thus this is usually true. They need complex organic molecules of carbon and nitrogen element for metabolism and biogenesis. The use of AM fungi and various organic inputs (seaweed extract, organic microbial fertilisers, fishmeal, crushed crab shells, and so on) not only helps to balance the micronutrient content in soil, but it also helps to enhance the population of important antibiotic producing actinomycetes. It is changing the soil to a disease suppressive condition over a comparatively short period. All useful microbes predominant in organic farming or natural farming ways can rely upon the ecosystem and environmental conditions. The ultimate goal is to pick microorganisms that are physiologically and ecologically compatible with one another, which can be introduced as single or consortium cultures into soil wherever their useful effects can be realized (Prasad et al., 2019; Prasad & Pandey, 2012; Prasad, 2015, 2017, 2020; Higa, 1994).

5.12 Conclusion and Future Prospects

With an intensive literature assessment, it is often all over that the AM fungi are utilized in biological science, agriculture, and forestry for numerous resolutions. A major deliberation in their application of valuable AM fungi to soils is that the improvement of their synergistic properties. AM fungi are to be effective once immunization into soil or agricultural crops. Various research analysis reports have already recognized the helpful role of AM fungi in improving plant growth and productivity even nerve-wracking environments. Therefore, during this manuscript, the prevailing information associated with the important role of AM fungi has been combined during a coherent manner for beneath standing of AM fungal dependent relationship with a spread of plants under stress environments. Mostly, AM fungi are principally mentioned as helpful entities for macro- and micronutrients uptake from soil ecosystem; but recently, it has been clearly pictured that plants inoculated with AM fungi can effectively combat numerous environmental such as salinity, drought, nutrient stress, alkali stress, cold stress, and extreme temperatures, and therefore helps increase yield in numerous crops. Encouragement of the use of AM fungus is critical for the long-term viability of modern international agriculture systems. AM fungal application in agricultural improvement can considerably scale back the utilization of artificial fertilizers, pesticides, insecticides, and alternative chemicals, thereby promoting biologically healthy agriculture. Truly AM fungi live in the soil facilitate plants to absorb most macro- and micronutrients for overall performances. AM fungi help plant growth, productivity, management of weeds, pests, and diseases and assist the plant to require increased essential energy sources. In return, plants donate their waste byproducts for the AM fungi to use as food. Phosphate, nitrogen, and potassium are necessary macronutrients for the expansion of plants. These compounds exist naturally within the surroundings environment; however, plants have a restricted ability to extract them. Phosphate plays a crucial role in crop stress tolerance, maturity, quality and directly or indirectly, in biological nitrogen fixation. AM fungal vaccinated plants boost growth and output, and they are frequently used to meet the increasing consumption demands of the world's population. These environments friendly technologies shall be extremely inspired due to their widespread uses in agriculture, forestry, floriculture, green land development, polluted land, and bioremediation and wasteland development. Typically, AM fungi elicited modulations within the tolerance mechanisms and also the cross-talk triggered to manage plant performance can facilitate improve crop productivity. It absolutely was all over that AM fungal biotechnology is cost effective, higher substitute of chemical fertilizers and better quality and productivity and nutritive crops. These approaches can be appropriate for agriculturalists and organic alimental crops growers with their low inputs farming practices. This technology has the potential to improve soil health while also preserving the environment for current and future generations.

References

Abdel Latef, A. A. (2011). Influence of arbuscular mycorrhizal fungi and copper on growth, accumulation of osmolyte, mineral nutrition and antioxidant enzyme activity of pepper (*Capsicum annuum L.*). *Mycorrhiza, 21*, 495–503.

Abdel Latef, A. A., & Chaoxing, H. (2011a). Arbuscular mycorrhizal influence on growth, photosynthetic pigments, osmotic adjustment and oxidative stress in tomato plants subjected to low temperature stress. *Acta Physiologiae Plantarum, 33*, 1217–1225.

Abdel Latef, A. A., & Chaoxing, H. (2011b). Effect of arbuscular mycorrhizal fungi on growth, mineral nutrition, antioxidant enzymes activity and fruit yield of tomato grown under salinity stress. *Scientia Horticulturae, 127*, 228–233.

Abdel Latef, A. A., & Chaoxing, H. (2014). Does the inoculations with *Glomus mosseae* improve salt tolerance pepper plants? *Plant Growth Regulation, 33*, 644–653.

Abdel Latef, A. A., & Miransari, M. (2014). The role of arbuscular mycorrhizal fungi in alleviation of salt stress. In *Use of microbes for the alleviation of soil stresses* (pp. 23–38). Springer.

Adesemoye, A., Torbert, H., & Kloepper, J. (2009). Plant growth-promoting Rhizobacteria allow reduced application rates of chemical fertilizers. *Microbial Ecology, 58*(4), 921–929.

Aguilera, P., Pablo, C., Fernando, B., & Fritz, O. (2014). Diversity of arbuscular mycorrhizal fungi associated with *Triticum aestivum L.* plants growing in an and soil with high aluminum level. *Agriculture, Ecosystems and Environment, 186*, 178–184.

Ahanger, M. A., & Agarwal, R. M. (2017). Potassium up-regulates antioxidant metabolism and alleviates growth inhibition under water and osmotic stress in wheat (*Triticum aestivum* L.). *Protoplasma, 254*(4), 1471–1486.

Ahanger, M. A., Tyagi, S. R., Wani, M. R., & Ahmad, P. (2014). Drought tolerance: role of organic osmolytes, growth regulators, and mineral nutrients. In P. Ahmad & M. R. Wani (Eds.), *Physiological mechanisms and adaptation strategies in plants under changing environment* (Vol. 1, pp. 25–55). Springer.

Ali, N., Masood, S., Mukhtar, T., Kamran, M. A., Rafique, M., & Munis, M. F. H. (2015). Differential effects of cadmium and chromium on growth, photosynthetic activity, and metal uptake of *Linumusitatissimum* in association with *Glomus intraradices*. *Environmental Monitoring and Assessment, 187*(6), 311.

Amiri, R., Ali, N., Nematollah, E., & Mohammad, R. S. (2017). Nutritional status, essential oil changes and water-use efficiency of rose geranium in response to arbuscular mycorrhizal fungi and water deficiency stress. *Symbiosis, 73*, 15–25.

Andrade, S. A. L., & Silveira, A. P. D. (2008). Mycorrhiza influence on maize development under Cd stress and P supply. *Brazilian Journal of Plant Physiology, 20*(1), 39–50.

Asrar, A. A., Abdel-Fattah, G. M., & Elhindi, K. M. (2012). Improving growth, flower yield, and water relations of snapdragon *Antirhinum majus L.* plants grown under well-watered and water-stress conditions using arbuscular mycorrhizal fungi. *Photosynthetica, 50*, 305–316.

Audet, P. (2014). Arbuscular mycorrhizal fungi and metal phytoremediation: Eco physiological complementarity irelation to environmental stress. In P. Ahmad & R. Raso (Eds.), *Emerging technologies and management of crop stress tolerance* (pp. 133–160). Academic.

Auge, R. M., Toler, H. D., & Saxton, A. M. (2014). Arbuscular mycorrhizal symbiosis and osmotic adjustment in response to NaCl stress: A meta-analysis. *Frontiers of Plant Science, 5*, 562.

Barea, J. M. (2000). Rhizosphere and mycorrhiza of field crops. In E. Balazs, E. Galante, J. M. Lynch, J. S. Schepers, J. P. Toutant, D. Werner, & P. A. Werry (Eds.), *Biological resource management connecting science and policy* (pp. 110–125). INRA Editions, Springer.

Barzana, G., Aroca, R., & Ruiz-Lozano, J. M. (2015). Localized and non-localized effects of arbuscular mycorrhizal symbiosis on accumulation of osmolytes and aquaporins and on antioxidant systems in maize plants subjected to total or partial root drying. *Plant, Cell & Environment, 38*, 1613–1627.

Bati, C. B., Santilli, E., & Lombardo, L. (2015). Effect of arbuscular mycorrhizal fungi on growth and on micronutrient and macronutrient uptake and allocation in olive plantlets growing under high total Mn levels. *Mycorrhiza, 25*(2), 97–108.

Battini, F., Gronlund, M., Agnolucci, M., Giovannetti, M., & Jakobsen, I. (2017). Facilitation of phosphorus uptake in maize plants by mycorrhizosphere bacteria. *Scientific Reports, 7*, 4686.

Bauddh, K., & Singh, R. P. (2012). Growth: Tolerance efficiency and phytoremediation potential of *Ricinus communis* (L.) and *Brassica juncea* (L.) in salinity and drought affected cadmium contaminated soil. *Ecotoxicology and Environmental Safety, 85*, 13–22.

Baylis, G. T. S. (1959). Effect of vesicular-arbuscular mycorrhizas on growth of *Griselinialittoralis* (Cornaceae). *New Phytologist, 58*, 274.

Bender, S. F., Wagg, C., & Van der Heijden, M. G. A. (2016). An underground revolution: Biodiversity and soil ecological engineering for agricultural sustainability. *Trends in Ecology and Evolution, 20*, 1–13.

Berruti, A., Lumini, E., Balestrini, R., & Bianciotto, V. (2016). Arbuscular mycorrhizal fungi as natural biofertilizers: Let's benefit from past successes. *Frontiers in Microbiology, 6*, 1559.

Birhane, E., Sterck, F., Fetene, M., Bongers, F., & Kuyper, T. (2012). Arbuscular mycorrhizal fungi enhance photosynthesis, water use efficiency, and growth of frankincense seedlings under pulsed water availability conditions. *Oecologia, 169*, 895–904.

Bitterlich, M., Franken, P., & Graefe, J. (2019). Atmospheric drought and low light impede mycorrhizal effects on leaf photosynthesis – A glasshouse study on tomato under naturally fluctuating environmental conditions. *Mycorrhiza, 29*, 13–28.

Bunn, R., Lekberg, Y., & Zabinski, C. (2009). Arbuscular mycorrhizal fungi ameliorate temperature stress in thermophilic plants. *Ecology, 90*(5), 1378–1388.

Calvo-Polanco, M., Sanchez-Romera, B., Aroca, R., Asins, M. J., Declerck, S., & Dodd, I. C. (2016). Exploring the use of recombinant inbred lines in combination with beneficial

microbial inoculants (AM fungus and PGPR) to improve drought stress tolerance in tomato. *Environmental and Experimental Botany, 131*, 47–57.
Cavagnaro, T. R., Bender, S. F., Asghari, H. R., & van der Heijden, M. G. A. (2015). The role of arbuscular mycorrhizas in reducing soil nutrient loss. *Trends in Plant Science, 20*, 283–290.
Chen, S., Jin, W., Liu, A., Zhang, S., Liu, D., & Wang, F. (2013). Arbuscular mycorrhizal fungi (AMF) increase growth and secondary metabolism in cucumber subjected to low temperature stress. *Scientia Horticulturae, 160*, 222–229.
Clark, R. B., & Zeto, S. K. (2000). Mineral acquisition by arbuscular mycorrhizal plants. *Journal of Plant Nutrition, 23*, 867–902.
Curl, E. A., & Truelove, B. (1986). *The rhizosphere* (p. 288). Springer.
De Andrade, S. A. L., Domingues, A. P., & Mazzafera, P. (2015). Photosynthesis is induced in rice plants that associate with arbuscular mycorrhizal fungi and are grown under arsenate and arsenite stress. *Chemosphere, 134*, 141–149.
Deacon, J. (2019). *The microbial world: Mycorrhizas*. bio.ed.ac.uk (archived). Archived from the original on 2018-04-27. Retrieved January 11, 2019.
Djodjic, F., Borling, K., & Bergstrom, L. (2004). Phosphorus leaching in relation to soil type and soil phosphorus content. *Journal of Environmental Quality, 33*, 678–684.
Evelin, H., Kapoor, R., & Giri, B. (2009). Arbuscular mycorrhizal fungi in alleviation of salt stress: A review. *Annals of Botany, 104*, 1263–1281.
Evelin, H., Giri, B., & Kapoor, R. (2012). Contribution of *Glomus intraradices* inoculation to nutrient acquisition and mitigation of ionic imbalance in NaCl-stressed *Trigonella foenum-graecum*. *Mycorrhiza, 22*, 203–217.
Gamalero, E., Lingua, G., Berta, G., & Glick, B. R. (2009). Beneficial role of plant growth promoting bacteria and arbuscular mycorrhizal fungi on plant responses to heavy metal stress. *Canadian Journal of Microbiology, 55*, 501–514.
Garg, N., & Chandel, S. (2012). Role of arbuscular mycorrhizal (AM) fungi on growth, cadmium uptake, osmolyte, and phytochelatin synthesis in *Cajanus cajan* (L.) Mill sp. under NaCl and Cd stresses. *Journal of Plant Growth Regulation, 31*(3), 292–308.
Gautam, S. P., & Prasad, K. (2001). VA mycorrhiza-importance and biotechnological application. In D. K. Maheshwari & R. C. Dubey (Eds.), *Innovative approaches in microbiology* (pp. 83–114). Bishon Singh Mahendra Pal Singh.
Gavito, M. E., Olsson, P. A., Rouhier, H., Medinapeñafiel, A., Jakobsen, I., & Bago, A. (2005). Temperature constraints on the growth and functioning of root organ cultures with arbuscular mycorrhizal fungi. *New Phytologist, 168*, 179–188.
Ghorchiani, M., Etesami, H., & Alikhani, H. A. (2018). Improvement of growth and yield of maize under water stress by co-inoculating an arbuscular mycorrhizal fungus and a plant growth promoting rhizobacteria together with phosphate fertilizers. *Agriculture, Ecosystems and Environment, 258*, 59–70.
Gomez-Bellot, M. J., Ortuno, M. F., Nortes, P. A., Vicente-Sanchez, J., Banon, S., & Sanchez Blanco, M. J. (2015). Mycorrhizal euonymus plants and reclaimed water: Biomass, water status and nutritional responses. *Scientia Horticulturae, 186*, 61–69.
Govindarajulu, M., Pfeffer, P. E., Jin, H. R., Abubaker, J., Douds, D. D., & Allen, J. W. (2005). Nitrogen transfer in the arbuscular mycorrhizal symbiosis. *Nature, 435*, 819–823.
Guether, M., Neuhauser, B., & Balestrini, R. (2009). A mycorrhizal-specific ammonium transporter from *Lotus japonicus* acquires nitrogen released by arbuscular mycorrhizal fungi. *Plant Physiology, 150*, 73–83.
Hameed, A., Dilfuza, E., Abd-Allah, E. F., Hashem, A., Kumar, A., & Ahmad, P. (2014). Salinity stress and arbuscular mycorrhizal symbiosis in plants. In M. Miransari (Ed.), *Use of microbes for the alleviation of soil stresses* (Vol. 1, pp. 139–159). Springer.
Hammer, E. C., Nasr, H., Pallon, J., Olsson, P. A., & Wallander, H. (2011). Elemental composition of arbuscular mycorrhizal fungi at high salinity. *Mycorrhiza, 21*(2), 117–129.
Harwani, D., Chaudhari, P., Dhaker, S., Prasad, K., & Mahna, S. K. (2009). Tripartite symbiotic association: Legume-rhizobia-mycorrhiza – A review. In D. K. Maheshwari & R. C. Dubey (Eds.),

Biotechnology for agricultural microorganisms – An agro-industry approach (pp. 406–435). I.K. International Pvt. Ltd.

Hasanuzzaman, M., Gill, S. S., & Fujita, M. (2013). Physiological role of nitric oxide in plants grown under adverse environmental conditions. In N. Tuteja & S. S. Gill (Eds.), *Plant acclimation to environmental stress*.

Hashem, A., Alqarawi, A. A., Radhakrishnan, R., Al-Arjani, A. F., Aldehaish, H. A., & Egamberdieva, D. (2018). Arbuscular mycorrhizal fungi regulate the oxidative system, hormones and ionic equilibrium to trigger salt stress tolerance in *Cucumis sativus* L. *Saudi Journal of Biological Sciences, 25*(6), 1102–1114.

Higa, T. (1994). Effective microorganisms: A new dimension for nature farming. In J. F. Parr, S. B. Hornick, & M. E. Simpson (Eds.), *Proceedings of the second international conference on Kyusei nature farming* (pp. 20–22). U.S. Department of Agriculture.

Hodge, A., & Fitter, H. (2010). Substantial nitrogen acquisition by arbuscular mycorrhizal fungi from organic material has implications for N cycling. *Proceedings of the National Academy of Sciences, 107*, 13754–13759.

Hodge, A., & Storer, K. (2015). Arbuscular mycorrhiza and nitrogen: Implications for individual plants through to ecosystems. *Plant and Soil, 386*, 1–19.

Janouskova, M., & Pavlikova, D. (2010). Cadmium immobilization in the rhizosphere of arbuscular mycorrhizal plants by the fungal extraradical mycelium. *Plant and Soil, 332*, 511–520.

Ji, L., Tan, W., & Chen, X. (2019). Arbuscular mycorrhizal mycelial networks and glomalin related soil protein increase soil aggregation in Calcaric Regosol under well-watered and drought stress conditions. *Soil and Tillage Research, 185*, 1–8.

Jiang, Y. N., Wang, W. X., Xie, Q. J., Liu, N., Liu, L. X., & Wang, D. P. (2017). Plants transfer lipids to sustain colonization by mutualistic mycorrhizal and parasitic fungi. *Science, 356*, 1172–1175.

Jin, H., Pfeffer, P. E., Douds, D. D., Piotrowski, E., Lammers, P. J., & Shachar- Hill, Y. (2005). The uptake, metabolism, transport and transfer of nitrogen in an arbuscular mycorrhizal symbiosis. *New Phytology, 168*, 687–696.

Johnson, N. C., Graham, J. H., & Smith, F. A. (1997). Functioning of mycorrhizal associations along the mutualism–parasitism continuum. *New Phytologist, 135*(4), 575–585.

Kabir, Z. (2005). Tillage or no-tillage: Impact on mycorrhizae. *Canadian Journal of Plant Science, 85*, 23–29.

Kanwal, S., Bano, A., & Malik, R. A. (2015). Effects of arbuscular mycorrhizal fungi on metals uptake, physiological and biochemical response of *Medicago sativa L.* with increasing Zn and Cd concentrations in soil. *American Journal of Plant Science, 6*, 2906–2923.

Kapoor, R., Evelin, H., Mathur, P., & Giri, B. (2013). Arbuscular mycorrhiza: Approaches for abiotic stress tolerance in crop plants for sustainable agriculture. In N. Tuteja & S. S. Gill (Eds.), *Plant acclimation to environmental stress* (pp. 359–401). LLC: Springer.

Kautz, T., Amelung, W., Ewert, F., Gaiser, T., Horn, R., & Jahn, R. (2013). Nutrient acquisition from arable sub soils temperate climates: A review. *Soil Biology and Biochemistry, 57*, 1003–1022.

Kawaguchi, M., & Minamisawa, K. (2010). Plant-microbe communications for symbiosis. *Plant Cell Physiology, 51*, 1377–1380.

Kohl, L., & van der Heijden, M. G. A. (2016). Arbuscular mycorrhizal fungal species differ in their effect on nutrient leaching. *Soil Biology and Biochemistry, 94*, 191–199.

Lehmann, A., & Rillig, M. C. (2015). Arbuscular mycorrhizal contribution to copper, manganese and iron nutrient concentrations in crops meta-analysis. *Soil Biology and Biochemistry, 81*, 147–158.

Lehmann, A., Veresoglou, S. D., Leifheit, E. F., & Rillig, M. C. (2014). Arbuscular mycorrhizal influence on zinc nutrition in crop plants: A meta-analysis. *Soil Biology and Biochemistry, 69*, 123–131.

Leifheit, E. F., Veresoglou, S. D., Lehmann, A., Morris, E. K., & Rillig, M. C. (2014). Multiple factors influence the role of arbuscular mycorrhizal fungi in soil aggregation-a meta-analysis. *Plant and Soil, 374*, 523–537.

Li, H., Smith, S. E., Holloway, R. E., Zhu, Y., & Smith, F. A. (2006). Arbuscular mycorrhizal fungi contribute to phosphorus uptake by wheat grown in a phosphorus-fixing soil even in the absence of positive growth responses. *New Phytologist, 172*(3), 536–543.

Li, X., Zeng, R., & Liao, H. (2016). Improving crop nutrient efficiency through root architecture modifications. *Journal of Integrative Plant Biology, 58*, 193–202.

Linquist, B., Van Groenigen, K. J., Adviento-Borbe, M. A., Pittelko, C., & Van Kessel, C. (2012). An agronomic assessment of greenhouse gas emissions from major cereal crops. *Global Change Biology, 18*, 194–209.

Liu, X., Song, Q., Tang, Y., Li, W., Xu, J., & Wu, J. (2013). Human health risk assessment of heavy metals in soil–vegetable system: A multi-medium analysis. *Science of the Total Environment, 463–464*, 530–540.

Luginbuehl, L. H., Menard, G. N., Kurup, S., Van Erp, H., Radhakrishnan, G. V., & Breakspear, A. (2017). Fatty acids in arbuscular mycorrhizal fungi are synthesized by the host plant. *Science, 356*, 1175–1178.

Mathur, S., Tomar, R. S., & Jajoo, A. (2019). Arbuscular mycorrhizal fungi (AMF) protect photosynthetic apparatus of wheat under drought stress. *Photosynthesis Research, 139*, 227–238.

Maya, M. A., & Matsubara, Y. (2013). Influence of arbuscular mycorrhiza on the growth and ant oxidative activity in Cyclamen under heat stress. *Mycorrhiza, 23*(5), 381–390.

Mayer, Z., Juhasz, A., Vo Trung, A., & Posta, K. (2019). Impact of arbuscular mycorrhizal fungi on some defense enzyme activities at early stage of maize (*Zea mays L.*) under different abiotic stresses. *Applied Ecology and Environmental Research, 17*(3), 6241–6253.

Meghavanshi, M. K., Prasad, K., Harwani, D., & Mahna, S. K. (2008). Response of soybean cultivars towards inoculation with three Arbuscular Mycorrhizal Fungi (AMF) and *Bradyrhizobium japonicum* in alluvial soil. *European Journal of Soil Biology, 44*, 316–323.

Meghavanshi, M. K., Prasad, K., & Mahna, S. K. (2010). Symbiotic potential, competitiveness and compatibility of indigenous *Bradyrhizobium japonicum* isolates to three soybean genotypes of two distinct agro-climatic regions of Rajasthan, India. *Saudi Journal of Biological Sciences, 17*, 303–310.

Miransari, M. (2017). Arbuscular mycorrhizal fungi and heavy metal tolerance in plants. In Q. S. Wu (Ed.), *Arbuscular mycorrhizas and stress tolerance of plants* (pp. 174–161). Springer.

Mitra, D., Navendra, U., Panneerselvam, U., Ansuman, S., Ganeshamurthy, A. N., & Divya, J. (2019). Role of mycorrhiza and its associated bacteria on plant growth promotion and nutrient management in sustainable agriculture. *International Journal of Life Sciences and Applied Sciences, 1*, 1–10.

Moghadam, H. R. T. (2016). Application of super absorbent polymer and ascorbic acid to mitigate deleterious effects of cadmium in wheat. *Pesquisa Agropecuaria Tropical, 6*(1), 9–18.

Mosse, B. (1962). The establishment of vesicular-arbuscular mycorrhiza under aseptic conditions. *Journal of General Microbiology, 27*, 509–520.

National Academy of Sciences. (1989). *Alternative agriculture. Committee on the role of alternative agriculture farming methods in modern production agriculture. National Research Council, Board on Agriculture* (p. 448). National Academy Press.

Nouri, E., Breuillinsessoms, F., Feller, U., & Reinhardt, D. (2015). Phosphorus and nitrogen regulate arbuscular mycorrhizal symbiosis in Petunia hybrid. *PLoS One, 9*, 90–841.

Ouziad, F., Hildebrandt, U., Schmelzer, E., & Bothe, H. (2005). Differential gene expressions in arbuscular mycorrhizal-colonized tomato grown under heavy metal stress. *Journal of Plant Physiology, 162*, 634–649.

Parr, J. F., & Hornick, S. B. (1992). Agricultural use of organic amendments: A historical perspective. *American Journal of Alternative Agriculture, 7*, 181–189.

Parr, J. F., & Hornick, S. B. (1994). *Assessment of the third international conference on Kyusei nature farming: Round table discussion by USDA Scientists, October 7, 1993*. Published by the Nature Farming Research and Development Foundation, Lompoc.
Pellegrino, E., & Bedini, S. (2014). Enhancing ecosystem services in sustainable agriculture: Bio fertilization and biofortification of chickpea (*Cicer arietinum* L.) by arbuscular mycorrhizal fungi. *Soil Biology and Biochemistry, 68*, 429–439.
Prasad, K. (1993). PhD thesis, B.R. Ambedkar Bihar University, Muzaffarpur, Bihar, India.
Prasad, K. (2000). Growth responses of Acacia nilotica (L.) Del. inoculated with *Rhizobium* and *Glomus fasciculatum* VAM fungi. *Journal of Tropical Forestry, 16*(1), 22–27.
Prasad, K. (2002a). Application of arbuscular mycorrhizae fungal biofertilizer on biomass yield and nutrient uptake of *Azadirachta indica*. In V. S. Hari Kumar (Ed.), *Recent trends in biotechnology* (pp. 73–78). Agrobios (India) Publication.
Prasad, K. (2002b). Effect of arbuscular mycorrhizae on biomass yield, uptake and translocation of nitrogen, phosphorus and potassium in *Azadirachta indica* L. In C. Manoharachary, D. K. Purohit, S. Ram Reedy, M. A. Singaracharya, & S. Girisham (Eds.), *Frontier of microbial biotechnology and plant pathology* (pp. 187–191). Scientific Publishers.
Prasad, K. (2006a). Impact of arbuscular mycorrhizal fungus (*Glomus fasciculatum*) and phosphate solubilizing bacterium (*Pseudomonas striata*) on growth and nutrient status of *Azadirachta indica* L. *Mycorrhiza News, 18*(2), 10–12.
Prasad, K. (2006b). Influence of indigenous VAM fungus (*Glomus fasciculatum*) and *Rhizobium* on growth, nutrient uptake and nodulation in *Acacia nilotica*. In A. Prakash & V. S. Mehrotra (Eds.), *Mycorrhiza* (pp. 139–144). Scientific Publishers (India).
Prasad, K. (2007). Improvement in in Bioavailability of Nutrition and Biomass production to *Azadirachta indic*a (Neem) by Arbuscular Mycorrhizal Symbiosis. *Journal of Basic and Applied Mycology, 6*(I& II), 96–101.
Prasad, K. (2010). Ectomycorrhizal symbiosis: Possibilities and prospects. In M. K. Rai & G. Kovices (Eds.), *Progress in Mycology* (pp. 290–308). Springer/Scientific Publisher.
Prasad, K. (2011a). Effect of *Glomus intraradices* AM fungi on the shoot dry matter, seed dry yield, and nitrogen and phosphorus uptake of soybean (*Glycine max* (L.) Merrill) inoculated with cultivar specific *Bradyrhizobium japonicum*. *Mycorrhiza News, 23*(3), 13–18.
Prasad, K. (2011b). Interaction between arbuscular Mycorrhizal fungus (*Glomus fasciculatum*) and rhizospheric fungi in *Saccharum officinarum* L. *Amravati University Research Journal, 5*, 53–65.
Prasad, K. (2013). Arbuscular mycorrhizal fungus plays a major role in agriculture and natural ecosystems to improve production in sustainable manner. In Jamaluddin & A. K. Singh (Eds.), *Microbes and sustainable plant productivity* (pp. 113–138). Scientific Publication.
Prasad, K. (2015). Biofertilizers: A new dimension for agriculture and environmental development to improve production in sustainable manner. *Journal of Basic and Applied Mycology, 11*(1& II), 5–13.
Prasad, K. (2017). Biology, diversity and promising role of mycorrhizal entophytes for green technology. In D. K. Maheshwari (Ed.), *Endophytes: Biology and biotechnology* (Series sustainable development and biodiversity 15) (Vol. 1, pp. 267–301). Springer.
Prasad, K. (2020). Positive importance of arbuscular mycorrhizal fungi for global sustainable agriculture and environment management for green technology. *Current Investigations in Agriculture and Current Research, 9*(2), 1182–1184.
Prasad, K., & Deploey, J. J. (1999). Incidence of arbuscular mycorrhizae and their effect on certain species of trees. *Journal of the Pennsylvania Academy of Science, 73*(3), 117–122.
Prasad, K., & Kaushik, S. (2004). Ecology, physiology, biochemistry and taxonomy of arbuscular mycorrhizal fungi. In S. P. Gautam, Y. K. Bansal, & A. K. Pandey (Eds.), *Biological diversity: Current trends* (Prof R.C. Rajak Festschrift volume) (pp. 134–141). Shree Publication & Distributors.

Prasad, K., & Meghavanshi, M. K. (2005). Interaction between indigenous *Glomus fasciculatum* (AM fungus) and Rhizobium and their stimulatory effect on growth, nutrient uptake and nodulation in *Acacia nilotica* (L.) *Del. Flora and Fauna, 11*(1), 51–56.

Prasad, K., & Pandey, A. K. (2012). Mycorrhizal symbiosis: A new dimension for agriculture and environmental development to improve production in sustainable manner. In D. J. Bagyaraj, T. KVBR, & H. K. Kehri (Eds.), *Microbial diversity and function* (pp. 389–402). New India Publishing Agency.

Prasad, K., & Rajak, R. C. (2000). Biotechnological application of mycorrhizae in reclamation of mined dumps. In M. K. Rai, A. Varma, & R. C. Rajak (Eds.), *Integrated management of plant resources* (pp. 283–292). Scientific Publishers (India).

Prasad, K., & Rajak, R. C. (2001). Microbes and wasteland management: Challenge ahead. In A. K. Roy & S. K. Varma (Eds.), *Wasteland management and environment* (pp. 27–39). Scientific Pub (India).

Prasad, K., & Rajak, R. C. (2002). Management of plant pests through biological entities: Possibilities and prospects. In P. C. Trivedi (Ed.), *Plant pest management* (pp. 75–99). Aavishkar Publishers.

Prasad, K., & Warke, R. V. (2018a). Arbuscular mycorrhizal fungi pay a promising role towards green technology and soil health management. In *Plant and soil health management: New challenges and opportunity* (pp. 191–192). ICAR – Indian Institute of Pules Research.

Prasad, K., & Warke, R. V. (2018b). Biology, diversity and promising role of microbial inoculations specially mycorrhizae in agriculture and natural ecosystems to improve productivity for sustainable development. In V. K. Vishnoi (Ed.), *Microbial inoculants for agriculture and environmental sustainability, National conference* (pp. 7–8). GKV.

Prasad, K., Meghavanshi, M. K., & Chaudhari, K. K. (2005). Arbuscular mycorrhizal symbiosis: An overview of research and extension needs. In M. Tiwari & S. C. Sati (Eds.), *The mycorrhizae: Diversity, ecology, and application* (pp. 87–101). Daya Publishing House.

Prasad, K., Meghavanshi, M. K., Harwani, D., Mahna, S. K., & Werner, D. (2005a). Synergistic effect of arbuscular mycorrhizal fungi and *Bradyrhizobium japonicum* on growth, yield and nutrient status of Soybean (*Glycine max* (L.) Merrill). *Anusandhan, 1*(1), 23–29.

Prasad, K., Meghavanshi, M. K., Harwani, D., Mahna, S. K., & Werner, D. (2005b). Synergistic effect of arbuscular mycorrhizal fungi and *Bradyrhizobium japonicum* on growth, yield and nutrient status of soybean [*Glycine max* (L.) Merrill]. *Anusandhan, 1*(1), 13–19.

Prasad, K., Warke, R. V., & Khadke, K. (2019). Management of soilborne pathogens to improve productivity of pulses using organic technologies for sustainable agriculture. *International Journal of Research and Analytical Reviews, 6*(2), 82–101.

Punamiya, P., Datta, R., Sarkar, D., Barber, S., Patel, M., & Da, P. (2010). Symbiotic role of Glomus mosseae in phytoextraction of lead in vetiver grass *Chrysopogonzizanioides L. Journal of Hazardous Materials, 177*, 465–474.

Rahimzadeh, S., & Pirzad, A. (2019). Pseudomonas and mycorrhizal fungi co-inoculation alter seed quality of flax under various water supply conditions. *Industrial Crops and Product, 129*, 518–524.

Rewari, R. B., & Tilak, K. V. B. R. (1988). Microbiology of pulses. In B. Baldev, S. Ramanujam, & H. K. Jain (Eds.), *Pulse crops* (pp. 373–411). Oxford & IBH.

Sale, V., Aguilera, P., Laczko, E., Mader, P., Berner, A., & Zihlmann, U. (2015). Impact of conservation tillage and organic farming on the diversity of arbuscular mycorrhizal fungi. *Soil Biology & Biochemistry, 84*, 38–52.

Schneider, F., Don, A., Hennings, I., Scmittman, O., & Seidel, S. J. (2017). The effect of deep tillage on crop yields – What do we really know? *Agriculture, Ecosystems & Environment, 174*, 193–204.

Schoumans, O. F. (2015). *Phosphorus leaching from soils: Process description, risk assessment and mitigation*. Available at: https://www.wur.nl/upload_mm/2/c/f/82da1d37-6fd7-4e67-a1d9 57848033f213_8412100377_ DissertationSchoumans 2015_ESG-versie.pdf

Shen, H., Christie, P., & Li, X. (2006). Uptake of zinc, cadmium and phosphorus by arbuscular mycorrhizal maize (*Zea mays, L.*) from a low available phosphorus calcareous soil spiked with zinc and cadmium. *Environmental Geochemistry and Health, 28*, 111.

Smith, S. E., & Read, D. J. (1997). *Mycorrhizal symbiosis* (p. 607). Academic.

Smith, S. E., & Smith, F. A. (2011). Roles of arbuscular mycorrhizas in plant nutrition and growth: New paradigms from cellular to ecosystem scales. *Annual Review of Plant Biology, 62*, 227–250.

Smith, S. E., Jakobsen, I., Grnlund, M., & Smith, F. A. (2011). Roles of arbuscular mycorrhizas in plant phosphorus nutrition: Interactions between pathways of phosphorus uptake in arbuscular mycorrhizal roots have important implications for understanding and manipulating plant phosphorus acquisition. *Plant Physiology, 156*, 1050–1057.

Sosa-Hernandez, M. A., Roy, J., Hempel, S., & Rillig, M. (2018). Evidence for subsoil specialization in arbuscular mycorrhizal fungi. *Frontiers in Ecology and Evolution, 6*, 67.

Souza, L. A., Andrade, S. A., Souza, S. C. R., & Schiavinato, M. A. (2012). Evaluation of mycorrhizal influence on the development and phytoremediation potential of *Canavalia gladiata* in Pb contaminated soils. *International Journal of Phytoremediation, 15*, 465–476.

Takacs, T., & Voros, I. (2003). Effect of metal non-adapted arbuscular mycorrhizal fungi on Cd, Ni and Zn uptake by ryegrass. *Acta Agronomica Hungarica, 51*, 347–354.

Tanaka, Y., & Yano, Y. (2005). Nitrogen delivery to maize via mycorrhizal hyphae depends on the form of N supplied. *Plant, Cell and Environment, 28*, 1247–1254.

Thirkell, T. J., Charters, M. D., Elliot, A. J., Sait, S. M., & Field, K. J. (2017). Are mycorrhizal fungi our sustainable saviour's considerations for achieving food security? *Journal of Ecology, 105*, 921–929.

Trappe, J. M. (1987). Phylogenetic and ecologic aspects of mycotrophy in the angiosperms from an evolutionary standpoint. In G. R. Safir (Ed.), *Ecophysiology of VA mycorrhizal plants*. CRC Press.

Turrini, A., Bedini, A., Loor, M. B., Santini, G., Sbrana, C., & Giovannetti, M. (2018). Local diversity of native arbuscular mycorrhizal symbionts differentially affects growth and nutrition of three crop plant species. *Biology and Fertility of Soils, 54*, 203–217.

van der Heijden, M. G. A. (2010). Mycorrhizal fungi reduce nutrient loss from model grassland ecosystems. *Ecology, 91*, 1163–1171.

Vessey, J. K. (2003). Plant growth promoting rhizobacteria as biofertilizers. *Plant and Soil, 255*(2), 571–586.

Wahid, A., Gelani, S., Ashraf, M., & Foolad, M. R. (2007). Heat tolerance in plants: An overview. *Environmental and Experimental Botany, 61*, 199–223.

Wang, Y., Jing, H., & Gao, Y. (2012). Arbuscular mycorrhizal colonization alters subcellular distribution and chemical forms of cadmium in *Medicago sativa L.* and resists cadmium toxicity. *PLoS One, 7*, 3161–3164.

Wang, Y., Wang, M., Li, Y., Wu, A., & Huang, J. (2018). Effects of arbuscular mycorrhizal fungi on growth and nitrogen uptake of *Chrysanthemum morifolium* under salt stress. *PLoS One, 13*(4), e0196408.

Warke, R. V., Prasad, K., & Khadke, K. (2018). Impact of arbuscular mycorrhizal fungi (ROC-WA329) and phosphate solubilizing bacteria (WA302) on *Zea Mays* cultivated at Sahara district of Maharashtra, India. In *Plant and soil health management: New challenges and opportunity* (p. 179). ICAR – Indian Institute of Pules Research.

Werner, D. (1998). II Kornerleguminosen: BiologischeStickst off-Fixierung. In Knollenun Wurzelfruchte, Korner-und Futterleguminosen ER, Keller H, Hanus und KU. Heyland (Hrgs.), *Handbuch De Pflanzenbaues* (*Band 3*, pp. 554–562). Verlag Eugen Ulmer.

Yang, G., Yang, X., Zhang, W., Wei, Y., Ge, G., & Lu, W. (2016). Arbuscular mycorrhizal fungi affect plant community structure under various nutrient conditions and stabilize the community productivity. *Oikos, 125*, 576–585.

Yin, N., Zhang, Z., Wang, L., & Qian, K. (2016). Variations in organic carbon, aggregation, and enzyme activities of gangue-fly ash-reconstructed soils with sludge and arbuscular mycorrhizal fungi during 6 years reclamation. *Environmental Science and Pollution Research, 23*(17), 17840–17849.

Yousaf, B., Liu, G., Wang, R., Imtiaz, M., Ziaur-Rehman, M., & Munir, M. A. M. (2016). Bioavailability evaluation, uptake of heavy metals and potential health risks via dietary exposure in urban-industrial areas. *Environmental Science and Pollution Research, 23*, 22443–22453.

Zaefarian, F., Rezvani, M., Ardakani, M. R., Rejali, F., & Miransari, M. (2013). Impact of mycorrhizae formation on the phosphorus and heavy-metal uptake of Alfalfa. *Communication in Soil Science and Plant Analysis, 44*, 1340–1352.

Zhang, F., Jia-Dong, H. E., Qiu-Dan, N. I., Qiang-Sheng, W. U., & Zou, Y. N. (2018). Enhancement of drought tolerance in trifoliate orange by mycorrhiza: Changes in root sucrose and proline metabolisms. *Notulae Botanicae Horti Agrobotanici, Cluj-Napoca, 46*, 270.

Zhu, X. C., Song, F. B., & Xu, H. W. (2010a). Arbuscular mycorrhizae improve low temperature stress in maize via alterations in host water status and photosynthesis. *Plant and Soil, 331*, 129–137.

Zhu, X. C., Song, F. B., & Xu, H. W. (2010b). Effects of arbuscular mycorrhizal fungi on photosynthetic characteristics of maize under low temperature stress. *Acta Ecologica Sinica, 21*, 470–475.

Zhu, X. C., Song, F. B., Liu, S. Q., Liu, T. D., & Zhou, X. (2012). Arbuscular mycorrhizae improves photosynthesis and water status of *Zea mays L.* under drought stress. *Plant Soil and Environment, 58*, 186–191.

Chapter 6
Mycometabolites in Industrial Applications with Emphasis on Bioherbicide Production

Puja Ray and Mayukh Ghosh

Contents

6.1 Introduction

Weeds are considered a perpetual menace leading to reduction of crop yield by limiting growth-associated resources (water, nutrient, sunlight, space, etc) and being alternate and alternative hosts of crop pathogens and insect pests (Kumar et al., 2021). Many weeds and nonnative invasive plant species, due to their large seed bank, fast growth rate, lack of natural control agents, allelopathic potential, etc., negatively impact native biodiversity and hence ecosystem goods and services

P. Ray (✉) · M. Ghosh
Multitrophic Interactions and Biocontrol Research Laboratory, Department of Life Sciences, Presidency University, Kolkata, India

A. C. Shukla (ed.), *Applied Mycology*, Fungal Biology,
https://doi.org/10.1007/978-3-030-90649-8_6

(Duncan et al., 2004). Present agricultural systems target to achieve maximum yield while minimizing the costs. This leads to several weed management options with more emphasis on using chemical herbicides, which are effective, fast, and relatively cheap means of weed control. In a recent study (Qi et al., 2020), it was observed that nontarget impacts of herbicides' resulted in changing the species composition, reduced the number of plant species, ecological adaptability of plants and thus reducing community diversity even at sublethal doses not only in the agricultural fields where herbicides are being applied but also in non-crop situations including field edges, fallow land, and other seminatural habitat. This herbicide further drifts into water bodies and protected areas causing a range of negative impacts on the ecosystems worldwide. Thus, owing to nontarget herbicide impact more emphasis is being made on alternate control strategies, including bioherbicides based on microorganisms (Ray et al., 2008).

Kingdom fungi are a diverse group of microorganisms with possibly over five million species (Blackwell, 2011), of which only a small fraction (5% approx.) have been identified. They are associated with some of the crucial ecological processes, like decomposition, recycling, and transportation of nutrients in different environments. Fungi produce a wide array of low molecular weight, non-nutritive metabolites that are exuded into the environment and play a key role in biotic interactions across species. The astounding structural diversity of these natural compounds has attracted the researchers globally to search for lead novel structures and find their possible utilization. The discovery of penicillin in 1928 by Alexander Fleming from the culture plate of bacterium *Staphylococcus* contaminated with the filamentous fungi *Penicillium rubens* has led to a revolutionized substantial research on fungi and their metabolites and simultaneous isolation of thousands of new fungal metabolites with a diverse range of biological and pharmacological effects. Many of these novel and natural bioactive compounds have potential bioprospecting applications in industries including pharmacy, cosmetics, nutraceutics, agriculture, etc., Some of the well-known mycotoxins include ergot alkaloids, aflatoxins, and fumonisins. There is an increasing interest in the isolation and identification of a range of unusual fungal metabolites with interesting and innovative mechanisms of action. Such compounds have been found to possess a range of properties including antidiabetic (H. Hussain et al., 2021a), anticancer(Singh & Datta, 2020), antioxidant (Ujam et al., 2021; Vitale et al., 2020), antimicrobial (Al-Fakih et al., 2019; Dalinova et al., 2020; Jakubczyk & Dussart, 2020), immunosuppressive (Duan et al., 2020; Ujam et al., 2021), and several other activities (Hyde et al., 2019). Recent studies by Nguyen et al. (2020) indicate *Monascuspurpureus* and its secondary metabolites are capable of reducing cholesterol content in vitro and hence may have a potential application in probiotics. Thus, kingdom fungi have demonstrated to be an exceptional but comparatively less explored source of biologically active compounds with diverse potential. Rising global concerns about environmental consequences of chemical compounds including pesticides have further boosted an era of research in using naturally occurring compounds for our everyday use through industrial production. In this chapter, we discuss the avenues of using fungal metabolites as bioherbicides against invasive weed species, their associated challenges, and way forward.

6.2 Ecological Role of Fungi and Their Metabolites

Due to wide differences in their chemical structures, the fungal metabolites have different ecological and environmental roles and mechanisms of action. Here we discuss some of the major roles played by fungi and their secondary metabolites in the ecosystems of their occurrence.

6.2.1 *Decomposition and Cycling of Organic and Inorganic Materials*

Fungi are well known for being decomposers in many ecosystems. Fungal secondary metabolites have been favoured by natural selection for their use in critical environmental conditions like intraspecific and interspecific interactions, growth and development, and abiotic stress (Rohlfs & Churchill, 2011). Coprophilous fungi produce secondary metabolites to not only reduce interspecific competition but also to metabolize very complex molecules (Sarrocco, 2016). They also make chemical changes to minerals and make them usable for plants and other organisms; the process is also known as bioweathering. Fungi also help in the breakdown of xenobiotics and detoxification of the environment (Gadd, 2007; Tang et al., 2013).

6.2.2 *Interactions with Other Organisms*

Fungi also have direct interactions with many different organisms like insects, molluscs, and various other organisms including many plant species, which can be mutualistic or antagonistic (Boddy et al., 2007; Masters & Brown, 1997). Many fungivore organisms obtain high amounts of nutrients from decomposing organic materials or directly from fungal mycelia (Maraun et al., 2003).

6.2.3 *Mutualistic Symbiosis with Plants*

Many fungi are associated with plant hosts in mutualistic symbiosis, where they increase fitness for each other's survival. Arbuscular mycorrhizal fungi help symbiont plants to accumulate more vitamins and minerals with antioxidant properties as well as help the plants in obtaining bioavailable nitrogen from mineral associated proteins, where the source of nitrogen is scarce (Baslam et al., 2013; Wang et al., 2020). *Curvularia* sp. was found to increase high temperature tolerance in *Dichanthelium lanuginosum* plant species (Redman et al., 2002). Peramine is a secondary metabolite produced by *Neotyphodium* sp. that deters herbivore insects from feeding on its host plants such as ryegrass (Tanaka et al., 2005).

6.2.4 Competitive Interactions with Different Organisms

There are various strategies to prevent insect competitors for the same resources as fungi by means of chemical warfare. Fungal secondary metabolites were found to reduce immune response in insects and rendered them susceptible to infectious diseases (Fiolka, 2008; Pal et al., 2007). Saprophagous insect larvae are strong competitors to fungi like *Aspergillus*, depending on the same food source. Mycotoxins like Kojic acid and Ochratoxin A were found to be effective in causing insect-specific responses in *Drosophila* spp. and larval death, which helped in the survival of the fungi (Rohlfs & Obmann, 2009; Trienens et al., 2010). *Ophiocordyceps unilateralis* fungus is known to manipulate behaviour of infected *Camponotusleonardi* ants to bite onto plants before the death of the ant at optimal zones for fungal growth (Andersen et al., 2009; Bekker, 2019).

6.3 Secondary Metabolites Biosynthesis

Secondary metabolites are generally synthesized from genes of a single cluster that usually contain one or more central genes for nonribosomal peptide synthetase (NRPS) or polyketide synthase (PKS) and other genes for modifications of the NRPs or PKs (Hertweck, 2009; Strieker et al., 2010; Yu & Keller, 2005).

The NRPSs are multinodular enzymes that catalyze the formation of nonribosomal peptides (NRPs) with very specific function of each domain in a series where each domain activates and adds a single monomer covalently to the substrate to form the NRP. So knowing the assembly and functions of the domains of the NRPs can help in determining the type of NRP to be formed (Fischbach & Walsh, 2006). The NRPs can be functionality divided into three domains: activation or adenylation (A) domain: activates the amino acids, thiolation, or peptidyl carrier protein (PCP) domain: binds the cofactor 4′-phosphopantetheine (4′PP), to which activated amino acids are covalently attached, and propagates the peptide chain being synthesized, condensation (C) domain: peptide bonds are formed, and at the terminal end there is a thioesterase (TE) domain: catalyzes peptide release by either hydrolysis or macrocyclization (Strieker et al., 2010).

PKSs also contain three main domains that are acyltransferase domain: selection and acylation of extender substrate molecules such as malonyl-CoA or methylmalonyl-CoA, acyl-carrier domain: loading extender molecules, and keto-acyl synthase domain: decarboxylative condensation of extender molecules (Brakhage, 2013).

6.3.1 Transcriptional Regulation of Secondary Metabolite Gene Clusters

The secondary metabolite gene clusters are regulated by complex interactions of transcription factors (Pan & Liu, 2018). The transcription factors for such gene clusters are mostly very specific to environmental conditions like carbon and nitrogen sources, pH, temperature, biofilm, light, hypoxic conditions, interspecific signalling, etc. (Bruns et al., 2010; Vödisch et al., 2011; Yin & Keller, 2011). There are two types of transcriptional signalling mechanisms, that is, global regulation: the transcription factor gene is not a part of the gene cluster and transcription factor is also involved in the regulation of other genes that are not part of secondary metabolite gene clusters; on the other hand, there is pathway-specific regulation: the transcription factor is very specific to the promoter sequence of genes from the cluster that are involved in the pathway that is required for the synthesis of the secondary metabolite, and the transcription factor gene may reside within the specific cluster itself or it can be a part of another gene cluster even on a different chromosome (Bergmann et al., 2010; Brakhage, 2013).

6.3.2 Signalling and Regulations for pH, Light Intensity, Iron Starvation, and Oxidative Stress During Production of Secondary Metabolites Are as Follows

I. *Ambient pH*

PacC transcription factor (contains zinc finger motifs) in *Aspergillus nidulans* is inactive in ambient acidic pH. In ambient alkaline pH, pal genes (pal A, B, C, F, H, and I) are expressed and the pal genes' products activate the PacC transcription factor by cleaving C-terminal domain and making it accessible to protease and then signalling protease cleaves PacC between residues 493 and 500. Active PacC works as transcriptional activator of alkaline expressed genes such as palD (alkaline phosphatase) and prtA (alkaline protease), and the penicillin biosynthesis genes N-(5-amino-5-carboxypentanoyl)-lcysteinyl-d-valine synthase (acvA)40 and isopenicillin N synthase (ipnA) by the interaction of zinc finger motifs to ipnA promoter element 5′GCCAAG3′ (Denison, 2000; Díez et al., 2002; Espeso & Penalva, 1996; Then Bergh & Brakhage, 1998).

II. *Light Intensity*

In *A. nidulans,* phytochrome FphA and other light sensor proteins LreA, LreB interact with LaeA protein in dark condition and nuclear localization of LaeA occurs. LaeA is a methyl transferase that removes methyl group from ninth lysine residue of histone H3 and heterochromatin protein 1 (HepA), and promotes euchromatinization, which positively regulates various secondary metabolites gene clusters like penicillin and sterigmatocystin by epigenetic modifications (Bayram et al., 2008; Sarikaya et al., 2010).

III. *Iron Starvation*

HapX is a basic region leucine zipper protein that is active in iron-deficient condition in *A. fumigatus* where it binds to CCAAT-binding complex (CBC) made of HapB, HapC, and HapE, and induces siderophore biosynthesis genes that contain a central NRPS gene (Brakhage, 2013; Furukawa et al., 2020).

IV. *Oxidative Stress*

In presence of H2O2, Yap1 is activated by forming disulfide bond with Gpx3 with the help of Ybp1 and the nuclear export signal at the C-terminal is inaccessible. The activated Yap1 binds to Yap1 response element (YRE) TTA(G/C)TAA of antioxidant genes like GSH1 (γ-glutamylcysteine synthetase), GPX2 (glutathione peroxidase), TRX2 (thioredoxin), TSA1 (thioredoxin peroxidase) and induces their transcription (Simaan et al., 2019).

6.4 Fungal Formulation–Based Early Bioherbicides

Bioherbicides are weed-control living agents, mostly microorganisms and plants or products derived from them. Mycoherbicides are weed control formulations derived from live fungi or their products. Early research and field success of bioherbicides, particularly mycoherbicides, suggested that a large number of spores or mycelial mass of native, target-specific fungi could be used to transform an ordinarily endemic pathogen into a severe epidemic (Templeton et al., 1979). Commercial mycoherbicides first appeared in the market in the United States in the early 1980s with Devine as the first registered bioherbicide, developed by the Abbott Laboratories, United States, for controlling the milkweed vine. It consisted of liquid formulation containing chlamydospores of fungi *Phytopthora pulmivora*, which was capable of infecting roots of the weed leading to wilting. This was followed by the release of several other products, including Collego and BioMal (TeBeest, 1991), and subsequently several others. Mycoherbicides have in particular attracted global attention due to their weed host specificity, low environmental impact, and mostly cost-effectiveness (Ash, 2010; Ray & Hill, 2013). Development of microbial herbicides can be especially beneficial against the herbicide-resistant weeds.

However, the efficacy of these products is often restricted as the growth, sporulation, dissipation of infection, disease development, and disease-causing potential of the pathogen can be overblown by daily changes in the environment under field conditions impacting weed management drastically(Ray & Hill, 2016). Several studies (Auld & Say, 1999; Hetherington et al., 2002; Zhang & Watson, 1997) show that the dew period requirement in terms of duration, temperature, timing, and frequency in most fungi is perhaps the predominant impediment to the development and deployment of mycoherbicides. Furthermore, after application, most fungal spores die within a short span of time: the infection process gets aborted midway due to this phenomenon. Also challenges due to increasing herbicide resistance in weeds (Bourdôt et al., 2007; Hussain et al., 2021b) and nontarget environmental effects of chemical herbicides have led to increasing research in finding their biological and eco-friendly alternatives (Bourdôt et al., 2007; Gurusubramanian et al., 2008; Ray et al., 2008).Thus, synthesis of novel agrochemicals like mycoherbicides has thus attracted much attention from researchers preferring eco-friendly pest control options and sustainable organic farming.

6.5 Fungal Metabolites as Bioherbicides

Fungal metabolites also provide a prosperous resource for developing environmentally viable and sustainable mycoherbicides. Decades of research and biotechnological advances have resulted in a surfeit of knowledge on extraction of interesting compounds through fractionation and characterization from a range of fungi with herbicidal potential (McLean, 1996; Ndam et al., 2014). The fairly high degradability of these microbial metabolites causes less environmental problems in the soil ecosystem or bioaccumulation of toxic compounds in the food. Thus, the primary goal of these extensive researches into these naturally existing microbial phytotoxins is to curtail the use of environmentally damaging chemical herbicides and facilitation of nonchemical farming, enhancing agricultural productivity and ecological systems in a sustainable way.

A few potential toxins of fungal origin exploited for the management of weeds include A-AL toxin from *Alternaria alternata* against prickly sida and range of other broadleaf weed species (Abbas et al., 1995; Meena & Samal, 2019), Fumonisins from *Fusarium moniliforme* against Hemp sesbania (Ismaiel & Papenbrock, 2015; Nelson et al., 1993), Ophiobolins from *Dreschslera oryzae* against sickle pod (Evidente et al., 2006; Fumio Sugawara et al., 1988), Pyrenophorol from *Drechslera avenae* against *Avena sterilis* L (Kastanias & Chrysayi-Tokousbalides, 2000) etc. A list of some interesting secondary metabolites has been given in Table 6.1. Zhao et al. (2021) while clarifying the role of phytotoxic substances of fungi *Bipolaris setariae* observed several ophiobolins associated with strong phytotoxicity toward green foxtail (*Setaria viridis*). Among these one of the ophiobolins (6-epi-Ophiobolin A) interestingly showed cytotoxicity against five kinds of human cancer cells. This study opens avenues for parallel development of potential compound both as herbicide as well as an antitumor drug.

Table 6.1 List of toxins of fungal origin exploited for the management of some weed

Toxin	Source fungal species	Target weed	References
A-AL toxin	*Alternaria alternata*	Prickly sida (*Sida spinosa*), *Lemna minor* L., Northern Jointvetch, Amaranthus retroflexus, *Xanthium strumarium*	Abbas et al. (1995)
Alternariol	*Alternaria alternata*	*Amaranthus retroflexus*	De Souza et al. (2013) and Tang et al. (2020)
Ascaulitoxin	*Ascochyta caulina*	*Chenopodium album* L.	Evidente et al. (1998)
AT-toxin	*Alternaria longipipes*	Tobacco (*Nicotiana tabacum)*	Nishimura and Kohmoto (1983)
Bipoloroxin	*Bipolariscyanodontis*	*Cynodondactylon*	Sugawara et al. (1985)
Brefeldin A	*Alternaria zinniae*	*Xanthium occidentale*	Vurro et al. (1998)
Colletotrichinin	*Colletotrichum tabacum*	*Nicotiana tabacum*	Duke et al. (1991)
Curvulin	*Dreschslera indica*	*Amaranthus spinosus L. and Portulaca oleraceea*	Kenfield et al. (1989)
Eremophilanes	*Bipolariscyanodontis*	Bermudagrass (*Cynodondactylon)*	Sugawara et al. (1985)
Fumonisins	*Fusarium moniliforme*	Anoda cristata, Sickle pod (*Cassia obtusifolia*), Hemp sesbania (*Sesbania exaltata*)	Nelson et al. (1993)
Maculosin	*Alternaria alternata*	Spotted Knapweed (*Centaurea maculosa*)	Meena and Samal (2019) and Stierle et al. (1988)
Ophiobolins	*Dreschsleraoryzae, D. sorghicola* *D. gigantea* *Bipolaris setariae*	*Sorghum*, Sickle pod (*Cassia obtusifolia*) crabgrass (*Digitariasanguinalis*) Green foxtail (*Setaria viridis* (L.) P.Beauv.)	Sugawara et al. (1988), Evidente et al. (2006), and Zhao et al. (2021)
Porriolide	*Alternaria porri*	Lettuce and Stoneleek seedling	Suemitsu et al. (1993)
Radicinin	*Cochliobolusaustraliensis* *Pyricularia grisea*	Buffelgrass (*Cenchrus ciliaris*)	Masi et al. (2019)
Tentotoxin	*Alternaria alternata*	Johnson grass (*Sorghum halepense*), Jimson weed (*Datura stramonium* L), *Sida spinosa*	Duke (1986)
Tentuazonic acid	*Alternaria alternata*	*Datura innoxia*	Umetsu et al. (1973)

(continued)

Table 6.1 (continued)

Toxin	Source fungal species	Target weed	References
Pyrenophorol	*Drechsleraavenae*	*Avenasterilis* L	Kastanias and Tokousbalides (2000)
Viridiol	*Gliocladium virens*	*Amaranthus retroflexus*	Jones et al. (1988)
Zinniol	*Alternaria cichorii*	Russian knapweed (*Acroptilon repens*)	Stierle et al. (1993)

6.6 Future Prospects of Fungal Metabolites as Bioherbicides and Entrepreneurship Possibilities

Numerous mycotoxins with phytotoxic potential with the possibility of being developed as bioherbicide have been identified over the last four decades. Most of the fungal metabolites have different modes of actions, biochemical reactions, and signalling mechanisms leading to the development of diseases symptoms and possible death in their respective host plants (Meena & Samal, 2019). These provide opportunities for elucidating new mechanisms of herbicide action not yet discovered by traditional pesticide discovery methods, thereby providing tools for combating environmental issues and herbicide resistance. Further commercial interest in these metabolites has mostly caused researchers to largely look into these natural compounds for structural elucidation and industrial application, while very little effort has been given to understand the causes and consequences or the primary role and impact of these metabolites. Also, several challenges exist in developing these biologically active metabolites into bioherbicides. One of the major challenges include dealing with biosynthesis of low quantities of the bioactive compound naturally or under the influence of growing conditions including, impact of media, and environmental factors like temperature (Rai et al., 2021). Thus, more studies shall be interesting to understand their ecological role in nature. However, rapid advent and magnitude of possible utilization of techniques, like metabolomics, meta-proteomics, meta-genomics, meta-transcriptomics, next-generation sequencing, etc., have vastly increased our ability to untap the infinite fungal resources. Though over the last few decades, metabolic engineering of fungi has led to understanding biosynthetic pathways leading to the production of large quantities of secondary metabolites for commercial application, yet an extensive scope of finding bioherbicidal agents still remains less explored. The continuing search for herbicidal agents from fungi is extremely necessary to find the possible ways to have safe and more effective weed management while reducing the use of chemical pesticides. With the advancement of expanding knowledge on fungi and their metabolites, formulation development research and field trials are needed. More researchers of the present and future generations need to take up these technological and scientific challenges for a pesticide pollution–free world.

6.7 Conclusion

The problem of invasive weeds in contemporary intensive agriculture under increasing anthropogenic population pressure and changing climatic conditions makes plant protection inevitable. Further aggravating environmental issues have ushered in extensive research needs into eco-friendly bio-based solutions to invasive weed management. Fungal diversity offers a rich natural resource with possibilities of providing for human needs, in a sustainable way, with immense possibilities. Fungal metabolites thus have promising potential as bioherbicides.

References

Abbas, H. K., Duke, S. O., Paul, R. N., Riley, R. T., & Tanaka, T. (1995). AAL-toxin, a potent natural herbicide which disrupts sphingolipid metabolism of plants. *Pesticide Science, 43*(3), 181–187. https://doi.org/10.1002/ps.2780430302

Al-Fakih, A. A., Qasem, W., & Almaqtri, A. (2019). Overview on antibacterial metabolites from terrestrial *Aspergillus* spp. *Mycology, 10*(4), 191–209. https://doi.org/10.1080/21501203.2019.1604576

Andersen, S. B., Gerritsma, S., Yusah, K. M., Mayntz, D., Hywel-Jones, N. L., Billen, J., Boomsma, J. J., & Hughes, D. P. (2009). The life of a dead ant: The expression of an adaptive extended phenotype. *The American Naturalist, 174*(3), 424–433.

Ash, G. J. (2010). The science, art and business of successful bioherbicides. *Biological Control, 52*(3), 230–240. https://doi.org/10.1016/j.biocontrol.2009.08.007

Auld, B. A., & Say, M. M. (1999). Comparison of isolates of Colletotrichum orbiculare from Argentina and Australia as potential bioherbicides for *Xanthium spinosum* in Australia. *Agriculture, Ecosystems and Environment, 72*(1), 53–58. https://doi.org/10.1016/S0167-8809(98)00161-3

Baslam, M., Garmendia, I., & Goicoechea, N. (2013). Enhanced accumulation of vitamins, nutraceuticals and minerals in lettuces associated with arbuscular mycorrhizal fungi (AMF): A question of interest for both vegetables and humans. *Agriculture, 3*(1), 188–209.

Bayram, O., Krappmann, S., Ni, M., Bok, J. W., Helmstaedt, K., Valerius, O., Braus-Stromeyer, S., Kwon, N. J., Keller, N. P., Yu, J. H., & Braus, G. H. (2008). VelB/VeA/LaeA complex coordinates light signal with fungal development and secondary metabolism. *Science (New York, N.Y.), 320*(5882), 1504–1506.

Bekker, C. (2019). Ophiocordyceps-ant interactions as an integrative model to understand the molecular basis of parasitic behavioural manipulation. *Current Opinion in Insect Science, 33*, 19–24.

Bergmann, S., Funk, A. N., Scherlach, K., Schroeckh, V., Shelest, E., Horn, U., Hertweck, C., & Brakhage, A. A. (2010). Activation of a silent fungal polyketide biosynthesis pathway through regulatory cross talk with a cryptic non-ribosomal peptide synthetase gene cluster. *Applied and Environmental Microbiology, 76*(24), 8143–8149.

Blackwell, M. (2011). The fungi: 1, 2, 3 ... 5.1 million species? *American Journal of Botany, 98*(3), 426–438. https://doi.org/10.3732/ajb.1000298

Boddy, L., Frankland, J., & Van West, P. (Eds.). (2007). *Ecology of saprotrophic basidiomycetes. Elsevier.*

Bourdôt, G. W., Fowler, S. V., Edwards, G. R., Kriticos, D. J., Kean, J. M., Rahman, A., & Parsons, A. J. (2007). Pastoral weeds in New Zealand: Status and potential solutions. *New Zealand Journal of Agricultural Research, 50*(2), 139–161. https://doi.org/10.1080/00288230709510288

Brakhage, A. A. (2013). Regulation of fungal secondary metabolism. *Nature Reviews Microbiology, 11*(1), 21–32.

Bruns, S., Seidler, M., Albrecht, D., Salvenmoser, S., Remme, N., Hertweck, C., Brakhage, A. A., Kniemeyer, O., & Müller, F. M. (2010). Functional genomic profiling of *Aspergillus fumigatus* biofilm reveals enhanced production of the mycotoxin gliotoxin. *Proteomics, 10*(17), 3097–3107.

Dalinova, A., Chisty, L., Kochura, D., Garnyuk, V., Petrova, M., Prokofieva, D., Yurchenko, A., Dubovik, V., Ivanov, A., Smirnov, S., Zolotarev, A., & Berestetskiy, A. (2020). Isolation and bioactivity of secondary metabolites from solid culture of the fungus, alternaria sonchi. *Biomolecules, 10*(1), 1–18. https://doi.org/10.3390/biom10010081

De Souza, G., Mithöfer, A., Daolio, C., Schneider, B., & Rodrigues-Filho, E. (2013). Identification of *Alternaria alternata* mycotoxins by LC-SPE-NMR and their cytotoxic effects to soybean (Glycine max) cell suspension culture. *Molecules, 18*(3), 2528–2538.

Denison, S. H. (2000). pH regulation of gene expression in fungi. *Fungal Genetics and Biology: FG & B, 29*(2), 61–71.

Díez, E., Álvaro, J., Espeso, E. A., Rainbow, L., Suárez, T., Tilburn, J., & Peñalva, M. Á. (2002). Activation of the *Aspergillus* PacC zinc finger transcription factor requires two proteolytic steps. *The EMBO Journal, 21*(6), 1350–1359.

Duan, X., Tan, X., Gu, L., Liu, J., Hao, X., Tao, L., Feng, H., Cao, Y., Shi, Z., Duan, Y., Deng, M., Chen, G., Qi, C., & Zhang, Y. (2020). New secondary metabolites with immunosuppressive activity from the phytopathogenic fungus *Bipolaris maydis*. *Bioorganic Chemistry, 99*, 103816. https://doi.org/10.1016/j.bioorg.2020.103816

Duke S. O. (1986). Microbial phytotoxins as herbicides-a perspective. In A. R. Putnam & C. S. Tang (Eds.), *The science of allelopathy* (pp. 287–304). Wiley.

Duke S. O., Abbas H. K., Boyette C. D., & Ghobara M. (1991). Microbial compounds with potential for herbicide use. In *Brighton. Crop protection conference, Weeds*, Vol. 1, pp. 155–164.

Duncan, C. A., Jachetta, J. J., Brown, M. L., Carrithers, V. F., Clark, J. K., Ditomaso, J. M., Lym, R. G., Mcdaniel, K. C., Renz, M. J., & Rice, P. M. (2004). Assessing the economic, environmental, and societal losses from invasive plants on rangeland and wildlands. *Weed Technology, 18*(sp1), 1411–1416. https://doi.org/10.1614/0890-037x(2004)018[1411:ateeas]2.0.co;2

Espeso, E. A., & Penalva, M. A. (1996). Three binding sites for the *Aspergillus nidulans* PacC zinc-finger transcription factor are necessary and sufficient for regulation by ambient pH of the isopenicillin N synthase gene promoter. *The Journal of Biological Chemistry, 271*, 28825–28830.

Evidente, A., Capasso, R., Cutignano, A., Taglialatela-Scafati, O., Vurro, M., Zonno, M. C., & Motta, A. (1998). Ascaulitoxin, a phytotoxic bis-amino acid N-glucoside from *Ascochyta caulina*. *Phytochemistry, 48*(7), 1131–1137.

Evidente, A., Andolfi, A., Cimmino, A., Vurro, M., Fracchiolla, M., & Charudattan, R. (2006). Herbicidal potential of ophiobolins produced by Drechslera gigantea. *Journal of Agricultural and Food Chemistry, 54*(5), 1779–1783. https://doi.org/10.1021/jf0528431

Fiolka, M. J. (2008). Immunosuppressive effect of cyclosporin A on insect humoral immune response. *Journal of Invertebrate Pathology, 98*(3), 287–292.

Fischbach, M. A., & Walsh, C. T. (2006). Assembly-line enzymology for polyketide and nonribosomal peptide antibiotics: Logic, machinery, and mechanisms. *Chemical Reviews, 106*(8), 3468–3496.

Furukawa, T., Scheven, M. T., Misslinger, M., Zhao, C., Hoefgen, S., Gsaller, F., Lau, J., Jöchl, C., Donaldson, I., Valiante, V., Brakhage, A. A., Bromley, M. J., Haas, H., & Hortschansky, P. (2020). The fungal CCAAT-binding complex and HapX display highly variable but evolutionary conserved synergetic promoter-specific DNA recognition. *Nucleic Acids Research, 48*(7), 3567–3590.

Gadd, G. M. (2007). Geomycology: Biogeochemical transformations of rocks, minerals, metals and radionuclides by fungi, bioweathering and bioremediation. *Mycological Research, 111*(Pt 1), 3–49.

Gurusubramanian, G., Rahman, A., Sarmah, M., Ray, S., & Bora, S. (2008). Pesticide usage pattern in tea ecosystem, their retrospects and alternative measures. *Journal of Environmental Biology, 29*(6), 813–826.

Hertweck, C. (2009). The biosynthetic logic of polyketide diversity. *Angewandte Chemie International Edition, 48*(26), 4688–4716.

Hetherington, S. D., Smith, H. E., Scanes, M. G., & Auld, B. A. (2002). Effects of some environmental conditions on the effectiveness of Drechslera avenacea (Curtis ex Cooke) Shoem.: A potential bioherbicidal organism for Avena fatua L. *Biological Control, 24*(2), 103–109. https://doi.org/10.1016/S1049-9644(02)00020-8

Hussain, A., Ding, X., Alariqi, M., Manghwar, H., Hui, F., Li, Y., Cheng, J., Wu, C., Cao, J., & Jin, S. (2021a). Herbicide resistance: Another hot agronomic trait for plant genome editing. *Plants, 10*(4), 1–24. https://doi.org/10.3390/plants10040621

Hussain, H., Nazir, M., Saleem, M., Al-Harrasi, A., Elizbit, & Green, I. R. (2021b). Fruitful decade of fungal metabolites as anti-diabetic agents from 2010 to 2019: Emphasis on α-glucosidase inhibitors. In *Phytochemistry reviews* (Vol. 20, Issue 1). Springer. https://doi.org/10.1007/s11101-020-09733-1

Hyde, K. D., Xu, J., Rapior, S., Jeewon, R., Lumyong, S., Niego, A. G. T., Abeywickrama, P. D., Aluthmuhandiram, J. V. S., Brahamanage, R. S., Brooks, S., Chaiyasen, A., Chethana, K. W. T., Chomnunti, P., Chepkirui, C., Chuankid, B., de Silva, N. I., Doilom, M., Faulds, C., Gentekaki, E., … Stadler, M. (2019). The amazing potential of fungi: 50 ways we can exploit fungi industrially. In *Fungal diversity* (Vol. 97, Issue 1, pp. 1–136). Springer. https://doi.org/10.1007/s13225-019-00430-9

Ismaiel, A. A., & Papenbrock, J. (2015). Mycotoxins: Producing fungi and mechanisms of phytotoxicity. In *Agriculture* (Vol. 5, Issue 3, pp. 493–537). Walter de Gruyter GmbH. https://doi.org/10.3390/agriculture5030492

Jakubczyk, D., & Dussart, F. (2020). Selected fungal natural products with antimicrobial properties. In *Molecules* (Vol. 25, Issue 4). MDPI AG. https://doi.org/10.3390/molecules25040911

Jones, R. W., Lanini, W. T., & Hancock, J. G. (1988). Plant growth response to the phytotoxin viridiol produced by the fungus gliocladium virens. *Weed Science, 36*(5), 683–687. https://doi.org/10.1017/s0043174500075627

Kastanias, M. A., & Chrysayi-Tokousbalides, M. (2000). Herbicidal potential of pyrenophorol isolated from a *Drechslera avenae* pathotype. *Pest Management Science, 56*(3), 227–232. https://doi.org/10.1002/(SICI)1526-4998(200003)56:3<227::AID-PS115>3.0.CO;2-A

Kenfield, D., Hallock, Y., Clardy, J., & Strobel, G. (1989). Curvulin and O-Methylcurvulinic acid: Phytotoxic metabolites of *Drechslera indica* which cause necroses on purslane and spiny amaranth. *Plant Science, 60*(1), 123–127. https://doi.org/10.1016/0168-9452(89)90052-6

Kumar, S., Bhowmick, M. K., & Ray, P. (2021). Weeds as alternate and alternative hosts of crop pests. *Indian Journal of Weed Science, 53*(1), 14–29. https://doi.org/10.5958/0974-8164.2021.00002.2

Maraun, M., Martens, H., Migge, S., Theenhaus, A., & Scheu, S. (2003). Adding to the 'enigma of soil animal diversity': Fungal feeders and saprophagous invertebrates prefer similar food substrates. *European Journal of Soil Biology, 39*, 85–95.

Masi, M., Freda, F., Sangermano, F., Calabrò, V., Cimmino, A., Cristofaro, M., … Evidente, A. (2019). Radicinin, a fungal phytotoxin as a target-specific bioherbicide for invasive buffelgrass (Cenchrus ciliaris) control. *Molecules, 24*(6), 1086.

Masters, G. J., & Brown, V. K. (1997). Host-plant mediated interactions between spatially separated herbivores: Effects on community structure. In *Multitrophic interactions in terrestrial systems: The 36th symposium of the British Ecological Society*, Royal Holloway College, University of London, 1995 (pp. 217–237). Blackwell Science Ltd.

McLean, M. (1996). The phytotoxicity of *Fusarium* metabolites: An update since 1989. *Mycopathologia, 133*(3), 163–179. https://doi.org/10.1007/BF02373024

Meena, M., & Samal, S. (2019). *Alternaria* host-specific (HSTs) toxins: An overview of chemical characterization, target sites, regulation and their toxic effects. In *Toxicology Reports* (Vol. 6, pp. 745–758). Elsevier Inc. https://doi.org/10.1016/j.toxrep.2019.06.021

Ndam, L. M., Mih, A. M., Fongod, A. G. N., Tening, A. S., Tonjock, R. K., Enang, J. E., & Fujii, Y. (2014). Phytochemical screening of the bioactive compounds in twenty (20) Cameroonian medicinal plants. *International Journal of Current Microbiology and Applied Sciences, 3*(12), 768–778.

Nelson, P. E., Desjardins, A. E., & Plattner, R. D. (1993). Fumonisins, mycotoxins produced by Fusarium species: Biology, chemistry, and significance. In *Annual review of phytopathology* (Vol. 31, pp. 233–252). Annual Reviews Inc. https://doi.org/10.1146/annurev.py.31.090193.001313

Nguyen, T. P. T., Garrahan, M. A., Nance, S. A., Seeger, C. E., & Wong, C. (2020). Assimilation of cholesterol by Monascus purpureus. *Journal of Fungi, 6*(4), 352. https://doi.org/10.3390/jof6040352

Nishimura, S., & Kohmoto, K. (1983). Host-specific toxins and chemical structures from *Alternaria* species. *Annual Review of Phytopathology, 21*(1), 87–116. https://doi.org/10.1146/annurev.py.21.090183.000511

Pal, S., St Leger, R. J., & Wu, L. P. (2007). Fungal peptide Destruxin A plays a specific role in suppressing the innate immune response in *Drosophila melanogaster*. *The Journal of Biological Chemistry, 282*(12), 8969–8977.

Pan, Y. Y., & Liu, G. (2018). Yi chuan = Hereditas, *40*(10), 874–887.

Qi, Y., Li, J., Guan, X., Yan, B., Fu, G., He, J., Du, L., Zhao, C., & Zhang, D. (2020). Effects of herbicides on non-target plant species diversity and the community composition of fallow fields in northern China. *Scientific Reports, 10*(1), 1–10. https://doi.org/10.1038/s41598-020-67025-2

Rai, M., Zimowska, B., Shinde, S., & Tres, M. V. (2021). Bioherbicidal potential of different species of Phoma: Opportunities and challenges. *Applied Microbiology and Biotechnology, 105*, 3009–3018. https://doi.org/10.1007/s00253-021-11234-w

Ray, P., & Hill, M. P. (2013). Microbial agents for control of aquatic weeds and their role in integrated management. *CAB Reviews: Perspectives in Agriculture, Veterinary Science, Nutrition and Natural Resources, 8*. https://doi.org/10.1079/PAVSNNR20128014

Ray, P., & Hill, M. P. (2016). More is not necessarily better: The interaction between insect population density and culture age of fungus on the control of invasive weed water hyacinth. *Hydrobiologia, 766*(1). https://doi.org/10.1007/s10750-015-2454-3

Ray, P., Sushilkumar, & Pandey, A. K. (2008). Deleterious effect of herbicides on waterhyacinth biocontrol agents Neochetina bruchi and Alternaria alternata. *Biocontrol Science and Technology, 18*(5). https://doi.org/10.1080/09583150802001734

Redman, R. S., Sheehan, K. B., Stout, R. G., Rodriguez, R. J., & Henson, J. M. (2002). Thermotolerance generated by plant/fungal symbiosis. *Science, 298*(5598), 1581–1581.

Rohlfs, M., & Churchill, A. C. (2011). Fungal secondary metabolites as modulators of interactions with insects and other arthropods. *Fungal Genetics and Biology, 48*(1), 23–34.

Rohlfs, M., & Obmann, B. (2009). Species-specific responses of dew fly larvae to mycotoxins. *Mycotoxin Research, 25*(2), 103–112.

Sarikaya Bayram, O., Bayram, O., Valerius, O., Park, H. S., Irniger, S., Gerke, J., Ni, M., Han, K. H., Yu, J. H., & Braus, G. H. (2010). LaeA control of velvet family regulatory proteins for light-dependent development and fungal cell-type specificity. *PLoS Genetics, 6*(12), e1001226.

Sarrocco, S. (2016). Dung-inhabiting fungi: A potential reservoir of novel secondary metabolites for the control of plant pathogens. *Pest Management Science, 72*(4), 643–652.

Simaan, H., Lev, S., & Horwitz, B. A. (2019). Oxidant-sensing pathways in the responses of fungal pathogens to chemical stress signals. *Frontiers in Microbiology, 10*, 567.

Singh, D. D., & Datta, M. (2020). A promising source of anticancer drug from fungal secondary metabolite. In *New and future developments in microbial biotechnology and bioengineering* (pp. 95–101). Elsevier. https://doi.org/10.1016/b978-0-12-821006-2.00008-x

Stierle, A. C., Cardellina, J. H., & Strobel, G. A. (1988). Maculosin, a host-specific phytotoxin for spotted knapweed from Alternaria alternata. *Proceedings of the National Academy of Sciences, 85*(21), 8008–8011. https://doi.org/10.1073/pnas.85.21.8008

Stierle, A., Hershenhorn, J., & Strobel, G. (1993). Zinniol-related phytotoxins from Alternaria cichorii. *Phytochemistry, 32*, 1145–1149.
Strieker, M., Tanović, A., & Marahiel, M. A. (2010). Nonribosomal peptide synthetases: Structures and dynamics. *Current Opinion in Structural Biology, 20*(2), 234–240.
Suemitsu, R., Ohnishi, K., Horiuchi, M., Morikawa, Y., Sakaki, Y., & Matsumoto, Y. (1993). Structure of porriolide, a new metabolite from alternaria porri. *Bioscience, Biotechnology, and Biochemistry, 57*(2), 334–335. https://doi.org/10.1271/bbb.57.334
Sugawara, F., Strobel, G., Fisher, L. E., Van Duyne, G., & Clardy, J. (1985). Bipolaroxin, a selective phytotoxin produced by bipolaris cynodontis. *Proceedings of the National Academy of Sciences of the United States of America, 82*(24), 8291–9294. https://www.jstor.org/stable/26574?seq=1
Sugawara, F., Takahashi, N., Strobel, G., Yun, C. H., Gray, G., Fu, Y., & Clardy, J. (1988). Some New Phytotoxic Ophiobolins Produced by Drechslera Oryzae. *Journal of Organic Chemistry, 53*(10), 2170–2172. https://doi.org/10.1021/jo00245a008
Tanaka, A., Tapper, B. A., Popay, A., Parker, E. J., & Scott, B. (2005). A symbiosis expressed non-ribosomal peptide synthetase from a mutualistic fungal endophyte of perennial ryegrass confers protection to the symbiotum from insect herbivory. *Molecular Microbiology, 57*(4), 1036–1050.
Tang, Y., Zeiner, C. A., Santelli, C. M., & Hansel, C. M. (2013). Fungal oxidative dissolution of the Mn (II)-bearing mineral rhodochrosite and the role of metabolites in manganese oxide formation. *Environmental Microbiology, 15*(4), 1063–1077.
Tang, J., Huang, L., Liu, Y., Toshmatov, Z., Zhang, C., & Shao, H. (2020). Two phytotoxins isolated from the pathogenic fungus of the invasive weed *Xanthium italicum*. *Chemistry & Biodiversity, 17*(4), e2000043.
TeBeest, D. O. (Ed.). (1991). *Microbial control of weeds*. Chapman and Hall, 284p.
Templeton, G. E., TeBeest, D. O., & Smith, R. J. (1979). Biological weed control with mycoherbicides. *Annual Review of Phytopathology, 17*(1), 301–310. https://doi.org/10.1146/annurev.py.17.090179.001505
Then Bergh, K., & Brakhage, A. A. (1998). Regulation of the *Aspergillus nidulans* penicillin biosynthesis gene acvA (pcbAB) by amino acids: Implication for involvement of transcription factor PACC. *Applied and Environmental Microbiology, 64*(3), 843–849.
Trienens, M., Keller, N. P., & Rohlfs, M. (2010). Fruit, flies and filamentous fungi–experimental analysis of animal–microbe competition using *Drosophila melanogaster* and *Aspergillus* mould as a model system. *Oikos, 119*(11), 1765–1775.
Ujam, N. T., Ajaghaku, D. L., Okoye, F. B. C., & Esimone, C. O. (2021). Antioxidant and immunosuppressive activities of extracts of endophytic fungi isolated from Psidium guajava and Newbouldia laevis. *Phytomedicine Plus, 1*(2), 100028. https://doi.org/10.1016/j.phyplu.2021.100028
Umetsu, N., Kaje, J., & Tamari, K. (1973). Isolation of tenuazonic acid from blast disease rice plant. *Agricultural and Biological Chemistry, 37*, 451–452.
Vitale, G. A., Coppola, D., Esposito, F. P., Buonocore, C., Ausuri, J., Tortorella, E., & de Pascale, D. (2020). Antioxidant molecules from marine fungi: Methodologies and perspectives. *Antioxidants, 9*(12), 1–35. https://doi.org/10.3390/antiox9121183
Vödisch, M., Scherlach, K., Winkler, R., Hertweck, C., Braun, H. P., Roth, M., Haas, H., Werner, E. R., Brakhage, A. A., & Kniemeyer, O. (2011). Analysis of the *Aspergillus fumigatus* proteome reveals metabolic changes and the activation of the pseurotin A biosynthesis gene cluster in response to hypoxia. *Journal of Proteome Research, 10*(5), 2508–2524.
Vurro, M., Evidente, A., Andolfi, A., Zonno, M. C., Giordano, F., & Motta, A. (1998). Brefeldin A and α, β-dehydrocurvularin, two phytotoxins from *Alternaria zinniae*, a biocontrol agent of *Xanthium occidentale*. *Plant Science, 138*(1), 67–79.
Wang, T., Tian, Z., Tunlid, A., & Persson, P. (2020). Nitrogen acquisition from mineral-associated proteins by an ectomycorrhizal fungus. *New Phytologist, 228*(2), 697–711.
Yin, W., & Keller, N. P. (2011). Transcriptional regulatory elements in fungal secondary metabolism. *Journal of Microbiology, 49*(3), 329–339.

Yu, J. H., & Keller, N. (2005). Regulation of secondary metabolism in filamentous fungi. *Annual Review of Phytopathology, 43*, 437–458.
Zhang, W. M., & Watson, A. K. (1997). Effect of dew period and temperature on the ability of *Exserohilum monoceras* to cause seedling mortality of *Echinochloa* species. *Plant Disease, 81*(6), 629–634. https://doi.org/10.1094/PDIS.1997.81.6.629
Zhao, X.-L., Niu, Y.-C., Deng, H., & Luo, D.-Q. (2021). Characterization and phytotoxicity of ophiobolins produced by *Bipolaris setariae*. *Mycoscience, 62*(1), 64–70.

Chapter 7
Lichenized Fungi as Significant Source of Pharmaceuticals: Possibilities and Limitations for Entrepreneurship Development

Musarrat Fatima and Mamta Bhat

Contents

7.1 Introduction

The practices related to the traditional use of plant-based medicines pertained to many parts of the world among which China is at the top followed by India and other nations (Liu, 2010). This could be attributed to their efficacy without any side effect; hence, the demand for herbal medicines at a high rate never declined. The trend continues at present and is expected to upsurge in the near future too (Verma & Singh, 2008).

Lichenized fungi or lichens are one such dual-natured biological entities composed of fungi and algae (cyanobacterium or green alga), known as mycobiont and photobiont, respectively. They are the rich repository of secondary metabolites, which are produced exclusively in them to perform special functions. Some of the important secondary metabolites reported in lichens are amino acid derivatives,

M. Fatima · M. Bhat (✉)
Department of Botany, BGSB University, Rajouri, Jammu and Kashmir, India

A. C. Shukla (ed.), *Applied Mycology*, Fungal Biology,
https://doi.org/10.1007/978-3-030-90649-8_7

sugar alcohols, aliphatic acids, macro lytic lactones, monocyclic aromatic compounds, quinines, chromones, xanthones, dibenzofurans, depsides, depsidones, dapsones, terpenoids, steroids, carotenoids, and diphenyl ethers (Bhattacharyya et al., 2016). These exhibit a range of pharmaceutical activities, including antimicrobial, antiproliferative, antioxidant, antiviral, anti-inflammatory, and further allelopathic, antiherbivore, and photoprotective activities (Goga et al., 2018; Wink, 2018).

7.2 Applications of Lichenized Fungi

7.2.1 Pharmaceuticals

Due to the presence of secondary metabolites, lichens are being explored in pharmaceutical industries. Streptomycin and 2-propanol are lichen derivatives isolated from *Parmotrema tinctorum*. These exhibit antibacterial and antifungal activities against human pathogenic bacterial strains and also against a wide range of fungi (Anjali et al., 2015).

Different solvent extracts of lichens (*Cladonia furcata, Parmelia caperata, Parmelia pertusa, Hypogymnia physodes, Umbilicaria polyphylla, Lasallia pustulata, Parmelia sulcata, Umbilicaria crustulosa and Umbilicaria cylindrica*) showed antibacterial and antifungal activities. However, antioxidant, antipyretic, antiviral, antitumor, and analgesic effects of these lichens have also been reported (Mitrović et al., 2011).

Chharila, an Indian lichenized drug composed of *Parmelia perlata*, *Parmelia perforata,* and *Parmelia sancti-angelii*, is reported to be effective against dyspepsia, spermatorrhoea, amenorrhoea, calculi, diseases of blood and heart, stomach disorders, enlarged spleen, bronchitis, bleeding piles, scabies, leprosy, excessive salivation, soreness of throat, tooth-ache, wounds, and general pain; it can also be used as astringent, resolvent, laxative, and carminative (Nayaka et al., 2010).

Usnea longissima is known for its wounds healing properties and is effective against tumor (Crockett et al., 2003).

In Brazil, since ancient times, *Parmelinella salacinifera*, *Heterodermia galactophylla* and *Parmotrema wrightii* were used to treat common digestive track problems (Londoño-Castañeda et al., 2017).

Usnea ghattensis shows antimicrobial activity against gram-negative and gram-positive pathogenic bacteria (*Staphylococcus aureus*, *Streptococcus faecalis*, *Bacillus cereus*, *Escherichia coli*, *Pseudomonas aeruginosa*, and *Salmonella typhimurium*). *Lecanora muralis*, *Peltigera polydactyla*, *Ramalina farinacea*, and *Xanthoria elegans* also possess antibacterial activity (Srivastava et al., 2013).

Lichen compounds such as Barbatic Acid, Barbatolic Acid, Caperatic Acid; Diffractaic Acid, Divaricatic Acid, Gyrophoric Acid, Hypostictic Acid, Lobaric Acid, Lecanoric Acid, Lichexanthone, Lobastin, Norticitic Acid, Olivetoric Acid,

Physodic Acid, Perlatolic Acid, Protolichesterinic Acid, Protocetraric Acid, Psoromic Acid, Potassium Usnate, Retigeric Acid, Ramalin, Salazanic Acid, Sekikaic Acid, Usnic Acid, and Vulpinic Acid show anticancer, antimicrobial, antioxidant, anti-inflammatory, antiproliferative and anti-diabetic properties (Solárová et al., 2020).

Evernia furfuracea used for medicinal purposes is reported to be effective against jaundice, rabies, and cough. *Cladonia*, *Evernia*, *Lobaria*, *Parmelia*, *Peltigera*, *Pertusaria*, *Physica*, *Rocella*, *Usnea*, and *Xanthoria* are used as herbal medicines (Zambare & Christopher, 2012).

Usnea barbata is effective against *Bacillus subtilis* and *Pseudomonas fluorescens*. Besides, it also possesses antioxidant properties (Bazarnova et al., 2018).

Anaptychia ciliaris, Cetrelia olivetorum, Lecanora muralis, Peltigera polydactyla, Peltigera praetextata, Ramalina farinacea, Rhizoplaca melanohpthalma, Umbilicaria vellea, Xanthoria elegans, Xanthoria parietina, Xanthoparmelia tinctina are known for their antibacterial potential against *Escherichia coli*, *Pseudomonas aeruginosa, Bacillus subtilis, Klebsiella pneumonia*, *Staphylococcus aureus*, and *Staphylococcus* (Karagoz et al., 2009).

Antimicrobial activity against some bacteria viz. *Escherichia coli*, *Pseudomonas aeruginosa, Enterococcus faecalis*, *Staphylococcus aureus*, and a fungus *Candida albicans* was shown by *Melanohalea exasperata, Physcia aipolia, Usnea florida, Usnea subfloridana and Xanthoria parietina*. Antimicrobial activity by lichens was reported both against gram positive and gram-negative strains (Çobanoğlu et al., 2016).

Usnic acid a derivative of *Cladonia* sp. is proved to be effective against lung cancer. On the other hand, many lichens are used as antibiotics against a number of bacteria (Shrestha & St. Clair, 2013).

A total of 153 active chemical constituents have been reported in 118 different species of genus *Ramalina*, which have antimicrobial, antioxidant, antiviral, antitumor, cytotoxic, anti-inflammatory, antifungal, fungicidal, anti-helminth, anti-Schistosoma, and larvicidal properties (Moreira et al., 2015).

7.2.2 *Pesticides*

Chemical pesticides are used to kill pests and insects to protect crops from destroying. On the other hand, many natural products have substituted these chemical-based pesticides without affecting soil, crops, and beneficial insects. Lichens are also being used for this purpose, and crop is treated either with the extract of single species or more than one species in different concentrations with different solvents. *Lecanora muralis* (Schreb.) Rabenh. *Letharia vulpina* (L.), Hue, and *Peltigera rufescens* are used as insecticides against *Sitophilus granarius* (Emsen et al., 2016).

7.2.3 Dyes

Dye extracted from lichens is used to color attires, food and also applied on body parts. Production of dye in lichens is due to many secondary metabolites present in them like gyrophoric, lecanoric acid, umbilicaric acids, usnic acid, atranorin, chloroatranorin, salazinic acid, and parietin. Some of the dye-producing lichens are *Flavopunctelia, Flavoparmelia, Cladonia, Parmelia, Umbilicaria, Xanthoria, Ochrolechia, Hyperphyscia, Hypogymnia, Dermatocarpon*, and *Parmotrema* (Shaheen et al., 2019).

7.2.4 Bioremediation

Many reports are available showing that the lichens are used for uptake of heavy metals and air pollutants from the atmosphere.

Numerous scientific records are available based on multiple uses of lichens (Table 7.1 and Fig. 7.1). This can be referred to while looking for new avenues to explore possibilities regarding the establishment of lichen-based enterprises.

7.3 Exploring Lichens for Entrepreneurship Development

The numerous benefits associated with lichens make them promising entities to be explored commercially. Entrepreneurship development focusing on these aspects of lichenized fungi is needed so as to tap these important bioresources.

As they are harvested from their natural habitat, their distribution in nature is already under threat. Therefore, lichen farming or cultivation can be a suitable option for business for which only a little space is required. It can be profitable with low capital investment. But to ensure good returns, both study and experience are required. Technical training both at national and international levels can be imparted to gain expertise. Then, it can even be perfect option for the unskilled as well.

For sustainable entrepreneurship, budgetary requirements need to be worked out, focusing clean and green products and target market such as domestic or export.

Certain limitations such as their slow growth rate poses a challenge, which needs to be worked upon. Ensuring safe environment for workers and development of harmless lichen products for end users also holds importance.

Genetic engineering techniques leading to industrial large-scale production of lichen-based natural products by modifying natural pathways to increase productivity to meet demands both globally as well as locally for lichen-based products.

However, researching the mainstream market and strategic use of these inconspicuous forms can definitely reap maximum benefits and fulfill demand for these pharmaceutical products in the market commercially.

Table 7.1 Bioactive metabolites and uses associated with lichenized fungi

S. no.	Lichen	Bioactive metabolites	Uses	References
1.	*Bacidia stipata*	Atranorin	Anticancer	Shrestha and St. Clair (2013)
2.	*Bryoria lactinea* (Nyl.) Brodo and D. Hawksw.	Fumaroprotocetraric acid	Biofuel	Shukla et al. (2014)
3.	*Bryoria* sp.		Immunosuppressive	Hwang et al. (2017)
4.	*Bulbothrix setschwanensis* (Zahlbr.) Hale	Salazinic acid	Biofuel	Shukla et al. (2014)
5.	*Cetraria braunsiana* (Mull. Arg.) W. Culb. and C. Culb.	Alectoronic and α-collatolic acid	Biofuel	Shukla et al. (2014)
6.	*Cladia retipora*	Usnic acid	Antiviral/ Cytotoxic	Perry et al. (1999)
7.	*Cladonia deformis*	Zeorin	Bioremediation	Hämäläinen et al. (2015) and Pawlik-Skowrońska and Bačkor (2011)
8.	*Cladonia furcata*	Depside, depsidones, dibenzofurane	Antibacterial/Antifungal	Mitrović et al. (2011) and Kosanic et al. (2018)
9.	*Cladonia lepidophora*	Usnic acid	Anticancer	Shrestha and St. Clair (2013)
10.	*Cladonia rangiferina* (L.) Nyl	Atranorin Fumarprotocetraric acid, lsousnic acid	Antibacterial	Smith (1932), Richardson (1974), Yoshimura et al. (1994), and Fraser (2006)
11.	*Cladonia substellata*	Usnic acid	Insecticidal	Sachin et al. (2018)
12.	*Cladonia sulphurina*	Usnic and Squamatic acid	Bioremediation	Hämäläinen et al. (2015) and Pasternak et al. (2015)
13.	*Cornicularia aculeata*	Protolichesterinic acid	Anticancer	Shrestha and St. Clair (2013)
14.	*Evernia divaricata* (L.) Ach.	β-pinene, α-pinene, limonene, α-phellandrene, camphene, and p-cymene	Antibacterial	Yuan et al. (2010) and Goga et al. (2018)

(continued)

Table 7.1 (continued)

S. no.	Lichen	Bioactive metabolites	Uses	References
15.	*Evernia prunastri* (L.) Ach	Atranorin, Chloroatranorin Evernic acid, Usnic acid	Astringent	Lopez et al. (2006) and Yoshimura et al. (1994)
16.	*Everniastrum nepalense* (Taylor) Hale	Salazinic and protolichesterinic acid	Biofuel	Shukla et al. (2014)
17.	*Flavopunctelia soredica* (Nyl.) Hale	Lecanoric acid	Biofuel	Shukla et al. (2014)
18.	*Heterodermia diademata* (Taylor) D. D. Awasthi	Zeorin	Biofuel	Shukla et al. (2014)
19.	*Heterodermia leucomela* (L.) Poelt	3,6-Dimethyl-2-hydroxy-4-methoxybenzoic acid, Zeorin, norstictic, salazinic acid, and triterpenoids	Insecticidal/biofuel	Shukla et al. (2014) and Sachin et al. (2018)
20.	*Hypogymnia physodes*	Depsidones, depsides, usnic acid	Antibacterial/antifungal	Mitrović et al. (2011) and Rankovic et al. (2009, 2014a, b)
21.	*Lasallia pustulata*	gyrophoric acid, arabitol, mannitol, and umbilicarin	Antibacterial/antifungal	Ranković et al. (2007) and Mitrović et al. (2011)
22.	*Lecanora muralis*	Phenolics and flavonoid	Antibacterial	Rankovic et al. (2014a) and Srivastava et al. (2013)
23.	*Lepraria tomentosa*	(+)-Usnic acid	Insecticidal	Sachin et al. (2018)
24.	*Leproloma sipmanianum*	3,6-dimethyl-2-hydroxy-4-methoxybenzoic acid	Insecticidal	Sachin et al. (2018)
25.	*Letharia vulpina*	Atranorin and vulpinic acid	Insecticidal	Sachin et al. (2018)
26.	*Loberia retigera* (Bory) Trev.	Triterpenoids and thelephoric acid	Biofuel	Shukla et al. (2014)
27.	*Ochrolechia deception*	Variolaric acid	Anticancer	Shrestha and St. Clair (2013)
28.	*Pseudocyphellaria homoeophylla*	Usnic acid	Antiviral/ Cytotoxic	Perry et al. (1999)
29.	*Parmelaria subthomsonii* D. D. Awasthi	Atranorin, alectoronic and α-collatolic acid	Biofuel	Shukla et al. (2014)

30.	*Parmelia caperata*	salazinic acid, usnic acid, depsidone, atranorin, chloroatranorin	Antibacterial/antifungal/antioxidant/cytotoxic	Manojlović et al. (2013) and Mitrović et al. (2011)
31.	*Parmelia petrusa*	atranorin and salazinic acid	Antibacterial/antifungal	Mitrović et al. (2011) and Rankovic et al. (2009)
32.	*Parmelia sulcata*	Salzinic acid	Antibacterial/antifungal	Mitrović et al. (2011) and Candan et al. (2014)
33.	*Parmelinella wallichiana* (Taylor) Elix and Hale	Salazinic and consalazinic acid	Biofuel	Shukla et al. (2014)
34.	*Peltigera polydactyla*	usnic acid, phenolic compounds, triterpenes, steroids, anthraquinones, depsides, depsidones, and dapsones	Antibacterial	Srivastava et al. (2013) and Kosanić and Ranković (2014)
35.	*Protousnea magellanica*	Diffractaic acid	Anticancer	Shrestha and St. Clair (2013)
36.	*Protousnea malacea*	Divaricatic acid	Anticancer	Shrestha and St. Clair (2013)
37.	*Pseudocyphellaria glabra*	Usnic acid	Antiviral/cytotoxic	Perry et al. (1999)
38.	*Psoroma dimorphum*	Vicanicin	Anticancer	Shrestha and St. Clair (2013)
39.	*Psoroma pallida*	Vicanicin	Anticancer	Shrestha and St. Clair (2013)
40.	*Pyxine consocians*	Cabraleadiol monoacetate, 4-O-methylcryptochlorophaeic acid, lichexanthone	Insecticidal	Sachin et al. (2018)
41.	*Ramalina conduplicans* Vain.	Usnic acid, sekikaic acid aggregate and salazinic acid	Biofuel	Shukla et al. (2014)
42.	*Ramalina farinacea*	usnic acid, phenolic compounds, triterpenes, steroids, anthraquinones, depsides, depsidones, and dapsones	Antibacterial	Srivastava et al. (2013) and Kosanić and Ranković (2014)
43.	*Rhizoplaca melanophthalma*	Protolichesterinic acid	Anticancer	Shrestha and St. Clair (2013)
44.	*Stereocaulon alpnium*	Lobaric acid	Anticancer	Shrestha and St. Clair (2013)

(continued)

Table 7.1 (continued)

S. no.	Lichen	Bioactive metabolites	Uses	References
45.	*Umbilicaria crustulosa*	methyl orsellinate, lecanoric acid, crustinic acid, gyrophoric acid and atranorin	Antibacterial/antifungal	Mitrović et al. (2011) and Zlatanović et al. (2017)
46.	*Umbilicaria cylindrica*	norstictic acid, methyl-β-orcinol carboxylate, ethyl haematommate, atranorin and usnic acid	Antibacterial/antifungal	Manojlovic et al. (2012)
47.	*Umbilicaria hirsuta*	Gyrophoric acid	Anticancer	Shrestha and St. Clair (2013)
48.	*Umbilicaria polyphylla*	Phenol, flavonoid	Antibacterial/antifungal	Kosanic and Rankovic (2011) and Mitrović et al. (2011)
49.	*Usnea ghattensis*	Usnic acid, norstictic acid	Antibacterial/antioxidant	Verma et al. (2008) and Srivastava et al. (2013)
50.	*Usnea longissima*	Usnic acid, barbatic acid, diffractaic acid, glutinol, longissiminone A, longissiminone B, salazinic acid, protocetraric acid, evernic acid, 4-O-demethyl-barbatic acid Diffractaic acid and usnic acid	Antibacterial/antifungal/antitumor/insecticidal	Sachin et al. (2018), Crockett et al. (2003), and Zugic et al. (2018)
51.	*Xanthoria elegans*	Anthraquinone	Antibacterial	Stocker-Wörgötter (2008) and Srivastava et al. (2013)
52.	*Xylopsora caradocensis*	–	Bioremediation	Hämäläinen et al. (2015)
53.	*Xylopsora friesii*	Friesiic acid	Bioremediation	Hämäläinen et al. (2015) and Elix (2005)

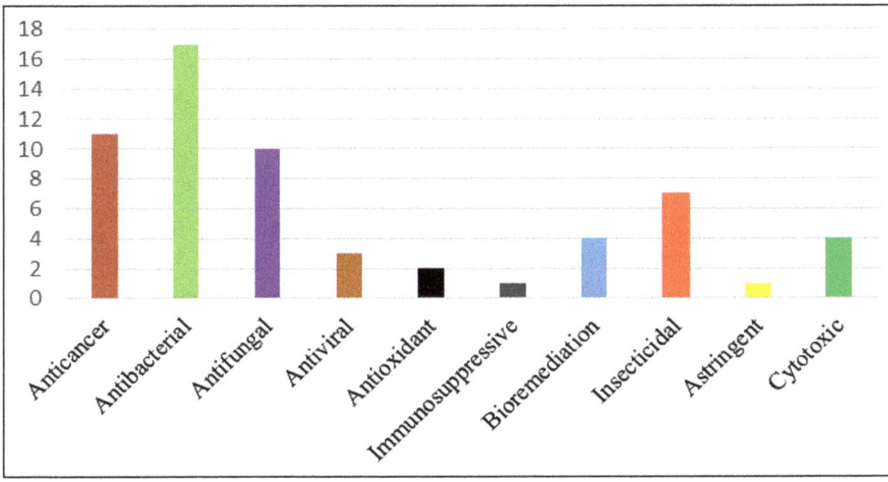

Fig. 7.1 Pharmacological uses of lichens

7.4 Conclusions

The multiple roles that lichenized fungi play in our environment are unmatched with the other living organisms. Every best effort must be done so as to harness their potential. The fundamental and applied science offers the best solution to the lichen-based economy. It opens new avenues and in the light of their contribution to 7 out of 17 United Nations Sustainable Development Goals, their role becomes more important for future, which needs to be realized, researched, tapped, and explored.

References

Anjali, D. B., Mohabe, S., Reddy, A. M., & Nayaka, S. (2015). Antimicrobial activities of 2-propanol crude extract from lichen *Parmotrema tinctorum* (Despr.ex. Nyl.) Hale, collected from Eastern Ghats, India. *Current Research in Environmental and Applied Mycology, 5*(3), 160–168.

Bazarnova, Y., Politaeva, N., & Lyskova, N. (2018). Research for the lichen *Usnea barbata* metabolites. *Zeitschrift Für Naturforschung C, 73*(7-8), 291–296.

Bhattacharyya, S., Deep, P. R., Singh, S., & Nayaka, B. (2016). Lichen secondary metabolites and its biological activity. *American Journal of Pharmatech Research., 6*(6), 28–44.

Candan, M., Yılmaz, M., Tay, T., Erdem, M., & Türk, A. O. (2014). Antimicrobial activity of extracts of the lichen *Parmelia sulcata* and its salazinic acid constituent. *Zeitschrift Für Naturforschung C, 62*(7–8), 619–621.

Çobanoğlu, G., Sesal, C., Açıkgöz, B., & Karaltı, I. (2016). Evaluation of antimicrobial activity of the lichens *Physcia aipolia, Xanthoria parietina, Usnea florida, Usnea subfloridana* and *Melanohalea exasperate*. *Modern Phytomorphology, 10*, 19–24.

Crockett, M., Kageyama, S., Homen, D., Lewis, C., Osborn, J., & Sander, L. (2003). *Antibacterial properties of Pacific Northwest Lichens*.

Elix, J. (2005). New species of sterile crustose lichens from Australasia. *ResearchGate, Mycotaxon, 94*, 219–224.
Emsen, B., Yildirim, E., & Aslan, A. (2016). Insecticidal activities of extracts of three lichen species on *Sitophilus granarius (L.)* (Coleoptera: Curculionidae). *Plant Protection Science, 51*(3), 155–161.
Fraser, M. H. (2006). *Ethnobotanical investigation of plants used for the treatment of type 2 diabetes by two cree communities in Quebec: quantitative comparisons and antioxidant evaluation*. M.Sc. thesis, McGill University.
Goga, M., Elecko, J., Marcincinova, M., Rucova, D., Backorova, M., & Backor, M. (2018). Lichen metabolites: An overview of some secondary metabolites and their biological potential. *Chemistry and Material Science*, 1–36.
Hämäläinen, A., Kouki, J., & Lõhmus, P. (2015). Potential biodiversity impacts of forest biofuel harvest: Lichen assemblages on stumps and slash of Scots pine. *Canadian Journal of Forest Research, 45*(10), 1239–1247.
Hwang, Y. H., Lee, S., Kang, K. Y., Hur, H., & Yee, S. T. (2017). Immunosuppressive effects of *Bryoria sp.* (Lichen-forming fungus) extracts via inhibition of CD8+ T-Cell proliferation and IL-2 production in CD4+ T Cells. *Journal of Microbiology and Biotechnology, 27*(6), 1189–1197.
Karagoz, A., Dogruoz, N., Zeybek, Z., & Aslan, A. (2009). Antibacterial activity of some lichen extracts. *Journal of Medicinal Plant Research, 3*(12), 1034–1039.
Kosanic, M., & Rankovic, B. (2011). Antioxidant and antimicrobial properties of some lichens and their constituents. *Journal of Medicinal Food, 14*(12), 1624–1630.
Kosanić, M., & Ranković, B. (2014). Lichen secondary metabolites as potential antibiotic agents. *Lichen Secondary Metabolites*, 81–104.
Kosanic, M., Ristic, S., Stanojkovic, T. P., & Manojlovic, N. (2018). Extracts of five *Cladonia sp.* Lichens as source of biologically active compounds. Farmacia, *66*(4), 644–651.
Liu, W. J. H. (2010). Introduction to traditional herbal medicines and their study. *Traditional Herbal Medicine Research Methods*, 1–26.
Londoño-Castañeda, P. A., Buril, M. L. L., Rego-Cunha, I. P., Silva, N. H., Honda, N. K., Pereira, E. C., & Andrade, L. H. C. (2017). Lichens used in the traditional medicine by the pankararu indigenous community, Pernambuco-Brazil. *C Biological Science, Global Journal of Science Frontier Research, 17*(4).
Lopez Eire, A., Cortes Gabaudan, F., Gutierrez Rodilla, B. M., & Vazquez de Benito, M. C. (2006). *Estudios y Traduccion*. Dioscorides. Ediciones Universida, Sobre los remedios medicinales. Manuscrito de salamanca.
Manojlovic, N. T., Vasiljevic, P. J., Maskovic, P. Z., Juskovic, M., & Bogdanovic-Dusanovic, G. (2012). Chemical composition, antioxidant, and antimicrobial activities of lichen *Umbilicaria cylindrica (L.) Delise* (Umbilicariaceae). *Evidence-Based Complementary and Alternative Medicine*, 1–8.
Manojlović, N., Ranković, B., Kosanić, M., Vasiljević, P., & Stanojković, T. (2013). Chemical composition of three *Parmelia* lichens and antioxidant, antimicrobial and cytotoxic activities of some their major metabolites. *Phytomedicine, 19*(13), 1166–1172.
Mitrović, T., Stamenković, S., Cvetković, V., Nikolić, M., Tošić, S., & Stojičić, D. (2011). Lichens as source of versatile bioactive compounds. *Biologica Nyssana, 2*(1), 1–6.
Moreira, A. S. N., Braz-Filho, R., Mussi-Dias, V., & Vieira, I. J. C. (2015). Chemistry and biological activity of Ramalina lichenized fungi. *Molecules, 20*(5), 8952–8987.
Nayaka, S., Upreti, D. K., & Khare, R. (2010). Medicinal lichens of India. In P.C. Trivedi (Ed.), *Drugs from plants*. Avishkar Publishers.
Pasternak, C., Chhoker, K., Kidwai, A., & Deduke, C. (2015). Secondary metabolites of *Cladonia* species in three community types. *Research Gate.*
Pawlik-Skowrońska, B., & Bačkor, M. (2011). Zn/Pb-Tolerant lichens with higher content of secondary metabolites produce less phytochelatins than specimens living in unpolluted habitats. *Environmental and Experimental Botany, 72*(1), 64–70.

Perry, N. B., Benn, M. H., Brennan, N. J., Burgess, E. J., Ellis, G., Galloway, D. J., & Tangney, R. S. (1999). Antimicrobial, antiviral and cytotoxic activity of New Zealand lichens. *The Lichenologist, 31*(06), 627–636.
Ranković, B., Mišić, M., & Sukdolak, S. (2007). Evaluation of antimicrobial activity of the *Lichens Lasallia pustulat, Parmelia sulcate, Umbilicaria crustulosa* and *Umbilicaria cylindrica. Microbiology, 76*(6), 723–727.
Ranković, B., Misic, M., & Sukdolak, S. (2009). Antimicrobial activity of extracts of the *lichens Cladonia furcata, Parmelia caperata, Parmelia pertusa, Hypogymnia physodes* and *Umbilicaria polyphylla. Versita, 64*(1), 53–58.
Ranković, B. R., Kosanic, M. M., & Stanojkovic, T. P. (2014a). Antioxidant, antimicrobial and anticancer activity of the *lichens Cladonia furcata, Lecanora atra* and *Lecanora muralis. BMC Complementary and Alternative Medicine, 11*(97).
Ranković, B., Kosanić, M., Manojlović, N., Rančić, A., & Stanojković, T. (2014b). Chemical composition of *Hypogymnia physodes Lichen* and biological activities of some its major metabolites. *Medicinal Chemistry Research, 23*, 408–416.
Richardson, D. H. S. (1974). *The vanishing lichens: their history and importance* (p. 231). Hafner.
Sachin, M. B., Mahalakshmi, S. N., & Prashith-Kekuda, T. R. (2018). Insecticidal efficacy of lichens and their metabolites-A mini review. *Journal of Applied Pharmaceutical Science, 8*(10), 159–164.
Shaheen, S., Iqbal, Z., & Hussain, M. (2019). First report of dye yielding potential and compounds of *Lichens*; A cultural heritage of Himalayan communities, Pakistan. *ResearchGate, Pakistan Journal of Botany, 51*(1).
Shrestha, G., & St. Clair, L. L. (2013). *Lichens*: A promising source of antibiotic and anticancer drugs. *Phytochemistry Reviews, 12*(1), 229–244.
Shukla, P., Upreti, D. K., & Tiwari, P. (2014). Assessment of dye yielding potential of Indian *Lichens. Indian Journal of Plant Sciences, 3*(1), 57–63.
Smith, H. H. (1932). Ethnobotany of the Ojibwe Indians. *Bulletin of Public Museum Milwaukee, 4*, 327–525.
Solárová, Z., Liskova, A., Samec, M., Kubatka, P., Büsselberg, D., & Solá, P. (2020). Anticancer potential of *Lichens'* secondary metabolites. *Biomolecules, 10*(87), 1–31.
Srivastava, P., Upreti, D. K., Dhole, T. N., Srivastava, A. K., & Nayak, M. K. (2013). Antimicrobial property of extracts of Indian *Lichen* against human pathogenic bacteria. In *Interdisciplinary perspectives on infectious diseases* (pp. 1–6). Hindawi Publishing Corporation.
Stocker-Wörgötter, E. (2008). Metabolic diversity of *Lichen*-forming ascomycetous fungi: culturing, polyketide and shikimate metabolite production, and PKS genes. *Natural Product Reports, 25*(1), 188–200.
Verma, S., & Singh, S. P. (2008). Current and future status of herbal medicines. *Veterinary World, 1*(11), 347–350.
Verma, N., Behera, B. C., & Makhija, U. (2008). Antioxidant and hepatoprotective activity of a *Lichen Usnea ghattensis* in vitro. *Applied Biochemistry and Biotechnology, 151*(2–3), 167–181.
Wink, M. (2018). Introduction: Biochemistry, physiology and ecological functions of secondary metabolites. *Annual Plant Reviews*, 1–19.
Yoshimura, I., Kinoshita, Y., Yamamoto, Y., Huneck, S., & Yamada, Y. (1994). Analysis of secondary metabolites from *Lichen* by high performance liquid chromatography with a photodiode array detector. *Phytochemical Analysis, 5*(4), 197–205.
Yuan, C., Zhang, X. J., Du, Y. D., Guo, Y. H., Sun, L. Y., & Ren, Q. (2010). Antibacterial compounds and their constituents of *Evernia divertica. Chemical Society of Pakistan, 32*(2), 189–193.
Zambare, V. P., & Christopher, L. P. (2012). Biopharmaceutical potential of *lichens. Pharmaceutical Biology, 50*(6), 778–798.

Zlatanović, I., Zrnzević, I., Jovanović, O., Stojanović, I., Petrović, G., & Stojanović, G. (2017). Chemical composition of *Umbilicaria crustulosa* and *U. cylindrica*. *Natural Product Communications, 12*(7).

Zugic, A., Tadic, V., & Kundakovic, T. (2018). Chemical composition and biological activities of the extracts and secondary metabolites of *Lichens* belonging to the genus *Usnea, Parmeliaceae*. *ResearhGate, Lekovite Sirovine, 38*, 68–80.

Chapter 8
Protocols for Extraction, Isolation, and Purification of Secondary Metabolites of Mushroom and Its Applications

Karina Afzal, Amritesh Chandra Shukla, and D. K. Srivastava

Contents

8.1 Introduction

Mushrooms, the macroscopic fungi, harbor miraculous potentials in their filamentous fruiting bodies that significantly contribute to the nutrition of both animal and plant kingdoms (Feeney et al., 2014). Thousands of secondary metabolites of agricultural, medicinal, nutritional, and industrial application have been identified from fungi (Keswani et al., 2014). Basidiomycetes contribute about 23% of the total natural bioactive metabolites known today (Bérdy, 2012). These mushroom-derived metabolites are indubitable stars of modern medicine unveiling massive numbers of

K. Afzal · A. C. Shukla (✉)
Biocontrol Laboratory, Department of Botany, University of Lucknow, Lucknow, Uttar Pradesh, India

D. K. Srivastava
Council of Science and Technology, Lucknow, Uttar Pradesh, India

A. C. Shukla (ed.), *Applied Mycology*, Fungal Biology,
https://doi.org/10.1007/978-3-030-90649-8_8

bioactivities like antifungal, antiviral, antibacterial, antitumor, and immunomodulating activities (Gargano et al., 2017).

These secondary metabolites are at present the topic of great research interest, although their part of extraction in phytochemical or biological researches presents specific challenges that are worth mentioning. Successful extraction and isolation of desired compounds begin with careful selection and preparation of mushroom samples. Further, literature review will be helpful to choose the appropriate protocol for a particular class of secondary metabolite or mushroom species. During the extraction process, it is vital to limit interference from components that may coextract with the target compounds and must take care to prevent decomposition of important metabolites and so on. This chapter presents an outline of the process of extraction, isolation, and identification of mushroom secondary metabolites. In addition, some bioactive secondary metabolites derived from *Ganoderma* spp., *Lentinula edodes,* and *Flammulina velutipe* are mentioned, along with their applications.

8.2 Protocol for Extraction, Isolation, and Purification of Secondary Metabolites of Mushroom

8.2.1 Selection, Collection, and Identification of Mushroom Material

The reproducibility of a phytochemical research is directly affected by the selection, collection, and identification of mushroom material. Any imprecision at this stage of an investigation might reduce the scientific value of the research when searching for a particular compound or metabolite. Specific secondary metabolites differ both qualitatively and quantitatively among members of single species, as well as among members of a population. Apart from the phytochemical considerations, norms related to intellectual property rights are to be taken care of when samples or extracts will cross international borders (Cantley, 1997).

8.2.2 Drying and Grinding

To begin with, mushrooms should be dried at temperatures below 30 °C in order to avoid decay of thermolabile compounds. Similarly, direct sunlight should be avoided to minimize the chemical transformations ensuing from exposure to ultraviolet radiation. When a material is required for research, it is recommended to extract it instantly using organic solvents, such as ethanol in order to deactivate enzymes present in the material. Small quantities of fungal material can be crushed using a mortar and pestle, or electric grinder. In the case of large quantities, milling can be best carried out in industrial-scale comminution apparatus. Proper grinding decreases the quantity of solvent required for extraction and improves the efficacy of extraction.

8.2.3 Extraction, Isolation, and Purification

Some conventional processes of extraction and isolation of secondary metabolites include Soxhlet extraction, hydro-distillation, maceration, pressing, percolation, and infusion. In order to isolate a specific compound in a purified form, it is essential to add chemicals while extraction. Most of the water-insoluble compounds are extracted with organic solvents, such as chloroform, methanol, or ethanol. However, the requirement of large quantities of solvents for efficient extraction along with the high temperatures, sometimes, causes degradation of target molecules with fractional loss of volatile compounds (Bubalo et al., 2018). Consequently, the advanced methods of extraction are envisioned to make use of nontoxic solvents, like ethanol, water, and carbon dioxide. Some most promising and competent extraction techniques include ultrasound-assisted extraction, microwave-assisted extraction, supercritical fluid extraction, subcritical water extraction, and accelerated solvent extraction (Roselló-Soto et al., 2016) (Fig. 8.1). The use of standard extraction techniques to obtain valuable extracts rich in secondary metabolites from different mushroom species has been expansively described by various workers (Ruthes et al., 2015; Kaewnarin et al., 2016; Boonsong et al., 2016). Kimatu et al. (2017) obtained a protein-rich extract with conventional techniques in order to produce mushroom protein hydrolysates. Hassainia et al. (2018) isolated chitin from the fruiting body of *A. bisporus* with alkaline treatment of 1 M NaOH solution at 80 °C for 2 h under optimum conditions. Maeng et al. (2016) prepared the phenolic-rich extracts of mushroom *Coriolus versicolor* means of microwave-assisted extraction and optimized a central composite experimental design.

Smiderle et al. (2017) compared the two powerful extraction techniques, microwave-assisted extractions and pressurized liquid extractions, to obtain β-d-glucans from *Pleurotus ostreatus* and *Ganoderma lucidum* fruiting bodies. Although both methods were efficient, quick, and easy, but its postextraction treatments required some extra separation steps. Another recent advancement in extraction strategies involves molecular imprinting. Hashim et al. (2016) applied this technique to extract ergosterol from *Ganoderma tsugae*. Moreover, they reported an indistinct increase in the ergosterol amount extracted from *G. tsugae* as compared to that obtained from conventional extraction techniques. Overall, representation of extraction, isolation, and identification of metabolites can be summarized as represented in Fig. 8.2.

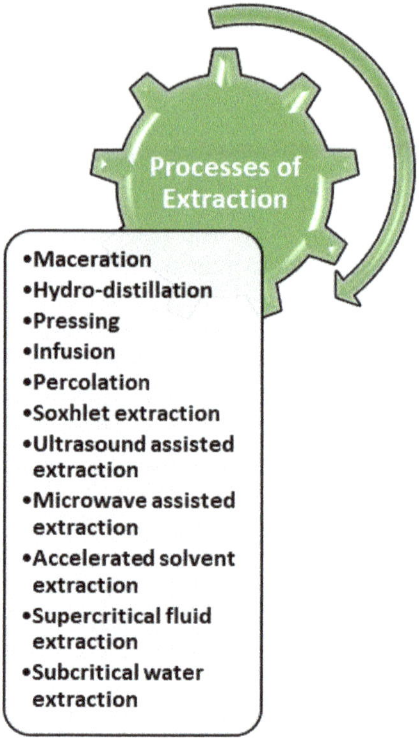

Fig. 8.1 Methods of extraction process. (Roselló-Soto et al., 2016)

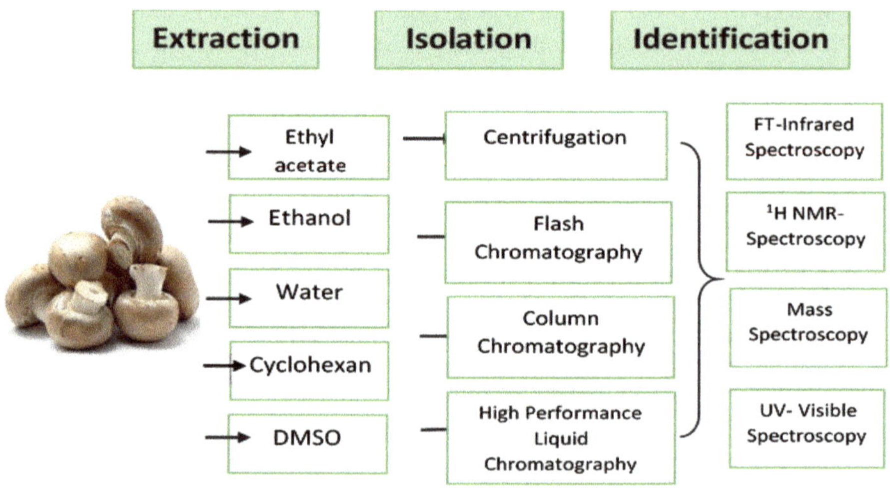

Fig. 8.2 Mushroom metabolites: extraction, isolation, and identification methods

8.2.3.1 Sequential Extraction Methods to Obtain Different Extracts from Shiitake Mushrooms (Morales et al., 2018)

A schematic outline of the extraction methods utilized to obtain different extracts from shiitake mushrooms is given in Fig. 8.3 (Morales et al., 2018):

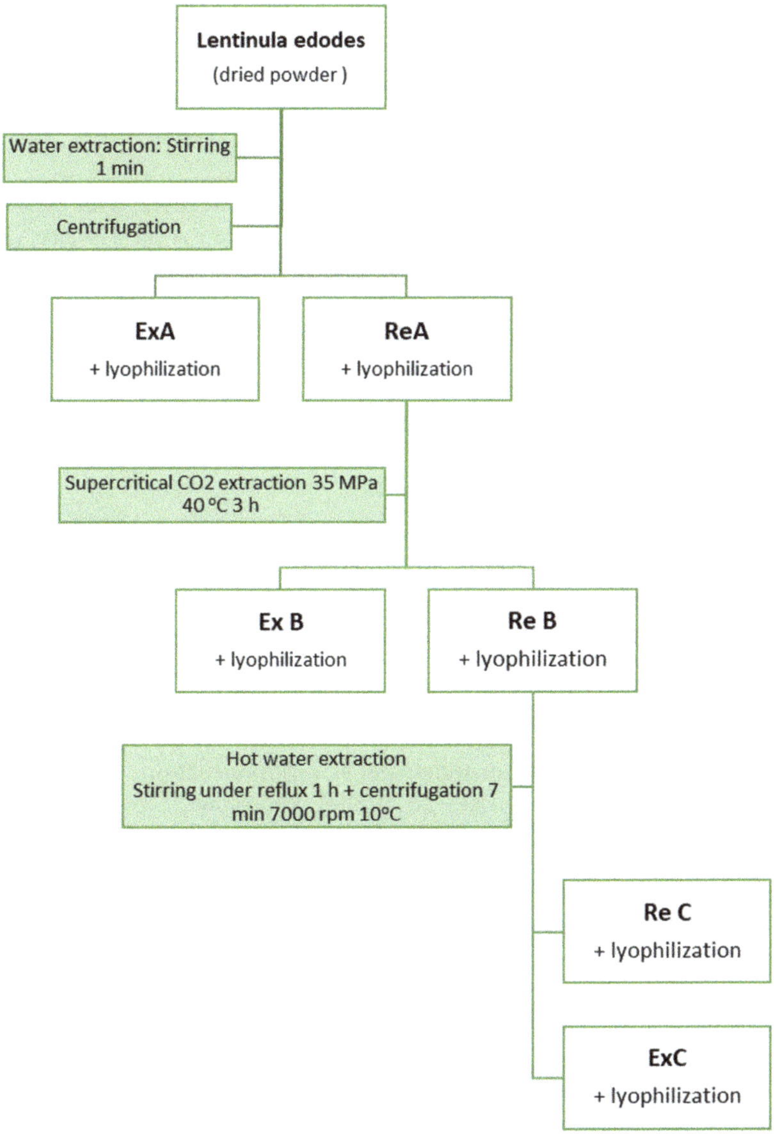

Fig. 8.3 Outline of the sequential extraction method utilized to obtain different extracts from shiitake mushrooms. (Morales et al., 2018).

Step 1. Collect mushroom sample and dry it.
Step 2. Mix mushroom powder with water (50 g/L) at room temperature and stir vigorously.
Step 3. Centrifuge the mixture, thus obtained (7 min, 7000 rpm, 10 °C) and separate the supernatant (ExA) from the residue (ReA). Now, freeze both fractions lyophilizer.
Step 4. Grind freeze-dried ReA and sieve until particles size <0.5 mm and submit it to
Step 5. Supercritical fluid extraction (SFE): Mis ReA (253 g) with 1.9 kg of 5 mm diameter stainless steel spheres (a ratio 1:1 (v/v) extract: spheres) in extraction cell.
Step 6. Collect the extracted material in two different separators.
Step 7. Precipitate both the extracted compounds.
Step 8. At the end of the extraction process, drag the fractions with ethanol and immediately submit them to concentration until dryness on a rotary vacuum evaporator.
Step 9. Dried extract (ExB) will be stored at −20 °C till further analysis and separate the nonextracted residual material (ReB) from steel spheres by sieving.
Step 10. Hot water extraction (98 °C) of ReB (100 g/L) in a flask under vigorous stirring.
Step 11. Separate the soluble (ExC) and nonsoluble (ReC) fractions by centrifugation and freeze-dry.

8.3 Extraction and Isolation of Metabolites from *Tapinella atrotomentosa*

Steps involved in the extraction and isolation of secondary metabolites from *T. atrotomentosa* can be summarized as follows (Béni et al., 2018):

Step 1. Collect the mushroom material, that is, *T. atrotomentosa.*
Step 2. Store the fruiting bodies at −20 °C until processing.
Step 3. Make Methanol extract (11.5 L) of the fruiting bodies of *T. atrotomentosa* (2 kg) and concentrate it.
Step 4. Dissolve the dry methanol extract (90.0 g) in 50% aqueous MeOH (600 mL).
Step 5. Perform solvent-solvent partition with n-hexane and chloroform (5 × 500 mL each) that will yield n-hexane, chloroform, and aqueous MeOH-soluble phases.
Step 6. Then evaporate the chloroform-soluble phase and roughly separate the residue (8.56 g) with flash column chromatography on silica gel column by means of n-hexane–acetone gradient system.
Step 7. Next, combine the obtained fractions on the bases of TLC into several fractions, for example, fractions I–VII in this case.
Step 8. Then, elute one fraction with n-hexane–acetone 85:15 and further separate it by multiple flash chromatography.

Step 9. The resulting two fractions can be finally purified with normal-phase HPLC by means of cyclohexane–isopropanol–water isocratic eluent system to result in compounds 1 and 2.

Step 10. Another fraction (fraction IV) will be subjected to flash column chromatography using n-hexane–acetone gradient system as mobile phase, and further subjected to fractions

Step 11. Further, purification of resulting fractions will be performed by HPLC that will lead to the isolation of compound 3 and so on (Fig. 8.4).

8.4 Extraction and Isolation of Secondary Metabolites from *Agaricus macrosporus*

Major steps involved in extraction and isolation of secondary metabolites from *Agaricus macrosporus* are summarizsed in Fig. 8.5. (Stadler et al., 2006)

Step 1. Collection of mushrooms.

Step 2. Propagation of mushroom in flask with culture media (for small-scale isolation) or in stirring fermenters (for large-scale isolation).

Step 3. Separate the mushroom mycelium from the culture broth by centrifugation.

Step 4. Filtration and extraction with acetone.

Step 5. Remove acetone in vacuo (ca. 40 °C, 250 mbar) to yield an aqueous residue.

Step 6. Dilute the resulting residue with water and subsequently extract it three times with EtOAc.

Step 7. Dry it over Na_2SO_4 and evaporate the combined organic phase in vacuo to yield oily residue.

Step 8. Divide this crude extract into portions and subject it to preparative HPLC and different compounds can be eluted at their respective retention times.

8.5 Secondary Metabolites of Some Mushrooms and Their Applications

8.5.1 Ganoderma *spp.*

Ganoderma, a medicinal mushroom, is splendidly known for its reservoir of abundant secondary metabolites and their biological significance (Sanodiya et al., 2009; Ríos et al., 2012; Baby et al., 2015). In context to Chinese medical manuscripts, various strains of *G. lucidum* have been medicinally applied in the treatment of chronic ailments such as cancer, diabetes, bronchitis, hypertension, arthritis, hepatopathy, insomnia, and nephritis (Fatmawati et al., 2010). Thus far, more than 200 secondary metabolites have been reported essentially from *G. lucidum* (Chen et al., 2012).

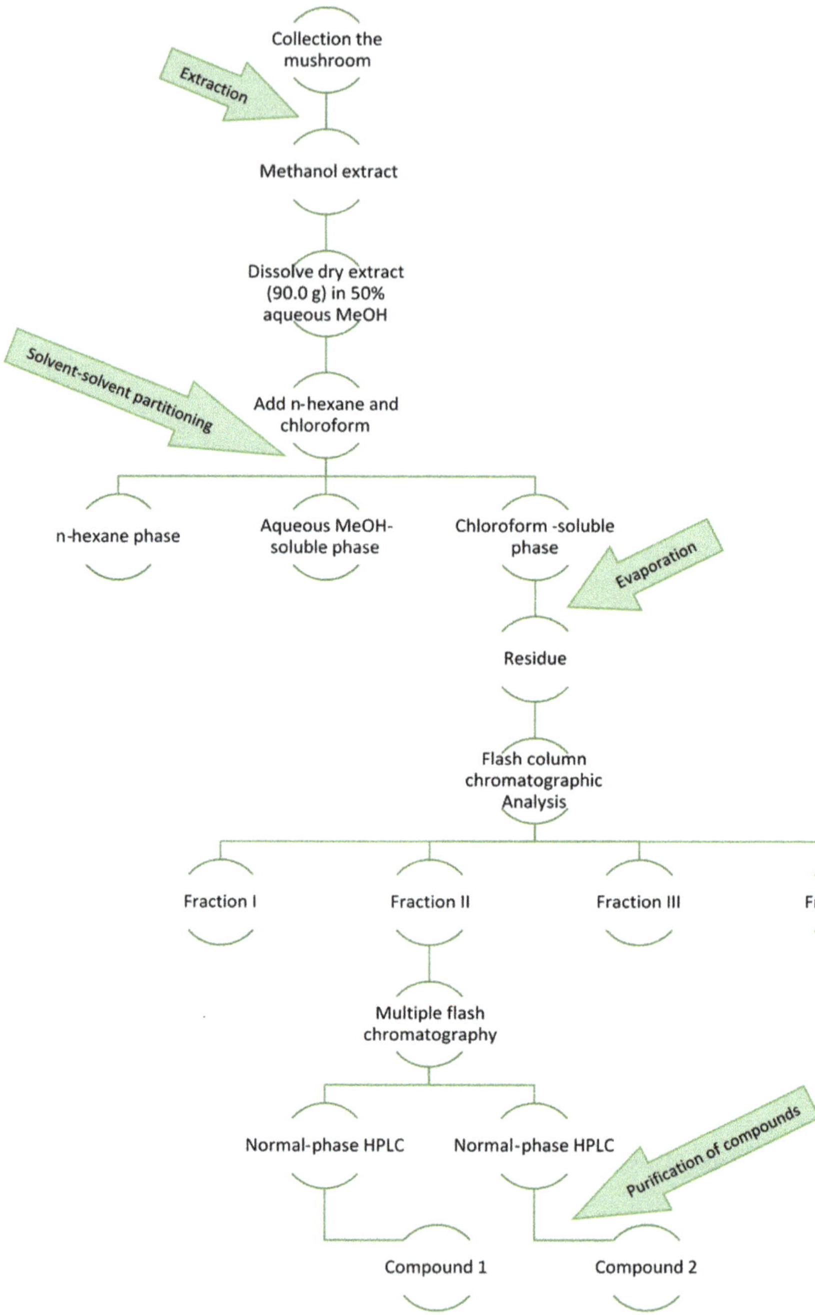

Fig 8.4 Schematic representation of extraction and isolation of metabolites from *Tapinella atrotomentosa.* (Beni et al., 2018)

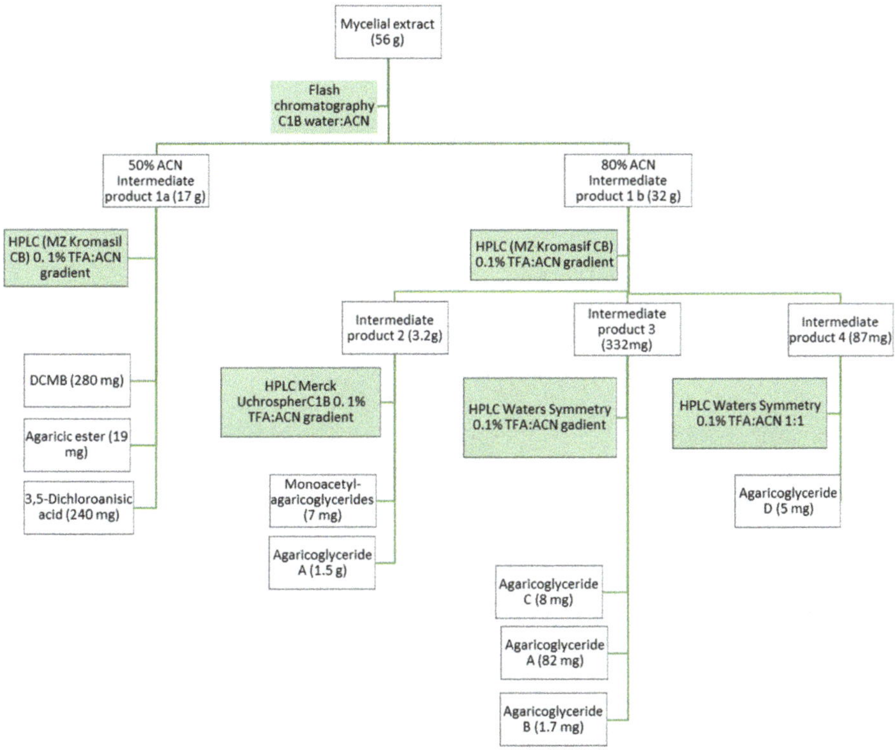

Fig. 8.5 Schematic representation of the isolation of secondary metabolites from mycelial crude extract of *Agaricus macrosporus* (200 litres scale; Q6/2 medium). (Stadler and Hoffmeister, 2015)

According to El Dine et al. (2008a), *Ganoderma colossum* is reported to have anti-HIV-1 protease activities. *Ganoderma applanatum* is very well known for its antitumor activities (Boh et al., 2000), antibacterial properties (Smania et al., 1999), inhibition of Epstein–Barr virus activation (Chairul & Hayashi, 1994), and inhibition of aldose reductase. Smania et al. (2007) reported *Ganoderma austral* to possess antimicrobial activities. Ngai and Ng (2004), described *Ganoderma capense* to be mitogenic and according to El Dine et al. (2008a), *Ganoderma colossum* was testified to embrace anti-HIV-1 protease activities.

Since the last 40 years, various phytochemical studies have led to the identification and isolation of more than 400 bioactive metabolites from several *Ganoderma* species. Some foremost secondary metabolites isolated include alkaloids, meroterpenoids, steroids, C30 lanostanes (aldehydes, ketones, alcohols, glycosides, esters, lactones), C15 sesquiterpenoids, C27 lanostanes (esters, alcohols, lactones), C24 and C25 lanostanes, C27 lanostanes (lucidenic acids), C30 lanostanes (ganoderic acids), farnesyl hydroquinones (meroterpenoids), C30 pentacyclic triterpenes, prenyl hydroquinone, benzopyran-4-one derivatives, benzofurans, and benzenoid derivatives (Baby et al., 2015) as represented in Fig. 8.6. Besides, some common

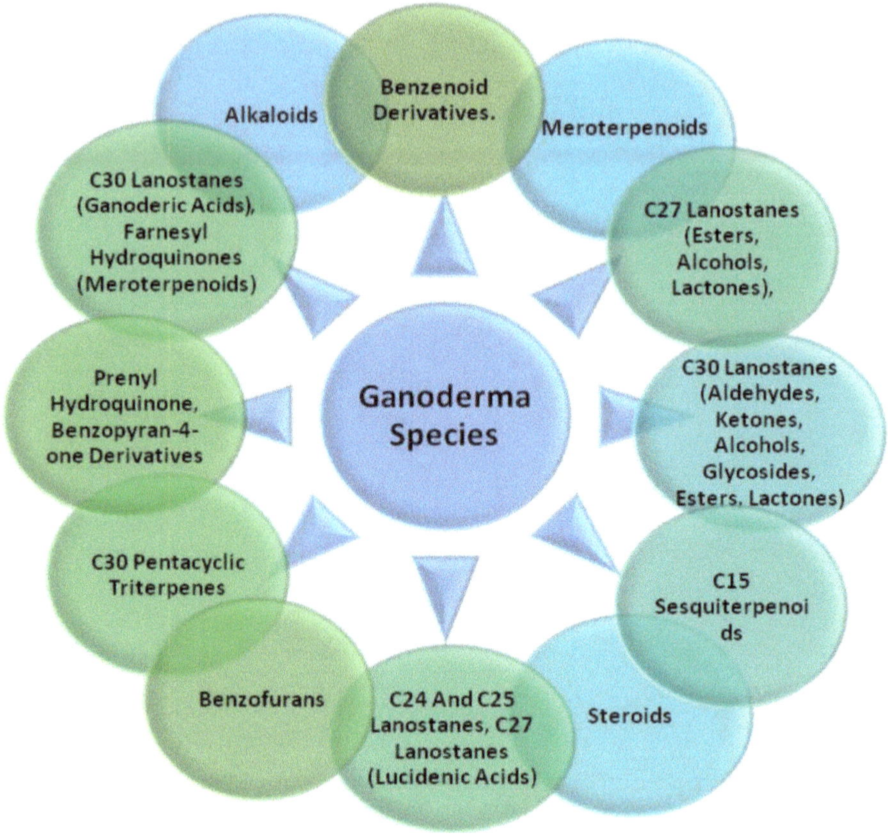

Fig. 8.6 Major groups of secondary metabolites found in *Ganoderma* spp

Ganoderma species that are exclusively subjected to phytochemical analysis include *Ganoderma annulare, Ganoderma australe, Ganoderma amboinense, Ganoderma applanatum, Ganoderma boninense Ganoderma colossum, Ganoderma capense Ganoderma concinna, Ganoderma cochlear, Ganoderma fornicatum, Ganoderma hainanense, Ganoderma lipsiense Ganoderma neo-japonicum, Ganoderma mastoporum, Ganoderma orbiforme, Ganoderma pfeifferi, Ganoderma resinaceum, Ganoderma sinense, Ganoderma tsugae, Ganoderma tropicum,* and *Ganoderma theaecolum.* Various secondary metabolites isolated from *Ganoderma* species are listed in Table 8.1 for better understanding.

Table 8.1 Secondary metabolites isolated from *Ganoderma* species

S. no	Compound name	Compound structure	*Ganoderma* species	References
1.	Ganoderic acid A	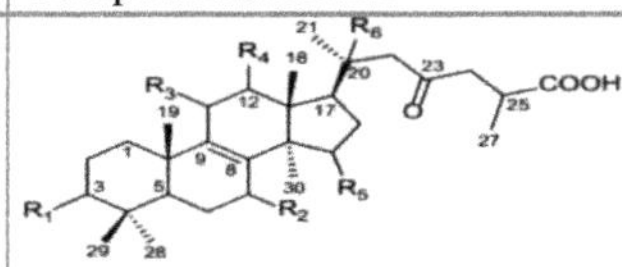 $R_1 = O$, $R_2 = \beta\text{-OH}$, $R_3 = O$, $R_4 = H$, $R_5 = \alpha\text{-OH}$, $R_6 = H$	*G. sinense*	Liu et al. (2012b)
2.	Ganoderic acid C1	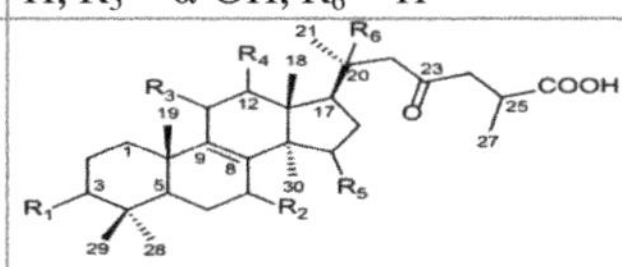 $R_1 = O$, $R_2 = \beta\text{-OH}$, $R_3 = O$, $R_4 = H$, $R_5 = O$, $R_6 = H$	*G. lucidum*	Seo et al. (2009)
3.	Ganoderic acid C2	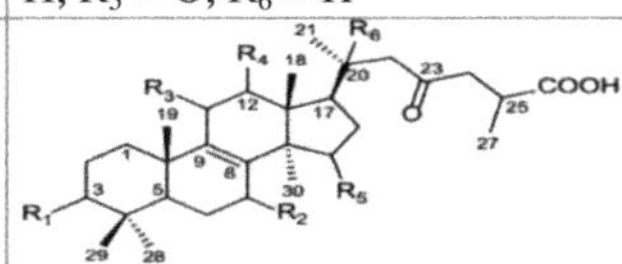 $R_1 = \beta\text{-OH}$, $R_2 = \beta\text{-OH}$, $R_3 = O$, $R_4 = H$, $R_5 = \alpha\text{-OH}$, $R_6 = H$	*G. lucidum*	Min et al. (2000)
4.	Ganoderic acid E	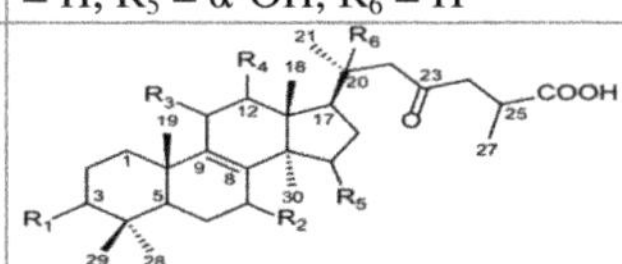 $R_1 = O$, $R_2 = O$, $R_3 = O$, $R_4 = H$, $R_5 = O$, $R_6 = H$	*G. sinense*	Liu et al. (2012b)
5.	Ganoderic acid F	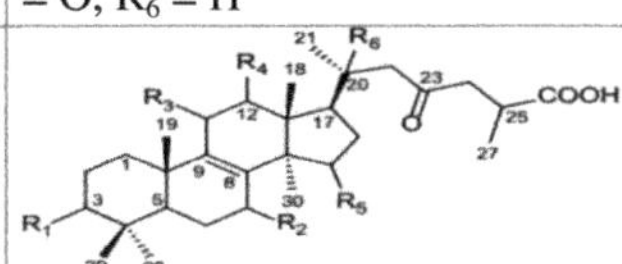 $R_1 = O$, $R_2 = O$, $R_3 = O$, $R_4 = \beta\text{-OAc}$, $R_5 = O$, $R_6 = H$	*G. amboinense*	Yang et al. (2012)
6.	Ganoderic acid AM1	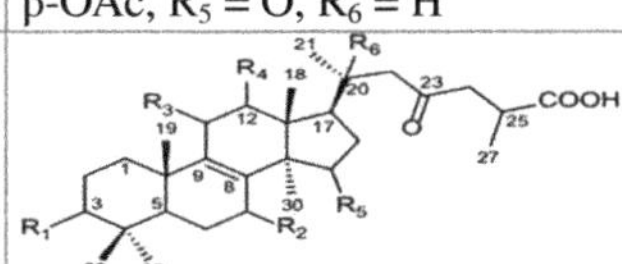 $R_1 = \beta\text{-OH}$, $R_2 = O$, $R_3 = O$, $R_4 = H$, $R_5 = O$, $R_6 = H$	*G. lucidum*	Cheng et al. (2010)
7.	Ganoderic acid AP3	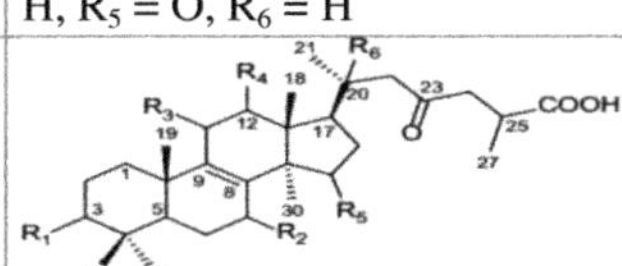 $R_1 = O$, $R_2 = O$, $R_3 = O$, $R_4 = H$, $R_5 = \alpha\text{-OH}$, $R_6 = \xi\text{-OH}$	*G. applanatum*	Wang and Liu (2008)

(continued)

Table 8.1 (continued)

S. no	Compound name	Compound structure	*Ganoderma* species	References
8.	Ganoderic acid Df	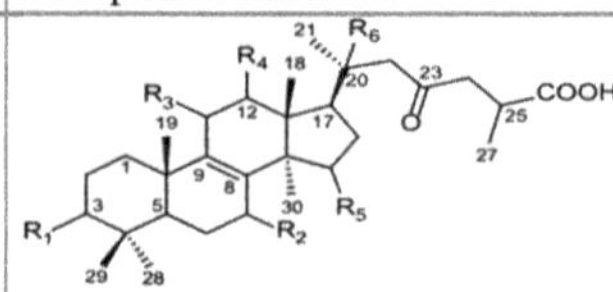 R_1 = O, R_2 = β-OH, R_3 = β-OH, R_4 = H, R_5 = O, R_6 = H	*G. lucidum*	Fatmawati et al. (2010)
9.	12-Hydroxy ganoderic acid C2	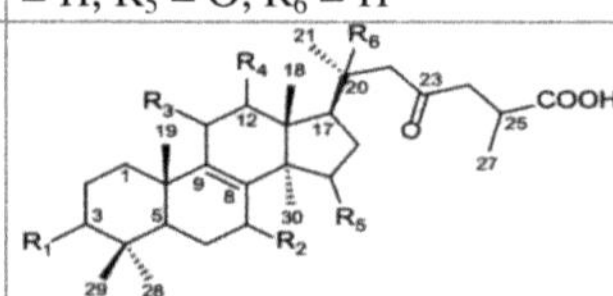 R_1 = β-OH, R_2 = β-OH, R_3 = O, R_4 = OH, R_5 = α-OH, R_6 = H	*G. lucidum*	Yang et al. (2007)
10.	20-Hydroxy ganoderic acid AM1	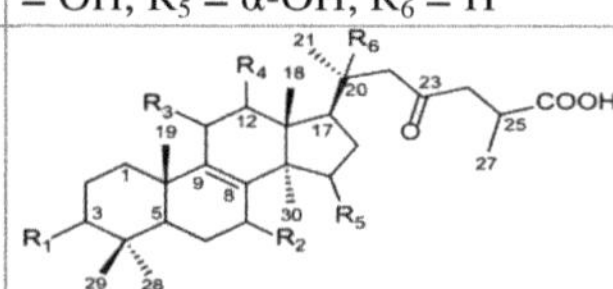 R_1 = β-OH, R_2 = O, R_3 = O, R_4 = H, R_5 = O, R_6 = ξ-OH	*G. theaecolum*	Liu et al. (2014)
11.	3-O-Acetyl ganoderic acid B	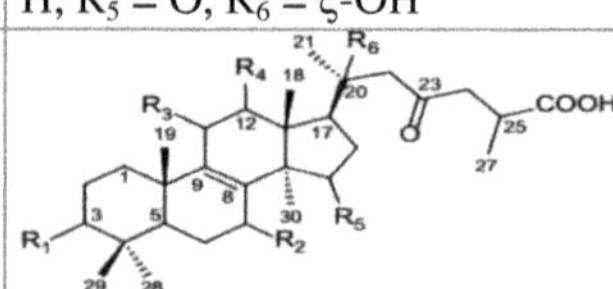 R_1 = β-OAc, R_2 = β-OH, R_3 = O, R_4 = H, R_5 = O, R_6 = H	*G. lucidum*	Li et al. (2009)
12.	3-O-Acetyl ganoderic acid H	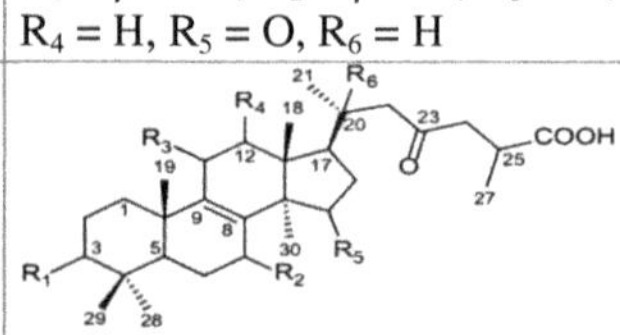 R_1 = β-OAc, R_2 = O, R_3 = O, R_4 = β-OAc, R_5 = O, R_6 = H	*G. lucidum*	Yang et al. (2007)
13.	3-O-Acetyl ganoderic acid K	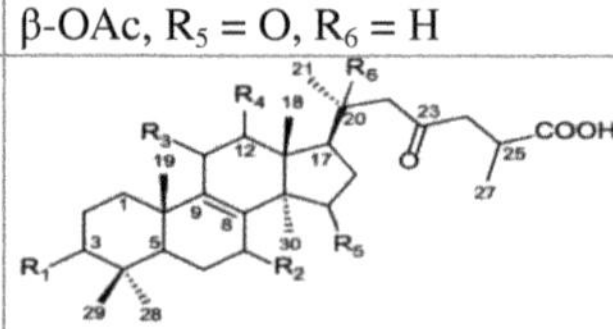 R_1 = β-OAc, R_2 = O, R_3 = O, R_4 = H, R_5 = α-OH, R_6 = H	*G. lucidum*	Li et al. (2009)
14.	12-Acetoxy ganoderic acid D	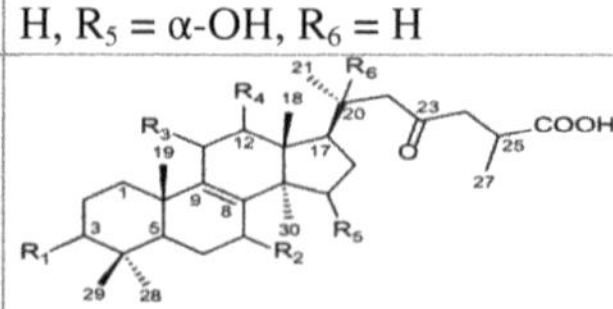 R_1 = O, R_2 = β-OH, R_3 = O, R_4 = OAc, R_5 = O, R_6 = H	*G. lucidum*	Yang et al. (2007)

(continued)

Table 8.1 (continued)

S. no	Compound name	Compound structure	*Ganoderma* species	References
15.	Ganolucidic acid B	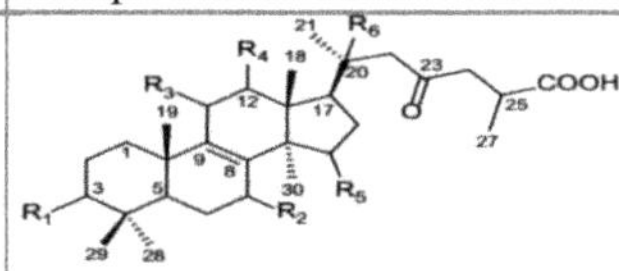 R_1 = β-OH, R_2 = H, R_3 = O, R_4 = H, R_5 = α-OH, R_6 = H	*G. sinense*	Liu et al. (2012b)
16.	12 b-Hydroxy-3,7,11,15,23-pentaoxo-5a-lanosta-8-en-26-oic acid	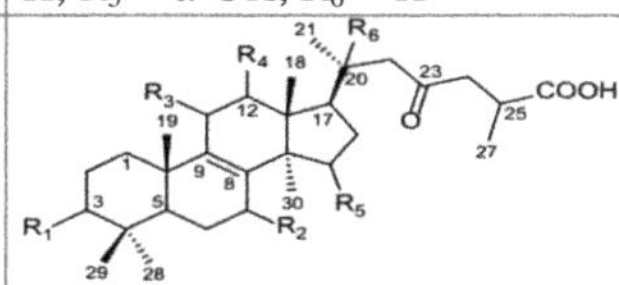 R_1 = O, R_2 = O, R_3 = O, R_4 = β-OH, R_5 = O, R_6 = H	*G. lucidum*	Cheng et al. (2010)
17.	12,15-Bis(acetyloxy)-3-hydroxy-7,11,23-trioxo-lanost-8-en-26-oic acid	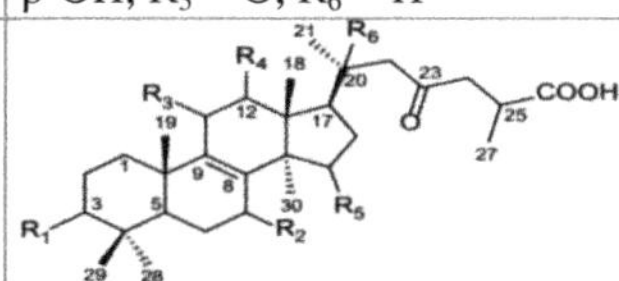 R_1 = OH, R_2 = O, R_3 = O, R_4 = OAc, R_5 = OAc, R_6 = H	*G. lucidum*	Yang et al. (2007)
18.	Ganoderic acid V	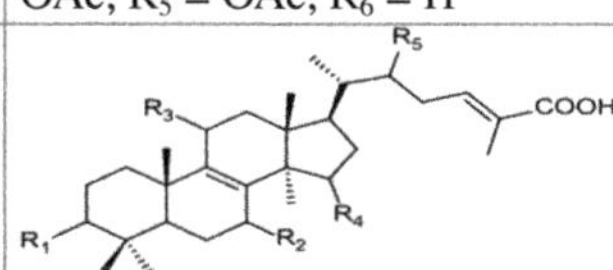 R_1 = O, R_2 = α-OH, R_3 = H, R_4 = α-OAc, R_5 = H	*G. orbiforme*	Isaka et al. (2013)
19.	Ganoderic acid GS-2	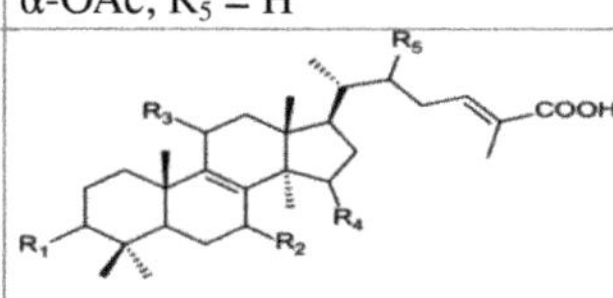 R_1 = O, R_2 = β-OH, R_3 = O, R_4 =α-OH, R_5 = H	*G. sinense*	Sato et al. (2009)
20.	Ganoderic acid GS-1	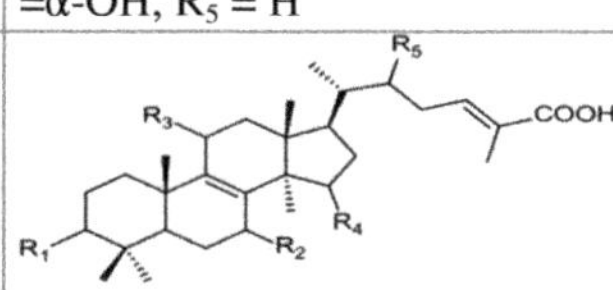 R_1 = O, R_2 = β-OH, R_3 = O, R_4 = O, R_5 = H	*G. sinense*	Sato et al. (2009)

8.5.2 Lentinula edodes *(Shittake Mushroom)*

Lentinula edodes has recently garnered international recognition because of its high nutritional value and due to the presence of invaluable bioactive compounds, although it has been cultivated across Asian countries from time immemorial (Royse et al., 2017; Raut, 2019). Studies related to Chinese traditional medicine reveal that Shittake mushroom had been working as therapeutic in the form of tonics to treat aging-associated sicknesses like pain and fatigue; moreover, they had beneficial effect on heart health and lung diseases (Money, 2016). Current research findings have elucidated the presence of high-molecular-weight compounds like peptides and polysaccharides (Finimundy et al. 2014) as well as small secondary metabolites (Emi Fukushima-Sakuno, 2020) that impart these mushrooms with health benefits avowed in traditional medicine. Nevertheless, some secondary metabolites with significant pharmacological possessions have been listed in Fig. 8.7 and are briefly explained below.

8.5.2.1 Polyacetylenes

L. edodes are found to be rich in with various polyacetylenes including octa-2,3-diene-5,7-diyne-1-ol (Lentinamycin), a compound with potent antimicrobial potential tested against *Trichoderma spp.* (Komemushi et al., 1996; Tokimoto & Komatsu, 1995). Another polyacetylene, 4-Hydroxyundeca-5,6-diene-8,10-diynoic acid (Cortinellin), has been isolated from a culture filtrate of *Cortinellus shiitake*, which is known for its antimicrobial action against mycobacteria, gram-positive as well as gram-negative bacteria, and fungi (MIC 1.6 μM) (Herrmann, 1962; Dembitsk & Maoka, 2007). Apart from these, several research findings have revealed the abundance of numerous polyacetylenes, as mentioned in Fig. 8.7 that possess potential bioactivity.

8.5.2.2 Sulfurous Compounds

Various sulfurous compounds occurring in *L. edodes* include 1,2,3,5,6-pentathiepane (Lenthionine), 1,2,4,6-tetrathiepane, 1,2,3,4,5,6-hexathiepane (Morita & Kobayashi, 1967), that embrace potent antimicrobial properties. A study recorded that lenthionine had inhibitory effect on liver damage and could be effectively used in treating thrombosis (Kumagai et al., 2013). Chen et al. (2015) also recognized several sulfur compounds in chloroform extracts of *L. edodes* including 1,3,5-Trithiane, 1,2,3,5-Tetrathiane, Dimethyl tetrasulfide Dimethyl disulfide, and others.

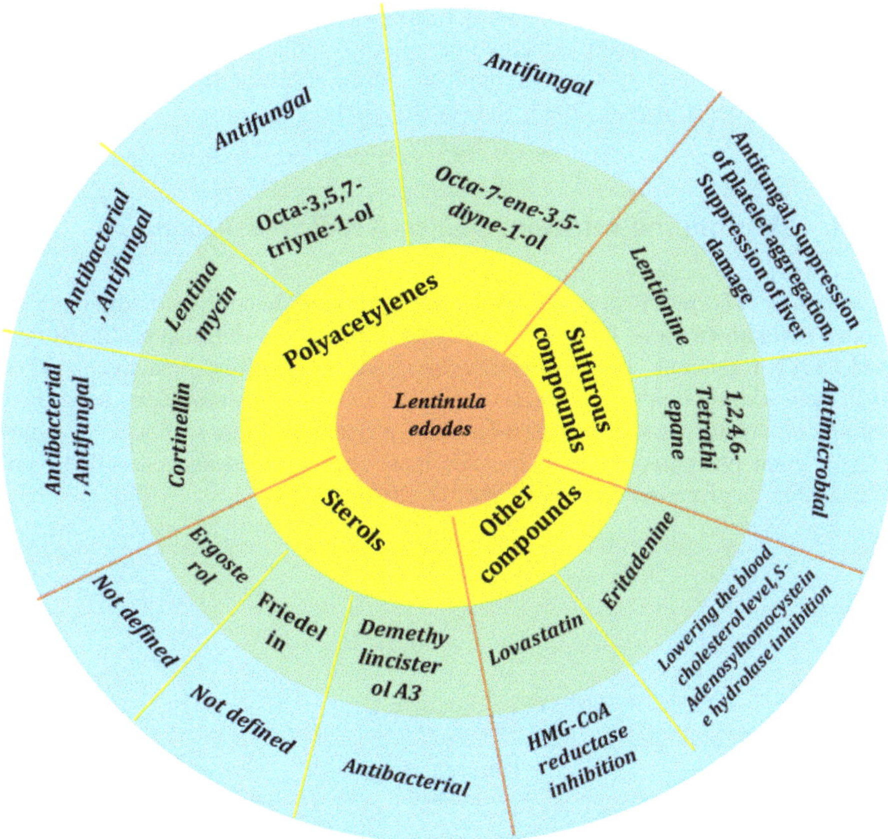

Fig. 8.7 Bioactive secondary metabolites and their related compounds from *Lentinula edodes*

8.5.2.3 Sterols

Ergosterol, a precursor of vitamin D2, is another group of abundant secondary metabolite occurring in L. edodes. However, various other terpenes and sterols have also been reported from *L. edodes*. This is known to possess Friedelin, Demethylincisterol A3, Ergosterol, Methylcholesta 7,22-diene-3β,5α,6β-triol, and, Methylcholesta 6,22-diene-3β,5α,8α -triol are some sterols isolated and identified from acetone extracts of fruiting bodies of L. edodes (Chen et al., 2015).

8.5.2.4 Other Compounds

Eritadenine, a nucleic acid derivative isolated from the fruiting body of *L. edodes*, is shown to inhibit angiotensin-converting enzyme and S-adenosyl-L-homocysteine hydrolase (Afrin et al., 2016). Various researches have also revealed the presence of lovastatin in many mushrooms including L. edodes (Kała et al., 2020). Lovastatin is an explicit inhibitor of hydroxymethylglutaryl-CoA reductase enzyme involved in catalysis of cholesterol synthesis in the liver; hence, it lowers the concentration of blood cholesterol (Lin et al., 2013; Kała et al., 2020).

8.5.3 Flammulina velutipe *(Golden Needle mushroom)*

*Flammulina velutipes***,** also called enokitake, is worldwide cultivated edible mushrooms, best known for its flavorsome taste and prodigious nutritional profile (Royse et al., 2017; Raut et al., 2019). Owing to the presence of diverse bioactive metabolites, they exhibit numerous pharmacological activities, like antioxidant, antitumor, immunomodulatory, and cholesterol-lowering properties. Many secondary metabolites of great pharmacologic significance have been isolated and classified from *F. velutipes*, as given in Fig. 8.8 (Tang et al., 2016).

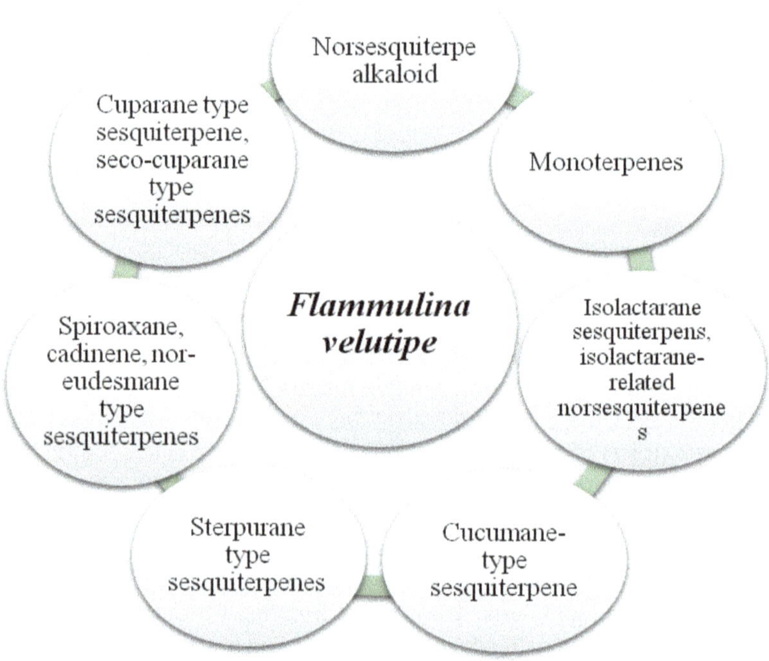

Fig. 8.8 *Flammulina. velutipes*: Secondary metabolites of pharmacologic significance

8.5.3.1 Cuparane Type Sesquiterpenes, Seco-Cuparane Type Sesquiterpenes

Flammulin velutipes are known to possess as sesquiterpenes with highly oxygenated cuparene. Ishikawa et al. isolated Enokipodins A, B, C, and D from a culture filtrate of F. velutipes. Tabuchi et al. studied the bioactivity of these three compounds in bioassays using liquid medium, where they were found to be antibacterial against B. subtilis (IC50 of 13–45 μM) (Tabuchi et al., 2020). Enokipodins B is also known to suppress spore germination of a few plant-pathogenic fungi (Tabuchi et al., 2020). Moreover, Enokipodins B and D also have strong antitumor activity and antiproliferative activity against several transformed cell lines (tsFT210, HL60, HeLa; IC50 of 4.9–18.7 μM) (Tabuchi et al., 2020). Furthermore, enokipodins E, F, G, H, I, and, J were also isolated from rice substrate fermented by *F. velutipes*. Enokipodin F, G, and I have shown antifungal activity against highly pathogenic fungi *Aspergillus fumigatus* (IC50 of 229–235 μM) (Wang et al., 2012). Flamvelutpenoids E & F (with cuparane skeleton) and flammufuranones A & B (secocuparane sesquiterpene) were also isolated from rice fermented *F. velutipes* (Tao et al., 2016).

8.5.3.2 Spiroaxane, Cadinene, Nor-Eudesmane-Type Sesquiterpenes

Flammuspirone A, 7,13,14-Trihydroxy-4-cadinen-15-oic acid methyl ester, Flammuspirone C and, 1,2,6,10-Tetrahydroxy-3,9-epoxy-14-nor-5(15)-eudesmane have been isolated from *F. velutipes* and they are known to show inhibitory activity against HMG-CoA reductase (Tao et al., 2016). Flammuspirone D, E, and H isolated from the same showed DPP4 inhibitory activity, with IC50 83.7, 70.9, and 79.7 respectively (Tao et al., 2016).

8.5.3.3 Sterpurane-Type Sesquiterpenes

Wang et al. (2012) isolated sterpuric acid and sterpurane sesquiterpenes, sterpurols A and B from *F. velutipes*. However, as per the literature, these three compounds have not shown any antimicrobial activity or antioxidative activity so far. Although sterpuric acid had shown weak antibacterial activity against MRSA and *B. subtilis* (Wang et al., 2012).

8.5.3.4 Cucumane-Type Sesquiterpene

Another cucumane-type sesquiterpene being isolated from rice fermented by *F. velutipes* is Flamulinol A. It is known to exhibit weak antibacterial activity against MRSA and B. subtilis (Wang et al., 2012).

8.5.3.5 Isolactarane Sesquiterpens, Isolactarane-Related Norsesquiterpenes

An isolactarane type sesquiterpene, flammulinolide, and a few isolactarane-related norsesquiterpenes, flammulinolides B–G have been reported from rice substrates fermented by *F. velutipes* (Wang et al., 2012). Flammulinolide A is known to be strongly cytotoxic against KB cells (IC50 of 3.9 μM) and moderately cytotoxic against HepG2 cells (IC50 of 34.7 μM), as per the literatures. Also, all of these compounds are known to possess antibacterial potentials against MRSA and *B. subtilis*.

8.5.3.6 Norsesquiterpe Alkaloid

A norsesquiterpe alkaloid, (R)-8-hydroxy-4,7,7-trimethyl-7,8-dihydrocyclopenta[e] isoindole-1,3 (2H, 6H) -dione has been reported from *F. velutipes* and it has shown to exhibit cytotoxicity against KB cells at an IC50 of 16.6 μM (Xu et al., 2013).

8.5.3.7 Monoterpenes

Some growth-promoting low-molecular-weight monoterpentriols have been successfully isolated from the fruiting body of *F. velutipes* viz., (1R,2R,4R,8S)-(-)-p-Menthane-2,8,9-triol and (1R,2R,4R,8R)-(-)-p-Menthane-2,8,9-triol (Hirai et al., 1998). Some important bioactive secondary metabolites isolated from *F. velutipes* have been listed in Table 8.2.

Table 8.2 Some important bioactive secondary metabolites isolated from *F. velutipes*

S. N.	Compound	Source	Bioactivity
1.	Enokipodin A	Liquid culture filtrate	Antibacterial, antifungal, antimalarial, cytotoxic
2.	Enokipodin B	Liquid culture filtrate	Antibacterial
3.	Enokipodin D	Solid culture substrate	Antifungal cytotoxic ntioxidative
4.	Enokipodin E	Solid culture substrate	Not known
5.	Enokipodin F	Solid culture substrate	Antifungal
6.	Enokipodin I	Solid culture substrate	Antibacterial, antifungal
7.	Enokipodin J	Solid culture substrate	antibacterial

(continued)

Table 8.2 (continued)

S. N.	Compound	Source	Bioactivity
8.	2,5-Cuparadiene-1,4-dione	Solid culture substrate	Cytotoxic antioxidative
9.	Flamvelutpenoid A	Solid culture substrate	Weak antibacterial
10.	Flammuspirone A	Solid culture substrate	HMG-CoA reductase inhibition
11.	1,2,6,10-Tetrahydroxy-3,9-epoxy-14-nor-5(15)-eudesmane		HMG-CoA reductase inhibition DPP-4 Inhibition
12.	7,13,14-Trihydroxy-4-cadinen-15-oic acid methyl ester		
13.	Flammuspirone D	Solid culture substrate	DPP-4 Inhibition
14.	Sterpuric acid	Solid culture substrate	Cytotoxic Weak, antibacterial
15.	Flammulinol A	Solid culture substrate	Weak antibacterial
16.	Flammulinolide A		Cytotoxic
17.	Flammulinolide B		Weak antibacterial
18.	(R)-8-Hydroxy-4,7,7-trimethyl-7,8-dihydrocyclopenta[e] isoindole1,3(2H,6H)-dion		Cytotoxic
19.	(1R,2R,4R,8S)-(-)-p-Menthane-2,8,9-triol	Fresh fruiting bodies	Growth promoting and inhibitory activity against stipe of enokitake
20.	(1R,2R,4R,8R)-(-)-p-Menthane-2,8,9-triol	Fresh fruiting bodies	Growth promoting and inhibitory activity against stipe of enokitake

References

Afrin, S., Rakib, M. A., Kim, B. H., Kim, J. O., & Ha, Y. L. (2016). Eritadenine from edible mushrooms inhibits activity of angiotensin converting enzyme in vitro. *Journal of Agricultural and Food Chemistry, 64*, 2263–2268.

Baby, S., Johnson, A. J., & Govindan, B. (2015). Secondary metabolites from *Ganoderma*. *Phytochemistry, 114*, 66–101.

Béni, Z., Dékány, M., Kovács, B., Csupor-Löffler, B., Zomborszki, Z. P., Kerekes, E., Szekeres, A., Urbán, E., Hohmann, J., & Ványolós, A. (2018). Bioactivity-guided isolation of antimicrobial and antioxidant metabolites from the mushroom *Tapinella atrotomentosa*. *Molecules: A Journal of Synthetic Chemistry and Natural Product Chemistry, 23*, 1082.

Bérdy, J. (2012). Thoughts and facts about antibiotics: Where we are now and where we are heading. *The Journal of Antibiotics, 65*, 385–395.

Boh, B., Hodzar, D., Dolnicar, D., Berovic, M., & Pohleven, F. (2000). Isolation and quantification of triterpenoid acids from *Ganoderma applanatum* of Istrianorigin. *Food Technology and Biotechnology, 38*, 11–18.

Boonsong, S., Klaypradit, W., & Wilaipun, P. (2016). Antioxidant activities of extracts from five edible mushrooms using different extractants. *Agriculture and Natural Resources, 50*, 89–97.

Bubalo, C. M., Vidović, S., Redovniković, I. R., & Jokić, S. (2018). New perspective in extraction of plant biologically active compounds by green solvents. *Food and Bioproducts Processing, 109*(52–73), 109. https://doi.org/10.1016/j.fbp.2018.03.001

Cantley, M. F. (1997). International instruments, intellectual property and collaborative exploitation of genetic resources. In S. Wrigley (Ed.), *Phytochemical diversity: A source of new industrial products* (pp. 141–157). *Royal Society of Chemistry*.

Chairul, S. M., & Hayashi, Y. (1994). Lanostanoid triterpenes from *Ganoderma applanatum*. *Phytochemistry, 35*, 1305–1308.

Chen, J., Wei, S. L., & Gao, K. (2015). Chemical constituents and antibacterial activities of compounds from *Lentinus edodes*. *Chemistry of Natural Compounds, 51*, 592–594.

Chen, S., Xu, J., Liu, C., Zhu, Y., Nelson, D. R., Zhou, S., Li, C., Wang, L., Guo, X., Sun, Y., Luo, H., Li, Y., Song, J., Henrissat, B., Levasseur, A., Qian, J., Li, J., Luo, X., Shi, L., … Sun, C. (2012). Genome sequence of the model medicinal mushroom Ganoderma lucidum. *Nature Communications, 3*, 913.

Cheng, C. R., Yue, Q. X., Wu, Z. Y., Song, X. Y., Tao, S. J., Wu, X. H., Xu, P. P., Liu, X., Guan, S. H., & Guo, D. A. (2010). Cytotoxic triterpenoids from *Ganoderma lucidum*. *Phytochemistry, 71*, 1579–1585.

Dembitsk, V. M., & Maoka, T. (2007). Allenic and cumulenic lipids. *Progress in Lipid Research, 46*, 328–375.

Fatmawati, S., Shimizu, K., & Kondo, R. (2010). Ganoderic acid Df, a new triterpenoid with aldose reductase activity from the fruiting body of *Ganoderma lucidum*. *Fitoterapia, 81*, 1033–1036.

Feeney, M. J., Miller, A. M., & Roupas, P. (2014). Mushrooms – Biologically distinct and nutritionally unique: Exploring a "third food kingdom". *Nutrition Today, 49*(6), 301–307.

Finimundy, T. C., Dillon, A. J. P., Henriques, J. A. P., & Ely, M. R. (2014). A review on general nutritional compounds and pharmacological properties of the *Lentinula edodes* mushroom. *Food and Nutrition Sciences, 5*, 1095–1105.

Fukushima-Sakuno, E. (2020). Bioactive small secondary metabolites from the mushrooms *Lentinula edodes* and *Flammulina velutipes*. *The Journal of Antibiotics, 73*. https://doi.org/10.1038/s41429-020-0354-x

Gargano, M. L., van Griensven, L. J. L. D., Isikhuemhen, O. S., Lindequist, U., Venturella, G., Wasser, S. P., & Zervakis, G. I. (2017). Medicinal mushrooms: Valuable biological resources of high exploitation potential. *Plant Biosystems, 151*, 548–565.

Hashim, S. N. N. S., Schwarz, L. J., Danylec, B., Mitri, K., Yang, Y., Boysen, R. I., & Hearn, M. T. W. (2016). Recovery of ergosterol from the medicinal mushroom, *Ganoderma tsugae, var. Janniae,* with a molecularly imprinted polymer derived from a cleavable monomer-template composite. *Journal of Chromatography A, 1468*, 1–9.

Hassainia, A., Satha, H., & Boufi, S. (2018). Chitin from *Agaricus bisporus:* Extraction and characterization. *International Journal of Biological Macromolecules, 117*, 1334–1342.

Herrmann, H. (1962). Cortinellin, eine antibiotisch wirkasame Substanz aus *Cortinellus shiitake*. *Naturwissenschaften, 49*, 542.

Hirai, Y., Ikeda, M., Murayama, T., & Ohata, T. (1998). New monoterpentriols from the fruiting body of *Flammulina velutipes*. *Bioscience, Biotechnology, and Biochemistry, 62*, 1364–1368.

Isaka, M., Chinthanom, P., Kongthong, S., Srichomthong, K., & Choeyklin, R. (2013). Lanostane triterpenes from cultures of the basidiomycete *Ganoderma orbiforme* BCC 22324. *Phytochemistry, 87*, 133–139.

Kaewnarin, K., Suwannarach, N., Kumla, J., & Lumyong, S. (2016). Phenolic profile of various wild edible mushroom extracts from Thailand and their antioxidant properties, anti-tyrosinase and hyperglycaemic inhibitory activities. *Journal of Functional Foods, 27*, 352–364.

Kała, K., Kryczyk-Poprawa, A., Rzewińska, A., & Muszyńska, B. (2020). Fruiting bodies of selected edible mushrooms as a potential source of lovastatin. *European Food Research and Technology, 246*, 713–722.

Keswani, C., Mishra, S., Sarma, B. K., Singh, S. P., & Singh, H. B. (2014). Unravelling the efficient applications of secondary metabolites of various *Trichoderma* spp. *Applied Microbiology and Biotechnology, 98*, 533–544.

Kimatu, B. M., Zhao, L., Biao, Y., Ma, G., Yang, W., Pei, F., & Hu, Q. (2017). Antioxidant potential of edible mushroom (*Agaricus bisporus*) protein hydrolysates and their ultrafiltration fractions. *Food Chemistry, 230*, 58–67.

Komemushi, S., Yamamoto, Y., & Fujita, T. (1996). Purification and identification of antimicrobial substances produced by *Lentinus edodes*. *Journal of Antibacterial and Antifungal Agents, 24*, 21–25.

Kumagai, H., Akao, M., & Masuda, H. (2013). *Hepatopathy inhibitor*. Japan Kokai Tokkyo Koho, JP2013103900A.

Li, Y.-Y., Mi, Z.-Y., Tang, Y., Wang, G., Li, D.-S., & Tang, Y. J. (2009). Lanostanoids isolated from *Ganoderma lucidum* mycelium cultured by submerged fermentation. *Helvetica Chimica Acta, 92*, 1586–1593.

Lin, S., Chen, Y., Yu, H., Barseghyan, G. S., Asatiani, M. D., & Wasser, S. P. (2013). Comparative study of contents of several bioactive components in fruiting bodies and mycelia of culinary-medicinal mushrooms. *International Journal of Medicinal Mushrooms, 15*, 315–323.

Liu, L. Y., Chen, H., Liu, C., Wang, H. Q., Kang, J., Li, Y., & Chen, R. Y. (2014). Triterpenoids of *Ganoderma theaecolum* and their hepatoprotective activities. *Fitoterapia, 98*, 254–259.

Maeng, J. H., Muhammad Shahbaz, H., Ameer, K., Jo, Y., & Kwon, J. H. (2016). Optimization of microwave-assisted extraction of bioactive compounds from *Coriolus versicolor* mushroom using response surface methodology. *Journal of Food Process Engineering, 40*. https://doi.org/10.1111/jfpe.12421

Money, N. P. (2016). Are mushrooms medicinal? *Fungal Biology, 120*, 449–453.

Morales, D., Piris, A. J., Ruiz-Rodriguez, A., Prodanov, M., & Soler-Rivas, C. (2018). Extraction of bioactive compounds against cardiovascular diseases from *Lentinula edodes* using a sequential extraction method. *Biotechnology Progress, 34*(3), 746–755.

Morita, K., & Kobayashi, S. (1967). Isolation, structure, and synthesis of lenthionine and its analogs. *Chemical & Pharmaceutical Bulletin, 15*, 988–993.

Raut, J. K. (2019). Current status, challenges and prospects of mushroom industry in Nepal. *International Journal of Agricultural Economics, 4*, 154–160.

Ríos, J. L., Andújar, I., Recio, M. C., & Giner, R. M. (2012). Lanostanoids from fungi: A group of potential anticancer compounds. *Journal of Natural Products, 75*, 2016–2044.

Roselló-Soto, E., Parniakov, O., Deng, Q., Patras, A., Koubaa, M., Grimi, N., & Barba, F. J. (2016). Application of non-conventional extraction methods: Toward a sustainable and green production of valuable compounds from mushrooms. *Food Engineering Reviews, 8*(2), 214–234.

Royse, D. J., Baars, J., & Tan, Q. (2017). Current overview of mushroom production in the world. In D. C. Zied & Pardo-Giménez (Eds.), *Edible and medicinal mushrooms: Technology and applications* (pp. 5–13). Wiley.

Ruthes, A. C., Smiderle, F. R., & Iacomini, M. (2015). D-glucans from edible mushrooms: A review on the extraction, purification and chemical characterization approaches. *Carbohydrate Polymers, 117*, 753–761.

Sanodiya, B. S., Thakur, G. S., Baghel, R. K., Prasad, G. B., & Bisen, P. S. (2009). *Ganoderma lucidum*: A potent pharmacological macrofungus. *Current Pharmaceutical Biotechnology, 10*, 717–742.

Sato, N., Zhang, Q., Ma, C. M., & Hattori, M. (2009). Anti-human immunodeficiency virus-1 protease activity of new lanostane-type triterpenoids from *Ganoderma sinense*. *Chemical & Pharmaceutical Bulletin, 57*, 1076–1080.

Seo, H. W., Hung, T. M., Na, M., Jung, H. J., Kim, J. C., Choi, J. S., Kim, J. H., Lee, H. K., Lee, I., Bae, K., Hattori, M., & Min, B. S. (2009). Steroids and triterpenes from the fruit bodies of *Ganoderma lucidum* and their anti-complement activity. *Archives of Pharmacal Research, 32*, 1573–1579.

Smania, A., Jr., Monache, F. D., Smania, E. F. A., & Cuneo, R. S. (1999). Antibacterial activity of steroidal compounds isolated from *Ganoderma applanatum* (Pers.) Pat. (Aphyllophoromycetideae) fruit body. *International Journal of Medicinal Mushrooms, 1*, 325–330.

Smiderle, F. R., Morales, D., Gil-Ramírez, A., de Jesus, L. I., Gilbert-López, B., Iacomini, M., & Soler-Rivas, C. (2017). Evaluation of microwave-assisted and pressurized liquid extractions to obtain β-d-glucans from mushrooms. *Carbohydrate Polymers, 156*, 165–174.

Stadler, M., Hellwig, V., Mayer-Bartschmid, A., Denzer, D., Wiese, B., & Burkhardt, N. (2006). Novel analgesic triglycerides from cultures of *Agaricus macrosporus* and other basidiomycetes as selective inhibitors of neurolysin. *The Journal of Antibiotics, 58*, 775–786. https://doi.org/10.1038/ja.2005.105

Stadler, M., & Hoffmeister, D. (2015). Fungal natural products—the mushroom perspective. *Frontiers in microbiology, 6*, 127.

Tabuchi, A., Fukushima-Sakuno, E., Osaki-Oka, K., Futamura, Y., Motoyama, T., Osada, H., et al. (2020). Productivity and bioactivity of enokipodins A–D of *Flammulina rossica* and *Flammulina velutipes*. *Bioscience, Biotechnology, and Biochemistry, 84*, 876–886.

Tang, C., Hoo, P. C., Tan, L. T., Pusparajah, P., Khan, T. M., Lee, L., et al. (2016). Golden needle mushroom: A culinary medicine with evidenced based biological activities and health promoting properties. *Frontiers in Pharmacology, 7*, Article 474.

Tao, Q., Ma, K., Yang, Y., Wang, K., Chen, B., Huang, Y., et al. (2016). Bioactive sesquiterpenes from the edible mushroom *Flammulina velutipes* and their biosynthetic pathway confirmed by genome analysis and chemical evidence. *The Journal of Organic Chemistry, 81*(98), 67–77.

Tokimoto, K., & Komatsu, M. (1995). Selection and breeding of Shiitake strains resistant to *Trichoderma* spp. *Canadian Journal of Botany, 73*(S1), 962–966.

Wang, F., & Liu, J. K. (2008). Highly oxygenated lanostane triterpenoids from the fungus *Ganoderma applanatum*. *Chemical & Pharmaceutical Bulletin, 56*, 1035–1037.

Wang, Y., Bao, L., Yang, X., Li, L., Li, S., Gao, H., et al. (2012). Bioactive sesquiterpenoids from the solid culture of the edible mushroom *Flammulina velutipes* growing on cooked rice. *Food Chemistry, 132*, 1346–1353.

Xu, Z. Y., Wu, Z. A., & Bi, K. S. (2013). A novel norsesquiterpene alkaloid from the mushroom-forming fungus *Flammulina velutipes*. *Chinese Chemical Letters, 24*, 57–58.

Yang, M., Wang, X. M., Guan, S. H., Xia, J. M., Sun, J. H., Guo, H., & Guo, D. A. (2007). Analysis of triterpenoids in *Ganoderma lucidum* using liquid chromatography coupled with electrospray ionization mass spectrometry. *Journal of the American Society for Mass Spectrometry, 18*, 927–939.

Yang, S. X., Yu, Z. C., Lu, Q. Q., Shi, W. Q., Laatsch, H., & Gao, J. M. (2012). Toxic lanostane triterpenes from the basidiomycete *Ganoderma amboinense*. *Phytochemistry Letters, 5*, 576–580.

Chapter 9
Fungi as Nutraceutical: Present to Future

Sibashish Kityania, Rajat Nath, Subrata Das, Deepa Nath, Jayanta Kumar Patra, and Anupam Das Talukdar

Contents

9.1 Introduction

"Nutraceutical" has been described as medicinally active or nutritional food supplement since the late 1980s (DeFelice, 1995). The term "nutraceutical" was coined by Stephen De Felice from two different branches of science as "nutrition" and "pharmaceutical" in 1989 (Maddi et al., 2007). "Nutraceuticals" means a combination of "nutrient" and "ceutical," where "nutrient" refers to functional foods having

S. Kityania · R. Nath · S. Das · A. D. Talukdar (✉)
Department of Life Science and Bioinformatics, Assam University, Silchar, Assam, India
e-mail: anupam@bioinfoaus.ac.in

D. Nath
Department of Botany, Guru Charan College, Silchar, Assam, India

J. K. Patra
Research Institute of Biotechnology and Medical Converged Science, Dongguk University-Seoul, Goyangsi, Republic of Korea

A. C. Shukla (ed.), *Applied Mycology*, Fungal Biology,
https://doi.org/10.1007/978-3-030-90649-8_9

medicinal value and "ceutical" refers to therapy (Borodina et al., 2020). Nutraceuticals generally have some pharmaceutical and antiaging properties and the capability of preventing or reducing symptoms of many long-term diseases. Nutraceuticals can also improve overall wellness, along with boosting immunity (Sharma, 2009). Nutraceutical food products are categorized into different subcategories like polyunsaturated fatty acids, probiotics, prebiotics, dietary fibers, and antioxidants. Industries also define any nontoxic functional food as nutraceuticals that have vital benefits on health issues and also have some potential bioactive compounds to treat/prevent many diseases (Das et al., 2012). These are food products having pharmaceutical properties. Nutraceuticals are used as supplements in developed countries as they are associated with many potential bioactive phytochemicals that also have health benefits (Al-Obaidi et al., 2021). There are various nutraceutical compounds also found in some higher plants, bacteria, and algae. There are many studies on functional foods to discover some bioactive compounds that can be used as a nutraceutical and also to make some beverages having great health benefits.

In nutraceutical development, fungi play an important role. Several species of fungi are nutritionally rich and are a good source of bioactive compounds that not only have nutritional values but also drug-like properties. Among these, mushrooms nowadays have become a hot topic among scientists searching for new therapeutic alternatives (Barros et al., 2008). Mushrooms are rapidly finding a place in our regular diet due to nutritional value, low fat, and high protein content (Barros et al., 2007). Other than having nutritional values, fungi also have antiviral, antibacterial, anticancer, immune-booster, and medicinal importance values (Fig. 9.1).

9.2 Fungi: As Nutraceuticals

Modern science considers and also uncovered some new fungi sources of natural nutraceuticals. There are various fungal species that are being used as medicine since a long time in different parts of the world. Modern scientific researchers have found many fungi species that are very rich sources of nutraceutical (Al-Obaidi, 2016). There are some species that have traditional medicinal values, including Chinese *Cordyceps* (Paterson, 2008) and *Ganoderma lucidum* (Wachtel-Galor et al., 2011). This includes mushrooms, a type of fungi having potential bioactive compounds with some functional properties like antioxidants (Kalaras et al., 2017), anticancer compounds (Ajith & Janardhanan, 2007), antibiotics (Kumar & Kaushik, 2012), immunomodulators (Mallard et al., 2019), hypocholesterolemics (Kim et al., 2009), and hypoglycemics (Giavasis, 2014). This also includes *Cordyceps*, a fungi with medicinal properties that possesses several bioactive molecules and is also used as traditional medicine, with cordycepin (Soltani et al., 2018). Cordycepin is a potential bioactive molecule due to some ability to fight diseases like cancer, cardiovascular disease, and diabetics, which makes it an important mushroom (Ashraf et al., 2020). Fungi like *Ganoderma* were also showing activity to prevent various kidney diseases and also used as traditional medicine. Recently, *Ganoderma* has

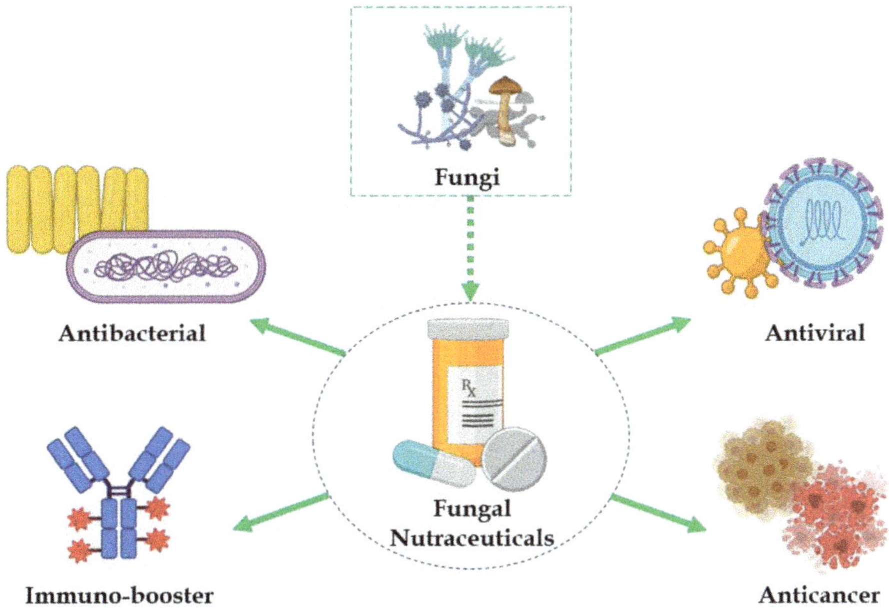

Fig. 9.1 Fungi as nutraceutical and its function

shown some activity having several bioactive compounds for different biological reactions through mass spectrometry studies and spectroscopy studies (Garuba et al., 2020). Research has shown some potential candidates of nutraceuticals that were derived from mushroom and can also be used to ease neurological disorders like mental dysfunction (dementia) and ataxia (Yadav et al., 2020). Beta-glucan is found in various fungal species, which is a common component for nutraceuticals (Kaushik et al., 2019). Cell walls of some plants like oats and barely possess the natural polysaccharide, that is, beta-glucan (Ciecierska et al., 2019). Beta-glucans also show some anticancerous properties that can promote apoptosis in cancer cells (Kobayashi et al., 2005). The antibacterial properties of some fungi are also advantageous to use fungal compounds in different nutraceutical industries. For example, cultured mycelia extracts of *Ophiocordyceps sinensis* has shown some antibacterial potential (Kaushik et al., 2019). Maitake and shiitake are well-known medicinal mushrooms with potential bioactive nutraceutical compounds; however, some other mushrooms like *Cordyceps, Agaricus,* and oyster mushrooms are also considered as potential species since the past decade. Some examples of nutraceutical mushrooms are as follows: *Grifola frondosa,* known as maitaka; *Ganoderma lucidum,* known as reishi; *Lentinula edodes,* known as shiitake; *Cordyceps sinensis,* known as *Cordyceps*; *Trametes versicolor* known as turkey tail; and *Hericium erinaceus,* known as lion's mane. The nutraceutical activity of various fungal species and their bioactive compounds are shown in Table 9.1. Structure of some bioactive compounds from fungal origin used as nutraceuticals shown in Fig 9.3. Future research on fungal nutraceuticals could highlight the natural nutraceuticals from traditionally

Table 9.1 List of different fungal species and its bioactive compounds as nutraceuticals

Sl. no.	Fungal species	Family	Bioactive compounds	Nutraceutical activity	References
1	*Clitocybe maxima*	Tricholomataceae	Quercetin, gallic acid	Antihyperglycemic and antioxidant, α-glucosidase inhibitory activity	Tsai et al. (2009)
2	*Catathelasma ventricosum*	Tricholomataceae	Quercetin, catechin, gallic acid, ergosterol	Antihyperglycemic and antioxidant, anticancer activity	Ohtsuka et al. (1973) and Liu et al. (2012)
3	*Stropharia rugosoannulata*	Strophariaceae	Quercetin	Antihyperglycemic and antioxidant	Liu et al. (2012)
4	*Craterellus cornucopioides*	Cantharellaceae	Quercetin, caffeic acid, gallic acid, ergosterol	Antihyperglycemic and antioxidant	Beluhan and Ranogajec (2011)
5	*Laccaria amethystina*	Hydnangiaceae	Quercetin, catechin	Antihyperglycemic and antioxidant, α-amylase inhibitory activity	Liu et al. (2012)
6	*Russula delica*	Russulaceae	Catechin, ascorbic acid, β-carotene	Antioxidant, antimicrobial activity	Yaltirak et al. (2009) and Kalač (2009)
7	*Russula griseocarnosa*	Russulaceae	Phenolics, ergosterol, β-carotene	Antioxidant	Chen et al. (2010)
8	*Boletus edulis*	Boletaceae	Ergothioneine, ergosterol, fungistero	Antioxidant	Ey et al. (2007)
9	*Ganoderma annulare*	Ganodermataceae	Applanoxidic acid A	Antioxidant, antifungal activity	Smania et al. (2003)
10	*Lentinula edodes (Berk.)*	Marasmiaceae	Oxalic acid, homogeneous polysaccharide	Antioxidant, antifungal activity, antitumor activity	Bender et al. (2003) and Ya (2017)
11	*Podaxis pistillaris (L.: Pers.) Morse*	Agaricaceae	Pipolythiopiperazine-2,5-diones	Antioxidant, antifungal activity	al Fatimi (2001)
12	*Ganoderma lucidum*	Ganodermataceae	Triterpenes, heteropolysaccharide	Antiviral agents, antitumor, antioxidative, antidecrepitude, antidiabetic activity	Lindequist et al. (2005), Zhang et al. (2007) and Xiao et al. (2017)
13	*Ganoderma pfeifferi*	Ganodermataceae	Ganodermadiol, lucidadiol, applanoxidic acid G	In vitro antiviral activity	Lindequist et al. (2005)
14	*Inonotus hispidus*	Hymenochaetaceae	Phenolic compounds, ergosterol	Antiviral agents	Ali et al. (2003)

15	*Collybia maculata*	Lyophyllaceae	Purine derivatives	Antiviral agents	Lindequist et al. (2005)
16	*Piptoporus betulinus (Bull.: Fr.) P. Karst.*	Fomitopsidaceae	Phenolics, flavonoids	Antiparasitic and antimicrobial agent, in vitro anticancer activity	Lemieszek et al. (2009)
17	*Schizophyllum* sp.	Schizophyllaceae	Schizophyllan (β-d-glucan)	Immunomodulation	c Ooi and Liu (2000)
18	*Sparassis* sp.	Sparassidaceae	Bioactiveβ-d-glucan, phenylderivatives, chalcones, sesquiterpenoids	Enhancement of the hematopoietic response and induction of cytokine production	Jiang et al. (2009) and Kimura (2013)
19	*Polyporus* sp.	Polyporaceae	Nucleotides, nucleosides, proteins, amino acids, vitamins, polysaccharides, triterpenoids	Very effective against urethral blockages	Wang et al. (2004)
20	*Pleurotus tuber-regium*	Pleurotaceae	b-D-glucan	Antioxidant, hepatoprotective, anti-breast cancer	Zhang et al. (2007) and Rahi and Malik (2016)
21	*Auricularia auricular*	Auriculariaceae	Glucan	Hyperglycemia, immunomodulating, antitumor	Rahi and Malik (2016)
22	*Schizophyllum commune*	Schizophyllaceae	Glucan, schizophyllan	Antitumor	Rahi and Malik (2016)
23	*Hericium erinaceus*	Hericiaceae	Heteroglycan, heteroglycan peptide	Hyperglycemia, immunomodulating, antitumor	Rahi and Malik (2016)
24	*Lentinus edodes*	Omphalotaceae	Mannogluca, polysaccharide-protein complex, glucan, lentinan	Immunomodulating, antitumor, antiviral	Rahi and Malik (2016)
25	*Sclerotinia sclerotiorum*	Sclerotiniaceae	Glucan, scleroglucan (SSG)	Antitumor	Rahi and Malik (2016)
26	*Polystictus versicolor*	Polyporaceae	Heteroglycan, glycopeptide, krestin (PSK)	Immunomodulating, antitumor, antiradiative, hyperglycemia, anti-inflammatory	Rahi and Malik (2016)

(continued)

Table 9.1 (continued)

Sl. no.	Fungal species	Family	Bioactive compounds	Nutraceutical activity	References
27	*Grifola frondosa*	Meripilaceae	Proteoglycan, glucan, galatomanna, heteroglycan, grifolan	Immunomodulating, antitumor, antiviral, hepatoprotective	Rahi and Malik (2016)
28	*Inonotus obliquus*	Hymenochaetaceae	Glucan	Antitumor, immunomodulating	Rahi and Malik (2016)
29	*Agaricus blazei*	Agaricaceae	Glucan, heteroglycan, glucan protein	Antitumor	Rahi and Malik (2016)
30	*Flammulina velutipes*	Physalacriaceae	Glucan-protein complex, heteropolysaccharide	Antitumor, anti-inflammatory, immunomodulating activity	Rahi and Malik (2016) and Feng et al. (2016)
31	*Polyporus umbellatus*	Polyporaceae	Glucan	Antitumor, immunomodulating	Rahi and Malik (2016)
32	*Clitopilus caespitosus*	Entolomataceae	Glucan	Antitumor	Rahi and Malik (2016)
33	*Trametes coriolus*	Polyporaceae	Polysaccharide, PSP-a glycopeptides	Antitumor and immunostimulant	and Rahi and Malik (2016)
34	*Tricholomopsis rutilans*	Tricholomataceae	Polysaccharides	Anticarcinogenic activity, antioxidative, anti-inflammatory	Rahi and Malik (2016)
35	*Volvariella* sp.	Pluteaceae	Polysaccharide	Cardiac tonic	Rahi and Malik (2016)
36	*Tremella fuciformis*	Tremellaceae	Polysaccharide	Hypocholesteric	Rahi and Malik (2016)
37	*Marasmius androsaceus*	Marasmiaceae	Polysaccharide	Analgesic/sedative effect	Rahi and Malik (2016)
38	*Cordyceps* sp.	Ophiocordycipitaceae	Glucan, heteroglycan	Antitumor, immunomodulating, antitumor, hyperglycemia	Rahi and Malik (2016)
39	*Coriolus versicolor*	Polyporaceae	Polysaccharides PSK and PSP	Anticancerous and antiviral effect on HIV and cytomegalovirus in vitro	Rahi and Malik (2016)
40	*Dictyophora indusiata*	Phallaceae	Heteroglycan, mannan, glucan	Antitumor, hyperlipidemia	Rahi and Malik (2016)
41	*Agaricus bisporus*	Agaricaceae	Heteropolysaccharide	Immunostimulatory and antitumor activities	Zhang et al. (2014)
42	*Pleurotus ostreatus*	Pleurotaceae	Homogeneous polysaccharide	Antitumor activity	Cao et al. (2015)

43	*Pleurotus eryngii*	Apiaceae	Heteropolysaccharide	Antitumor activity	Ren et al. (2016)
44	*Calocybe indica*	Lyophyllaceae	Heteropolysaccharide	Antioxidant and antiaging activities	Govindan et al. (2016)
45	*Dictyophora indusiata*	Phallaceae	Vitamin C (55)	Antioxidant	Liu et al. (2019)
46	*Pleurotus ostreatus*	Pleurotaceae	Resveratrol	Antioxidant	Koutrotsios et al. (2017)
47	*Aspergillus oryzae*	Trichocomaceae	Agmatine (28)	Neurological benefits	Akasaka and Fujiwara (2020)
48	*Mucor circinelloides*	Mucoraceae	γ-Linolenic acid	Anti-inflammatory	Chan et al. (2018)
49	*Rhizopus oryzae*	Mucoraceae	Lipase (Lipopan F)	Decreases in glycemic response	Huang et al. (2020)
50	*Mortierella alpina*	Mortierellaceae	Arachidonic acid	Enhancement of immunity	Mamani et al. (2019)

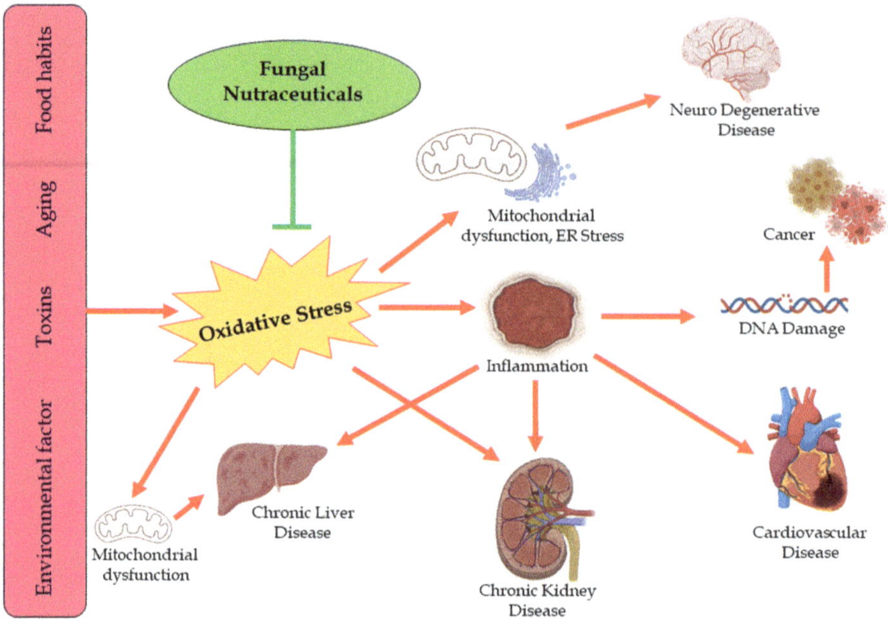

Fig. 9.2 Fungi as a nutraceutical and its function on curing different diseases

used remedies referring to fungi and also can reintroduce the metabolites as potential health supplements in modern pharmaceuticals or nutraceutical industries as potential disease-controlling agents (Fig. 9.2).

9.3 Nutraceutical Production: Using Fungi as Elicitors

Some issues regarding nutraceutical production are the time and cost for its multiple extraction process can be alleviated through advanced techniques of metabolic and some genetic approaches (Liu et al., 2017). Nutraceuticals can also be found in many plants and that co-culturing with their native endophytic fungus can increase yields of nutraceutical production. Therefore, a fungus acts as an elicitor that can influence some metabolic changes to produce the target product and also enhance its production. For example, cell cultures of *Taxus chinensis* were co-cultured with *Aspergillus niger* to increase production of paclitaxel (Wang et al., 2001). There are some other fungal elicitors for compounds like glycyrrhizin (Karwasara et al., 2010) and ajmalicine (Namdeo et al., 2002) that were also studied. Some pathogenic fungi were used as elicitors for nutraceutical production, for example, *Alternaria panax* Whetz, elicitor for *Panax ginseng* (Hao et al., 2020). Recent studies show that *Penicillium* sp. YJM-2013 activates the transcription factor expression, which promotes the increment of ginsenoside production (Wang et al., 2020). Recent studies

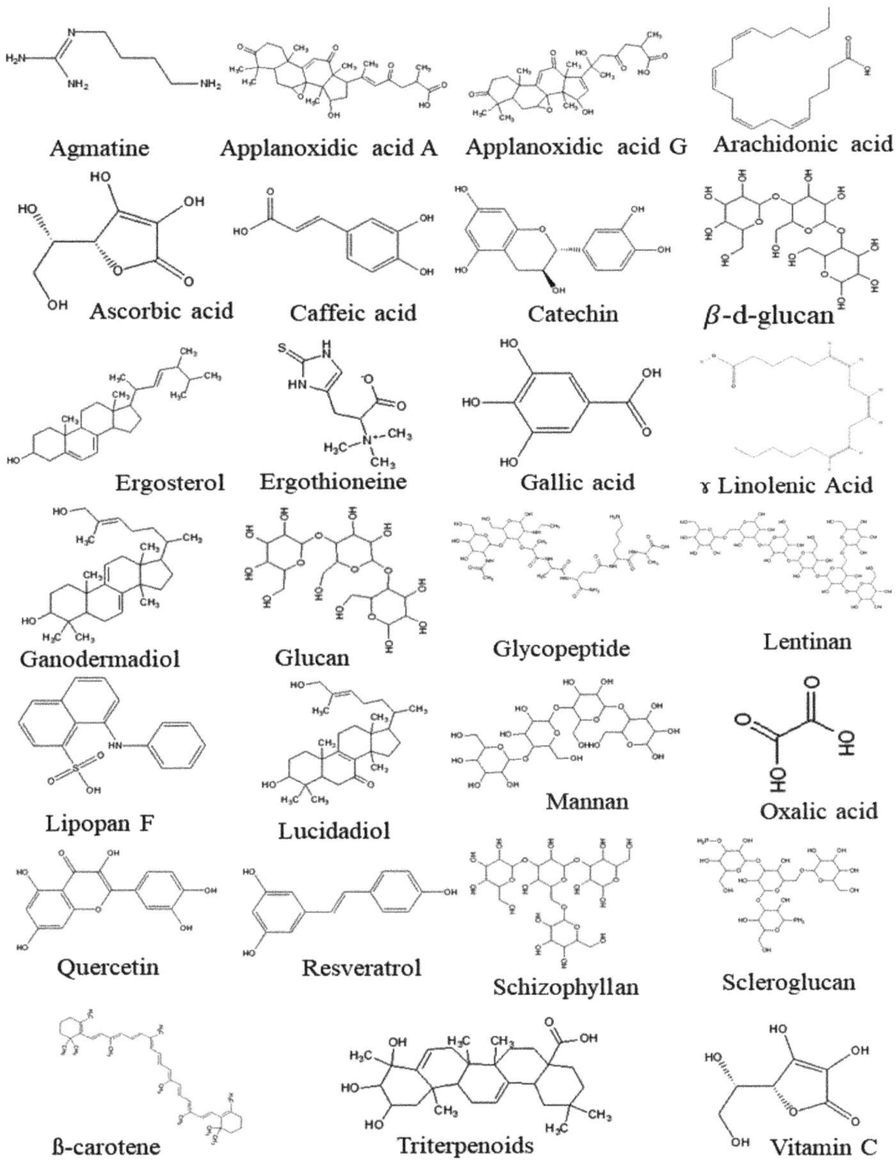

Fig. 9.3 Bioactive compounds from fungal origin

have found that the role of fungi is not only as nutraceuticals but also as a mediator in eliciting nutraceutical production from some other organisms also.

9.4 Nutraceutical Discovery: Prospects of Fungi

So many species of fungi are found on earth that can have advantages to be used as novel nutraceuticals and also as potential food supplements. Few species of fungi have been discovered till now, where it is reported that only 7% were screened for discovering some active bioactive compounds (Hawksworth, 2004). A study found 1526 out of 10,000 fungi in the screening of bioactivity level, and only a few compounds from their extract have certain bioactivity (Hoeksma et al., 2019). Studies were also carried out to discover some novel nutraceuticals from fungi species that were from marine environment and the Antarctic (Nichols et al., 2002). The *Parmeliaceae* family produces some metabolites and proteins that have promising nutraceutical activity found in the Arctic and Antarctic regions (González-Burgos et al., 2019). This will lead to the discovery of some novel nutraceutical compounds and bioactive phytochemicals of fungal specie that live in extreme and remote ecological areas. Future prospects of nutraceutical research originating from fungi will highlight the traditional medicinal values of fungi, along with modern approaches in the pharmaceutical/nutraceutical industry. The aim of these types of researches is to introduce some new health supplements using those metabolites (Zhang et al., 2020).

9.5 Entrepreneurship Aspects of Fungal Nutraceutical Industry

The branches of fungal nutraceuticals, the mushrooms, are potential elements and also have global demands. The global demands of various potential mushrooms having pharmaceutical and nutraceutical benefits will lead to establishment of industries and development of various new nutraceuticals. The application of biotechnological and advanced nutrigenomics increases the relevance of nutraceutical industries to produce quality food supplements and increases the range of scope of natural nutraceuticals in potential remedies for human health. Some nutraceutical products like vitamin complexes and inorganic nutrients and also some food ingredients are taken as supplements in many developed countries, for example, there were 47% men and 50% women taking those supplements in the United States (Bas, 2017). The nutraceutical industry is nowadays leading as a functional food supplement producer in developed countries, and hence it opens new wings of entrepreneurship possibilities in natural nutraceutical development. Companies that are involved in this field are also concerned about the intellectual property rights, along with entrepreneurship knowledge (BAS, 2016). Nutraceuticals act as necessary food and also help develop the global market of fungal nutraceuticals. The Western medicinal science is looking for the most significant medicinal mushrooms that have the preventive as well as adjuvant therapeutic properties. The global market of mushrooms is quite huge for the production of nutraceuticals from mushrooms, and

Asia alone has nearly about $2 billion-valued industries for the production of nutraceuticals from medicinal mushrooms. The publicized market for only *Agaricus* became a potential one with nearly $450 million in retail in Japan and also for the mushrooms having a rich traditional medicine value in Japan and China craving the attention of scientific researchers to develop them. Mushroom farming is an easy and very profitable agri-business for entrepreneurs with a limited investment and less infrastructure to start. Recent data regarding mushroom cultivation and production market show a value of $35 billion in 2015, and between 2016 and 2021 the value of the market will be about $60 billion in 2021. Some countries like the United States, Spain, the Netherlands, Poland, China, Italy, Canada, and the United Kingdom are leading in the field of mushroom producers (Raman et al., 2018). In India, cultivation of different mushrooms is growing and has become an alternative income source for many entrepreneurs. In the recent times, mushroom have become the potential source of nutraceuticals that can help treat chronic diseases.

The global market of fungal nutraceuticals is mainly drives by the increasing demands of mushrooms, and hence mushroom cultivation and production in a large scale may be a suitable entrepreneur opportunity that leads to successful entrepreneurs in nutraceutical industries. Global mushroom demands and production rates estimated a value of US$ 16.7 billion in 2020, which will reach a value of US$ 20.4 billion by 2025 (Report: Mushroom Cultivation Market 2020). The innovation and development of nutraceutical products is important to emphasize on the entrepreneurship knowledge, academics, and also intellectual property rights. It is also seen that some researchers do the entrepreneurship practices along with their scientific research that creates major posts in industries and also in journals or scientific community. Nowadays, the nutraceutical industries are leading in healthy food supplement production and bio-pharmaceutical applications that can help the entrepreneurs to grow and establish new opportunities in the field of fungal nutraceuticals (Fig. 9.4). Entrepreneurship can also have an impact on the social and economic growth of the country.

9.6 Conclusion and Future Prospective of Nutraceuticals

Issues associated with nutraceutical development regarding the efficiency have been raised. Nutraceuticals are commonly used as supplements and sold with optimizing dosage. The delivery system and optimizing dosage are very important parameters for nutraceutical efficacy. There are some studies regarding the specificity of long-term effect and also for the dose optimization for positive health impact. It is quite difficult to find the exact bioactive nutrient responsible for acclaimed benefits due to their complex chemical constituents. The rising health challenges due to increasing population rate have made nutraceuticals a potential health-related food supplement obtained from different fungal sources. The higher incidences of different diseases in aging population increase the need of nutraceuticals. A study has shown some that invertebrate models can promote the beneficial areas of using nutraceuticals in

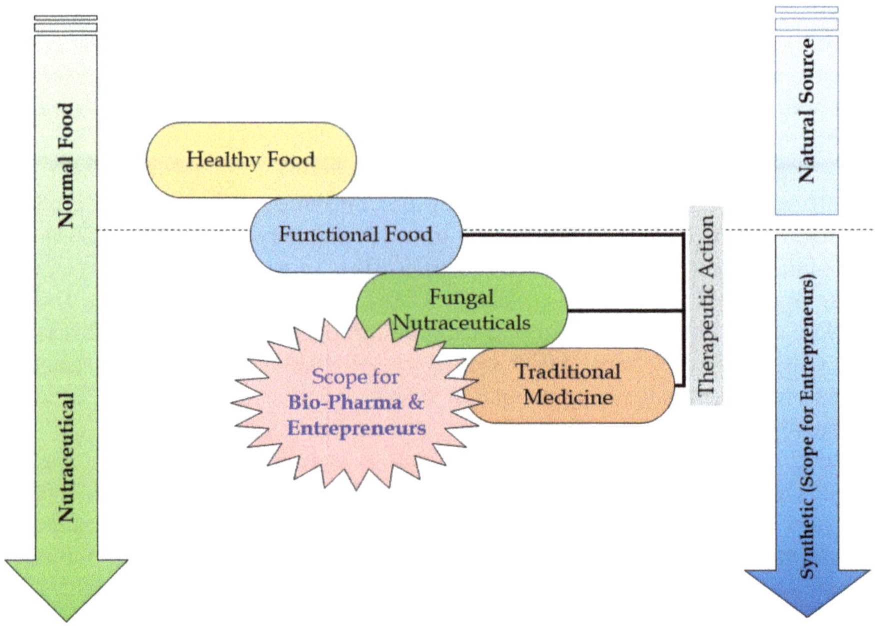

Fig. 9.4 Scopes for entrepreneurs

healthy aging, reducing risk of cancer and inflammation (Dong et al., 2012). The advanced prospective of nutraceutical development can apply in many fields of supplement manufacturing industries. The production time and cost can greatly reduce through the advanced technology of enzyme engineering, coupled with gene/protein discovery. Synthesis of nutraceuticals, the time, and cost can be reduced with the help of advanced DNA-sequencing and gene-editing methods that also open a new avenue for nutraceutical development (Goss et al., 2012). Nutraceutical developmental studies also include research on increasing the bioavailability as its bioavailability profile differs due to diverse chemical nature and structure (Santini et al., 2017). Nutraceutical research for developing a better delivery system for its potential activity also opens a new era of nutraceutical application as a health supplement. Some specialized nano-delivery methods are also used to cure cancer by nutraceuticals (Nair et al., 2010). Modern researches have shown the confidence and trust on nutraceuticals as naturally occurring food/supplements for the treatment of different chronic diseases. Some mushrooms had proved to be very active and effective nutraceutical sources having potential and numerous serious ailments to treat different diseases. Fungal nutraceuticals nowadays attract much attention in the global market for producing potential health supplements and also for their medicinal properties.

References

Ajith, T. A., & Janardhanan, K. K. (2007). Indian medicinal mushrooms as a source of antioxidant and antitumor agents. *Journal of Clinical Biochemistry and Nutrition, 40*, 157–162.

Akasaka, N., & Fujiwara, S. (2020). The therapeutic and nutraceutical potential of agmatine, and its enhanced production using *Aspergillus oryzae*. *Amino Acids, 52*, 181–197.

Al Fatimi, M. A. M. (2001). Isolierung und Charakterisierung antibiotisch wirksamer Verbindungen aus Ganoderma pfeifferi Bres. und aus Podaxis pistillaris (L.: Pers.) Morse. *Zeitschrift für Pflanzenphysiologie, 83*(1), 69–80.

Ali, N. A., Mothana, R., Lesnau, A., Pilgrim, H., & Lindequist, U. (2003). Antiviral activity *of Inonotus hispidus*. *Fitoterapia, 74*, 483–485.

Al-Obaidi, J. R. (2016). Proteomics of edible mushrooms: A mini-review. *Electrophoresis, 37*, 1257–1263.

Al-Obaidi, J. R., Alobaidi, K. H., Al-Taie, B. S., Wee, D. H.-S., Hussain, H., Jambari, N. N., Ahmad-Kamil, E., & Ariffin, N. S. (2021). Uncovering prospective role and applications of existing and new nutraceuticals from bacterial, fungal, algal and cyanobacterial, and plant sources. *Sustainability, 13*, 3671.

Ashraf, S. A., Elkhalifa, A. E. O., Siddiqui, A. J., Patel, M., Awadelkareem, A. M., Snoussi, M., Ashraf, M. S., Adnan, M., & Hadi, S. (2020). Cordycepin for health and wellbeing: A potent bioactive metabolite of an entomopathogenic *cordyceps* medicinal fungus and its nutraceutical and therapeutic potential. *Molecules, 25*, 2735.

Barros, L., Baptista, P., Correia, D. M., Casal, S., Oliveira, B., & Ferreira, I. C. J. F. C. (2007). Fatty acid and sugar compositions, and nutritional value of five wild edible mushrooms from Northeast Portugal. *Food Chemistry, 105*(1), 140–145.

Barros, L., Cruz, T., Baptista, P., Estevinho, L. M., Ferreira, I. C. J. F., & Toxicology, C. (2008). Wild and commercial mushrooms as source of nutrients and nutraceuticals. *Food and Chemical Toxicology, 46*, 2742–2747.

Bas, T. G. (2016). Entrepreneurship and scientific research in the nutraceutical industry. The concentration and dispersion intellectual property. *International Journal of Applied Management Science, 2*, 21–26.

Bas, T. G. (2017). Nutraceutical industry with the collaboration of biotechnology and nutrigenomics engineering: The significance of intellectual property in the entrepreneurship and scientific research ecosystems. Comparative Approaches to Biotechnology Development and Use in Developed and Emerging Nations. *IGI Global*.

Beluhan, S., & Ranogajec, A. (2011). Chemical composition and non-volatile components of *Croatian* wild edible mushrooms. *Food Chemistry, 124*, 1076–1082.

Bender, S., Dumitrache-Anghel, C. N., Backhaus, J., Christie, G., Cross, R. F., Lonergan, G. T., & Baker, W. L. (2003). A case for caution in assessing the antibiotic activity of extracts of culinary-medicinal Shiitake mushroom [*Lentinus edodes* (Berk.) Singer] (Agaricomycetideae). *International Journal of Medicinal Mushrooms, 5*, 1–6.

Borodina, I., Kenny, L. C., Mccarthy, C. M., Paramasivan, K., Pretorius, E., Roberts, T. J., Van Der Hoek, S. A., & Kell, D. B. (2020). The biology of ergothioneine, an antioxidant nutraceutical. *Nutrition Research Reviews, 33*, 190–217.

C Ooi, V. E., & Liu, F. (2000). Immunomodulation and anti-cancer activity of polysaccharide-protein complexes. *Current Medicinal Chemistry, 7*, 715–729.

Cao, X. Y., Liu, J. L., Yang, W., Hou, X., & Li, Q. J. (2015). Antitumor activity of polysaccharide extracted from *Pleurotus ostreatus* mycelia against gastric cancer in vitro and in vivo. *Molecular Medicine Reports, 12*, 2383–2389.

Chan, L. G., Cohen, J. L., Ozturk, G., Hennebelle, M., Taha, A. Y., & De Moura Bell, J. M. L. (2018). Bioconversion of cheese whey permeate into fungal oil by *Mucor circinelloides*. *Journal of Biological Engineering, 12*, 1–14.

Chen, X.-H., Xia, L.-X., Zhou, H.-B., & Qiu, G.-Z. (2010). Chemical composition and antioxidant activities of *Russula griseocarnosa sp. nov. Journal of Agricultural and Food Chemistry, 58*, 6966–6971.
Ciecierska, A., Drywien, M., Hamulka, J., & Sadkowski, T. (2019). Nutraceutical functions of beta-glucans in human nutrition. *Roczniki Państwowego Zakładu Higieny, 70(4)*, 315–324.
Das, L., Bhaumik, E., Raychaudhuri, U., & Chakraborty, R. (2012). Role of nutraceuticals in human health. *Journal of Food Science and Technology, 49*, 173–183.
Defelice, S. L. (1995). The nutraceutical revolution: Its impact on food industry R and D. *Trends in Food Science and Technology, 6*, 59–61.
Dong, Y., Guha, S., Sun, X., Cao, M., Wang, X., & Zou, S. (2012). Nutraceutical interventions for promoting healthy aging in invertebrate models. *Oxidative Medicine and Cellular Longevity, 2012*, 718491. https://doi.org/10.1155/2012/718491
Ey, J., Schömig, E., & Taubert, D. (2007). Dietary sources and antioxidant effects of ergothioneine. *Journal of Agricultural and Food Chemistry, 55*, 6466–6474.
Feng, T., Jia, W., Wang, W., Lin, C.-C., Fan, H., Zhang, J., & Bao, H.-Y. (2016). Structural characterization and immunological activities of a novel water-soluble polysaccharide from the fruiting bodies of culinary-medicinal winter mushroom, *Flammulina velutipes* (Agaricomycetes). *International Journal of Medicinal Mushrooms, 18*, 807–819.
Garuba, T., Olahan, G., Lateef, A., Alaya, R., Awolowo, M., & Sulyman, A. (2020). Proximate composition and chemical profiles of Reishi mushroom (*Ganoderma lucidum* (Curt: Fr.) Karst). *Journal of Scientific Research, 12*, 103–110.
Giavasis, I. (2014). Bioactive fungal polysaccharides as potential functional ingredients in food and nutraceuticals. *Current Opinion in Biotechnology, 26*, 162–173.
González-Burgos, E., Fernández-Moriano, C., & Gómez-Serranillos, M. P. (2019). Current knowledge on *Parmelia* genus: Ecological interest, phytochemistry, biological activities and therapeutic potential. *Phytochemistry, 165*, 112051.
Goss, R. J., Shankar, S., & Abou Fayad, A. (2012). The generation of "unnatural" products: Synthetic biology meets synthetic chemistry. *Natural Product Reports, 29*, 870–889.
Govindan, S., Johnson, E. E. R., Christopher, J., Shanmugam, J., Thirumalairaj, V., & Gopalan, J. (2016). Antioxidant and anti-aging activities of polysaccharides from *Calocybe indica var.* APK2. *Experimental and Toxicologic Pathology, 68*, 329–334.
Hao, Y.-J., An, X.-L., Sun, H.-D., Piao, X.-C., Gao, R., & Lian, M.-L. (2020). Ginsenoside synthesis of adventitious roots in *Panax ginseng* is promoted by fungal suspension homogenate of *Alternaria panax* and regulated by several signaling molecules. *Industrial Crops and Products, 150*, 112414.
Hawksworth, D. L. (2004). Fungal diversity and its implications for genetic resource collections. *Studies in Mycology, 50*, 9–18.
Hoeksma, J., Misset, T., Wever, C., Kemmink, J., Kruijtzer, J., Versluis, K., Liskamp, R. M., Boons, G. J., Heck, A. J., & Boekhout, T. (2019). A new perspective on fungal metabolites: Identification of bioactive compounds from fungi using zebra fish embryogenesis as read-out. *Scientific Reports, 9*, 1–16.
Huang, Z., Brennan, C. S., Zheng, H., Mohan, M. S., Stipkovits, L., Liu, W., Kulasiri, D., Guan, W., Zhao, H., & Liu, J. (2020). The effects of fungal lipase-treated milk lipids on bread making. *LWT, 128*, 109455.
Jiang, M.-Y., Zhang, L., Dong, Z.-J., & Liu, J.-K. (2009). Two new metabolites from basidiomycete *Sparassis crispa. Zeitschrift für Naturforschung B, 64*, 1087–1089.
Kalač, P. (2009). Chemical composition and nutritional value of European species of wild growing mushrooms: A review. *Food chemistry, 113*, 9–16.
Kalaras, M. D., Richie, J. P., Calcagnotto, A., & Beelman, R. B. (2017). Mushrooms: A rich source of the antioxidants ergothioneine and glutathione. *Food Chemistry, 233*, 429–433.
Karwasara, V. S., Jain, R., Tomar, P., & Dixit, V. (2010). Elicitation as yield enhancement strategy for glycyrrhizin production by cell cultures of *Abrus precatorius Linn. In Vitro Cellular and Developmental Biology-Plant, 46*, 354–362.

Kaushik, V., Arya, A., Sindhu, A., & Singh, A. (2019). Identification, optimization of culture conditions, and bioactive potential of Chinese caterpillar mushroom *Ophiocordyceps sinensis* (Ascomycetes) mycelium isolated from fruiting body. *International Journal of Medicinal Mushrooms, 21*, 931–942.

Kim, S.-Y., Chung, S.-I., Nam, S.-H., & Kang, M.-Y. (2009). Cholesterol lowering action and antioxidant status improving efficacy of noodles made from unmarketable oak mushroom (*Lentinus edodes*) in high cholesterol fed rats. *Journal of the Korean Society for Applied Biological Chemistry, 52*, 207–212.

Kimura, T. (2013). Natural products and biological activity of the pharmacologically active cauliflower mushroom Sparassis crispa. *BioMed Research International, 2013(11)*, 1–9.

Kobayashi, H., Yoshida, R., Kanada, Y., Fukuda, Y., Yagyu, T., Inagaki, K., Kondo, T., Kurita, N., Suzuki, M., & Kanayama, N. (2005). Suppressing effects of daily oral supplementation of beta-glucan extracted from *Agaricus blazei Murill* on spontaneous and peritoneal disseminated metastasis in mouse model. *Journal of Cancer Research and Clinical Oncology, 131*, 527–538.

Koutrotsios, G., Kalogeropoulos, N., Stathopoulos, P., Kaliora, A. C., & Zervakis, G. I. (2017). Bioactive compounds and antioxidant activity exhibit high intraspecific variability in *Pleurotus ostreatus* mushrooms and correlate well with cultivation performance parameters. *World Journal of Microbiology and Biotechnology, 33*, 98.

Kumar, S., & Kaushik, N. (2012). Metabolites of endophytic fungi as novel source of bio fungicide: A review. *Phytochemistry Reviews, 11*, 507–522.

Lemieszek, M. K., Langner, E., Kaczor, J., Kandefer-Szerszen, M., Sanecka, B., Mazurkiewicz, W., & Rzeski, W. (2009). Anticancer effect of fraction isolated from medicinal Birch polypore mushroom, Piptoporus betulinus (Bull.: Fr.) P. Karst. (Aphyllophoromycetideae): In vitro studies. *International Journal of Medicinal Mushrooms, 11*, 351–364.

Lindequist, U., Niedermeyer, T. H., & Jülich, W.-D. (2005). The pharmacological potential of mushrooms. *Evidence-Based Complementary and Alternative Medicine, 2*, 285–299.

Liu, Y.-T., Sun, J., Luo, Z.-Y., Rao, S.-Q., Su, Y.-J., Xu, R.-R., & Yang, Y.-J. (2012). Chemical composition of five wild edible mushrooms collected from Southwest China and their antihyperglycemic and antioxidant activity. *Food and Chemical Toxicology, 50*, 1238–1244.

Liu, L., Guan, N., Li, J., Shin, H.-D., Du, G., & Chen, J. (2017). Development of GRAS strains for nutraceutical production using systems and synthetic biology approaches: Advances and prospects. *Critical Reviews in Biotechnology, 37*, 139–150.

Liu, D., Chen, Y.-Q., Xiao, X.-W., Zhong, R.-T., Yang, C.-F., Liu, B., & Zhao, C. (2019). Nutrient properties and nuclear magnetic resonance-based metabonomic analysis of macro fungi. *Foods, 8*, 397.

Maddi, V., Aragade, P., Digge, V., & Nitalikar, M. (2007). Phcog Rev.: Short review importance of nutraceuticals in health management. *Pharmacognosy Reviews, 1*, 377–379.

Mallard, B., Leach, D. N., Wohlmuth, H., & Tiralongo, J. (2019). Synergistic immuno-modulatory activity in human macrophages of a medicinal mushroom formulation consisting of *Reishi, Shiitake* and *Maitake*. *PloS One, 14*, e0224740.

Mamani, L. D. G., Magalhães, A. I., Jr., Ruan, Z., De Carvalho, J. C., & Soccol, C. R. (2019). Industrial production, patent landscape, and market trends of arachidonic acid-rich oil of *Mortierella alpina*. *Biotechnology Research and Innovation, 3*, 103–119.

Nair, H. B., Sung, B., Yadav, V. R., Kannappan, R., Chaturvedi, M. M., & Aggarwal, B. B. (2010). Delivery of antiinflammatory nutraceuticals by nanoparticles for the prevention and treatment of cancer. *Biochemical Pharmacology, 80*, 1833–1843.

Namdeo, A., Patil, S., & Fulzele, D. P. (2002). Influence of fungal elicitors on production of ajmalicine by cell cultures of Catharanthus roseus. *Biotechnology Progress, 18*, 159–162.

Nichols, D. S., Sanderson, K., Buia, A., Van De Kamp, J., Holloway, P., Bowman, J. P., Smith, M., Mancuso-Nichols, C., Nichols, P., & Mcmeekin, T. (2002). Bioprospecting and biotechnology in Antarctica. In *The Antarctic: Past, present and future* (Antarctic CRC Research Report, 28).

Ohtsuka, S., Ueno, S., Yoshikumi, C., Hirose, F., Ohmura, Y., Wada, T., Fujii, T., & Takahashi, E. (1973). *Polysaccharides having an anticarcinogenic effect and a method of producing them from species of Basidiomycetes*. UK Patent, 1331513, 02.

Paterson, R. R. M. (2008). Cordyceps–a traditional Chinese medicine and another fungal therapeutic biofactory? *Phytochemistry, 69*, 1469–1495.

Rahi, D. K., & Malik, D. (2016). Diversity of mushrooms and their metabolites of nutraceutical and therapeutic significance. *Journal of Mycology, 2016(1-2)*, 1–18.

Raman, J., Lee, S.-K., Im, J.-H., Oh, M.-J., Oh, Y.-L., & Jang, K.-Y. (2018). Current prospects of mushroom production and industrial growth in India. *Journal of Mushroom, 16*, 239–249.

Ren, D., Wang, N., Guo, J., Yuan, L., & Yang, X. (2016). Chemical characterization of *Pleurotus eryngii* polysaccharide and its tumor-inhibitory effects against human hepatoblastoma HepG-2 cells. *Carbohydrate Polymers, 138*, 123–133.

Santini, A., Tenore, G. C., & Novellino, E. (2017). Nutraceuticals: A paradigm of proactive medicine. *European Journal of Pharmaceutical Sciences, 96*, 53–61.

Sharma, R. (2009). Nutraceuticals and nutraceutical supplementation criteria in cancer: A literature survey. *The Open Nutraceuticals Journal, 2*, 92–106.

Smania, E., Delle Monache, F., Smania, A., Jr., Yunes, R., & Cuneo, R. (2003). Antifungal activity of sterols and triterpenes isolated from Ganoderma annulare. *Fitoterapia, 74*, 375–377.

Soltani, M., Abd Malek, R., Elmarzugi, N. A., Mahomoodally, M. F., Uy, D., Leng, O. M., & El-Enshasy, H. A. (2018). Cordycepin: A biotherapeutic molecule from medicinal mushroom. In *Biology of macro fungi* (pp. 319–349). Springer.

Tsai, S.-Y., Huang, S.-J., Lo, S.-H., Wu, T.-P., Lian, P.-Y., & Mau, J.-L. (2009). Flavour components and antioxidant properties of several cultivated mushrooms. *Food Chemistry, 113*, 578–584.

Wachtel-Galor, S., Yuen, J., Buswell, J. A., & Benzie, I. F. (2011). Ganoderma lucidum (Lingzhi or Reishi). In *Herbal medicine: Biomolecular and clinical aspects* (2nd ed.). CRC Press/Taylor and Francis.

Wang, C., Wu, J., & Mei, X. (2001). Enhancement of taxol production and excretion in *Taxus chinensis* cell culture by fungal elicitation and medium renewal. *Applied Microbiology and Biotechnology, 55*, 404–410.

Wang, H., Ng, T., & Liu, Q. (2004). Alveolarin, a novel antifungal polypeptide from the wild mushroom *Polyporus alveolaris*. *Peptides, 25*, 693–696.

Wang, S.-H., Liang, W.-X., Lu, J., Yao, L., Wang, J., & Gao, W.-Y. (2020). *Penicillium sp.* YJM-2013 induces ginsenosides biosynthesis in Panax ginseng adventitious roots by inducing plant resistance responses. *Chinese Herbal Medicines, 12*, 257–264.

Xiao, C., Wu, Q., Zhang, J., Xie, Y., Cai, W., & Tan, J. (2017). Antidiabetic activity of *Ganoderma lucidum* polysaccharides F31 down-regulated hepatic glucose regulatory enzymes in diabetic mice. *Journal of Ethnopharmacology, 196*, 47–57.

Ya, G. (2017). A *Lentinus edodes* polysaccharide induces mitochondrial-mediated apoptosis in human cervical carcinoma HeLa cells. *International Journal of Biological Macromolecules, 103*, 676–682.

Yadav, S. K., Ir, R., Jeewon, R., Doble, M., Hyde, K. D., Kaliappan, I., Jeyaraman, R., Reddi, R. N., Krishnan, J., & Li, M. (2020). A mechanistic review on medicinal mushrooms-derived bioactive compounds: Potential mycotherapy candidates for alleviating neurological disorders. *Planta Medica, 86*(16), 1161–1175.

Yaltirak, T., Aslim, B., Ozturk, S., & Alli, H. (2009). Antimicrobial and antioxidant activities of *Russula delica Fr. Food and Chemical Toxicology, 47*, 2052–2056.

Zhang, M., Cui, S., Cheung, P., & Wang, Q. (2007). Antitumor polysaccharides from mushrooms: A review on their isolation process, structural characteristics and antitumor activity. *Trends in Food Science and Technology, 18*, 4–19.

Zhang, Y., Ma, G., Fang, L., Wang, L., & Xie, J. (2014). The immunostimulatory and anti-tumor activities of polysaccharide from *Agaricus bisporus* (brown). *Journal of Food and Nutrition Research, 2*, 122–126.

Zhang, L., Fasoyin, O. E., Molnár, I., & Xu, Y. (2020). Secondary metabolites from hypocrealean entomopathogenic fungi: Novel bioactive compounds. *Natural Product Reports, 37*, 1181–1206.

Chapter 10
Potential Application of Edible Mushrooms in Nutrition-Medical Sector and Baking Industries

Deepali Koreti, A. Kosre, A. Kumar, S. K. Jadhav, and N. K. Chandrawanshi

Contents

D. Koreti · A. Kosre · S. K. Jadhav · N. K. Chandrawanshi (✉)
School of Studies in Biotechnology, Pt. Ravishankar Shukla University, Raipur, Chhattisgarh, India

A. Kumar
Department of Biotechnology, Sant Gahira Guru Vishwavidyalaya, Sarguja Ambikapur, Chhattisgarh, India

A. C. Shukla (ed.), *Applied Mycology*, Fungal Biology,
https://doi.org/10.1007/978-3-030-90649-8_10

10.1 Introduction

Mushroom belongs to the assembly of a macro-fungus with a characteristic fruiting body and morphology. It can be either epigeous or hypogenous. It belongs to Basidiomycetes and is fleshy and edible by nature (Chang, 2008). Mushrooms and their cultivation have a long history in human health and dietary food. It has been valued for its edible and medicinal properties for humankind. Humans used mushrooms as a food source even before they understood the use of other microorganisms that have been discovered later. It is utilized as a food source and medicine source since a very long and ancient time (Gupta et al., 2019). In another sense, the word mushroom is also known as the fruiting body or basidiocarp or sporophores. It has desirable flavors with high nutritive value and is mostly consumed by several countries all over the world. It is a good source of proteins, carbohydrates, vitamins, as well as minerals, dietary fiber, amino acids, etc. It is also reported for its diverse range of secondary metabolites compound, which is the biologically active agent and used as therapeutic purposes (Abdelshafy et al., 2021). Various researches reported these secondary metabolites having medicinal value and also good candidate for utilization as functional foods and used as an indirect cure of diseases.

However, wild mushrooms, edible mushrooms, and medicinal mushrooms are the three main categories in the mushroom industry that are supported by their respective international bodies or forums that have helped to bring each category to the forefront of international attention and show their positive contribution to human welfare. For many years, mushrooms have been used as a food because of their flavor and nutritional properties; of approximately 1.5 million species, there were 69,000 fungal species described and only 14,000 are recognized as mushroom species. Currently, there are still many unknown species (approximately 126,000) yet to be discovered. The growth of mushroom can be distinguished into two phases: the reproductive phase (fruiting body) and the vegetative phase (mycelia) (Ukwuru et al., 2018). There are at least 12,000 species of fungi identified as mushrooms, of which 2000 species are edible (Chang, 1999). In the overall cultivated mushrooms, about 35 species have been commercially cultivated among which 20 are cultivated on an industrial level. According to the order of cultivation patterns, the following are the mushroom species: *Agaricus bisporus, Lentinus edodes*, *Pleurotus* species, *Auricula auricula*, *Flammulina velutipes, Volvariella volvacea,* etc. (Chang, 2008). Mushrooms and their extracts have long been used in folk medicine and food due to their low calorific value and pleasant taste and are also reported to have beneficial biological activities, thus now found applications in nutraceutical and pharmaceutical products (Giavasis, 2014). The higher market growth rate of mushrooms is due to their increasing acceptability for nutraceutical and medicinal purposes. The nutraceutical sector growth has increased extremely in the past several years due to the functional characterization of a wide variety of bioactive compounds (Valverde et al., 2015). Among others, mushrooms in the Asian continent have been consumed largely in China, India, and Japan due to their traditional food and nutritional value (Boels et al., 2014). Besides, the current scientific research has also confirmed their beneficial effects mainly as dietary supplements, antioxidants, antitumor and immunemodulator agents, and are also used as stamina and vitality supplement, etc. (Pan et al., 2018). Moreover, Europe is considered to own the largest market for mushrooms and its demand has also increased in the United States in the past years (Zong et al., 2012).

Recently, vast studies and research have been conducted worldwide fto increase the interest of mushrooms as a processed food both directly and indirectly (Lin et al., 2008). Mushrooms are added in the products directly as an ingredient or used indirectly as a source of fermentation such as vine making. According to Singh et al. (2010), 55% of the produced mushrooms are used in processed form or canned form because of their short shelf life, and only 45% are consumed in their fresh form. Mushroom is also known as a functional source by containing bioactive compounds, which can be consumed directly in fresh form and processed form, and by isolating particulate compound forms. In functional food industries, mushroom powder is frequently used for making food products or fortifying the food products by using mushroom supplementation. Food supplementation by mushrooms gives a new way to generate the opportunity of employment as well as cure various diseases easily and sustainably. Functional food by mushroom also offers chip source of nutrient for nutrient-deficient or malnutritious population all over the world. This chapter covers the important bioactive compounds of edible mushrooms, their nutraceutical properties, and the application of edible mushrooms in food sectors such as bakery industries.

10.2 Edible Mushrooms and Their Bioactive Compounds

Major bioactive compounds in fungal groups are known as mycochemicals, which can be found in the form of their cell wall components like proteins and polysaccharides or as secondary metabolites like phenolic compounds, terpenes, steroids, etc. (Patel & Goyal, 2012; Gupta et al., 2019). These compounds are naturally found in the mycelium as well as fruiting bodies of mushroom, and their concentration and efficacy depend on the type of mushroom, the substrate for growth, substrate composition, growth conditions, developmental stage, culture or postharvest conditions, storage and cooking procedures, etc. (Enshasy & Hatti-Kaul, 2013; Guillamon et al., 2010). Besides their pharmacological features, mushrooms contain functional metabolites and are used as functional food also beneficial in our diet (Khatun et al., 2012). Mushroom bioactive compounds are categorized into two categories: primary and secondary metabolites. Mushroom secondary metabolites include vitamins, polyphenols, alkaloids, terpenoids, sesquiterpenes, lactones, sterols, nucleotide analogs, glycoproteins, and polysaccharides (β-glucans). Still, new proteins are being discovered from a mushroom that can further be used for the development of new drugs. Some of the identified bioactive compounds are laccase, lectins, lignin-degrading enzymes, proteases, protease inhibitors, ribosome-inactivating proteins, hydrophobins, etc. (Erjavec et al., 2012). Mushrooms contain all essential amino acids required for the human body; also, they are an excellent resource of vitamins and minerals (Barros et al., 2008).

Numerous scientists investigated the preparation methods and structural characterizations of the molecules obtained from mushrooms bioactive compounds. These molecules mainly belong to polysaccharides, proteins, terpenes, phenolic compounds, unsaturated fatty acids, etc. These mushroom bioactive compounds are responsible for the development of efficient foods resources (Valverde et al., 2015).

10.2.1 Polysaccharides

Polysaccharides are polymeric carbohydrates that are widely distributed in animals, plants, and microorganisms. Various studies reported that mushroom polysaccharides are used as potential pharmacological agents with different bioactivities (Rodrigues Barbosa et al., 2019; Bai et al., 2019; Cheng et al., 2018; Rathore et al., 2019; Ruiz-Herrera & Ortiz-Castellanos, 2019). The polysaccharides found in the genus *Pleurotus* species have specific characteristics related to the chemical structures of glucans, which are varying degrees of bonds and conformations (Singdevsachan et al., 2016). Polysaccharides are formed in numerous types in the mushrooms cells and either can be simpler ones (monosaccharides) or they can be of complex form (oligo- and polysaccharides). Some edible mushroom polysaccharides are reported for the developed as functional food substances such as the schizophyllan from *Schizophyllum commune*, lentinan from *L. edodes,* pleuran from *Pleurotus* species, calocyban from *Calocyb*e species, all of which are proved and considered as excellent representatives of D-glucans. Simple polysaccharides such as glucose, fructose, galactose, xylose, mannose, fucose, rhamnose, arabinose, trehalose, and mannitol are the most common polysaccharides found in mushrooms. It has been reported that antitumor, anti-inflammatory, and immunomodulatory activities of mushroom polysaccharides are most important health benefits for human body. Mushroom polysaccharide can activate dendritic cells, monocytes, natural killer cells, neutrophil, cytotoxic macrophages, and different cytokines (Wasser, 2011; Chandrawanshi et al., 2017). Bhandari et al. (2020) reported another group of peptide called bioactive peptides (BAPs) is small fragments (contain 2–20 amino acids) of proteins that provide some health benefits. Most of the BAPs contain >20 AAs (Ryan et al., 2011). Depending on their amino acid composition and confirmation, they perform various biological activities (Bechaux et al., 2019). Because of their size, these BAPs could be easily absorbed by the intestine and directly produce local effects in the digestive tract, or they also enter the circulatory system and show their physiological effects (Erdmann et al., 2008). According to Daliri et al. (2017) and Montesano et al. (2020), the biological function of the bioactive proteins has been credited to their encoded BAPs and can be released without losing the bioactivities. Peptide shows improved bioactivity as compared to their parent proteins (Udenigwe & Aluko, 2012). Because mushroom protein contains it, it also can be an ideal source for the discovery of BAPs. β-Glucans are biologically active molecules or glucose polymers found in yeast, grain, alga, and fungus. β-Glucans are polysaccharides composed of multiple linear and branched D-glucose units linked by β-1-3 and 1-6 bonds. The bioactivity of polysaccharides is significantly influenced by physicochemical modifications such as the degree of branching and addition of substituent groups (sulfates, selenates) (Li et al., 2015). *Pleurotus* species mushrooms are an excellent source of crude fiber (10.2%) and particularly for β-glucans content (25.9%), which can be widely associated with prebiotic and anticholesterol, anticancer, and immunomodulator properties (Aida et al., 2009; Zhu et al., 2015). *P. citrinopileatus* and *P. ostreatus* have been reported for their highest crude fiber and β-glucan content, which are 20.7 and 50%, respectively. It can be used as drugs and is also known as a biological response modifier. β-Glucans, being pleuran, are more studied and recognized in *Pleurotus* genus mushrooms having bioactivity in humans and natural immunostimulant (Imunoglukan P4H®) (Bergendiova et al., 2011; Jesenak et al., 2013). Sulfated polysaccharide shows increased solubility, antiviral, anticancer, anticoagulant, and antioxidant activities, etc. It has anti-infective, antitumor, antiviral, and immunomodulatory activities as also wound-healing properties (Pretus et al., 1991; Williams et al., 1996; Majtan Jesenak, 2018).

10.2.2 Proteins and Peptides

Edible mushrooms are also an excellent source of bioactive proteins and peptides, which possess different biological activities. Numerous proteins and peptides are identified and characterized as well as their mechanisms of action are well established by researchers. But there are many other unidentified proteins, so fungi represent a relatively unexplored source of novel proteins. Similarly, it is also an endless source of novel peptides. The studies revealed that dry mushrooms contain 228–249 g (g/kg of mushroom) protein, which is much higher than other protein sources (Petrovska, 2020; Ma et al., 2018).

10.2.3 Terpenes

Chemically terpenes are the cluster of volatile unsaturated hydrocarbons and are classified as monoterpenoids, diterpenoids, triterpenoids, sesquiterpenoids, etc. Researchers have found and isolated various types of sesquiterpenoids from mushrooms, including drimane, bisabolane, aristolane, spiro, cuparene, lactarane, nardosinone, fomannosane, sterpurane, etc. Most of the triterpenoid compounds are lanostane type, which is isolated from mushrooms. Moreover, the mushroom terpenes have been proved to have numerous health benefits, such as antioxidant, antiviral, and anticancer activities (Klaus et al., 2017).

10.2.4 *Phenols*

Phenols, also called phenolic compounds, have one aromatic ring (C6) at least and one or more (OH) groups. It is found in the form of small simple molecules to complicated polymers (Michalak, 2006). Different kinds of phenolic compounds are identified in edible mushroom extract such as phenolic acids, flavonoids, fannins, tocopherols, etc. (Abdelshafy et al., 2021). Studies revealed that the antioxidant activity is the most significant bioactivity shown by phenolic compounds. It is well known as free radical inhibitors, metal inactivators, peroxide decomposers, and oxygen scavengers (Heleno et al., 2012). It is well known that phenolic compounds are active biomolecular and protect against several degenerative disorders such as brain dysfunction, cardiovascular diseases, and aging (Finimundy et al., 2013).

10.3 Application of Bioactive Compounds from the Edible Mushroom in the Nutraceutical Sector

Bioactive compounds provide medicinal or health benefits like prevention and treatment of human disease (Rathee et al., 2012). These bioactive compounds can act as immunomodulatory, anticarcinogenic, antiviral, antioxidant, and anti-inflammatory agents (Badalyan, 2014). Nutraceuticals are defined as biologically active molecules that have both nutritional as well as pharmacological potential. Nutraceuticals are present in various foods and are natural bioactive or chemical compounds. They provide nutrient source for the human body by also having health-promoting, disease–curing, or disease-prevention properties. Nutraceuticals included lipids, vitamins, carbohydrates, proteins, minerals, secondary metabolites, etc. (Sachdeva et al., 2020). Nutraceutical industries are rising rapidly, but are still in their initial phase. It gives new opportunities for employment as well as is good for human welfare. Approximately $379.061 billion in the year 2017 has accounted for the global nutraceuticals market and will be expected to grow up to $734.601 billion by 2026 (Sharma et al., 2013). Currently, some of the nutraceuticals are available in the market, and some are popular worldwide, such as RoxR, an energy drink (Rox America), ProteinexR, a protein supplement (Pfizer Ltd.), Snapple-a-day™, a meal replacement beverage (Snapple Beverage Group), etc.

For effective and successful nutraceuticals industry development, research is needed and continues to rise. In recent times, there are more than hundreds of mushroom species for which a variety of bioactive properties are detected. Mushrooms are consumed as concentrate powder as well as in extracted (in hot water) form. Currently, in the food markets and pharmaceutical industries, a large number of mushroom products and its isolate compounds exist and are represented as "nutriceutics" or "functional food or additives.". The extracts can be used as a drink, freeze-dried or spray-dried to form granular powders, which allow for easier handling, packaging, transportation, and consumption (Mizuno et al., 1995). These liquid concentrates and dried-powdered extracts or fruiting body powder can be placed in a capsule that is considered and utilized as dietary supplements or nutraceuticals. Researchers also reported that isolated bioactive substances can be used for fortification of common food products (Shamtsyan, 2016). Bioactive compounds isolated from edible mushrooms and their biological activities are shown in Table 10.1.

10.3.1 Antioxidants

Antioxidants are substances or compounds that play a very vital role in protecting human body from the free radicals generated by oxidative damages. Therefore, it becomes very essential for food and food supplements in the diet with compounds that are rich in antioxidants. It has been well reported that mushrooms extract alone and isolated polysaccharides have antioxidant activity and can decrease the production of oxygen-free radicals (Wasser & Weis, 1999). A variety of edible mushrooms are reported to possess antioxidant activity. It is generally accepted that extracts of fungi contain many components, each of which has its specific biological effects (Kozarski et al., 2011; Chandrawanshi et al., 2018). Flavonoids, glycosides, phenolics, tocopherols, polysaccharides, carotenoids, ergothioneine, ascorbic acid, etc., are some of the antioxidant compounds found in fruiting bodies and mycelium of mushrooms (Chen et al., 2012). Water-soluble glucans, and other glucans from *A. bisporus* as well as glucans from *Auricularia auricula*, *Flammulina velutipes*, *Ganoderma lucidum*, and *L. edodes*, had notable antioxidant capacity and free radical scavenging potential, thus reported as novel natural antioxidant source (Boh & Berivic, 2007).

10.3.2 Cholesterol-Lowering Effect

Cholesterol has been divided into two major categories: low-density lipoprotein and very-low-density lipoprotein, which is formally called "bad" cholesterol, and high-density lipoprotein, called "good" cholesterol. Hypercholesterolemia is the condition that increases the risk of cardiovascular diseases that cause risk for human health.

High levels of circulating cholesterol in the blood cause deposits inside blood vessels, and this deposit generates another disease called arteriosclerosis. A major rate-limiting step in the biosynthetic pathway for cholesterol formation is at the level of the microsomal enzyme 3-hydroxy-3-methylglutaryl-coenzymeA reductase (HMG-CoA reductase) that catalyzes the reductions of HMG-CoA into mevalonate (Bobek et al., 1991). It was reported that some mushrooms produce compounds that inhibit cholesterol syntheses by inhibiting specific enzyme in biosynthesis pathway. Some species of the genus *Pleurotus* are reported for the production of mevinolin

Table 10.1 Identified mushroom bioactive compounds and their bioactivity

Bioactive compounds	Name of mushroom	Compound name	Part	Identified bioactivity	References
Proteins and peptides	*Pleurotus ostreatus*	Lectin	Fruiting body	Immunomodulatory activity	He et al. (2017)
	Pleurotus ostreatus	Lectin	Mycelium	Immunomodulatory	Devi et al. (2013)
	Mushrooms	Fungal immunomodulatory proteins (FIPs)	Fruiting body	Useful in tumor immunotherapy, suppress tumor invasion and metastasis	Li et al. (2019)
	Hypsizygus marmoreus	Ribosome-inactivating proteins (RIPs)	Fruiting body	Used in antiviral and cancer therapies, antifungal	Wong et al. (2008) and Lam and Ng (2001)
	Agaricus bisporus	Tyrosinases	Fruiting body	Involved in the synthesis of melanin, used as biomarker for vitiligo	Zehtab et al. (2000)
	Amanita hemibapha	Ribonucleases (RNases)	Fruiting body	Antiproliferative activity/antitumor.	Sekete et al. (2012)
	Agaricus bisporus	Lectin	Fruiting body	Immunomodulatory activity, reducing the innate and adaptive responses	Ditamo et al. (2016)
	Pleurotus eryngii	Laccases	Fruiting body	Antiviral activity, inhibit the activity of HIV-1 reverse transcriptase	Wang and Ng (2006)
	Pleurotus cornucopiae	Laccases	Fruiting body	Antiviral and antitumor activities, inhibited the activity of HIV-I reverse transcriptase, and proliferation of the hepatoma cells HepG2 and the breast cancer cells	Wu et al. (2014)

Polysaccharides	*Pleurotus eryngii*	Heteropolysaccharide	Fruiting body	Hypolipidemic and hypoglycemic activities, increase the level cholesterol and liver glycogen of high-density lipoprotein	Chen et al. (2016)
	Pleurotus tuber-regium	*β*-d-glucan	Mycelium	Hepatoprotective, antibreast cancer	Zhang et al. (2007)
	Auricularia auricular	Glucan	Fruiting body	Hyperglycemia, immunomodulating, antitumor, anti-inflammatory, antiradiative	Zhang et al. (2007)
	Lentinus edodes	Mannoglucan	Culture broth, fruiting body	Polysaccharide-protein complex, glucan, lentinan immunomodulating, antitumor, antiviral	Zhang et al. (2007)
	Pleurotus ostreatus	Glycoprotein	Fruiting body	Antitumor, hyperglycemia, antioxidant	Zhang et al. (2007)
	Lentinus edodes	Lentinan	Fruiting body	Antigastric cancer	Oba et al. (2009)
	Agaricus bisporus	α-Glucans	Juice from fruiting body	Anti-hypercholesterolemia	Volman et al. (2010)
	Agaricus bisporus	Heteropolysaccharide	Mushroom wastewater	Hepatoprotective, against CCl4-induced hepatic injury	Huang et al. (2016)
	Agaricus bisporus	Homogeneous polysaccharide	Powder from fruiting body	Antitumor activity, inducing apoptosis by the mitochondrial death pathway	Pires et al. (2017)
	Pleurotus ostreatus	Heteropolysaccharide	Powder from fruiting body	Antidiabetic activity, activate GSK3 phosphorylation and GLUT4 translocation	Zhang et al. (2016a, b)
	Lentinula edodes	Heteropolysaccharide	Dried fruiting body	Antitumor activity	Wang et al. (2015)
	Pleurotus ostreatus	Heteropolysaccharide	Ultrasonic-assisted extracted of fruiting body	Alleviate cognitive impairment activity	Zhang et al. (2016a, b)
	Pleurotus eryngii	Heteropolysaccharide	Hot water extracted of fruiting body	Antitumor activity and stimulate the production of intracellular ROS of HepG-2 cells	Ren et al. (2016)
	Pleurotus sajor-caju	Lovastatin polysaccharide homopolysaccharides	Fruiting body	Lower cholesterol, prevents cardiovascular disorders	Lakhanpal and Rana (2005)
	Cordyceps sinensis	Heteropolysaccharides, cordycepin	Fruiting body	Hypoglycemic activity, anti depressant activity, cellular health properties	Lakhanpal and Rana (2005)

(continued)

Table 10.1 (continued)

Bioactive compounds		Name of mushroom	Compound name	Part	Identified bioactivity	References
Phenols	**Phenolic acids**	*Amanita crocea*	Cinnamic, p-hydroxybenzoic, p-coumaric	Fruiting body	Antioxidant, enzyme inhibition, and antimutagenic properties	Alkan et al. (2020)
		Agaricus brasiliensis	Gallic, ferulic, benzoic, p-hydroxybenzoic, gentisic, trans-cinnamic, p-coumaric, fumaric acids, catechol	Fruiting body	Antioxidant and antimicrobial activities	Bach et al. (2019)
		Agaricus subrufescens	p-Coumaric	Fruiting body	Antioxidant activity	Ferrari et al. (2021)
		Agrocybe aegerita	Sinapic, chlorogenic, Gallic, protocatechuic, ferulic acids	Fruiting body	Antiangiogenic	Lin et al. (2017)
		Calocybe indica	Gallic, vanillin, protocatechuic acid, naringin, naringenin, homogentisic acid, hesperetin	Fruiting body	Antioxidant and antityrosinase activities	Alam et al. (2019)
		Pleurotus pulmonarius	Caffeic, gallic, and ferulic acids	Mycelium	Antioxidant activity	Contato et al. (2020)

	Flavonoids	*Laccaria amethystea* and *Laccaria ventricosum*	Quercetin and catechin	Powder from fruiting body	Antihyperglycemic and antioxidant activity	Liu et al. (2012)
		Lactarius indigo	Myricetin, procyanidin, quercetin-hexoside, isoquercetin-hexoside, and kaempferol	Fruiting body	Antioxidants	Yahia et al. (2017)
		Lentinus edodes	Flavonoids rutin and quercetin	Fruiting body	Antioxidant activity	Xiaokang et al. (2020)
		Pleurotus citrinopileatus	Rutin and quercetin	Ultrasound and ethanolic extracted from fruiting body	Antioxidants	Gogoi et al. (2019)
	Tannins	*Pleurotus tuber-regium*	–	Powder from fruiting body	Antioxidant activity	Akindahunsi and Oyetayo (2006)
		Astraeus hygrometricus	Catechin	Mushroom flours	Antioxidants	Pavithra et al. (2016)
	Tocopherols	*Pleurotus ostreatus*	α-Tocopherol	Ethanolic extracted of from fruiting body	Antioxidants	Jayakumar et al. (2009)
		Morchella esculenta	Catechol, p-coumaric acid, rutin, hyperoside, quercetin, tocopherol, and ellagic acid	Dried fruiting bodies	Antioxidants	Wagay et al. (2019)
		Russula delica, Boletus badius, and *Agaricus bisporus*	α-Tocopherol	Methanolic extracts of from fruiting body	Antioxidants	Elmastas et al. (2007)
		Pleurotus ostreatus and *Pleurotus eryngii*	Tocopherol	Mushroom mycelia	Sources of lipophilic antioxidants	Bouzgarrou et al. (2018)

(lovastatin), which is known as the inhibitor of the HMG-CoA reductase and used for the treatment of hypocholesteremia.

Another compound, eritadenine, a compound extracted from *L. edodes* mushroom, is able to lower blood serum cholesterol level. It was reported that the addition of mushrooms' dried fruit bodies or submerge mycelium in a high cholesterol diet can effectively reduce cholesterol accumulation in the blood serum and liver. It was observed that mushroom diet in the experimental rats redistributed cholesterol in favor of HLDL, reduced VLDL and LDL production, and reduced HMG-CoA reductase activity and cholesterol absorption in the liver and blood serum (Popov et al., 2008).

10.3.3 Antiviral Activity

According to Pradeep et al. (2019), antiviral is an agent that kills a virus or it can suppress the replication and therefore inhibits its propagation. Antiviral compounds interfere with viral replication and its multiplication as a host. These are isolated from either natural sources like plants, bacteria, and fungi or can be chemical synthesis (Kulkarni & Sanghai, 2014). Many researchers reported the antiviral effects of *Pleurotus* mushroom from the whole extracts as well as isolated specific compounds. It is reported (Fan et al., 2006) that the extract from both mycelia and fruiting bodies showed various bioactivities such as antiviral, antibacterial, hematological, antitumor, hepatoprotective and hypotensive effects, etc. The isolated active compounds (small molecules) from mushroom played a direct role such as acting as inhibitors of viral enzymes and nucleic acids synthesis, and inhibit the absorption and uptake of viruses into mammalian cells. Polysaccharides and other complex molecules played indirect antiviral effects (Brandt & Piraino, 2000). These bioactive compounds acted either directly (exhibited especially by smaller molecules) or indirectly (the immune-stimulating polysaccharides or other complex molecules). β-Glucans had been potentially used by many viruses such as; infectious hematopoietic necrosis virus), spring viremia of carp virus, dengue virus, herpes simplex virus type 1 (HSV-1), human immunodeficiency virus, influenza virus, tobacco mosaic virus, etc. But only a few studies reported its human and animal clinical trials, and has been conducted for the treatment or prevention of viral infections (Adotey et al., 2011). β-Glucans can work directly through the inhibition and/or disruption of virus particles or indirectly by enhancing the immune system. β-Glucan's action is mediated by its several receptors such as the dectin-1 receptor, toll-like receptors (TLR 2, 4, and 6), complement receptor 3 (CR3), scavenger receptor, and lactosylceramide, among which dectin-1 receptor is the most important, which is highly expressed in many immunocompetent cells like neutrophils, eosinophils, macrophages, dendritic cells, monocytes, several T lymphocytes, etc. (Legentil et al., 2015).

10.3.4 Antiobesity

Obesity is a group of disorders in which body mass index is more than 30 kg/m^2. This causes enhancement of body fat deposited in the adipose tissue, which can cause harmful health effects. There are various complications associated with obesity such as diabetes, cardiovascular diseases (CVDs), pulmonary diseases, cancer, osteoarthritis, etc. (Ganesan & Xu, 2018). Obesity is caused by several factors or reasons like high food intake, sedentary lifestyles, lack of physical activity, and a genetic predisposition. Excess stored fat and high lipid content in the plasma in obese persons was observed. Adipose tissue is an organ that helps to maintain energy balance in the body, and in the case of obesity high fat accumulated within causes unusual progress of white adipose tissue (Kim, 2011). A balanced diet, exercise, and surgical intervention are some primary managements of obesity. Several therapeutic drugs are also available for the treatment of obesity, but these are associated with other health complications (Kennett & Clifton, 2010). Obesity also causes and generates other health complications. Obesity is a metabolic disorder, so there is a need for treatment that is associated with daily routine or that can say in the form of food. Edible mushrooms have been well documented for their numerous bioactive compounds and can be used as a functional food. It can directly or indirectly be used for the cure of obesity. Its hypocholesterolemic effects are connected to CVD (diminishes CVD)-related lipid metabolism, anti-inflammatory properties, and the prevention of oxidative stress with platelet agglutination (Sun et al., 2007). It is observed that the cholesterol-lowering effect might be due to a decrease in VLDL and decrease in the catalytic functions of HMG-CoA reductase and amplification of the rate of cholesterol catabolism (Bobek et al., 1991). It is reported that the polysaccharide from *Tremella fuciformis* prevented the variation of 3T3-L1 adipocytes by decreasing the expression of mRNA and suggesting that the possible role of the polysaccharide act as an antiobesity prebiotic. Cheskin et al. (2008) and Poddar et al. (2013) reported long-term and short-term clinical studies with obese or diabetic participants. They evaluated the impact of substituting 20% of high-energy beef with 20% to low-energy white button mushrooms in their diet, and the results showed that the mushroom regime consumers had lesser BMI, decreased belly circumference, and increased satiety without diminishing palatability. This study concluded that the consumption of *A. bisporus* has shown potential as an antidiabetic, antiobesity, etc.

10.4 Mushroom-Based Bakery Products and Their Benefits as a Functional Food

Fortification or enrichment is defined as the addition or enhancement of desired nutrients elements in developing foodstuff products. Fortification is an alternative and effective approach to prevent deficiency of one or more nutrients in certain specific groups and population (Codex Alimentarius, 1987). A fortification strategy is used for making functional foods in which food is fortified with bioactive metabolites. The demand and research related to "functional foods" have recently attracted increasing attention worldwide. By containing various metabolites, mushrooms are considered as a popular functional food. According to the Institute of Medicine's Food and Nutrition Board, "functional foods" are defined as foods or dietary components that provide health benefits beyond basic nutrition (Ferreira et al., 2017). Bioactive compound type and quantity in mushrooms depend upon the mushroom species, environment, and substrate in which they grow. Similarly, its functional activity depends upon the type and concentration of bioactive compounds (Ferreira et al., 2017). Till now there are hundreds of mushroom species that have been identified and extensively studied for their biological properties and potential for the prevention and treatment of several human diseases. Recently, there is a gradual increase in interest in the field of functional food from mushrooms by both direct as well as indirect use (Lin et al., 2008). It has been reported that 55% of the total mushroom produced is used in a processed form and 45% are consumed in fresh form (Moon & Lo, 2013; Singh et al., 2010). At present, advanced lifestyle demands have grown interest in the formulation and production of healthy and tasty food items that are rich in nutrients. Edible mushrooms can be used an alternative and healthy substitute for animal protein because they contain high protein as well as all the essential amino acids and vitamin D (Abigail et al., 2020). The approximate compositions of some edible mushrooms are shown in Table 10.2.

They have reported high content of sulfur amino acids and glutamic acids that provides meaty test or umami taste to mushrooms (Kim, 2011). Mushrooms are well known for their unique taste and texture flavor that differ from other food sources (Alemu, 2014). Their global and economic value is now increasing rapidly due to food as well as their medicinal and nutritional values. There are several

Table 10.2 Protein, carbohydrate, fat, ash, and moister content of some edible mushrooms

Mushroom	Moister content	Carbohydrate content	Protein	Fat	Ash	Reference
Agaricus bisporus	91.27 ± 0.45	73.80 ± 6.53	14.09 ± 0.23	2.18 ± 0.34	9.74 ± 1.95	Reis et al. (2012)
Pleurotus ostreatus	89.17 ± 2.12	85.87 ± 19.21	7.02 ± 0.55	1.39 ± 0.18	5.72 ± 0.74	Reis et al. (2012)
Pleurotus eryngii	82.59 ± 0.36	78.60 ± 0.75	2.09 ± 0.01	4.36 ± 0.14	14.95 ± 0.91	Reis et al. (2014)
Pleurotus sajor-caju	88.00 ± 0.00	69.05 ± 0.00	12.37 ± 0.00	1.80 ± 0.00	5.78 ± 0.00	Mahamud et al. (2012)
Pleurotus pulmonarius	89.17 ± 0.00	76.11 ± 0.00	30.30 ± 0.00	1.10 ± 0.00	13.21 ± 0.00	Rana et al. (2015)
Lentinus edodes	90 ± 91.80	67.5 ± 78.0	13.4 ± 17.50	4.9 ± 8.00	3.7 ± 7.00	Crisan and Sands (1998)

mushroom species that have been identified and cultivated worldwide. As a form of food source, mushroom is used in various forms and has generated various industrial opportunities (Waktola & Temesgen, 2018). Bakery products are one of the widely consumed food products and have a big market and income-generating industries. Bakery products are well known for their varied tastes, long shelf-life, and low cost, all these properties making it very popular in the food industry. Bakery products such as biscuits, bread, cookies, cakes, muffins, and snacks are popular products and consumed all over the world. New techniques and methods have evolved with the progress of bakery industries. Recently, bakery industry has also started making products that provide additional health benefits like fortified with specific nutrient(s) by using food sources. Mushrooms are one of the best nutrient-rich sources and can be a good alternative option for bakery products' enrichment with nutrients (Salehi, 2019), as displayed in Table 10.3. For enrichment or making bakery products edible, mushrooms are used in powder form. It provides additional protein, vitamins, mineral polyphenols, and crude fiber in bakery products. Recently, various researchers have tried to formulate and produce functional bakery products with mushrooms without the defect of their quality characteristics and sensory attributes (Lin et al., 2008). It has been reported that properties like color, physicochemical, and texture of bakery products were affected by the replacement of wheat flour with mushroom powder, and it was observed that addition of 4–10% mushroom powder in bakery products showed the best sensory attribute. Rai and Arumuganathan (2008) prepared and analyzed various mushroom powder value-added products like biscuits, cakes, nuggets, soup powder, noodles, pickles, candies, ketchup, etc. There are several studies already present in literature that show mushrooms are potentially used for making bakery products, as shown in Table 10.3, such as bread, biscuits, and cakes. By adding mushroom powder in bakery products, it provides valuable products that are highly nutritious and can be a good source of functional food that can be consumed by a wide group of consumers.

There are some steps in the production of mushroom-based bakery products including the processing of harvested mushrooms, and their preservation and addition in the bakery product. Because of the perishable and short shelf-life of harvested mushrooms, it is processed after harvesting. Therefore, preservation and drying methods are applied. Dried mushrooms as easy and more valuable for mixing or making bakery products. Mushroom contains various metabolites; therefore, a correct drying method is applied for the drying process. Different drying methods are reported by various scientists (Giri & Prasad, 2007; Ghanbarian et al., 2016; Salehi et al., 2016; Salehi, 2019), such as sun drying, hot air, freeze drying, infrared vacuum, and dehydration being some drying methods applied according to mushroom species and physiochemical condition. For the selection of a particular drying method, it is also considered that it does not give unwanted aroma or color to the final product. Before making bakery products, mushroom must be prepared in powder form by using an efficient method (Ibrahium & Hegazy, 2014; Salehi et al., 2016). The work is based on different drying methods on shiitake mushrooms and achieves good color, improves rehydration and better aroma, etc. (Wang et al., 2015). Salehi et al. (2017) studied the drying kinetics of *A. bisporus* mushroom,

Table 10.3 Mushroom species used in bakery products and their supplementation percentage

Bakery product	Mushroom	Quality of the product	Supplementation (%)	References
Bread	*Pleurotus ostreatus*	Increased protein, fiber and ash	25	Azeez et al. (2018)
	Pleurotus ostreatus and *Calocybe indica*	Increased protein, fiber and ash	–	Oyetayo and Oyedeji (2017)
	Pleurotus sajor-caju	Increased nutrients	5	Mahamud et al. (2012)
	Lentinula edodes	Increased nutrients and ash	5	Lin et al. (2008)
	Pleurotus pulmonarius	Increased protein, fiber, and ash	10	Okafor et al. (2012)
	Lentinula edodes	Less microbial counts and increased nutrients	5	Yen et al. (2011)
	Agaricus blazei, Antrodia camphorata, Hericium erinaceus, Phellinus linteus	Increased nutrients and ash, etc.	5	Ulziijargal et al. (2013)
	Oyster mushroom	Increased protein content	6	Majeed et al. (2017)
	Shiitake, porcini, and white button mushroom	Increased antioxidant capacity	–	Lu et al. (2021)
	Lentinus tuber-regium	Increased nutrients and ash	–	Lee et al. (2004)
Cookie and biscuits	*Pleurotus eryngii*	Improve the quality and antioxidant potential	3	Kim et al. (2010)
	Lentinus edodes	Increase nutrients content	–	Singh et al. (2016)
	Pleurotus sajor-caju	Nutritious and glycemic index	8	Ng et al. (2017)
	Agaricus bisporus	Increase nutrients content	20	Kumar & Barmanray (2007)
	Pleurotus sajor-caju	Increased nutrients	10	Bello et al. (2017)
	Pleurotus pulmonarius	Increased protein, fiber, and minerals	20	Ibrahium and Hegazy (2014)
	Calocybe indica	Increased protein and minerals	10	Rathore et al. (2010)
	Pleurotus ostreatus	Higher protein	25	Cornelia and Chandra (2019)
	Tricholoma matsutake	Antifatigue effect	–	Ning et al. (2020)

(continued)

Table 10.3 (continued)

Bakery product	Mushroom	Quality of the product	Supplementation (%)	References
Muffin	*Lentinus edodes*	Increased protein and fiber	–	Kim and Joo (2012)
Cake	*Agaricus bisporus*	Increased the protein and ash	10	Salehi et al. (2016)
	Pleurotus sajor-caju	Enhance essential nutritional components	4	Aishah and Wan Rosli (2013)
	Pleurotus ostreatus	Increased the protein and ash	15	Singh and Thakur (2016)
	Pleurotus ostreatus	Higher protein and Fat	15	Sheikh et al. (2010)
3D snacks	*Agaricus bisporus*	Useful for personalized nutrition	20	Keerthana et al. (2020)
Cheese spreads	*Pleurotus ostreatus*	Increased nutrients composition and mineral	2	Khider et al. (2017)

with combined IR vacuum drying. After drying, mushrooms are ground and made powder. Various scientists reported the nutritional and chemical composition of edible mushrooms in both fresh as well as dried forms.

10.5 Bakery Product Development from Mushroom as New Ways for Entrepreneurship

10.5.1 Bread

Bread is traditionally made by the process of fermentation of protein-rich sources. Salehi et al. (2016) reported that *A. bisporus* contains high-quality protein and can be a good source for making bread. Okamura-matsui et al. (2003) mixed mushroom powder with wheat flour and baked the mixture to make bread and determined the effect of mushroom supplementation on the functional properties of bread. In has been recorded that 10% addition of mushrooms (*Grifola frondosa*, *Hypsizygus marmoreus,* and *Pholiota nameko*) decreased the loaf volume of bread. Similarly, Lin et al. (2008) reported the substitution of *L. edodes* stipe for 2–7% of wheat flour for bread making and observed that the 5% shiitake stripe bread showed a higher fiber content with no interference with the bread's specific volume. Likewise, when a specific compound is isolated from mushroom like 5% of fungal chitin from shiitake mushrooms is mixed with wheat flour, it shows all similar to normal wheat bread sensory results. It reported slow moisture loss and fewer microbial counts during storage as compared to wheat bread (Yen et al., 2011). It is also reported that increasing mushroom content significantly increases nutrient content in bakery products. Aishah and Wan Rosli (2013) formulated paratha flat bread, rice porridge, and cake with the dried oyster powder and analyzed the increased percentage of moisture, ash, and protein with respect to oyster powder content. Mahamud et al. (2012) studied wheat bread fortification by oyster mushroom at three levels – 5%, 10%, and 15% – and concluded that 5% supplementation gave better texture and acceptability. It is also observed the increased nutritional composition of bread compared to control bread with mushroom powder. Ulziijargal et al. (2013) also reported that mushroom mycelia (5%) are used for the supplementation of bread and observed that mycelium-supplemented bread showed higher umami taste intensity than control white bread. Lu et al. (2021) worked with three different species of mushrooms powder (shiitake, porcini, and white button mushroom) for wheat bread supplementation and investigated starch (gelatinization and digestibility) and its antioxidant capacities. This study reported that total starch contents decrease and increase phenolic contents of the bread with increased mushroom powder supplementation. Mushroom supplementation enhanced antioxidants capacity that was observed in the DPPH radical scavenging assay and oxygen radical absorbance ability of bread.

10.5.2 Cake

A natural fiber from natural products gives good texture and firmness to the cake or can say that quality and quantity of fiber make cakes better. Quality makes a good sensorial attribute and obtained positive acceptance by consumers. It has been observed by Jeong and Shim (2004) and Salehi et al. (2016) that the physicochemical properties of sponge cake are changed by the addition of mushroom powder. In cake making, the addition of 4% powdered oyster mushroom gives higher scores for softness and flavor attributes. It is also a cost-effective way for protein and nutrient enrichment in normal wheat bread. Fibers in cakes decreased the volume and increased the firmness, or all the changes observed in cake textures were dependent upon the fiber content in the used mushroom powder. Salehi et al. (2016) supplemented cake with *A. bisporus* powder and studied the rheological, textural, physicochemical, and sensory attributes prepared of cake batters as well as cake. In this report, it was observed that increasing the percentage of *A. bisporus* powder significantly increased the protein and ash content in cake. This study observed that the viscosity and volume of cake batters, cohesiveness, and springiness of baked cakes increased with addition of mushroom powder, but it was also observed that it reversed value in some properties like consistency, hardness, gumminess, chewiness, density, and crumb color indexes. The result indicates that 10% mushroom powder is most accepted in the sensorial evaluation test. Another studiey by Sheikh et al. (2010), designated to the effect of mushroom powder addition on the quality of cake, found that 15% of mushroom addition gave the best acceptability.

10.5.3 Biscuits

Mushroom is renowned for its unique flavor and low-fat content. It is successfully utilized for making biscuits and cookies (Aishah & Wan Rosli, 2013). Mushrooms such as *L. edodes* and *P. eryngii* powder have been used for muffins and cookies making by Kim and Joo (2012) and Kim et al. (2010). Although the optimum amount of mushroom additive for product quality differs with the type of baked products and the selection of mushrooms used. The addition of mushrooms is a promising way to produce a healthy product with compatible quality characteristics to wheat flour bakery products (Moon & Lo, 2013). They formed biscuits by the substitution of wheat flour at three levels (10, 20, and 30%) with *P. pulmonarius* mushroom powder along with sweet potato flour (Ibrahium & Hegazy, 2014). This study reported that supplemented biscuits with 10 or 20% of mushroom powder and sweet potato flour exhibited a good sensory attribute. It determined the supplementation effect of oyster mushroom powder at three levels (0, 4, 8 and 12%) on the nutritional values, thermal and pasting properties, microstructure, in vitro starch digestibility, and sensorial properties of biscuits (Ng et al., 2017). They found significant increasing in the level of protein, total dietary fiber, ash, and β-glucan, with the 12% mushroom powder supplementation, etc. Another study by Singh and Thakur (2016) observed improved nutrient, physicochemical, and sensorial properties of biscuits by the supplementation of shiitake mushroom powder. This experiment also observed that the crude protein content of biscuits increases with the addition of shiitake mushroom powder. Biscuit is not much healthier for daily use because of its high carbohydrates, calories, and fat contents, but also due to its low vitamin, fiber, and mineral element contents and also having only 6–7% protein, etc. So, by using the fortification of biscuits by using mushroom powder it makes its more protein-rich source and healthier for daily use bases (Farzana & Mohajan, 2015; Serrem et al., 2011). They fortified soy flour to wheat flour with *P. ostreatus* powder and studied the effect on nutritional properties and sensory quality of biscuits (Farzana & Mohajan, 2015). A similar experiment was performed by Bello et al., 2017, who analyzed the nutrient composition and sensory characteristics of biscuits containing oyster mushroom (OM) powder. It was found that the protein content increased with mushroom powder addition. It was also observed that minerals such as K and Na are the main mineral elements in the biscuit samples and Ca, Mg, P, Fe, Cu, Zn, and Mn composition increases with mushroom powder addition. They studied functional, rheological, and physicochemical properties of biscuits that are made by gluten-free dough supplemented with fermented and unfermented polysaccharide flour of mushroom (Sulieman et al., 2019). The result signifies that *A. bisporus* polysaccharide flours in gluten-free dough formulation, having a plentiful amount of minerals, protein, essential amino acids, and rich amount of proteins, etc. It for the first time reported the development of mushroom-based fiber-enriched snacks by using 3D food printing technology. In this technique, they used *A. bisporus* mushroom powder at different percentages (5, 10, 15, 20, and 25% w/w) along with different printing speed. They found that the 20% mushroom powder formulation printed at 800 mm/min printing speeds using the 1.28 mm nozzle with 0.383 g/min flow rate gave the best results (Keerthana et al., 2020). These printed snacks are further processed in microwaves and found that the spice-flavored snacks have an excellent sensory acceptance.

10.6 Concluding Remarks and Future Prospects

The mushrooms are utilized since ancient times as a food source as well as a medicinal source. They have several nutrients and bioactive compounds; they are a well-known natural source that can be used in many different fields of human welfare. Continuous rise in the population and its advanced lifestyle raise the demand of food and medicine. All these consequences lead to generate such type of functional food that overcomes this problem and gives both food and medicine source in a single-step manner. Therefore, edible mushrooms are the most potential candidate that successfully utilized both these criteria. It is well-established research that the mushrooms are used as a functional food and are the source of bioactive compounds; furthermore, researches are still needed for unidentified mushroom metabolites. It is well known for its nutritional and bioactive compound and also can be preserved and utilized in food industries. Food fortification with mushroom extracts offers great potential for a future generation with very high importance of functional food.

There are various products made by mushrooms and utilized worldwide, mushroom-supplemented bakery products being one of them. Bakery products are the easy and most utilized food stuff all over the world, and its supplementation by using mushroom makes it more nutritious. These supplemented bakery products are easily consumed as a form of functional food. The supplementation of bakery products by mushrooms can help improve the nutritional requirements and status of the underdeveloped as well as developing countries. By increasing awareness of the various properties of mushroom and increasing utilization of mushroom in bakery industries, it can generate rise in mushroom demand. These mushroom demands can be fulfilled by increasing mushroom cultivation. Mushroom cultivation can give a contribution to sustainable development as well as provide a livelihood for rural populations.

Employment is a bigger challenge for the rural population. The majority of the rural population mostly depend on agriculture activity (farming), and farmers are still using traditional cultivation methods that result in low productivity. The agricultural activity produces a high number of by-products such as rice straw, wheat straw, stumps, barley straw, husk, sorghum stalks, grass seed straw, flax coil seed, corn stalks, sugarcane bagasse, etc. This agro-waste was combusted for disposal, which increases carbon emissions and harms the environment. Mushroom cultivation is the best option to utilize this agro-waste and make value-added products. In future, mushroom cultivation can be the best alternative source of employment generation in rural and backward areas. Also, mushroom cultivation is a better technique for generating income, providing employment to the rural population, and making their livelihood improvement to rural population. In the current situation, we require a food source that provides additional nutrition, and food supplementation by mushrooms is considered as a better and quick option of functional food. Mushroom cultivation and its utilization of food sector are helpful for defending malnutrition to the local population. It is an easy and cost-effective technique that provides a new way for farming, providing functional food to the nutrient-deficient population.

References

Abdelshafy, A. M., Belwal, T., Liang, Z., Wang, L., Li, D., Luo, Z., & Li, L. (2021). A comprehensive review on phenolic compounds from edible mushrooms: Occurrence, biological activity, application and future prospective. *Critical Reviews in Food Science and Nutrition*. https://doi.org/10.1080/10408398.2021.1898335

Abigail, G., Mario, C., Carolina, L., Clarisse, N., Araceli, L., Rosa, R., Juan, C., & Ruth, B. (2020). Edible mushrooms as a novel protein source for functional foods. *Food & Function Journal*. https://doi.org/10.1039/D0FO01746A

Adotey, G., Quarcoo, A., Holliday, J. C., Fofie, S., & Saaka, B. (2011). Effect of immunomodulating and antiviral agent of medicinal mushrooms (immune assist 24/7TM) on CD4+ T-lymphocyte counts of HIV-infected patients. *International Journal of Medicinal Mushrooms, 13*, 109–113.

Aida, F. M. N. A., Shuhaimi, M., Yazid, M., & Maaruf, A. G. (2009). Mushroom as a potential source of prebiotics: A review. *Trends in Food Science and Technology, 20*, 567–575.

Aishah, M., & Wan Rosli, W. (2013). The effect of addition of *Oyster* mushroom (*Pleurotus sajor-caju*) on nutrient composition and sensory acceptation of selected wheat and rice-based products. *International Food Research Journal, 20*, 183–188.

Akindahunsi, A. A., & Oyetayo, F. L. (2006). Nutrient and antinutrient distribution of edible mushroom, *Pleurotustuberregium* (fries) singer. *LWT Food Science Technology, 39*, 548–553.

Alam, N., Sikder, M. M., Karim, M. A., & Amin, S. M. R. (2019). Antioxidant and anti-tyrosinase activities of milky white mushroom. *Bangladesh Journal of Botany, 48*, 1065–1073.

Alemu, F. (2014). Cultivation of *Lentinus edodes* on teff straw (agricultural residue) at Dilla University, Ethiopia. *Applied Microbiology, 1*, 49–59.

Alkan, S., Uysal, A., Kasik, G., Vlaisavljevic, S., Berezni, S., & Zengin, G. (2020). Chemical characterization, antioxidant, enzyme inhibition and antimutagenic properties of eight mushroom species: A comparative study. *Journal of Fungi, 6*, 166.

Azeez, L., Adedokun, S., Adeoti, A., & Babalola. (2018). Quality characteristics of fortified bread produced from cassava and mushroom flours. *Journal of Food Processing & Technology, 9*, 9–13.

Bach, F., Zielinski, A. A. F., Helm, C. V., Maciel, G. M., Pedro, A. C., Stafussa, A. P., Avila, S., & Haminiuk, C. W. I. (2019). Bio compounds of edible mushrooms: *In vitro antioxidant and antimicrobial activities*. *LWT Food Science Technology, 107*, 214–220.

Badalyan, S. M. (2014). Potential of mushroom bioactive molecules to develop healthcare biotech products. In *Proceedings of the 8th international conference on mushroom biology and mushroom products (ICMBMP8)* (pp. 373–378).

Bai, J., Ren, Y., Li, Y., Fan, M., Qian, H., Wang, L., & Rao, Z. (2019). Physiological functionalities and mechanisms of β-glucans. *Trends in Food Science and Technology, 88*, 57–66.

Barros, L., Cruz, T., Baptista, P., Estevinho, L. M., & Ferreira, I. C. (2008). Wild and commercial mushrooms as source of nutrients and nutraceuticals. *Food and Chemical Toxicology, 46*, 2742–2747.

Bechaux, J., Gatellier, P., Le Page, J. F., Drillet, Y., & Sante-Lhoutellier, V. (2019). A comprehensive review of bioactive peptides obtained from animal byproducts and their applications. *Food & Function, 10*, 6244–6266.

Bello, M., Oluwamukomi, M. O., & Enujiugha, V. N. (2017). Nutrient composition and sensory properties of biscuit from mushroom-wheat composite flours. *Archives of Current Research International, 9*, 1–11.

Bergendiova, K., Tibenska, E., & Majtan, J. (2011). Pleuran (β-glucan from *Pleurotus ostreatus*) supplementation, cellular immune response and respiratory tract infections in athletes. *European Journal of Applied Physiology, 111*, 2033–2040.

Bhandari, D., Rafiq, S., Gat, Y., Gat, P., Waghmare, R., & Kumar, V. (2020). A review on bioactive peptides: Physiological functions, bioavailability and safety. *International Journal of Peptide Research and Therapeutics, 26*, 139–150.

Bobek, P., Ginter, E., Jurcovicova, M., Ozdin, L., & Mekinova, D. (1991). Cholesterol lowering effect of the mushroom *Pleurotus ostreatus* in hereditary hypercholesterolemic rats. *Annals of Nutrition and Metabolism Journal, 35*, 191–195.

Boels, D., Landreau, A., Bruneau, C., Garnier, R., Pulce, C., & Labadie, M. (2014). *Shiitake* dermatitis recorded by French poison control centers – New case series with clinical observations. *Clinical Toxicology, 52*, 625–628.

Boh, B., & Berivic, M. (2007). *Grifola frondosa* (Diks.: Fr.) S.F Gray: (*Maitake* mushroom): Medicinal properties, active compounds, and biotechnological cultivation. *International Journal of Medicinal Mushrooms, 9*, 89–108.

Bouzgarrou, C., Amara, K., Reis, F. S., Barreira, J. C. M., Skhiri, F., Chatti, N., Martins, A., Barros, L., & Ferreira, I. (2018). Incorporation of tocopherol-rich extracts from mushroom mycelia into yogurt. *Food & Function, 9*, 3166–3172.

Brandt, C. R., & Piraino, F. (2000). Mushroom's antivirals. Recent research and development. *Antimicrobial Agents and Chemotherapy, 4*, 11–26.

C. Codex Alimentarius. (1987). "General principles for the addition of essential nutrients to foods CAC/GL 09–1987," Jt. FAO/WHO Food Stand. Program. Codex Aliment. *Commission*, 13–16.

Chandrawanshi, N. K., Tandiya, D. K., & Jadhav, S. K. (2017). Nutraceutical properties evaluation of *Schizophyllum commune*. *Indian Journal of Science Research, 13*, 57–62.

Chandrawanshi, N. K., Tandiya, D. K., & Jadhav, S. K. (2018). Determination of antioxidant and antidiabetic activities of polar solvent extracts of *Daedaleopsis confragosa* (Bolton) J. Schröt. *Research Journal of Pharmacy and Technology, 11*, 5623–5630.

Chang, S. T. (1999). Global impact of edible and medicinal mushrooms on human welfare in the 21st century: Non-green revolution. *International Journal of Medicinal Mushrooms, 1*, 1–7.

Chang, S. T. (2008). Overview of mushroom cultivation and utilization as functional foods. In P. C. K. Cheung (Ed.), *Mushrooms as functional foods* (pp. 1–33). Wiley.

Chen, S. Y., Ho, K. J., Hsieh, Y. J., Wang, L. T., & Mau, J. L. (2012). Contents of lovastatin, γ-amino butyric acid and ergothioneine in mushroom fruiting bodies and mycelia. *LWT Food Science Technology, 47*, 274–278.

Chen, L., Zhang, Y., & Sha, O. (2016). Hypolipidemic and hypoglycemic activities of polysaccharide from *Pleurotus eryngii*in Kunming mice. *International Journal of Biological Macromolecules, 93*, 1206–1209.

Cheng, L., Wang, Y., He, X., & Wei, X. (2018). Preparation, structural characterization and bioactivities of se-containing polysaccharide: A review. *International Journal of Biological Macromolecules, 120*, 82–92.

Cheskin, L. J., Davis, L. M., Lipsky, L. M., Mitola, A. H., Lycan, T., Mitchell, V., Mickle, B., & Adkins, E. (2008). Lack of energy compensation over 4 days when white button mushrooms are substituted for beef. *Appetite, 51*, 50–57.

Contato, A. G., Inacio, F. D., de Araujo, C. A. V., Brugnari, T., Maciel, G. M., Haminiuk, C. W. I., Bracht, A., Peralta, R. M., & de Souza, C. G. M. (2020). Comparison between the aqueous extracts of mycelium and basidioma of the edible mushroom *Pleurotus pulmonarius*: Chemical composition and antioxidant analysis. *Journal of Food Measurement and Characterization, 14*, 830–837.

Cornelia, M., & Chandra, J. (2019). Utilization of white oyster mushroom powder (*Pleurotus ostreatus*) in the making of biscuit as emergency food product. *EurAsian Journal of BioSciences, 13*, 1859–1866.

Crisan, E. V., & Sands, A. (1998). *The biology and cultivation of edible mushrooms* (p. 137). In S. T. Chang and W. A. Hayes (Eds) Academic.

Daliri, E. B. M., Oh, D. H., & Lee, B. H. (2017). Bioactive peptides. *Food, 6*, 32.

Devi, K. S. P., Roy, B., Patra, P., Sahoo, B., Islam, S. S., & Maiti, T. K. (2013). Characterization and lectin microarray of an immunomodulatory heteroglycan from *Pleurotus ostreatus* mycelia. *Carbohydrate Polymers, 94*, 857–865.

Ditamo, Y., Rupil, L. L., & Sendra, V. G. (2016). *In vivo* immunomodulatory effect of the lectin from edible mushroom *Agaricus bisporus*. *Food & Function, 7*, 262–269.

El Enshasy, H. A., & Hatti-Kaul, R. (2013). Mushroom immunomodulators: Unique molecules with unlimited applications. *Trends in Biotechnology, 31*, 1–10.

Elmastas, M., Isildak, O., Turkekul, I., & Temur, N. (2007). Determination of antioxidant activity and antioxidant compounds in wild edible mushrooms. *Journal of Food Composition and Analysis, 20*, 337–345.

Erdmann, K., Cheung, B. W. Y., & Schroder, H. (2008). The possible roles of food-derived bioactive peptides in reducing the risk of cardiovascular disease. *The Journal of Nutritional Biochemistry, 19*, 643–654.

Erjavec, J., Kos, J., Ravnikar, M., Dreo, T., & Saboti, J. (2012). Proteins of higher fungi from forest to application. *Trends in Biotechnology, 30*, 259–273.

Fan, L., Pan, H., Soccol, A. T., Pandey, A., & Soccol, C. R. (2006). Advances in mushroom research in the last decade. *Food Technology and Biotechnology, 44*, 303–311.

Farzana, T., & Mohajan, S. (2015). Effect of incorporation of soy flour to wheat flour on nutritional and sensory quality of biscuits fortified with mushroom. *Food Science & Nutrition, 3*, 363–369.

Ferrari, S., Azevedo De Oliveira, A. B. G., Mannochio Russo, H., Carvalho Bertozo, L., Silva Bolzani, V., Cunha Zied, D., Farias Ximenes, V., & Zeraik, M. L. (2021). *Pleurotusostreatus* and *Agaricus subrufescens*: Investigation of chemical composition and antioxidant properties of these mushrooms cultivated with different handmade and commercial supplements. *International Journal of Food Science, 56*, 452–460.

Ferreira, I. C. F. R., Fernandes, A., & Heleno, S. A. (2017). *Chemical, nutritional, and bioactive potential of mushrooms edible and medicinal mushrooms: Technology and applications* (1st edn). In D. C. Zied & A. Pardo-Gimenez (Eds) Wiley.

Finimundy, T., Gambato, G., & Fontana, R. (2013). Aqueous extracts of *Lentinula edodes* and *Pleurotus sajor-caju* exhibit high antioxidant capability and promising in vitro antitumor activity. *Nutrition Research, 33*, 76–84.

Ganesan, K., & Xu, B. (2018). Anti-obesity effects of medicinal and edible mushrooms. *Molecules, 23*, 2880.

Ghanbarian, D., BaraaniDastjerdi, M., & Torki-Harchegani, M. (2016). Mass transfer characteristics of bisporus mushroom (*Agaricus bisporus*) slices during convective hot air drying. *Heat and Mass Transfer, 52*, 1081–1088.

Giavasis, I. (2014). Bioactive fungal polysaccharides as potential functional ingredients in food and nutraceuticals. *Current Opinion in Biotechnology, 26*, 162–173.

Giri, S., & Prasad, S. (2007). Drying kinetics and rehydration characteristics of microwave-vacuum and convective hot-air dried mushrooms. *Journal of Food Engineering, 78*, 512–521.

Gogoi, P., Chutia, P., Singh, P., & Mahanta, C. L. (2019). Effect of optimized ultrasound-assisted aqueous and ethanolic extraction of *Pleurotus citrinopileatus* mushroom on total phenol, flavonoids and antioxidant properties. *Journal of Food Process Engineering, 42*, 13172.
Guillamon, S., Garcia-Lafuente, A., & Lozano, M. (2010). Edible mushrooms: Role in the prevention of cardiovascular diseases. *Fitoterapia, 81*, 715–723.
Gupta, S., Summuna, B., Gupta, M., & Annepu, S. K. (2019). Edible mushrooms: Cultivation, bioactive molecules, and health benefits. In J. M. Merillon & K. G. Ramawat (Eds.), *Bioactive molecules in food*. Springer. https://doi.org/10.1007/978-3-319-54528-8_86-1
He, M., Su, D., & Liu, Q. (2017). Mushroom lectin overcomes hepatitis B virus tolerance via TLR6 signaling. *Scientific Reports, 7*, 581–584.
Heleno, S. A., Barros, L., & Martins, A. (2012). Phenolic, polysaccharidic, and lipidic fractions of mushrooms from Northeastern Portugal: Chemical compounds with antioxidant properties. *Journal of Agricultural and Food Chemistry, 60*, 4634–4640.
Huang, J., Ou, Y., & Yew, T. W. D. (2016). Hepatoprotective effects of polysaccharide isolated from *Agaricus bisporus* industrial wastewater against CCl4-inducedhepatic injury in mice. *International Journal of Biological Macromolecules, 82*, 678–685.
Ibrahium, M., & Hegazy, A. (2014). Effect of replacement of wheat flour with mushroom powder and sweet potato flour on nutritional composition and sensory characteristics of biscuits. *Current Science International, 3*, 26–33.
Jayakumar, T., Thomas, P. A., & Geraldine, P. (2009). In-vitro antioxidant activities of an ethanolic extract of the oyster mushroom, *Pleurotus ostreatus*. *Innovative Food Science and Emerging Technologies, 10*, 228–234.
Jeong, C. H., & Shim, K. H. (2004). Quality characteristics of sponge cakes with addition of *Pleurotuseryngii* mushroom powders. *Journal of the Korean Society of Food Science and Nutrition, 33*, 716–722.
Jesenak, M., Majtan, J., Rennerova, Z., Kyselovic, J., Banovcin, P., & Hrubisko, M. (2013). Immunomodulatory effect of pleuran (*β*-glucan from *Pleurotus ostreatus*) in children with recurrent respiratory tract infections. *International Immunopharmacology, 15*, 395–399.
Keerthana, K., Anukiruthika, T., Moses, J. A., & Anandharamakrishnan, C. (2020). Development of fiber-enriched 3D printed snacks from alternative foods: A study on button mushroom. *Journal of Food Engineering, 287*, 110–116.
Kennett, G. A., & Clifton, P. G. (2010). New approaches to the pharmacological treatment of obesity: Can they break through the efficacy barrier. *Pharmacology, Biochemistry, and Behavior, 97*, 63–83.
Khatun, S., Islam, A., Cakilcioglu, U., & Chatterjee, N. C. (2012). Research on mushroom as a potential source of nutraceuticals: A review on Indian perspective. *American Journal of Experimental Agriculture, 2*, 47–73.
Khider, M., Seoudi, O., & Abdelaliem, Y. F. (2017). Functional processed cheese spreads with high nutritional value as supplemented with fresh and dried mushrooms. *International Journal of Food Sciences and Nutrition, 6*, 45–52.
Kim, K. (2011). Bioproduction of mushroom mycelium of *Agaricus bisporus* by commercial submerged fermentation for the production of meat analogue. *Society of Chemical Industry, 91*, 1561–1568.
Kim, B. R., & Joo, N. M. (2012). Optimization of sweet rice muffin processing prepared with oak mushroom (*Lentinus edodes*) powder. *Journal of The Korean Society of Food Culture, 27*, 202–210.
Kim, Y. J., Jung, I. K., & Kwak, E. J. (2010). Quality characteristics and antioxidant activities of cookies added with *Pleurotuseryngii* powder. *Korean Journal of Food Science and Technology, 42*, 183–189.
Klaus, A., Kozarski, M., & Vunduk, J. (2017). Antibacterial and antifungal potential of wild basidiomycete mushroom *Ganoderma applanatum*. *Lekovite Sirovine, 36*, 37–46.
Kozarski, M., Klaus, A., Niksic, M., Jakovljevic, D., Helsper, J. P. F. G., & Van Griensven, L. J. D. (2011). Antioxidative and immunomodulating activities of polysaccharide extracts of

the medicinal mushrooms *Agaricus bisporus*, *Agaricus brasiliensis*, *Ganoderma lucidum* and *Phellinus linteus*. *Food Chemistry, 129*, 1667–1675.

Kulkarni, S., & Sanghai, N. (2014). Screening of antiviral compounds from plants-A review. *Journal of Pharmacy Research, 8*, 1050–1058.

Kumar, K., & Barmanray, A. (2007). Nutritional evaluation and storage studies of button mushroom powder fortified biscuits. *Proteins, 96*, 325.

Lakhanpal, T. N., & Rana, M. (2005). Medicinal and nutraceutical genetic resources of mushrooms. *Plant Genetic Resources, 3*, 288–303.

Lam, T. B., & Ng, S. K. (2001). Hypsin, a novel thermostable ribosome-inactivating protein with antifungal and antiproliferative activities from fruiting bodies of the edible mushroom *Hypsizigusmarmoreus*. *Biochemical and Biophysical Research Communications, 285*, 1071–1075.

Lee, M., Kyung, K., & Chang, H. (2004). Effect of mushroom (*Lentinus tuber-regium*) powder on the bread making properties of wheat flour. *Korean Journal of Food Science and Technology, 36*, 32–37.

Legentil, L., Paris, F., & Ballet, C. (2015). Molecular interactions of β-(1⟶3)-glucans with their receptors. *Molecules, 20*, 9745–9766.

Li, S., Xiong, Q., Lai, X., Li, X., Wan, M., Zhang, J., & Li, Y. (2015). Molecular modification of polysaccharides and resulting bioactivities. *Comprehensive Reviews in Food Science and Food Safety, 15*, 237–250.

Li, X. Y., Zheng, Q. Z., & Zhou, Y. Z. (2019). Fungal immunomodulatory proteins: Characteristic, potential antitumor activities and their molecular mechanisms. *Drug Discovery Today, 24*, 307–314.

Lin, L. Y., Tseng, Y. H., Li, R. C., & Mau, J. L. (2008). Quality of *Shiitake* stipe bread. *Journal of Food Processing & Preservation, 32*, 1002–1015.

Lin, S., Ching, L. T., Lam, K., & Cheung, P. C. K. (2017). Anti-angiogenic effect of water extract from the fruiting body of *Agrocybeaegerita*. *LWT Food Science Technology, 75*, 155–163.

Liu, Y. T., Sun, J., Luo, Z. Y., Rao, S. Q., Su, Y. J., Xu, R. R., & Yang, Y. J. (2012). Chemical composition of five wild edible mushrooms collected from Southwest China and their antihyperglycemic and antioxidant activity. *Food and Chemical Toxicology, 50*, 1238–1244.

Lu, X., Brennan, M. A., Guan, W., Zange, J., Yuan, L., & Brennan, C. S. (2021). Enhancing the nutritional properties of bread by incorporating mushroom bioactive compounds: The manipulation of thepre-dictiveglycaemic response and the phenolic properties. *Food, 10*, 731.

Ma, G., Yang, W., Zhao, L., Pei, F., Fang, D., & Hu, Q. (2018). A critical review on the health promoting effects of mushrooms nutraceuticals. *Food Science and Human Wellness, 7*, 125–133.

Mahamud, M. M., Shirshir, M. R. I., & Hasan, M. R. (2012). Fortification of wheat bread using mushroom powder. *Bangladesh Research Publications Journal, 7*, 60–68.

Majeed, M., Khan, M. U., Owaid, M. N., Khan, M. R., Shariati, M. A., Igor, P., & Ntsefong, G. N. (2017). Development of oyster mushroom powder and its effects on physicochemical and rheological properties of bakery products. *Journal of Microbiology, Biotechnology and Food Sciences, 6*, 1221–1227.

Majtan Jesenak, M. (2018). β-Glucans: Multi-functional modulator of wound healing. *Molecules, 23*, 806.

Michalak, A. (2006). Phenolic compounds and their antioxidant activity in plants growing under heavy metal stress. *Polish Journal of Environmental Studies, 15*, 523–530.

Mizuno, T., Sakai, T., & Chihara, G. (1995). Health foods and medicinal usage of mushrooms. *Food Reviews International, 11*, 69–81.

Montesano, D., Gallo, M., Blasi, F., & Cossignani, L. (2020). Biopeptides from vegetable proteins: New scientific evidences. *Current Opinion in Food Science, 31*, 31–37.

Moon, B., & Lo, Y. M. (2013). Conventional and novel applications of edible mushrooms in today's food industry. *Journal of Food Processing and Preservation, 38*, 2146–2153.

Ng, S. H., Robert, S. D., Ahmad, W. A. N. W., & Ishak, W. R. W. (2017). Incorporation of dietary fiber-rich oyster mushroom (*Pleurotussajor-caju*) powder improves postprandial glycemic response by interfering with starch granule structure and starch digestibility of biscuit. *Food Chemistry, 227*, 358–368.

Ning, M., Hongling, T., Hengjun, D., Liyan, Z., Qiuhui, H., & Hang, X. (2020). Antifatigue effect of functional cookies fortified with mushroom powder (*Tricholoma matsutake*) in mice. *Journal of Food Science*. https://doi.org/10.1111/1750-3841.15510

Oba, K., Kobayashi, M., & Matsui, T. (2009). Individual patient based meta-analysis of lentinan for unresectable/recurrent gastric cancer. *Anticancer Research, 29*, 2739–2745.

Okafor, J., Okafor, G., Ozumba, A., & Elemo, G. (2012). Quality characteristics of bread made from wheat and Nigerian oyster mushroom (*Pleurotusplumonarius*) powder. *Pakistan Journal of Nutrition, 11*, 5–10.

Okamura-matsui, T., Tomada, T., Fukuda, S., & Ohsugi, M. (2003). Discovery of alcohol dehydrogenases from mushrooms and application to alcoholic beverages. *Journal of Molecular Catalysis B: Enzymatic, 23*, 133–144.

Oyetayo, V. O., & Oyedeji, R. R. (2017). Proximate and mineral composition of bread fortified with mushroom (*Plerotusostreatus* and *Calocybe indica*). *Microbiology Research Journal International, 19*, 1–9.

Pan, W. J., Ding, Q. Y., Wang, Y., Wang, D. D., Lu, Y. M., & Yang, W. W. (2018). A bioactive polysaccharide TLH-3 isolated from *Tricholomalobayense* protects against oxidative stress-induced premature senescence in cells and mice. *Journal of Functional Foods, 42*, 159–170.

Patel, S., & Goyal, A. (2012). Recent developments in mushrooms as anti-cancer therapeutics: A review. *3Biotech, 2*, 1–15.

Pavithra, M., Sridhar, K. R., Greeshma, A. A., & Tomita-Yokotani, K. (2016). Bioactive potential of the wild mushroom *Astraeus hygrometricus* in South-west India. *Mycology, 7*, 191–202.

Petrovska, B. B. (2020). Protein fraction in edible Macedonian mushrooms. *European Food Research and Technology, 212*, 469–472.

Pires, R. A., Ruthes, A. C., & Cadena, S. M. S. C. (2017). Cytotoxic effect of a Mann galactoglucan extracted from *Agaricus bisporus* on HepG2 cells. *Carbohydrate Polymers, 170*, 33–42.

Poddar, K. H., Ames, M., Hsin-Jen, C., Feeney, M. J., Wang, Y., & Cheskin, L. J. (2013). Positive effect of mushrooms substituted for meat on body weight, body composition, and health parameters. A 1-year randomized clinical trial. *Appetite, 71*, 379–387.

Popov, A., Panchenko, A., Denisova, N., Petrischev, N., & Shamtsyan, M. (2008). Hypocholesterolic effect of some higher basidiomycetes. In G. E. Zaikov (Ed.), *Research progress in biotechnology* (pp. 53–58). Nova Science Publishers, Inc.

Pradeep, P., Vidya, M., & Feraz, M. F. (2019). Antiviral potency of mushroom constituents. https://doi.org/10.1007/978-981-13-6382-5_10.

Pretus, H. A., Ensley, H. E., McNamee, R. B., Jones, E. L., Browder, I. W., & Williams, D. L. (1991). Isolation, physicochemical characterization and preclinical efficacy evaluation of soluble scleroglucan. *The Journal of Pharmacology and Experimental Therapeutics, 257*, 500–510.

Rai, R. D., & Arumuganathan, T. (2008). Post harvest technology of mushrooms. *Indian Council of Agricultural Research*, Chambaghat, Solan –173 213 (HP), India, 68–69.

Rana, N., Vaidya, D., Sharma, S., & Chauhan, N. (2015). Chemical profile and amino acids composition of edible mushroom. *International Journal of Agriculture Environment and Biotechnology, 8*, 675–679.

Rathee, S., Rathee, D., & Rathee, D. (2012). Mushrooms as therapeutic agents. *Brazilian Journal of Pharmacognosy, 22*, 459–474.

Rathore, H., Sehwag, S., Prasad, S., & Sharma, S. (2010). Technological, nutritional, functional and sensorial attributes of the cookies fortified with *Calocybe indica* mushroom. *Journal of Food Measurement and Characterization, 13*, 976–987.

Rathore, H., Prasad, S., Kapri, M., Tiwari, A., & Sharma, S. (2019). Medicinal importance of mushroom mycelium: Mechanisms and applications. *Journal of Functional Foods, 56*, 182–193.

Reis, F. S., Barros, L., Martins, A., & Ferreira, I. C. F. R. (2012). Chemical composition and nutritional value of the most widely appreciated cultivated mushrooms: An inter-species comparative study. *Food and Chemical Toxicology, 50*, 191–197.
Reis, F. S., Barros, L., Sousa, M. J., Martins, A., & Ferreira, I. C. F. R. (2014). Analytical methods applied to the chemical characterization and antioxidant properties of three wild edible mushroom species from northeastern Portugal. *Food Analytical Methods, 7*, 645–652.
Ren, D., Wang, N., & Guo, J. (2016). Chemical characterization of *Pleurotuseryngii* polysaccharide and its tumor-inhibitory effects against human hepato blastoma HepG-2 cells. *Carbohydrate Polymers, 138*, 123–133.
Rodrigues Barbosa, J., dos Santos Freitas, M. M., da Silva Martins, L. H., & de Carvalho, R. N. (2019). Polysaccharides of mushroom *Pleurotus* spp: New extraction techniques, biological activities and development of new technologies. *Carbohydrate Polymers*. https://doi.org/10.1016/j.carbpol.2019.115550
Ruiz-Herrera, J., & Ortiz-Castellanos, L. (2019). Cell wall glucans of fungi. A review. *The Cell Surface, 5*, 100022. https://doi.org/10.1016/j.tcsw.2019.100022
Ryan, J. T., Ross, R. P., Bolton, D., Fitzgerald, G. F., & Stanton, C. (2011). Bioactive peptides from muscle sources: Meat and fish. *Nutrients, 3*, 765–791.
Sachdeva, V., Roy, A., & Bharadvaja, N. (2020). Current prospects of nutraceuticals: A review. *Current Pharmaceutical Biotechnology, 21*, 884–896.
Salehi, F. (2019). Characterization of different mushrooms powder and its application in bakery products: A review. *International Journal of Food Properties, 22*, 1375–1385.
Salehi, F., Kashaninejad, M., Asadi, F., & Najafi, A. (2016). Improvement of quality attributes of sponge cake using infrared dried button mushroom. *Journal of Food Science and Technology, 53*, 1418–1420.
Salehi, F., Kashaninejad, M., & Jafarianlari, A. (2017). Drying kinetics and characteristics of combined infrared-vacuum drying of button mushroom slices. *Heat and Mass Transfer, 53*, 1751–1759.
Sekete, M., Ma, D., Wang, B., Wang, H., & Ng, T. (2012). First biochemical characterization of a novel ribonuclease from wild mushroom *Amanita hemibapha*. *Springer Plus, 1*, 79.
Serrem, C. A., de Kock, H. L., & Taylor, J. R. N. (2011). Nutritional quality, sensory quality and consumer acceptability of *Sorghum* and bread wheat biscuits fortified with defatted soy flour. *International Journal of Food Science & Technology, 46*, 74–83.
Shamtsyan, M. (2016). Potential to develop functional food products from mushroom bioactive compounds. *Journal of Hygienic Engineering and Design, 15*, 1–9.
Sharma, A., Kumar, P., Sharma, P., & Shrivastav, B. A. (2013). Comparative study of regulatory registration procedure of nutraceuticals in India, Canada and Australia. *International Journal of Pharmaceutical Quality Assurance, 4*, 61–66.
Sheikh, M., Kumar, A., Islam, M., & Mahomud, M. (2010). The effects of mushroom powder on the quality of cake. *Progressive Agriculture, 21*, 205–214.
Singdevsachan, S. K., Auroshree, P., Mishra, J., Baliyarsingh, B., Tayung, K., & Thatoi, H. (2016). Mushroom polysaccharides as potential prebiotics with their antitumor and immunomodulating properties: A review. *Bioactive Carbohydrates and Dietary Fibre, 7*, 1–14.
Singh, K., & Thakur, M. (2016). Formulation, organoleptic and nutritional evaluation of value-added baked product incorporating oyster mushrooms (*Pleurotus ostearus*) powder. *International Journal of Food Sciences and Nutrition, 1*, 16–20.
Singh, P., Langowski, H. C., Wani, A. A., & Saengerlaub, S. (2010). Recent advances in extending the shelf life of fresh *Agaricus* mushrooms: A review. *Journal of the Science of Food and Agriculture, 90*, 1393–1402.
Singh, J., Sindhu, S. C., Sindhu, A., & Yadav, A. (2016). Development and evaluation of value-added biscuits from dehydrated shiitake (*Lentinus edodes*) mushroom. *International Journal of Current Research, 8*, 27155–27159.
Sulieman, A. A., Zhu, K. X., Peng, W., Hassan, H. A., Obadi, M., Siddeeg, A., & Zhou, H. M. (2019). Rheological and quality characteristics of composite gluten-free dough and

biscuits supplemented with fermented and unfermented *Agaricus bisporus* polysaccharide flour. *Food Chemistry, 271*, 193–203.

Sun, Q., Ma, J., Campos, H., Hankinson, S. E., Manson, J. E., Stampfer, M. J., Rexrode, K. M., Willett, W. C., & Hu, F. B. (2007). A prospective study of trans fatty acids in erythrocytes and risk of coronary heart disease. *Circulation, 115*, 1858–1865.

Udenigwe, C. C., & Aluko, R. E. (2012). Food protein-derived bioactive peptides: Production, processing, and potential health benefits. *Journal of Food Science, 77*, 11–24.

Ukwuru, M. U., Muritala, A., & Eze, L. U. (2018). Edible and non-edible wild mushroom: Nutrition, toxicity and strategies for recognition. *Journal of Clinical Nutrition and Metabolism, 2*, 2.

Ulziijargal, E., Yang, J. H., Lin, L. Y., Chen, C. P., & Mau, J. L. (2013). Quality of bread supplemented with mushroom mycelia. *Food Chemistry, 138*, 70–76.

Valverde, M. E., Hernandez-Perez, T., Paredes-Lopez, O., Valverde, M. E., Hernandez-Perez, T., & Paredes-Lopez, O. (2015). Edible mushrooms: Improving human health and promoting quality life. *International Journal of Microbiology, 2015*, 1–15.

Volman, J. J., Mensink, R. P., Van Griensven, L. J. L. D., & Plat, J. (2010). Effects of α-glucans from *Agaricus bisporus* on *ex vivo* cytokine production by LPS and PHA-stimulated PBMCs; a placebo-controlled study in slightly hypercholesterolemic subjects. *European Journal of Clinical Nutrition, 64*, 720–726.

Wagay, J. A., Nayik, G. A., Wani, S. A., Mir, R. A., Ahmad, M. A., Rahman, Q. I., & Vyas, D. (2019). Phenolic profiling and antioxidant capacity of *Morchella esculenta* L. by chemical and electrochemical methods at multiwall carbon nanotube paste electrode. *Journal of Food Measurement and Characterization, 13*, 1805–1819.

Waktola, G., & Temesgen, T. (2018). Application of mushroom as food and medicine. *Advances in Biotechnology & Microbiology, 11*, 555817.

Wang, H., & Ng, T. (2006). Purification of a laccase from fruiting bodies of the mushroom *Pleurotuseryngii*. *Applied Microbiology and Biotechnology, 69*, 521–525.

Wang, H. C., Zhang, M., & Adhikari, B. (2015). Drying of *Shiitake* mushroom by combining freeze-drying and mid-infrared radiation. *Food and Bioproducts Processing, 94*, 507–517.

Wasser, S. P. (2011). Current findings, future trends, and unsolved problems in studies of medicinal mushrooms. *Applied Microbiology and Biotechnology, 89*, 1323–1332.

Wasser, S. P., & Weis, A. L. (1999). Medicinal properties of substances occurring in higher basidiomycete mushrooms: Current perspective. *International Journal of Medicinal Mushrooms, 1*, 31–62.

Williams, D. L., Mueller, A., & Browder, W. (1996). Glucan-based macrophage stimulators. *Clinical Immunotherapeutics, 5*, 392–399.

Wong, J. H., Wang, H. X., & Ng, T. B. (2008). Marmorin, a new ribosome inactivating protein with antiproliferative and HIV-1 reverse transcriptase inhibitory activities from the mushroom *Hypsizigusmarmoreus*. *Applied Microbiology and Biotechnology, 81*, 669–674.

Wu, X., Huang, C., & Chen, Q. (2014). A novel laccase with inhibitory activity towards HIV-I reverse transcriptase and antiproliferative effects on tumorcells from the fermentation broth of mushroom *Pleurotuscornucopiae*. *Biomedical Chromatography, 28*, 548–553.

Xiaokang, W., Brunton, N. P., Lyng, J. G., Harrison, S. M., Carpes, S. T., & Papoutsis, K. (2020). Volatile and non-volatile compounds of *Shiitake* mushrooms treated with pulsed light after twenty-four-hour storage at different conditions. *Food Bioscience, 36*, 100619.

Yahia, E. M., GutierrezOrozco, F., & Moreno-Perez, M. A. (2017). Identification of phenolic compounds by liquid chromatography mass spectrometry in seventeen species of wild mushrooms in Central Mexico and determination of their antioxidant activity and bioactive compounds. *Food Chemistry, 226*, 14–22.

Yen, M. T., Yang, J. H., Tsenh, Y. H., Li, R. C., & Mau, J. L. (2011). Quality of fungal chitin bread. *Journal of Food Processing & Preservation, 35*, 708–713.

Zehtab, T., Yazdanparast, R., & Rafieii, S. (2000). Prevention of experimental autoimmune vitiligo by oral administration of mushroom tyrosinase. *Iranian Biomedical Journal, 4*, 31–36.

Zhang, M., Cui, S. W., Cheung, P. C. K., & Wang, Q. (2007). Antitumor polysaccharides from mushrooms: A review on their isolation process, structural characteristics and antitumor activity. *Trends in Food Science & Technology, 18*, 4–19.

Zhang, Y., Hu, T., & Zhou, H. (2016a). Antidiabetic effect of polysaccharides from *Pleurotusostreatus* in streptozotocin-induced diabetic rats. *International Journal of Biological Macromolecules, 83*, 126–132.

Zhang, Y., Yang, X., & Jin, G. (2016b). Polysaccharides from *Pleurotusostreatus* alleviates cognitive impairment in a rat model of Alzheimer's disease. *International Journal of Biological Macromolecules, 92*, 935–941.

Zhu, F., Du, B., Bian, Z., & Xu, B. (2015). *β*-Glucans from edible and medicinal mushrooms, characteristics, physicochemical and biological activities. *Journal of Food Composition and Analysis, 41*, 165–173.

Zong, A., Cao, H., & Wang, F. (2012). Anticancer polysaccharides from natural resources: A review of recent research. *Carbohydrate Polymers, 90*, 1395–1410.

Chapter 11
Fungi in Pharmaceuticals and Production of Antibiotics

Bandita Dutta, Dibyajit Lahiri, Moupriya Nag, Sougata Ghosh, Ankita Dey, and Rina Rani Ray

Contents

B. Dutta · A. Dey
Department of Biotechnology, Maulana Abul Kalam Azad University of Technology, Haringhata, West Bengal, India

D. Lahiri · M. Nag
Department of Biotechnology, University of Engineering & Management, Kolkata, West Bengal, India

S. Ghosh · R. R. Ray (✉)
Department of Microbiology, School of Science, RK University, Rajkot, Gujarat, India

A. C. Shukla (ed.), *Applied Mycology*, Fungal Biology,
https://doi.org/10.1007/978-3-030-90649-8_11

11.1 Introduction

The contribution of microbes in the enhancement of human well-being and health is immense. The natural compounds that are obtained from the microbes not only helped in treating diseases but also helped in enhancing the average life expectancy in human. The invention of penicillin from the mold *Penicillium notatum* by Alexander Fleming in the year 1928 showed that microbial species are the storehouse of a large number of natural compounds that play important role in acting against various types of disease-causing pathogens. Fungi are one of the most important among those five kingdoms as they are essential for the good growth of the plants, different crops through the nutrient cycle by degrading the cellulose and lignin (Pointing et al., 2001). In spite of having numerous negative aspects, fungi possesses biotechnological potentials for industrial exploitation. The large use of antibiotics since its discovery has resulted in the development of various types of resistant microbes by the mechanism of evolutional selection pressure driven by the antibiotics (Davies & Davies, 2010). Thus, classic therapeutics is being rendered ineffective in various types of infections along with resulting in morbidity due to the condition of multidrug resistance (MDR). The natural products produced by the primary and secondary metabolism of microbes have a broad range of therapeutic applications, which is further modified and then amplified through biotechnology for the betterment of society. Microorganisms are the most abundant sources of diverse bioactive metabolites. The pharmaceutical industries have extended their screening program to focus the metabolites from the microbes, effective toward different types of disease areas. Fungi are one of the groups of microbes capable of producing a diverse array of secondary metabolites that have been used in the pharmaceutical industry on a large scale. As the result of incorporation of the screening process, some most important products are immunosuppressants, antitumor drugs, hypocholesterolemia drugs, enzyme inhibitors, growth stimulants, insecticides, and herbicides. This chapter focuses on the use of the fungal metabolites in various types of drug development targeting diseases at large having a special focus on the development of antibiotics with the present-day need.

11.2 Different Roles of Fungi in Pharmaceuticals

Since the discovery of antibiotics in the 1940s, millions of lives were saved by antibiotics. Due to the excessive use of antibiotics, many of the bacteria develop their own protection against antibiotics and become antibiotic resistant. This can lead to the development of lethal infections that can be diminished by the alternative strategy. The discovery and development of the new antibiotic can accomplish the urge of the alternative strategy in order to reduce the lethal infections.

In 2018, the scientific community celebrated the ninetieth anniversary of the discovery of penicillin by Sir Alexander Fleming, from which the era of te antibiotic chemotherapy has started. After the discovery of penicillin in 1928, numerous antibiotic compounds like cephalosporins, fusidic acid, and pleuromutilin were derived from the fungi, which are effective against several diseases.

11.2.1 Antibacterial Antibiotics

Antibiotics are the group of chemical substances that are produced by one set of microbial cells used for the purpose of inhibiting other microorganisms. When an organism is able to produce specific secondary metabolites, which can destroy the competing organisms present in the same habitat, then that secondary metabolite is considered as antibiotics. It is evident that most of the filamentous fungi produce antibiotic compounds (Bills et al., 2009).

In the post-antibiotic era, there is a drastic change in the number and percentages of multidrug-resistant organisms and antibiotics. The inverse relationship between the drug-resistant pathogens, which is increased with the decreased number of therapeutic agents, leads to the development of new antibiotics in order to combat the pathogenicity (Cooper & Shlaes, 2011). The loss of interest of the pharmaceutical companies in research and development on the natural product gave rise to consequences like the scarcity of the antibiotics in the treatment of the multidrug-resistant human pathogenic bacteria. After two decades of this crisis, the discovery of the new antibiotics has increased, which is basically nothing but the chemical modification of the old compounds with a known mode of action. There are few compounds under the pipeline that are advanced and yet to be used in the pharmaceutical sectors.

For example, fermentation-derived compound mutilins from the basidiomycete *Clitopilus passeckerianus* is considered as a new antibacterial drug. For skin infections, several derivatives are under clinical trials, among which a derivative retapamulin has been discovered to be used as the topical antibiotic agent. However, it is difficult to culture basidiomycete on a large scale due to its slow growth rate and low yields. It has been discovered that a transfer of the biosynthetic genes into a fast-growing heterologous host *Aspergillus* can enhance the production of secondary metabolites like pleuromutilin (Bailey et al., 2016). In the middle of twentieth century, around 22% of the known 12,000 antibiotics were produced by the filamentous fungi, among which the β-lactam antibiotics including penicillin, cephalosporin, clavulanic acid, and carbapenems are the most important antibiotic class (Berdy, 1995).

Due to the immense application of those antibiotics, the production needs to be increased to reach the optimum goal. 1,3-Diaminopropane (1,3-DAP) (secreted by *P. chrysogenum* and *Acremonium chrysogenum*) and spermidine increase transcription level of penicillin biosynthetic genes pcbAB, pcbC and penDE to stimulate penicillin G production (Martín, 2012). laeA is the global regulator that regulates expression of secondary metabolism genes via heterochromatin reorganization and gets stimulated via the mechanism. Also, the cephamycin production and spermidine activity in *Amycolatopsis lactamdrans* get stimulated by 1,3-DAP. Due to the increase of antibiotic resistance among fungi and bacteria, novel antibiotics are in dire need. One example of the new and approved antibiotic is ceftobiprole, which is a modified version of cephalosporin, is active against methicillin-resistant S. *aureus,* and is not hydrolyzed by β-lactamases from Gram-positive (Shang, 2011). Another example of antibiotic that acts as an antifungal agent and also the first discovered inhibitor of fatty acid biosynthesis cerulenin is produced by *Acremonium caerelens.* The active site of the nucleophilic cystine of the ketosynthase enzyme of fatty acid synthetase through epoxide ring opening gets alkylated and inactivated by this antibiotic (Vance, 1972).

In a pathogenic fungi *Candida albicans,* poly(A) polymerase gets inhibited by an antifungal product parnafungin from *Fusarium layarum.* Most of the Gram-negative bacteria are resistant to the carbapenem antibiotics, mainly imipenem and meropenem, which give rise to the major antibiotic problem due to the occurrence of New Delhi metallo-β-lactamase (NDM-1), the extended spectrum metallo-β-lactamase. NDM-1 and another metallo-β-lactamase VIM-2 are inhibited by the aspergillomarasmine (AMA) from *Aspergillus versicolor* (King, 2014). AMA is the amended version of meropenem that inhibits metalloproteinases of the bacteria carrying NDM and VIM metallo-β-lactamase (King, 2014).

11.2.2 Antimycotics and Fungicides

Apart from the multidrug-resistant pathogenic bacteria, there are a number of pathogenic fungi whose resistivity is also high. Several efficient compounds are available in the market for antimycotic chemotherapy. One of those compounds is griseofulvin, which is widely used. Echinocandins (pneumocandin B_0) is the new class of antimycotic that has been launched to the market (Denning, 2002). The PKS-NRPS hybrid gene cluster regulates the biosynthesis of highly complex lipopeptides. The significant knowledge about the molecular mechanism leads to incisive manipulation of the production process, which can result in the synthesis of the new derivative by using a diverse group of fungi for the large-scale fermentation of the subsequent biosynthesis of those gene clusters (Chen et al., 2013). It was evidenced that Dothideomycetes, Eurotiomycetes, and Leotiomycetes are different classes of the producer organism Ascomycota having the high homology of the gene cluster due to horizontal gene transfer during evolution (Yue et al., 2015). A substantial number of developmental projects for the discovery of the novel antifungal compounds is under pipeline. Enfumafungin is one of the antifungal drugs originated from the fungal endophyte *Hormonema* spp. used to treat human disease after 15 years of its first discovery (Pelaez et al., 2000). Rediscovery of the old compounds can accomplish the urge of the novel antimycotics and fungicides, which have their significant effectivity against human fungal pathogens. Very few of those compounds have toxic effect toward human, but most of them have strong antifungal effects without any cytotoxicity. For example, favolon is the rediscovered form of strobilurins, which is produced by the invasive basidiomycete *Favolaschia calocera* (Chepkirui et al., 2016), and sporothriolodes is from xylarialean fungus *Hypoxylon monticulosum,* which is now classified under the new genus *Hypomontagnella* (*H. monticulosa*) (Surup et al., 2014; Lambert et al., 2019).

11.2.3 Biofilm Inhibitor

The discovery of the new antibiotics is mainly aimed to combat the infectious diseases caused by both the fungal and bacterial pathogens, which is possible through the inhibition of the biofilm formation. It has been reported that the fungal metabolites have the potency to inhibit the biofilm formation by interfering with a quorum sensing mechanism or by destroying preformed biofilms (Abraham & Estrela, 2016). A research showed that a small molecular derivative of edible mushroom *Coprinus comatus*, coprinuslactone, acts against the *Pseudomonas aeruginosa* biofilms (de Carvalho et al., 2016). The biofilm of the Gram-positive bacteria *Staphylococcus aureus* is inhibited by roussoellenic acid from *Roussoella* sp. (Phukhamsakda et al., 2018) and microporenic acid A from *Kenyan basidiomycete* (Chepkirui et al., 2018). A human pathogenic yeast *Candida albicans* can also be destroyed by microporenic acid A. The efficacy of an antibiotic is enhanced by the biofilm inhibitor that can increase penetration through the biofilm. Though some fungal antibiotics are available and have their efficacy toward the infections, a vast range of fungi and their applications in novel antibiotics are still underexplored.

11.2.4 Antimalarial Agent

In spite of tropical and subtropical areas, worldwide malaria is one of the major causes of illness and death. Around 500 million new cases are detected every year where 1.5 million, people especially pregnant women and children, die due to this. The two major drugs quinine from the bark of the Cinchona tree and artemisinin from the Chinese herb *Artemisia annua* have been used successfully against malaria. In spite of being a new drug, artemisinin is used extensively due to the side effects of the quinine like arrhythmia, thrombocythemia, and cinchonism. Endoperoxide sesquiterpene lactone artemisinin is also effective against multidrug-resistant *Plasmodium falciparum.* Genetic engineering has been implemented as very low amount of artemisinin is obtained from the *A. annua*, 0.01–1% of the weight of the dried leaves. A genetically engineered *S. cerevisiae* strain, which is able to produce 100 mg l−1 of artemisinic acid, has been developed by the Keasling group in Berkeley, California, where further artemisinic acid is chemically converted to artemisinin and artemisinic acid production increased by 1 million fold. Also, by the engineered *S. cerevisiae* the artemisinin precursor amorpha-4, 11-diene is made at 40 g/L (Liu et al., 2013).

11.2.5 Anticancer Agents

After the cardiovascular disease, the second leading cause of mortality is cancer, where nearly 9.6 million cancer-related death cases had been reported. With the loss of growth factors, the uncontrolled cell division and cell proliferation are termed as "cancer." The presence of cancerous cell is referred to as malignancy, which can invade nearby and destroy tissues by suppressing the human immune system. Several treatment processes like chemotherapy, radiation therapy, surgery, and immunotherapy have been administered according to the developmental stage of cancer disease. A wide range of cytotoxic agents are used to suppress cancerous tumor cells. Advanced cancer treatment therapy is based on regulating the gene or protein of the specific cancer cells by the drugs. Though the most used drug albeit has originated from plant and bacteria, some important natural products can be obtained from fungi for the development of novel drugs, and in most cases, those drugs are at clinical or preclinical stages. Irofulven is a semi-synthetic derivative of the natural toxin illudin S, isolated from *Omphalotus illudens* (Chin et al., 2006). Irofulven has antitumor activity, which has been evaluated at the phase I and phase II clinical trials against brain, central nervous system, breast, blood, colon, sarcoma, prostate, lungs, and ovarian and pancreas cancers (Alexandre et al., 2004). Irofulven interacts with the DNA replication complexes and inhibits the DNA synthesis to stop the abnormal cell divisions (Walser & Heinstein, 1973). A tetracyclic diterpene aphidicolin was isolated from *Cephalosporium aphidicola* (Bucknall et al., 1973), competes with DNA polymerase α, δ, and ε enzymes for specific binding sites, and has the antiviral and antimitotic properties (Crosetto et al., 2013). There are other anticancer lead compounds derived from the fungi, among which leptosins derived from *Leptosphaeria* sp. have the antitumor activity in mouse embryos (Pejin et al., 2013). Some natural compounds like β-glucans, palmarumycin, and spiropressione A are found on fungi cell wall and also have the anticancer potency, which is yet to be explored.

The fungal and plant secondary metabolites taxol (paclitaxel isolated from pacific yew tree, *Taxus brevifolia*) and camptothecin have showed marked result in the clinical trial of anticancer drug. Taxol is a steroidal diterpene alkaloid with tetracycline ring and N-benzoylphenyl isoserine side chain that inhibits the mammalian cancer cells from dividing by promoting tubulin polymerization and interfering microtubule breakdown during cell division. The taxol was approved against the refractory ovarian cancer in 1992, and further it was used to treat breast cancer and advanced Kaposi's sarcoma (Newman & Cragg, 2007). Certain plants (angiosperms) and fungi (endophytic), *Entrophospora infrequens,* produce a modified indole alkaloid, camptothecin, which is used as an important antitumor agent and also used to treat colon cancer, unusual activity of lungs, ovarian, and uterine cancers (Amna, 2006).

11.2.6 Antidiabetic Agents

Diabetes mellitus is a chronic metabolic disorder, where insulin is not produced or effectively used in the body. Diabetic patients have high sugar of glucose in their body to the insulin imbalance. There are two types of diabetes, type 1 (insulin dependent) and type 2 (noninsulin dependent). In case of type 1 diabetes patients, due to the malfunctioning of the pancreatic beta cells, insulin is not produced sufficiently. Insulin needs to be injected from outside of the body. Worldwide, around 5–10% people, mostly children and adolescent patients, are type 1 diabetes patients (Meier et al., 2005). Rest 90–95% patients suffer from type 2 diabetes, where their body is not able to produce sufficient insulin or cannot effectively metabolize it. Around 7% of the world's elder population are affected by this disease. In 2017, the largest number of diabetic patients, around 114 million, was recorded in China. Roughly 73 million diabetic patients were recorded in India, and 30 million were recorded in the United States.

Untreated diabetic patients can be the sufferer of blindness, kidney failure, depression, cardiovascular diseases, cancer, and even death. The most common complications that are attributed to the diabetes are retinopathy and neuropathy (De Silva et al., 2016).

For the treatment or prevention of type 2 diabetes, Basidiomycota such as *Agaricus bisporus, Cyclocybe aegerita, C. cylindracea,* and *Tremella fuciformis* are used as alternative medicine as they contain very less amount of the digestible carbohydrate, which can avert high glucose intake of patients (Poucheret et al., 2006). It has been reported that the extracts of *Inocutis levis* have the ability to increase insulin resistance and sensitivity and also control blood glucose levels by increasing glucose uptake in tissues (Ehsanifard et al., 2017). Hyperglycemia (when blood glucose level is high) and hyperinsulinemia (when blood insulin level is high) both along with type 2 diabetes can be treated by *Grifola frondosa* (Poucheret et al., 2006). Some antidiabetic medicinal products like *Ophiocordyceps sinensis* capsules, SX-Fraction (enhance insulin sensitivity), Reishi-Max capsules, and Tremella are made with medicinal mushrooms, used to decrease blood glucose level of type 2 diabetes and reduce blood pressure and body weight (Li et al., 2004).

11.2.7 Improving Nerve Function

Millions of people are affected worldwide by the most common Alzheimer's, Huntington's, and Parkinson's neurodegenerative diseases. To improve nerve functioning, the discovery and development of the neuroactive compounds from the medicinal mushroom are studied to a great extent. Earlier it has been discovered that *Antrodia camphorate, Ganoderma* spp., *Hericium erinaceus, Lignosus rhinocerotis,* and *Pleurotus giganteus* can enhance the peripheral nervous system. Most of the neurodegenerative diseases occur due to the disappearance of nerve growth factor as it is one of the important materials for the adult brain survival, maintenance, and regeneration of specific neuronal populations. Since the past 20 years, scientists are trying to discover some drug components from the fungi that can easily penetrate the blood–brain barrier and induce production of nerve growth factor. *Hericium erinaceus*, the medical mushroom, is mainly used to discover potential neuroactive compounds for the prevention of the neurodegenerative diseases. Hericenones and erinacines, the two terpenoid group of this medical mushroom, stimulate the nerve growth factor synthesis via TrkA/Erk1/2 pathway. Apart from inducing nerve growth factor, this medicinal mushroom also provides the immune support by improving digestive function and enhancing anti-inflammatory and antioxidant activities. It was demonstrated that hericenone A and erinacines C can induce synthesis of nerve factor both in vivo and in vitro (Thongbai et al., 2015). Cyathane diterpenoids, derived from *Cyathus* species, is reported as nerve growth factor enhancer, where the similar effect was observed in the cyathanes from the mycorrhizal basidiomycetes (genus *Sarcodon*) fruit bodies (Bai et al., 2015).

11.2.8 Cardiovascular Disease Control and Cholesterol-Lowering Agent

Heart and blood vessel diseases are termed as "cardiovascular disease." which can cause the average death of 17.3 million people per year due to the elevated level of plasma cholesterol, by which arteries get clogged due to fat deposition. Around 30% of the cholesterol comes from diet and rest 70% is synthesized in the human liver. Only healthy diet is not able to control the cholesterol level in most cases, which leads to the high blood cholesterol level causing atherosclerosis, a chronic disease due to the accumulation of atheromatous plaque within the arterial wall leading toward stenosis and ischemia. Scientists came across that plasma cholesterol level can get reduced by the inhibition of de novo synthesis of cholesterol (Miller, 2001). The reduction of cholesterol is possible by inhibiting the rate-determining step where 3-hydroxy-3-methylglutaryl-coenzyme A (HMG-CoA) is reduced to mevalonate by HMG-CoA reductase (Brown et al., 1976).

Fungal secondary metabolites, hypolipidemic drugs statins are the HMD-CoA reductase inhibitor, have two different moieties: hexahydro-naphthalene system and β-hydroxylactone system. Compactin was isolated from *Penicillium brevicompactum* and has the antibiotic as well as hypocholesterolemic activity. Statin lowers the cholesterol level by inhibiting the rate-limiting enzyme 2-hydroxymethyl glutaryl-CoA of mevalonate pathway of cholesterol biosynthesis. This inhibition mechanism of the enzyme stimulates low-density lipoprotein receptors in the liver, resulting in the augmentation of low-density lipoprotein clearing from the bloodstream in order to decrease the cholesterol level. Hence, the risk of cardiovascular disease is reduced with the reduction of total plasma cholesterol by 20–40% (Endo et al., 1976).

Mevinolin is another HMG-CoA reductase inhibitor that was isolated from *Aspergillus terreus7* (Alberts et al., 1980). In medicinal mushroom, certain molecules are present that are able to modify cholesterol absorption, metabolism, and modulate gene expression of cholesterol hemeostasis (Gil-Ramırez et al., 2016). Some other cholesterol-emic product like compactin (from *Penicillium citrinum*) and lovastatin (from *Monascus rubra* and *Aspergillus terreus*) had been developed at the same time, which has strong medical use with the commercial success.

11.2.9 Antiviral Agents

Virus can cause a serious global epidemic through the health and mortality-deliberating diseases due to insufficient or unavailability of the antiviral chemotherapy or vaccines. Some drug-resistant viral strains are responsible for the current virus-related pandemic due to the loss of drug efficacy. Hence, to combat this viral epidemic by controlling viral infections, a new strategy related to the discovery of the natural drug leads needs to be devised. There are several natural compounds screened for the antiviral activity but are yet to reach the market; one of those compounds is plethora, isolated from fungi. Scientists are more focused on discovering natural compounds that can exhibit potent activity against human pathogenic viruses.

11.2.9.1 Inhibitory Natural Product from Fungi Against Human Immunodeficiency Virus

The discovery of the anti-HIV drug is based on virus entry, reverse transcription, and integration. The discovery of the viral entry blocker is based on the HIV interactions with proteins. HIV-1 protease, which is a key enzyme for replication and maturation of HIV-1 virus, can be inhibited by bis-indolyl quinone, hinnuliquinone isolated from *Quercus coccifera* (Singh et al., 2004). At micromolar concentration, altertoxins I–III and V (oxidized perylenes) from *Alternaria tenuissima* inhibit HIV-1 replication (Bashyal et al., 2014). A strong anti-HIV activity was observed by the dimeric tetrahydroxanthone and penicillixanthone A, from *Aspergillus fumigatus,* as they are able to inhibit CCR5-tropic HIV-1 SF162 and CXCR4-tropic HIV NL4-3 (Tan et al., 2017). Anti-HIV-1 activity is also observed by malformin C, from marine-derived A. *niger* (Zhou et al., 2015). Phenalone and cytochalasin derivatives of endophytic Aspergillus sp. CPCC 400735 showed anti-HIV activity (Pang et al., 2017). Moderate anti-HIV-1 replication was observed in C8166 cells by novel sesquiterpenoids from *Paraconiothyrium brasiliense*, whereas significant HIV-1 activity was observed by pupukeanane sesquiterpenoid chloropupukeannolide A from *Pestalotiopsis fici* (Liu et al., 2010). The discovery of anti-HIV drug depends on the inhibition of three consecutive functions (RNA reverse transcription to DNA, degradation of RNA template by RNase H, and duplication of the remaining DNA strand) controlled by HIV reverse transcription. *Stachybotrys chartarum*-derived phenylspirodrimane metabolite Stachybosin D has the inhibitory effects on HIV-1 replication by targeting reverse transcriptase in wildtype HIV-1 and NNRTIs-resistant strains (Ma et al., 2013). Along with protease and reverse transcriptase, integrase is one of the important proteins encoded by HIV-1. It was reported that several compounds like equisetin, phomasetin, integracins, epiphiobolins C and K, hispidin, caffeic acid, naphtho-y-pyrones, xanthoyiricatins, etc. have the inhibitory activity against integrase.

11.2.9.2 Inhibitory Natural Products from Fungi Against Influenza Virus

Among all the influenza viruses, H1N1 and H3N2 are the most targeted ones for the development of the drug. IC_{50} value determines the moderate to high effectivity of drug molecules against H1N1 and H3N2 viruses. The terpenoid isolate of *Stachybotrys*, stachyflin, has the modest activity against influenza A (H1N1) virus with IC_{50} value of 39×10^{-3} μM (Minagawa et al., 2002). Low micromolar activity against both the influenza viruses was observed by y-pyrone isoasteltoxin from *Aspergillus ochraceopetaliformis* ($IC_{50} = 0.23$ μM) (Wang et al., 2016). Y-pyrone is a derivative asteltoxin E from *Aspergillus* sp. active against H1N1 and H3N2 (Tian et al., 2016). H3N2 influenza A and B virus replication was inhibited by the metabolite of *Gliocladium* sp. aureonitol, with an EC_{50} value 100 nM, via influenza hemagglutination suppression (Sacramento et al., 2015)

11.2.9.3 Inhibitory Natural Products from Fungi Against Herpes Simplex Virus (HSV)

There are no such drug molecules against HSV-1 and HSV-2 pathogenic viruses available in the fungal metabolite library till date. Very few drug molecules had been reported with anti-HSV activity (Huang et al., 2017). *Scytalidium* sp. fungus derived five lipopeptides that have moderate anti-HSV-1 and anti-HSV-2 activities, which are dose and time dependent (Rowley et al., 2003). The diphenyletherglycoside cordyol C from *Cordyceps* sp. exhibited significant anti-HSV-1 activity (Bunyapaiboonsri et al., 2011).

11.2.9.4 Inhibitory Natural Products from Fungi Against Hepatitis Virus

Vanitaracin A is a novel tricyclic polyketide fungal derivative that inhibits the viral entry process. This compound can inhibit all HBV genotypes (A–D) through interacting viral bile acid transport pathway (Kaneko et al., 2015).

Through the inhibition of the Akt activity and depletion of autophagic genes LC3 and p62, the HBV-X replication is decreased by epipolythiodiocopiperazine derivate, 11'-deoxyverticillin (Wu et al., 2015). From nematode-trapping basidiomycete

Hohenbuehelia grisea, anti-HCV agent 4-hydroxypleurogrisein has been discovered (Sandargo et al., 2018). Meroterpenoid rhodatin from rare basidiomycete *Rhodotus palmatus* has the notable anti-HCV activity (Sandargo et al., 2019). Since the viral diseases are increasing day by day, the discovery of the novel antiviral molecules is gaining importance. Most of the natural compounds with nanomolar activity are at the pipeline of the drug discovery.

11.2.10 Immunosuppressive and Immunomodulatory Agents from Fungi

Immunosuppression is a therapeutic process that accounts for preventing the immunity of various patients from defending against organ transplant and tissue transplant. The entire process takes place in the absence of any type of modern medicines, presence of drugs that are immunosuppressive, any type of modern medicine related to transplantation heart, liver, and kidney. Immunosuppressants also possess the ability to control many manifestations of some autoimmune diseases and allergic conditions. In these drugs, majority of them specially devote themselves to some specific biochemical pathways, which are very important for the proper functioning of the defense mechanism of human immunity in response to invading pathogens or microorganisms coming from outside. This is done by selective inhibition of the signal transduction cascade of pathways or the immunocompetent lymphocytes. These cascade pathways or cascade of steps undergo regulation of the cytokine transcription process. This results in the hospitalization and antibiotics treatment of the patients undergoing immunosuppressants for preventing the occurrence of infection in them. Few of the noted immunosuppressive drugs are simply some natural products that are artificially formed by the application of biotechnology that is by fermentation of fungi and bacteria. The compounds tacrolimus and sirolimus are isolated from actinobacteria. Moreover, mycophenolate and cyclosporine mofetil are secondary metabolites formed by fungi. But it is also known that penicillin is considered one of the oldest natural antibiotics available. The first antibiotic invented and extracted from fungi in its crystalline form is a meroterpenoid that is formed by *Penicillium* species, which consists of *P. brevicompactum* and *P. roqueforti* (Del-Cid et al., 2016), and then undergo their process of biosynthesis that is elucidated recently in the other species. Because of different reasons, the compound did not reach its clinical development phase as an antifungal agent and antibacterial agent. So basically, the main purpose as an immunosuppressant is to become an active component of many marketed drugs like those of Myfortic® and CellCept® and to become evident. This compound undergoes selective inhibition to form inosine monophosphate dehydrogenase (IMPDH). Along with this, it also produces an enzyme that is important for the process of biosynthesis of guanosine nucleotides into the cells of mammals. This is because this enzyme is needed more inside the B-lymphocytes and T-lymphocytes than in other cell types (Allison & Eugui, 2000).

Also, the compound and its isotypes, along with the isoform in the lymphocytes, possess more sensitivity to mycophenolic acid, the drug that has more efficient cytostatic effect on lymphocytes than on the rest of the type of cells, and by this entire process, human immune system suppression takes place.

11.2.11 Traditional Chinese Medicine

Traditional Chinese medicine (TCM), scattered in Japan and other Asian countries, has been used since thousands of yearsy. Use of natural remedies (mainly herbal medicines) is the main objective of TCM for disease therapy. The first Chinese medicine was recorded in the eleventh-century BC (1100–1001 BC), and throughout the previous centuries combination of conceptual theories and practical experiences helped to evolve the TCM. Compared to Western medicine, TCM has more holistic ideology in the pharmaceutical field. "Protector herb" or the Chinese medicinal mushrooms are being used for enhancement and maintenance of good health. Majority of Chinese medicines are plant-derived and in very few macrofungi are used. Several bioactive compounds are extracted from the fungal sources, are considered to have the potency against various diseases like cancer, diabetes, and cardiovascular and neurodegenerative diseases, and are used in fungal TCM. Due to the unavailability of a proper description, there are thorough debates regarding the utility and therapeutic applications of TCM in the modern healthcare sectors. Western medicine is more focused to develop drug from a single substance depending upon strong selectivity and pharmacological activity for the therapy of disease. TCM is a mixture of compounds that are isolated from plants, fungi, and other organisms, possessing the ability to obstruct the TCM to fit into the modern healthcare. Still there are many evidences that claim the TCM to be approved in the healthcare system (Tang et al., 2018).

11.3 Pharmaceutical Uses of Fungal Metabolites

Fungi have an important role in concurrent drug discovery and pharmaceuticals. Some fungal metabolites have a remarkable activity against the pathogenic organisms.

11.3.1 Cephalosporin

A broad spectrum of antibiotics, cephalosporins, is usually used to treat penicillin-allergic patients or against the infections resistant to penicillin. Since the 1950s, a large number of cephalosporins have been used as pharmaceutical.

Cephalosporin was isolated from the fungus *Cephalosporium acremonium,* which has potent antibiotic activity against both the Gram-positive and Gram-negative bacteria. Though cephalosporin N and cephalosporin C are closely related compounds, cephalosporin has the greater beta-lactamase withstanding ability by possessing low toxicity (Crawford et al., 1952).

11.3.2 Coumarins

Fungal metabolite dicoumarol-derived warfarin acts as an anticoagulant and anti-thrombotic. Dicoumarol is a hemorrhagic agent from which anticoagulant is formed by fungal oxidation of coumarins to 4-hydroxycoumarins along with its subsequent coupling with formaldehyde.

11.3.3 Cyclosporin

A wide range of biological activities like antifungal, antibacterial, and immunosuppressive activities were found in the cyclic peptide cyclosporin. Use of this drug was approved in 1983, and now it has wide application as first-line drug in the prevention of organ transplantation rejection. Cyclosporin A was isolated from the fermentation broth of *Tolypocladium inflatum*, which showed low antifungal but high immunosuppressive activity (Stähelin, 1996).

11.3.4 Echinocandins

An antifungal agent echinocandin inhibits fungal b-1-3glucan synthesis, which affects fungal cell wall (Saravolatz et al., 2003), mainly used to treat invasive fungal infections from *Candida* and *Aspergillus.* Due to the undesired hemolytic activity, semisynthetic investigations were performed to overcome this shortcoming. Pneumocandin B_0 and echinocandin B are the two precursors used in pharmaceutical. The modification of hexapeptide increases the antimycotic activity of pneumocandin $B_{0.}$

11.3.5 Ergot Alkaloids

The ergot alkaloid has the dopamine receptor agonist abilities to treat Parkinson's disease. The two fungal families Clavicipitaceae (e.g., *Claviceps* and *Neotyphodium*) and Trichocomaceae (e.g., *Aspergillus* and *Penicillium*) are responsible for the production of ergot alkaloids. The use of ergot started from 1582 according to the records. This ergot alkaloid is also used as an antimigraine drug. Ergotamine and caffeine combination proved to be best for migraine treatment. It was found that ergometrine from *Claviceps* sp. can induce uterine contraction for which it is used in front-line pharmaceutical to treat postpartum hemorrhage (Wallwey & Li, 2011).

11.3.6 Fingolimod

Fingolimod is an immunomodulatory compound of fungal metabolites that is used as oral first-line treatment for multiple sclerosis. The antibiotic activity was observed in the compound myriocin, isolated from *Myriococcum albomyces,* a thermophilic microbe (Kluepfel et al., 1972). Though this compound has antibiotic activity, a significant result was not observed in the clinical study.

11.4 Biotechnology in Sustainable Production of Antibiotics

The discovery of the highly pure form of antibacterial product from specific species is done through extraction, followed by screening of the antibacterial activity and isolation process. This whole process of natural product discovery is slow, monotonous, laborious, and inept. Biotechnology contributes to the advancement of this antibiotic discovery process by controlled miniaturization for increased screening throughput. “System Duetz” (Duetz et al., 2000) and “BioLector” (Samorski et al., 2005) are used as specialized small-scale fermentation system under the controlled biotechnological approach (Fig. 11.1).

11.4.1 Biotechnological Processes-Associated Production

The most crucial step for drug development is production of natural products, which is achieved by different biotechnological approaches. Full fermentative processes using the natural producer, semi-synthetic approaches, and heterologous production in genetically modified hosts are the best three approaches of production.

11.4.1.1 Production Using Full Fermentative Processes

Filamentous fungal biotechnology is implemented for the industrial-scale fermentation of the antibiotic products (Ng et al., 2015). Most of the antibiotic production strategies were established shake flask (Erlenmeyer flasks, EMF) level, with successful scale-up. To obtain sufficient amount of isolated compounds, large-scale

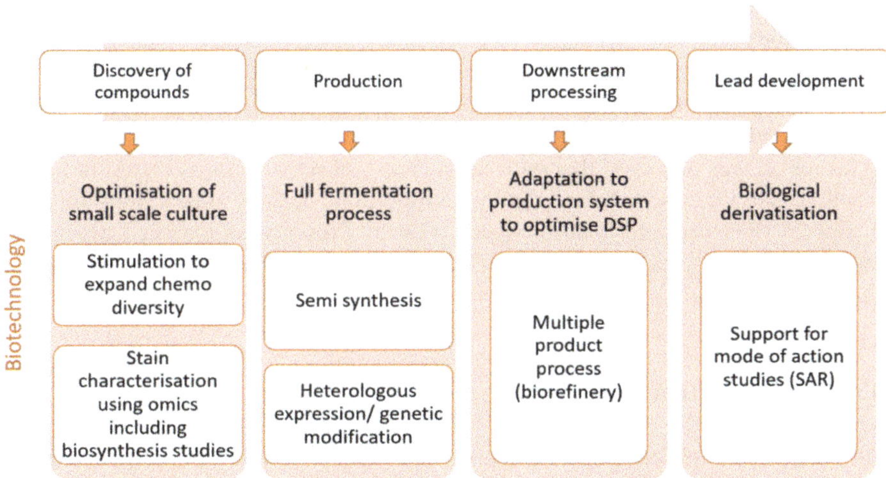

Fig. 11.1 Biotechnological approach in the production of antibiotics from fungi

fermentation is performed in multiple EMF frequently. The transfer of cultures from EMF to bioreactors under controlled conditions is required for the successful and economically expedient scale-up. Successful scale-up processes have some strategies where operating factors like dissolved oxygen, carbon dioxide, pH, temperature, aeration, and foam have influence over antibiotic production. In fungal biotechnology, stirred tank reactors (STR), moving bed, and solid-state system are most commonly used, and terrestrial fungi have been proved to be technically and economically feasible for large-scale production (Riley et al., 2000). Production of antibiotics from fungi is also dependent on the successful transfer into STR, where the controlled process can increase yield by a factor of 100. One of the examples of this process is production of tetramic acid compound ascosetin and lindgomycin by arctic fungus. Application of some specific experimental condition like high pressure for the cultivation of obligate fungi stimulates the expression of gene cluster for new antibiotic production. It was observed that 15 L of 3, 5-dihydroxydecanoic polyester exophilin A had been obtained from the 20-L culture of *Exophiala pisciphila*, member of "black yeast" isolated from the sponge *Mycale adhaerens* in a glass-bottle fermenter under controlled pH (Doshida et al., 1996), where the antibacterial activity increases from 6 days of the cultivation and reaches the highest after 10 days. Carbon and nitrogen also played an important role in compound production for terrestrial bioprocess (Xu et al., 2012). Along with pH and alternation of mycelial morphology, the adaptation of carbon to nitrogen ratio increases the production of macrocyclic polyester calcaride A from *Calcarisporium* sp. by 200-fold (Tamminen et al., 2015). Adding to the physicochemical parameters' morphology is one of the important parameters, which influences biosynthesis of the product. From dispersed filaments to highly dense networks of mycelia, filamentous fungi have the ability to grow in various morphological appearances (Papagianni, 2004).

Morphology has a great impact on product formation through the fermentation process. Unique metabolic mechanism was observed in the presence of salt concentration in the halotolerant marine fungi, though the metabolite production rate in the presence of high salt concentration is slow. Production of ortho-quinone obioninene from marine fungi *Leptosphaeria oraemaris* has been observed in varying salt concentration. It was observed that with the increase of seawater concentration fungal growth is also increased and maximum antimicrobial activity appeared in the presence of 25–50% seawater (Masuma et al., 2001). Presence of specific salts in seawater has the impact on the osmotic effect on the cells and are sensitive toward the antibiotic production.

11.4.1.2 Production Using Semi-Synthesis

Full fermentative process is not economically feasible due to the limitation of the complex chemical synthesis; semi-synthesis is an alternative way in product development. For the stereocomplex molecules, the optically pure molecule can be obtained by sterosensitivity of the enzymatic conversions. Lactam antibiotic cephalosporin C, with increased antibacterial activity, was developed from *Acremonium chrysogenum* by semi-synthetic approach (Kück et al., 2014). Biotechnologically produced cephalosporin C is used as the precursor for the synthesis of two-thirds of commercial cephalosporin. Implementation of mutagenesis and genetic engineering improves the yield of cephalosporin C from *A. chrysogenum* in a fed-batch fermentation (Elander, 2003). The chemistry behind the full or partial fermentation process of synthesis and broad biological screening process provides the access to the synthetic analogous for structure–activity relationship (SAR) studies. One example of chemical synthesis of fungal antibiotics is corollosporine (phthalide derivative) from *Corollospora maritima* (Liberra et al., 1998). Apart from organic synthesis, total synthesis of the antibiotic was also reported, where corollosporine from *Aplidium* sp. and pestalone from *Pestalotia* sp. were subjected under total synthesis.

11.4.1.3 Genetic, Metabolic Engineering, and Heterologous System for Production

In the case of the unknown and uncultivable, microorganism's metagenomics and genome mining have been to understand the hidden chemodiversity. Transfection of the DNA from the environment to the host cell may lead to the production of the new product. Use of different molecular techniques is an alternative approach to express the silent gene cluster and manipulate the biosynthetic pathway in order to optimize production process (Xiong et al., 2013). Heterologous host is an important tool for industrial-scale bioreactor production and also helps in transient process for cultivation condition. The multiple regulatory cascade and network are responsible for the antibiotic production, where different genetic approaches to regulate genes can activate antibiotic production. Different strategies have been implemented to increase the yield of *Saccharomyces cerevisiae* and change its product specificity along with heterologous protein production (Bhadury et al., 2006). The use of metabolic engineering has been reported for the production of novel sesterterpenoid from marine fungi *Fusarium heterosporum* and *Aspergillus versicolor* (Wang et al., 2003). Another example of using metabolic engineering is optimization in fermentation and high-yield stain improvement to obtain 30 g/L cephalosporin C from *A. chrysogenum* in fed batch fermentation (Elander, 2003). Strain improvement through mutagenesis and genetic engineering leads to the homologous cloning of the many genes involved in the biosynthetic pathway of cephalosporin synthesis, and this enables the increase of the flux of production of metabolites from primary to secondary. All the genetic and metabolic engineering are based on the significant knowledge of the omics techniques, where the RNA sequencing reveals the transcription state of a cell along with full genome sequence.

11.4.2 Role of Biotechnology in Downstream Processing

Several steps like separation followed by cell disruption, capture, concentration, extraction, purification, polishing, and formulation are involved in downstream processing (DSP), which is the second main part of the production process. DSP is the most expensive and ineffective part of the bioprocess, which is put down in biotechnological process development. It is mainly used for the commercialization purpose, where a particular batch of antibiotic is being processed before coming to the market. No standard applications of DSP are suitable for all chemical classes. A broad spectrum of downstream tools is used in the biotechnological processes, which can be applied to the marine biotechnology in antibiotic production.

11.4.3 Role of Biotechnology in Lead Development

Lead development is the late stage of discovery pipeline where biotechnology methods also can be used. In lead development, medicinal chemistry has been focused in order to generate products with improved properties like better bioavailability and metabolization with the SAR studies. Mode of action studies during lead development helps in the improvement of the therapeutic application. A small deviation in nature can affect biodiversity more than biotechnology, which was evident by the naturally occurring macrocyclic and linear polyester calcarides from a marine *Calcarisporium* sp. Though *S. epidermidis* and *X. campestris* are inhibited by all macrocyclic calcarides, neither activity was observed by closely related linear polyesters below a MIC of 100 μM (Silber et al., 2013). The occurrence of different derivatives from marine fungi has enzymatic as well as ecological relevance. Bioconversions of biological derivatives are possible through enzyme treatment in biotechnological processes. For example, laccase-catalyzed amination generates corollosporine derivative (Mikolasch et al., 2008) and cephalosporin C is enzymatically converted into 7-amino cephalosporanic acid (Barber et al., 2004). For the production of the deacetyl intermediate enzyme, catalyzed de-esterification method is used widely on industrial scale (Table 11.1).

Table 11.1 Different fungi, their antibacterial activity, and biotechnological approaches

Fungal origin	Approaches in the field of biotechnology	Activity of antibiotics	Class of the compound	References
The fungi are usually obligate in nature living within deep sea	The cultivation is scaled 20–100 L under high pressure High-pressure cultivation scaling 20–100 L	Acts against broad groups of microbial cells	Not determined	Xu et al. (2012)
Hypoxylon oceanicum	Various optimized conditions help in the enhancement of titer values	*Xanthomonas campestris, Propionibacterium acnes, Staphylococcus epidermidis*	Lipodepsipeptide and macrocyclic polylactones	Abbanat et al. (1998)
Halichondria panicea	Helps in the increase in yield with decreased time of productivity	*P. acnes, S. aureus, S. epidermidis, Septoria tritici, X. campestris*	Tetramic acid and ascosetin	Wu et al. (2015)
Cladosporium sp.	Transfer to STR from EMF enhances the productivity	*Pseudoalteromonas piscicida, Vibrio harveyi, Ruegeria* sp. *Loktanella hongkongensis, M. luteus, Rhodovulum* sp.	Bis(2-ethylhexyl) phthalate, phthalate *	Qi et al. (2008)
Ascochyta sp.	Medium optimization is done in small scale	The two-membrane bacterial regulatory system helps in the process of regulation	Spirodioxynaph-thalene Ascochytatin	Kanoh et al. (2008)
Calcarisporium sp.	Biological derivatization is performed by various types of biological optimizations The yield is dependent on C/N ratio	Macrocyclic compounds: *X. campestris S. epidermidis*, linear polyesters	Macrocyclic, linear polyesters and calcarides A–E	Silber et al. (2013)

(continued)

Table 11.1 (continued)

Fungal origin	Approaches in the field of biotechnology	Activity of antibiotics	Class of the compound	References
Unidentified marine biofilm-producing fungi	EMF scaling	It possesses antibiofilm and antibacterial activity against *S. aureus*, *S. haemolyticus*, *Vibrio* sp., *Micrococcus* sp.	Cyclo-(Pro-Phe), diketopiperazine	Quian et al. (2006)
Aspergillus chrysogenum	The genetically engineered organisms helps in reducing the formation of the by-products	They act against broad groups of microbial species	β-Lactam cephalosporin,	Posch et al. (2013)
Cephalosporium chrysogenum	Mutagenesis-associated modification of DNA	They act against broad groups of microbial species		Kensy et al. (2009)
Ampelomyces sp.	Scaling in EMF	*Vibrio* sp., *S. aureus*, *Pseudoalteromonas* sp., *Micrococcus* sp.	3-Chloro-2,5-dihydroxy benzyl alcohol, benzene derivative	Samorski et al. (2005)
Porteresia coarctata, *Penicillium chrysogenum*	Enhancement of yield take place by alteration of carbon and nitrogen source	*Vibrio cholera*	Diketopiperazine Chrysogenazine	Posch et al. (2013)
Halosarpheia sp.	Reprogramming of the biosynthetic pathway	*S. aureus, Yersinia enterocolitica, Clostridium perfringens, P. aeruginosa, E. coli, Salmonella enterica, Shigella dysenteriae*	Cyclodepsipeptides, Enniatins	Duetz et al. (2000)
Corollospora maritima	Salt-dependent enzyme-associated treatment of biological derivatives	*S. epidermidis, Candida maltosa, Bacillus subtilis, Escherichia coli, P. aeruginosa, S. aureus*	Phthalide derivatives and Corollosporine derivatives	Bills et al. (2008)
Phakellia fusca, *Aspergillus terreus*	Agitation, inoculation, rate of aeration, pH, speed of agitation, and nutrient feeding need to be optimized	*B. subtilis*	(+)-Terrein, cyclopentenone	Wang et al. (2003)

(continued)

Table 11.1 (continued)

Fungal origin	Approaches in the field of biotechnology	Activity of antibiotics	Class of the compound	References
Mycale adhaerens, *Exophiala pisciphila*	Transfer from EMF to STR	*S. aureus*, *E. faecalis*, *E. facium*	3,5-Dihydroxy-decanoic polyester, Exophilin A	Xiong et al. (2009)
Halichondria panicea	Adaptation of medium Transfer from EMF to STR enhances the yield	*S. epidermidis*, *P. acnes*, *S. tritici*, *S. aureus*, *X. campestris*	Tetramic acid lindgomycin,	Mikolasch et al. (2008)
Aspergillus versicolor, *Fusarium heterosporum*	Metabolic engineering	They act against broad groups of microbial species	Sesterterpenoid	Wang et al. (2003)
Leptosphaeria oraemaris	Salinity affects the production of antibiotics	–	Ortho-quinone, obioninene	Miller and Savard (1989)
Arthrinium saccharicola	Systematic manipulation of culture conditions	*Vibrio vulnificus*, *Pseudoalteromonas spongiae*	Not determined	Miao et al. (2006)

11.5 Conclusion

The use of microbial species in human need has been observed for more than 100 years. Fungi proved to be one of the most beneficial microbial organisms due to a large content of biomass being associated with them. They are the storehouse of large numbers of natural compounds that have acted efficiently for curing various types of diseases. Thus, the compounds associated with fungi can be an alternative to the conventional drugs after detailed investigations.

References

Abbanat, D., Leighton, M., Maiese, W., Jones, E. B., Pearce, C., & Greenstein, M. (1998). Cell wall active antifungal compounds produced by the marine fungus *Hypoxylon oceanicum* LL-15G256. I. Taxonomy and fermentation. *The Journal of Antibiotics, 51*, 296–302.

Abraham, W. R., & Estrela, A. B. (2016). Fungal metabolites for the control of biofilm infections. *Agriculture, 6*, 37.

Alberts, A. W., Chen, J., Kuron, G., Hunt, V., et al. (1980). Mevinolin: A highly potent competitive inhibitor of hydroxymethyl glutaryl coenzyme A reductase and a cholesterol-lowering agent. *Proceedings of the National Academy of Science United States of America, 77*, 3957–3961.

Alexandre, J., Raymond, E., Kaci, M. O., Brain, E. C., et al. (2004). Phase I and pharmacokinetic study of irofulven administered weekly or biweekly in advanced solid tumor patients. *Clinical Cancer Research, 10*, 3377–3385.

Allison, A. C., & Eugui, E. M. (2000). *Mycophenolate mofetil* and its mechanisms of action. *Immunopharmacology, 47*, 85–118.

Amna, T. (2006). Bioreactor studies on the endophytic fungus *Entrophospora* for the production of an anticancer alkaloid camptothecin. *Canadian Journal of Microbiology, 52*, 189–196.

Bai, R., Zhang, C. C., Yin, X., Wei, J., & Gao, J. M. (2015). Striatoids A-F, cyathane diterpenoids with neurotrophic activity from cultures of the fungus *Cyathus striatus*. *Journal of Natural Products, 78*, 783–788.

Bailey, A. M., Alberti, F., Kilaru, S., Collins, C. M., et al. (2016). Identification and manipulation of the pleuromutilin gene cluster from *Clitopilus passeckerianus* for increased rapid antibiotic production. *Scientific Reports, 6*, 25202.

Barber, M. S., Giesecke, U., Reichert, A., & Minas, W. (2004). Industrial enzymatic production of cephalosporin-basedbeta-lactams. *Advances in Biochemical Engineering/Biotechnology, 88*, 179–215.

Bashyal, B., Wellensiek, B., Ramakrishnan, R., Faeth, S. H., Ahmad, N., & Gunatilaka, A. A. (2014). Altertoxins with potent anti-HIV activity from *Alternaria tenuissima* QUE1Se, a fungal endophyte of *Quercus emoryi*. *Bioorganic & Medicinal Chemistry, 22*, 6112–6116.

Berdy, J. (1995). Are actinomycetes exhausted as a source of secondary metabolites? *Proceedings of 9th International Symposium on the Biology of Actinomycetes, Part 1*.

Bhadury, P., Mohammad, B. T., & Wright, P. C. (2006). The current status of natural products from marine fungi and their potential as anti-infective agents. *Journal of Industrial Microbiology & Biotechnology, 2006*(33), 325–337.

Bills, G. F., Platas, G., Fillola, A., Jimenez, M. R., Collado, J., Vicente, F., Martin, J., Gonzalez, A., Bur-Zimmermann, J., Tormo, J. R., et al. (2008). Enhancement of antibiotic and secondary metabolite detection from filamentous fungi by growth on nutritional arrays. *Journal of Applied Microbiology, 104*, 1644–1658.

Bills, G. F., Martın, J., Collado, J., Platas, G., et al. (2009). Measuring the distribution and diversity of antibiosis and secondary metabolites in the filamentous fungi. *SIM News, 59*, 133–147.

Brown, A. G., Smale, T. C., King, T. J., Hasenkamp, R., & Thompson, R. H. (1976). Crystal and molecular structure of compactin, a new antifungal metabolite from *Penicillium brevicompactum*. *Journal of the Chemical Society, 11*, 1165–1170.

Bucknall, R. A., Moores, H., Simms, R., & Hesp, B. (1973). Antiviral effects of aphidicolin, a new antibiotic produced by *Cephalosporium aphidicola*. *Antimicrobial Agents and Chemotherapy, 4*, 294–298.

Bunyapaiboonsri, T., Yoiprommarat, S., Srisanoh, U., Choowong, W., et al. (2011). Isariotins G-J from cultures of the Lepidoptera pathogenic fungus *Isaria tenuipes*. *Phytochemistry Letters, 4*, 283–286.

Chen, M. C., Lai, J. N., Chen, P. C., & Wang, J. D. (2013). Concurrent use of conventional drugs with Chinese herbal products in Taiwan: A population-based study. *Journal of Traditional and Complementary Medicine, 3*, 256–262.

Chepkirui, C., Richter, C., Matasyoh, J. C., & Stadler, M. (2016). Monochlorinated calocerins A-D and 9-oxostrobilurin derivatives from the basidiomycete *Favolaschia calocera*. *Phytochemistry, 132*, 95–101.

Chepkirui, C., Cheng, T., Matasyoh, J., Decock, C., & Stadler, M. (2018). An unprecedented spiro [furan-2,1' indene]-3-one derivative and other nematicidal and antimicrobial metabolites from *Sanghuangporus* sp. (Hymenochaetaceae, Basidiomycota) collected in Kenya. *Phytochemistry Letters, 25*, 141–146.

Chin, Y. W., Balunas, M. J., Chai, H. B., & Kinghorn, A. D. (2006). Drug discovery from natural sources. *The AAPS Journal, 8*, 239–253.

Cooper, M. A., & Shlaes, D. (2011). Fix the antibiotics pipeline. *Nature, 472*, 32.

Crawford, K., Heatley, N. G., Boyd, P. F., Hale, C. W., Kelly, B. K., Miller, G. A., & Smith, N. (1952). Antibiotic production by a species of *Cephalosporium*. *Journal of General Microbiology, 6*, 47. https://doi.org/10.1099/00221287-6-1-2-47

Crosetto, N., Mitra, A., Silva, M. J., Bienko, M., et al. (2013). Nucleotideresolution DNA double-strand breaks mapping by next-generation sequencing. *Nature Methods, 10*, 361–365.

Davies, J., & Davies, D. (2010). Origins and evolution of antibiotic resistance. *MMBR, 74*, 417–433.

de Carvalho, M. P., Gulotta, G., do Amaral, M. W., Lunsdorf, H., et al. (2016). Coprinuslactone protects the edible mushroom *Coprinus comatus* against biofilm infections by blocking both quorum sensing and Mur A. *Environmental Microbiology, 18*, 4254–4264.

De Silva, N. I., Lumyong, S., Hyde, K. D., Bulgakov, T., et al. (2016). Mycosphere essays 9: Defining biotrophs and hemibiotrophs. *Mycosphere, 7*, 545–559.

Del-Cid, A., Gil-Duran, C., Vaca, I., Rojas-Aedo, J. F., Garcıa-Rico, R. O., Levican, G., & Chavez, R. (2016). Identification and functional analysis of the mycophenolic acid gene cluster of *Penicillium roqueforti*. *PLoS One, 11*(1), e0147047.

Denning, D. W. (2002). Echinocandins: A new class of antifungals. *The Journal of Antimicrobial Chemotherapy, 49*, 889–891.

Doshida, J., Hasegawa, H., Onuki, H., Shimidzu, N., & Exophilin, A. (1996). A new antibiotic from a marine microorganism *Exophilia pisciphila*. *The Journal of Antibiotics, 49*, 1105–1109.

Duetz, W. A., Ruedi, L., Hermann, R., O'Connor, K., Buchs, J., & Witholt, B. (2000). Methods for intense aeration, growth, storage, and replication of bacterial strains in microtiter plates. *Applied and Environmental Microbiology, 66*, 2641–2646.

Ehsanifard, Z., Mir-Mohammadrezaei, F., Safarzadeh, A., & Ghobad-Nejhad, M. (2017). Aqueous extract of *Inocutis levis* improves insulin resistance and glucose tolerance in high sucrose-fed Wistar rats. *Journal of Herbmed Pharmacology, 6*, 160–164.

Elander, R. P. (2003). Industrial production of lactam antibiotics. *Applied Microbiology and Biotechnology, 61*, 385–392.

Endo, A., Kuroda, M., & Tsujita, Y. (1976). ML-236A, ML-236B, and ML- 236C, new inhibitors of cholesterogenesis produced by *Penicillium citrinum*. *The Journal of Antibiotics, 29*, 1346–1348.

Gil-Ramırez, A., Caz, V., Smiderle, F. R., & Martin-Hernandez, R. (2016). Water-soluble compounds from *Lentinula edodes* influencing the HMG-CoA reductase activity and the expression of genes involved in the cholesterol metabolism. *Journal of Agricultural and Food Chemistry, 64*, 1910–1920.

Huang, M. Y., Lin, K. H., Lu, C. C., Chen, L. R., et al. (2017). The intensity of blue light-emitting diodes influences the antioxidant properties and sugar content of oyster mushrooms (*Lentinus sajor-caju*). *Scientia Horticulturae, 218*, 8–13.

Kaneko, M., Watashi, K., Kamisuki, S., Matsunaga, H., et al. (2015). A novel tricyclic polyketide, vanitaracin A, specifically inhibits the entry of hepatitis B and D viruses by targeting sodium taurocholate co transporting polypeptide. *Journal of Virology, 89*, 11945–11953.

Kanoh, K., Okada, A., Adachi, K., Imagawa, H., Nishizawa, M., Matsuda, S., Shizuri, Y., & Utsumi, R. (2008). Ascochytatin, a novel bioactive spirodioxynaphthalene metabolite produced by the marine-derived fungus, *Ascochyta* sp. NGB4. *The Journal of Antibiotics, 61*, 142–148.

Kensy, F., Zang, E., Faulhammer, C., Tan, R. K., & Buchs, J. (2009). Validation of a high-throughput fermentation system based on online monitoring of biomass and fluorescence in continuously shaken microtiter plates. *Microbial Cell Factories, 8*, 31.

King, A. M. (2014). Aspergillomarasmine A overcomes metallo-β-lactamase antibiotic resistance. *Nature, 510*, 503–506.

Kluepfel, D., Bagli, J., Baker, H., Charest, M.-P., Kudelski, A., Sehgal, S. N., & Vézina, C. (1972). Myriocin, a new antifungal antibiotic from *Myriococcum* albomyces. *The Journal of Antibiotics, 25*(2), 109–115.

Kück, U., Bloemendal, S., & Teichert, I. (2014). Putting fungi to work: Harvesting a cornucopia of drugs, toxins, and antibiotics. *PLoS One, 10*, e1003950.

Lambert, C., Wendt, L., Hladki, A. I., Stadler, M., & Sir, E. B. (2019). *Hypomontagnella* (Hypoxylaceae): A new genus segregated from *Hypoxylon* by a polyphasic taxonomic approach. *Mycological Progress, 18*, 187–201.

Li, W. L., Zheng, H. C., Bukuru, J., & De Kimpe, N. (2004). Natural medicines used in the traditional Chinese medical system for therapy of diabetes mellitus. *Journal of Ethnopharmacology, 92*, 1–21.

Liberra, K., Jansen, R., & Lindequist, U. (1998). Corollosporine, a new phtalide derivative from the marine fungus *Corollospora maritima* Werderm. 1069. *Pharmazie, 53*, 578–581.

Liu, L., Gao, H., Chen, X., Cai, X., et al. (2010). Brasilamides A-D: Sesquiterpenoids from the plant endophytic fungus *Paraconiothyrium brasiliense*. *European Journal of Organic Chemistry, 17*, 3302–3306.

Liu, L., Redden, H., & Alper, H. S. (2013). Frontiers of yeast metabolic engineering: Diversifying beyond ethanol and Saccharomyces. *Current Opinion in Biotechnology, 24*, 1023–1030.

Ma, X., Li, L., Zhu, T., Ba, M., et al. (2013). Phenylspirodrimanes with anti- HIV activity from the sponge-derived fungus *Stachybotrys chartarum* MXHX73. *Journal of Natural Products, 76*(12), 2298–2306.

Martín, J. (2012). The inducers 1,3-diaminopropane and spermidine produce a drastic increase in the expression of the penicillin biosynthetic genes for prolonged time, mediated by the LaeA regulator. *Fungal Genetics and Biology, 49*, 1004–1013.

Masuma, R., Yamaguchi, Y., Noumi, M., Omura, S., & Namikoshi, M. (2001). Effect of sea water concentration on hyphal growth and antimicrobial metabolite production in marine fungi. *Mycoscience, 42*, 455–459.

Meier, J. J., Bhushan, A., Butler, A. E., Rizza, R. A., & Butler, P. C. (2005). Sustained beta cell apoptosis in patients with long-standing type 1 diabetes: Indirect evidence for islet regeneration? *Diabetologia, 48*, 2221–2228.

Miao, L., Kwong, T. F., & Qian, P. Y. (2006). Effect of culture conditions on mycelial growth, antibacterial activity, and metabolite profiles of the marine-derived fungus *Arthrinium* c.f. *saccharicola*. *Applied Microbiology and Biotechnology, 72*, 1063–1073.

Mikolasch, A., Hessel, S., Salazar, M. G., Neumann, H., Manda, K., Gordes, D., Schmidt, E., Thurow, K., Hammer, E., Lindequist, U., et al. (2008). Synthesis of new *N*-analogous corollosporine derivatives with antibacterial activity by laccase-catalyzed amination. *Chemical & Pharmaceutical Bulletin, 56*, 781–786.

Miller, S. J. (2001). Emerging mechanisms for secondary cardioprotective effects of statins. *Cardiovascular Research, 52*, 5–7.

Miller, J. D., & Savard, M. E. (1989). Antibiotic activity of the marine fungus *Leptosphaeria oraemaris*. *Proceedings of the Nova Scotian Institute of Science, 39*, 51–58.

Minagawa, K., Kouzuki, S., Yoshimoto, J., Kawamura, Y., et al. (2002). Stachyflin and acetylstachyflin, novel anti-influenza a virus substances, produced by *Stachybotrys* sp. RF-7260. I. Isolation, structure elucidation and biological activities. *The Journal of Antibiotics, 55*, 155–164.

Newman, D. J., & Cragg, G. M. (2007). Natural products as sources of new drugs over the last 25 years. *Journal of Natural Products, 70*, 461–477.

Ng, T. B., Cheung, R. C., Wong, J. H., Bekhit, A. A., & Bekhit Ael, D. (2015). Antibacterial products of marine organisms. *Applied Microbiology and Biotechnology, 99*, 4145–4173.

Pang, X., Zhao, J., Fang, X., & Zhang, T. (2017). Metabolites from the plant endophytic fungus *Aspergillus* sp. CPCC 400735 and their anti-HIV activities. *Journal of Natural Products, 80*, 2595–2601.

Papagianni, M. (2004). Fungal morphology and metabolite production in submerged mycelial processes. *Biotechnology Advances, 22*, 189–259.

Pejin, B. K., Jovanovic, K., Mojovic, M. G., & Savic, A. (2013). New and highly potent antitumor natural products from marine-derived fungi: Covering the period from 2003 to 2012. *Current Topics in Medicinal Chemistry, 13*, 2745–2766.

Pelaez, F., Cabello, A., Platas, G., Dıez, M. T., et al. (2000). The discovery of enfumafungin, a novel antifungal compound produced by an endophytic *Hormonema* species biological activity and taxonomy of the producing organisms. *Systematic and Applied Microbiology, 23*, 333–343.
Phukhamsakda, C., Macabeo, A. P. G., Yuyama, K., Hyde, K. D., & Stadler, M. (2018). Biofilm inhibitory abscisic acid derivatives from the plant-associated Dothideomycete fungus, *Roussoella* sp. *Molecules, 23*, 2190.
Pointing, S. B., Pelling, A. L., Smith, G. J. D., Hyde, K. D., & Reddy, C. A. (2001). Screening of basidiomycetes and xylariaceous fungi for lignin peroxidase and laccase gene-specific sequences. *Mycological Research, 109*, 115–124.
Posch, A. E., Herwig, C., & Spadiut, O. (2013). Science-based bioprocess design for filamentous fungi. *Trends in Biotechnology, 31*, 37–44.
Poucheret, P., Fons, F., & Rapior, S. (2006). Biological and pharmacological activity of higher fungi: 20-year retrospective analysis. *Cryptogamie Mycologie, 27*, 311–333.
Qi, S.-H., Xu, Y., Xiong, H.-R., Qian, P.-Y., & Zhang, S. (2008). Antifouling and antibacterial compounds from a marine fungus *Cladosporium* sp. F14. *World Journal of Microbiology and Biotechnology, 25*, 399–406.
Quian, P.-Y., Li, Y., Kwong, F. N., Yang, L. H., & Dobretsov, S. V. (2006). *Use of marine fungus originated compounds as antifouling agents*. U.S. Patent US2006/0147410 A1.
Riley, G. L., Tucker, K. G., Paul, G. C., & Thomas, C. R. (2000). Effect of biomass concentration and mycelial morphology on fermentation broth rheology. *Biotechnology and Bioengineering, 68*, 160–172.
Rowley, D. C., Kelly, S., Kauffman, C. A., Jensen, P. R., & Fenical, W. (2003). Halovirs A-E, new antiviral agents from a marine-derived fungus of the genus *Scytalidium*. *Bioorganic & Medicinal Chemistry, 11*, 4263–4274.
Sacramento, C. Q., Marttorelli, A., Fintelman-Rodrigues, N., de Freitas, C. S., et al. (2015). Aureonitol, a fungi-derived tetrahydrofuran, inhibits influenza replication by targeting its surface glycoprotein hemagglutinin. *PLoS One, 10*, e0139236.
Samorski, M., Müller-Newen, G., & Buchs, J. (2005). Quasi-continuous combined scattered light and fluorescence measurements: A novel measurement technique for shaken microtiter plates. *Biotechnology and Bioengineering, 92*, 61–68.
Sandargo, B., Thongbai, B., Padutya, D., Steinmann, E., et al. (2018). Antiviral 4-hydroxypleurogrisein and antimicrobial pleurotin derivatives from cultures of nematophagous basidiomycete *Hohenbuehelia grisea*. *Molecules, 23*, 2697.
Sandargo, B., Michehl, M., Praditya, D., Steinmann, E., Stadler, M., & Surup, F. (2019). Antiviral meroterpenoid rhodatin and sesquiterpenoids rhodocoranes A-E from the wrinkled peach mushroom, *Rhodotus palmatus*. *Organic Letters, 21*, 3286–3289.
Saravolatz, L. D., Deresinski, S. C., & Stevens, D. A. (2003). Caspofungin. *Clinical Infectious Diseases, 36*(11), 1445–1457. https://doi.org/10.1086/375080
Shang, S. (2011). Activities of TMC207, rifampin, and pyrazinamide against Mycobacterium tuberculosis infection in guinea pigs. *Antimicrobial Agents and Chemotherapy, 55*, 124–131.
Silber, J., Ohlendorf, B., Labes, A., Erhard, A., & Imhoff, J. F. (2013). Calcarides A–E, antibacterial macrocyclic and linear polyesters from a *Calcarisporium* strain. *Marine Drugs, 11*, 3309–3323.
Singh, S. B., Ondeyka, J. G., Tsipouras, N., Ruby, C., et al. (2004). Hinnuliquinone, a C2-symmetric dimeric non-peptide fungal metabolite inhibitor of HIV-1 protease. *Biochemical and Biophysical Research Communications, 324*, 108–113.
Stähelin, H. F. (1996). The history of cyclosporin A (Sandimmune®) revisited: Another point of view. *Experientia, 52*(1), 5–13. https://doi.org/10.1007/bf01922409
Surup, F., Kuhnert, E., Lehmann, E., Heitkamper, S., et al. (2014). Sporothriolide derivatives as chemotaxonomic markers for *Hypoxylon monticulosum*. *Mycology, 5*, 110–119.
Tamminen, A., Wang, Y., & Wiebe, M. G. (2015). Production of calcaride A by *Calcarisporium* sp. in shaken flasks and stirred bioreactors. *Marine Drugs, 13*, 3992–4005.

Tan, S., Yang, B., Liu, J., Xun, T., et al. (2017). Penicillixanthone A, a marine-derived dual-coreceptor antagonist as anti-HIV-1 agent. *Natural Product Research, 19*, 1–5.
Tang, H., Huang, W., Ma, J., & Liu, L. (2018). SWOT analysis and revelation in traditional Chinese medicine internationalization. *Chinese Medicine, 13*, 1–9.
Thongbai, B., Rapior, S., Hyde, K. D., Wittstein, K., & Stadler, M. (2015). *Hericium erinaceus*, an amazing medicinal mushroom. *Mycological Progress, 14*, 1–23.
Tian, Y., Lin, X., Wang, Z., Zhou, X., et al. (2016). Asteltoxins with antiviral activities from the marine sponge-derived fungus *Aspergillus* sp. SCSIO XWS02F40. *Molecules, 21*, 34/1–34/10.
Vance, D. (1972). Inhibition of fatty acid synthetases by the antibiotic cerulenin. *Biochemical and Biophysical Research Communications, 48*, 649–656.
Wallwey, C., & Li, S.-M. (2011). Ergot alkaloids: Structure diversity, biosynthetic gene clusters and functional proof of biosynthetic genes. *Natural Product Reports, 28*(3), 496–510. https://doi.org/10.1039/c0np00060d
Walser, J., & Heinstein, P. F. (1973). Mode of action of illudin S. *Antimicrobial Agents and Chemotherapy, 3*, 357–363.
Wang, G. Y., Laidlaw, R. D., Marshall, J., & Keasling, J. D. (2003). Metabolic engineering of fungal secondary metabolic pathways. In Z. Q. An (Ed.), *Handbook of industrial mycology* (p. 10016). Marcel Dekker.
Wang, J., Wei, X., Qin, X., Tian, X., et al. (2016). Antiviral merosesquiterpenoids produced by the antarctic fungus *Aspergillus ochraceopetaliformis* SCSIO 05702. *Journal of Natural Products, 79*, 59–65.
Wu, B., Wiese, J., Labes, A., Kramer, A., Schmaljohann, R., & Imhoff, J. F. (2015). Lindgomycin, an unusual antibiotic polyketide from a marine fungus of the Lindgomycetaceae. *Marine Drugs, 13*, 4617–4632.
Xiong, H., Qi, S., Xu, Y., Miao, L., & Qian, P. Y. (2009). Antibiotic and antifouling compound production by the marine-derived fungus *Cladosporium* sp. F14. *Journal of Hydro-environment Research, 2*, 264–270.
Xiong, Z. Q., Wang, J. F., Hao, Y. Y., & Wang, Y. (2013). Recent advances in the discovery and development of marine microbial natural products. *Marine Drugs, 11*, 700–717.
Xu, B., Yin, Y., Zhang, F., Li, Z., & Wang, L. (2012). Operating conditions optimization for (+)-terrein production in a stirred bioreactor by *Aspergillus terreus* strain PF-26 from marine sponge *Phakellia fusca*. *Bioprocess and Biosystems Engineering, 35*, 1651–1655.
Yue, Q., Chen, L., Zhang, X., Li, K., et al. (2015). Evolution of chemical diversity in the echinocandin lipopeptide antifungal metabolites. *Eukaryotic Cell, 14*(7), 76.
Zhou, H., Li, L., Wang, W., Che, Q., et al. (2015). Chrodrimanins I and J from the Antarctic moss-derived fungus *Penicillium funiculosum* GWT2-24. *Journal of Natural Products, 78*, 1442–1445.

Chapter 12
Optimizing Physical Parameters for Amylase Production Using *Aspergillus niger* and Ammonium Molasses Medium

Moitrayee Banerjee Chakraborty, Saranga Ranjan Patgiri, Mandira Sarma, and Hrishikesh Upadhyaya

Contents

12.1 Introduction

Amylases are a class of hydrolytic enzymes that converts starch to sugar like, maltose, etc. α-Amylases (EC 3.2.1.1) are starch-degrading enzymes capable of hydrolyzing internal α-1,4-glycosidic linkages in polysaccharide starch, resulting in the production of short-chain dextrin or low molecular weight carbohydrates (Sindhu et al., 2017; de Souza & de Oliveira Magalhães, 2010). Global market for industrial

M. B. Chakraborty (✉) · S. R. Patgiri · M. Sarma · H. Upadhyaya
Department of Botany, Cotton University, Guwahati, Assam, India

A. C. Shukla (ed.), *Applied Mycology*, Fungal Biology,
https://doi.org/10.1007/978-3-030-90649-8_12

enzymes was estimated at $4.2 billion in 2014 and expected to develop at a compound annual growth rate (CAGR) of approximately 7% over the period from 2015 to 2020 (Singh et al., 2016). α-Amylase covers about 30% shares of the total enzyme market (Paul et al., 2021). Amylase enzymes are used in diverse industrial sectors like food, pharmaceuticals, textile, and detergents. In fermentation-based industries like alcohol, etc., amylase is used in breaking starchy substrates to fermentable sugars. Growing demand for processed food, increasing demand for bioethanol due to rising environmental concerns, new technological innovations like dish washer, which require dish-washing detergents, and growing research and developmental activities are some of the factors driving the growth of the amylase market share.

Amylase enzyme is produced by plants, animals, and microorganisms. However, enzymes from fungal and bacterial sources have dominated applications in industrial sectors (Rana et al., 2013). Microorganisms are favored sources for industrial enzymes due to easy availability and fast growth rate. Genetic changes using recombinant DNA technology can easily be done on microbial cells for elevated enzyme production and scientific development (Illanes et al., 2012). Microbial sources of amylase are also preferred due to cost-effectiveness, consistency, and ease of process modification. Bacterial amylase is mostly obtained from *Bacillus* genus such as *B. stearothermophilus*, *B. subtilis*, *B. cereus*, *B. licheniformis*, and *B. amyloliquefaciens* that are isolated and screened for amylase production (Sivaramakrishnan et al., 2006, as cited in Elmansy et al., 2018). Among the fungi, *Aspergillus niger*, *Aspergillus oryzae*, *Thermomyces lanuginosus*, *Penicillium expansum*, and many species of *Mucor* are capable of producing amylase enzyme (El-Fallal et al., 2012).

For amylase production, different methods like solid-state fermentation (SSF) or submerged fermentation can be used. Solid-state fermentation process occurring in the absence or near-absence of free water has various advantages, such as low energy requirement, high product concentration, and little wastewater production, and is environmentally friendly compared with submerged fermentation. But a major disadvantage of SSF is thermal conduction or heat removal. Submerged fermentation (SMF) technology has the advantages of short period, low cost, high yield, easier purification of products, and easier control of the process.

Thus, the growing demand for amylase enzyme and the possibility of using cheap substrates make production of amylase a potential avenue for entrepreneurship. For economic competitiveness, it becomes important to standardize optimum parameters like media composition, temperature, pH, and time of incubation. Although there are many reports of optimum parameters for high yield, there is no consensus over the optimum parameters. So, this study attempts to determine the optimum range of pH and temperature that can help fungal amylase production entrepreneurship.

12.2 Materials and Methods

12.2.1 Isolation of the Microorganism

Aspergillus niger, used for amylase production, was isolated from soil sample by the method of serial dilution and plating on PDA media. Black fungal colonies were observed under microscope using cotton blue lactophenol stain and identified using handbook and pure cultured by streaking.

12.2.2 Screening

Screening of amylase activity by the fungi was done using starch agar medium. Starch agar medium was prepared using starch, 20 g; KNO_3, 1 g; K_2HPO_4, 0.5 g; $MgSO_4 \cdot 7H_2O$, 0.5 g; NaCl, 0.5 g; $FeSO_4 \cdot 7H_2O$, 0.01 g; agar, 15 g; water to 1 L and sterilized (Shirling & Gottlieb, 1966). The solidified starch agar plates were inoculated with a loop full of *Aspergillus niger* pure culture and incubated at 30 °C for 48 h. Iodine solution was poured over the plates. A colorless zone surrounding the fungal colony indicated conversion of starch to maltose by the fungal amylase.

12.2.3 Inoculation

Aspergillus niger from pure culture was streaked out and mixed in 5.0 ml of sterile deionized water. Then, from the spore suspension, 1.0 ml was inoculated to fresh sterilized potato dextrose broth medium for development of working culture and incubated at 30 ± 1 °C up to 7 days to attain about 5.0×10^8 spores ml^{-1}. For the entire study, 1×10^7 spores ml^{-1} was used as inoculum to carry out the fermentation process (Sethi et al., 2013). The sterile media was allowed to cool and inoculated with spore suspension inside laminar air flow chamber and incubated according to experimental design.

12.2.4 Medium

Composition of medium used for microbial fermentation is very critical because it influences the output of the process. Various carbon sources like molasses, whey, grains, and agricultural or industrial wastes are used with the objective of cutting down cost of production, and also addresses the environmental issue of waste management. Cane molasses are often used in industrial microbiology as carbon source. In cane molasses, sugar content is diluted to about 25% sugar level by adding double-distilled water. The molasses solution, after adding 35 ml of 1N H_2SO_4 per liter, is boiled for half an hour, cooled, neutralized with lime water (CaO), and left to stand overnight for clarification (Panda et al., 1984). Neutralized molasses solution was treated with 2% (w/v) tricalcium phosphate (TCPH followed by autoclaving at 105 °C for 5 min. The mixture is cooled and centrifuged at 3000 rpm for 15 min. The insoluble matter is discarded. TCPH treatment reduces considerably the metal loads of molasses compared to untreated molasses. (Kundu et al., 1984). Nitrogen is another important structural and functional component of all living organisms and plays a very important role in economic feasibility of any industrial fermentation process. Different organic nitrogen sources like peptone, tryptone, gelatin, yeast extract, beef extract, and inorganic nitrogen sources like ammonium nitrate, ammonium sulfate, ammonium chloride, and sodium nitrate are frequently used in microbial media. Although very good yield of organic acids, amylase, etc. has been reported by using organic nitrogen sources, because of their high cost, inorganic nitrogen sources like ammonium sulfate or ammonium nitrate are preferred for industrial production to lower the cost of production. The inorganic source ammonium nitrate (0.1%) was found suitable to produce high amount of enzyme and showed increased enzymatic activity by *Bacillus subtilis* (Deb et al., 2013). Amylase production by *Aspergillus* has been found to be higher using ammonium sulfate as an inorganic nitrogen source as compared to ammonium nitrate (Oshoma et al., 2010).

12.2.5 Extraction

Extraction of the crude enzyme was done by centrifugation of the fermented media at 2000 rpm (revolution per minute) for 5 min, supernatant collected and filtered off using Whatman No.1 filter paper. The filtrate was used as crude enzyme extract (Oyeleke et al., 2010)

12.2.6 Influence of Initial pH, Temperature, and Incubation Time

Amylase activity was checked at different initial pH values ranges 5.0–11.0 and temperature range 30–60 °C. 250 mL flask containing 50 mL of enzyme production medium was inoculated with 10 (% v/v) inoculum from 24-h-old seeds culture and was incubated in shaking incubator at 37 °C for 60 h, while, in case of temperature optimization, pH was adjusted to 7.0 and other conditions remained unchanged. Amylase activity was determined from culture broth obtained after centrifugation (12,100 for 10 min at 10 °C). To observe the optimum incubation time, the inoculated media with initial pH 7 were incubated at 37 °C for (24–96) h.

12.2.7 Enzyme Assay

The α-amylase activity was assayed by 3,5-dinitrosalicylic acid (DNS) procedure using 1% soluble starch (Bernfeld, 1955). Dinitro salicyclic acid required for enzyme assay is prepared by dissolving 1 g of DNS in 50 ml of distilled water. To this solution, about 30 g of sodium potassium tartarate tetrahydrate is added in small lots, and the solution turns milky yellow in color. To this, 20 ml of 2N NaOH is added, which turns the solution to transparent orange yellow color. The final volume is made to 100 ml with the distilled water and stored in an amber-colored bottle.

To prepare a standard graph of maltose, a stock solution is prepared using 50 mg of maltose in 50 ml of water. From the stock solution, 10 ml of solution is dissolved with 40 ml of distilled water to prepare the working solution. From the working solution, different amounts, that is,, 0.2, 0.4, 0.6, 0.8, and 1 ml, are taken in five test tubes, and one test tube is kept blank. The final volume of all the test tubes is made 1 ml by adding distilled water, and then 1 ml of DNS is added to each test tube. The contents in the test tubes are heated in a boiling water bath for 5 minutes. The test tubes are cooled and 10 ml of distilled water is added to each test tube. Optical density of each test tube is determined by a spectrophotometer at 540 nm (Table 12.1). A graph is plotted with the amount of maltose on X-axis vs. OD at 540 nm on Y-axis (Fig. 12.1).

For estimating enzyme activity, 1% starch solution is prepared by using 1 gram of starch in 100 ml of distilled water and 1 ml of enzyme extract maintained at different conditions under experimental design. In one of the test tube, to be treated as blank, 1 ml of distilled water is added instead of enzyme. All the test tubes are incubated for 10 minutes at 30 °C. Then, 1 ml of DNS was added and incubated in boiling water bath for 5 min. The test tubes are then cooled and 10 ml of distilled water is added and mixed well. The optical density of the sample is then determined against blank at 540 nanometers. All assays were performed in duplicate and each reading was taken three times (Chakraborty et al., 2018).

Table 12.1 Spectrophotometric absorbance shown by maltose at different concentrations

Maltose in micrograms	Average optical density at 540 nm
0	0
40	0.039
80	0.119
120	0.138
160	0.194
200	0.226

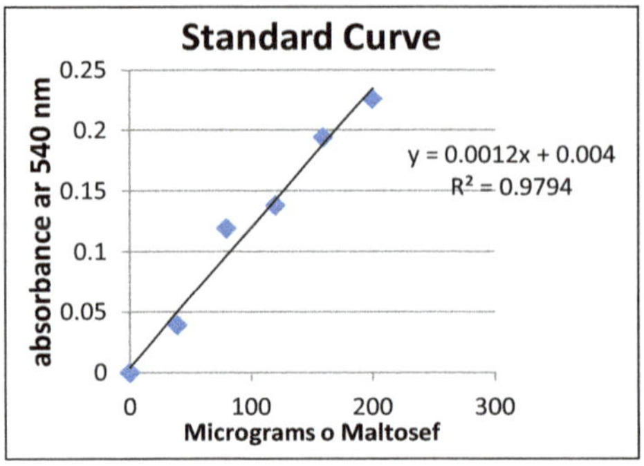

In the standard graph, $\mathbf{y = mx + b}$

y = Absorbance, X = concentration of maltose in the working solution,

slope (m) = **0.0012** and b = 0.004

$\mathbf{y = mx + b}$, So, $\mathbf{X = (y - b)/m}$

Fig. 12.1 Standard maltose curve

Table 12.2 Effect of pH on the production of amylase by *Aspergillus niger*

Effect of initial pH		
pH	Absorbance	μg of maltose units per ml of enzymes
4.5	0.189	154.16
5	0.527	435.83
5.5	0.766	635.00
6	0.829	687.50
6.5	1.14	946.67
7	0.794	658.33
8	0.686	568.33

12.3 Results

12.3.1 Screening

Starch agar medium inoculated with *Aspergillus niger* when flooded with iodine, after incubation, showed a clear zone, indicating amylase activity on starch.

12.3.2 Effect of Initial pH

The effect of pH on amylase production is shown in Table 12.2. Of the range of pH 4.5–8, maximum enzyme activity per ml of enzyme was observed at pH 6.5. The amount of maltose produced by the enzyme was found to be 946.67 μg/ml of enzyme extract. Enzyme activity is seen to rise gradually from pH 4.5 to 6.5 and again declines at pH 7 and 8 (Fig. 12.2).

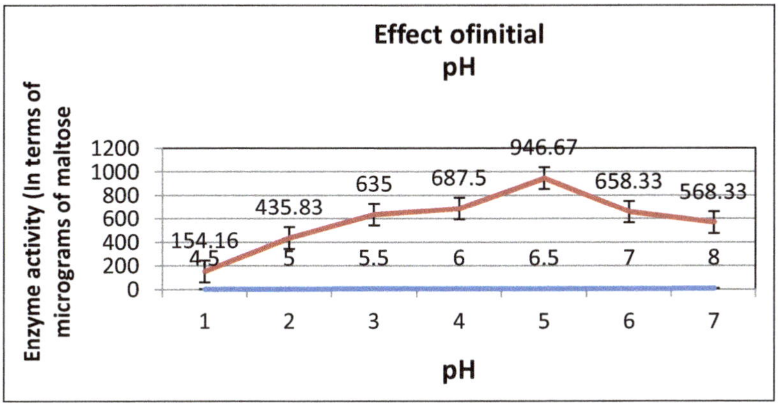

Fig. 12.2 The amount of maltose released by amylase increases from pH 4 to 6.5 and then starts declining

Table 12.3 Effect of initial temperature on the production of amylase by *Aspergillus niger*

Effect of initial temperature		
Temperature (°C)	Absorbance	μg of maltose units
25	0.522	431.67
30	0.732	606.67
35	1.984	1650
40	2.173	1807.5
45	1.276	1506250

12.3.3 Effect of Initial Temperature

Table 12.3 shows the effect of temperature of incubation on amylase production. Amylase production observed in terms of amylase activity on starch releasing maltose was found to be 431.67 μg/ml of enzyme extract when incubated at 25 °C and increased to maximum of 1807.5 μg/ml of enzyme extract at 40 °C. Further rise of temperature to 45 °C decreased enzyme activity 1060 μg/ml of enzyme extract (Fig. 12.3).

12.3.4 Effect of Incubation Period

The effect of incubation time on amylase production is shown in Table 12.4. It is seen that 24 h after inoculation, amylase activity in terms of maltose produced is 58.33 μg/ml of enzyme extract. The production increased to a maximum of 126.33 μg/ml of enzyme extract after 96 h of incubation and started declining with further incubation (Fig. 12.4).

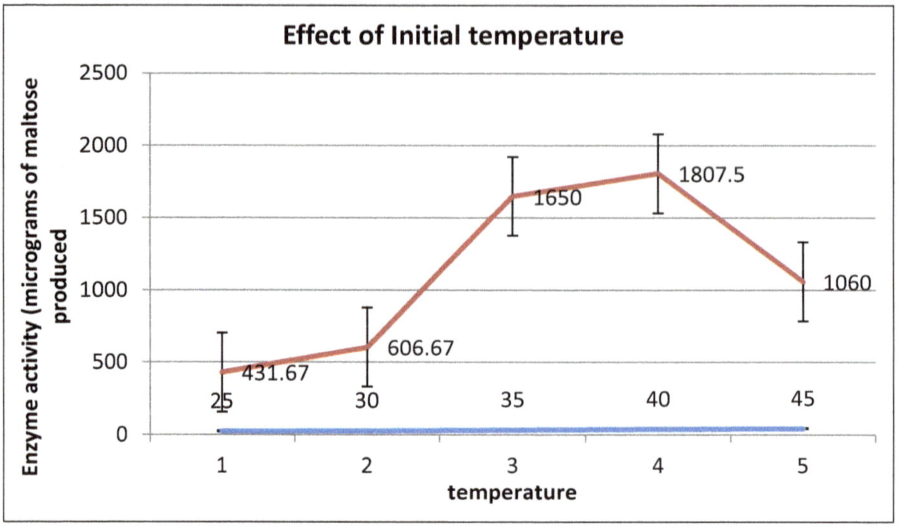

Fig. 12.3 Maltose concentration increases with temperature till 40 °C and then starts declining

Table 12.4 Effect of incubation time on the production of amylase by *Aspergillus niger*

Incubation time (h)	Absorbance	μg of maltose units
24	0.074	58.33
48	0.092	73.330
72	0.098	78.33
96	1.52	126.33
120	1.39	115.5
144	1.08	89.67

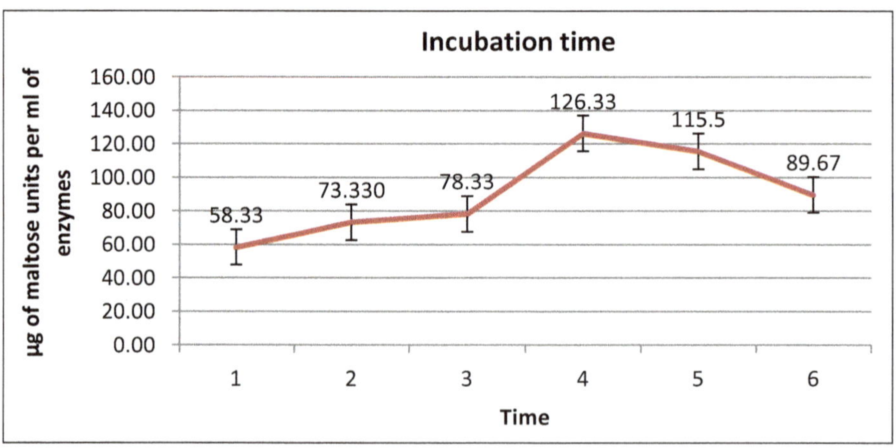

Fig. 12.4 The amount of maltose produced by maltose increases from 1-day incubation till 4 days and declines with further incubation

12.4 Discussion

Growing demands for amylase enzymes by different carbohydrate-based industries have opened up opportunities for entrepreneurship for setting small- and medium-scale industries for producing amylase enzyme at low-input cost. Startups using microbes and fermentation include MycoTechnology, which turns pea and rice protein fermented with shiitake mycelium, or fungi, into food ingredients, and Cargill produces fish food from naturally occurring microbes that use methane as an energy source (Financial Time, 2019).

To be economically competent, process optimization is very important. Effect of various physical parameters on α-amylase production was investigated in this study. pH is a very important parameter that affects any biochemical process by affecting the active sites of the enzymes involved in the process. The pH optima for production of amylase by submerged fermentation has mostly been reported as slightly acidic to neutral. In this study, the optimum pH for amylase production by A niger under submerged foundation was found to be 6.5. According to other studies, the optimum pH was found to vary.

Optimum pH for amylase production was found to be 6 in various studies (Behailu & Abebe, 2018; Wang et al., 2016; Singh et al., 2014; Ahmed et al., 2020). Monga et al. (2011) reported 4.6 pH as optima for amylase production.

Temperature of incubation for amylase production has been found to affect amylase production significantly. In this study, the temperature optima for α-amylase production by *A. niger* was found to be 40 °C. In different studies, a range of optimum values of incubation temperature for amylase production has been reported. Ahmed et al. (2020) reported 30 °C, while Sethi et al. (2016) reported range of 27–36 °C and 45 °C (Behailu & Abebe, 2018).

It has been established in growing volume of studies that production of enzyme and also metabolites vary at different points of incubation. In this study, maximum amylase production observed in terms of amylase activity was observed after 96 h of incubation similar to the findings of Sethi et al., 2016. Optimum incubation time reported in other studies was 72 h (Behailu & Abebe, 2018), 144 h (Singh et al. 2014), and 120 h (Ahmed et al., 2020; Monga et al., 2011). Although most of the studies have reported the optimum parameters in a close range, but there is a need for further studies on specific conditions to optimize the process.

References

Ahmed, N. E., El Shamy, A. R., & Awad, H. M. (2020). Optimization and immobilization of amylase produced by *Aspergillus terreus* using pomegranate peel waste. *Bulletin of the National Research Centre, 44*, 109. https://doi.org/10.1186/s42269-020-00363-

Behailu, A., & Abebe, G. (2018). Isolation, production and characterization of amylase enzyme using the isolate Aspergillus niger FAB-211. *International Journal of Biotechnology and Molecular Biology Research, 9*(2), 7–14. https://doi.org/10.5897/ijbmbr2018.0289

Bernfeld, P. (1955). Amylases, a and (3. Pages 149–158 in Methods in Enzymology. Vol. 1. S. P. Colo-wick and N. O. Kaplan, ed. Academic Press, New York, NY.
Chakraborty, M. B., Patgiri, S. R., Rahman, A. S., Dasgupta, A., & Pegu, G. (2018). Production, assay and optimisation of amylase by submerged fermentation using Aspergillus niger. *Annals of Plant Sciences, 6*, 2357–2360.
Deb, P., Talukdar, S. A., Mohsina, K., Sarker, P. K., & Sayem, S. A. (2013). Production and partial characterization of extracellular amylase enzyme from Bacillus amyloliquefaciens P-001. *SpringerPlus, 2*(1), 154. https://doi.org/10.1186/2193-1801-2-154
de Souza, P. M., & de Oliveira Magalhães, P. (2010). Application of microbial α-amylase in industry – A review. *Brazilian Journal of Microbiology: [Publication of the Brazilian Society for Microbiology], 41*(4), 850–861. https://doi.org/10.1590/S1517-83822010000400004
El-Fallal, A., Abou, M., El-Sayed, A., & Omar, N. (2012). Starch and microbial α-amylases: From concepts to biotechnological applications. *In Carbohydrates – Comprehensive Studies on Glycobiology and Glycotechnology*. https://doi.org/10.5772/51571
Elmansy, E. A., Asker, M. S., El-Kady, E. M., et al. (2018). Production and optimization of α-amylase from thermo-halophilic bacteria isolated from different local marine environments. *Bulletin of the National Research Centre, 42*, 31. https://doi.org/10.1186/s42269-018-0033-2
Emiko Terazono. (2019). Entrepreneurs hope microbes hold the key to a food revolution'. *The Financial Times*. https://www.ft.com/content/c6eabf26-d49e-11e9-8367-807ebd53ab77
Illanes, A., Cauerhff, A., Wilson, L., et al. (2012). Recent trends in biocatalysis engineering. *Bioresource Technology, 115*, 48–57.
Kundu, S., Panda, T., Majumdar, S. K., Guha, B., & Bandyopadhyay, K. K. (1984). Pretreatment of Indian cane molasses for increased production of citric acid. *Biotechnology & Bioengineering, 26*, 1114–1121. https://doi.org/10.1002/bit.260260915
Monga, M., Goyal, M., Kalra, K. L., & Soni, G. (2011). Production and stabilization of amylases from *Aspergillus niger*. *Mycosphere, 2*, 129–134.
Oshoma, C. E., Imarhiagbe, E. E., Ikenebomeh, M. J., & Eigbaredon, H. E. (2010). Nitrogen supplements effect on amylase production by Aspergillus niger using cassava whey medium. *African Journal of Biotechnology, 9*(5), 682–686.
Oyeleke, S. B., Egwim, E. C., & Auta, S. H. (2010). Screening of *Aspergillus flavus* and *Aspergillus fumigatus* strains for extracellular protease enzyme production. *Journal of Microbiology and Antimicrobial, 2*, 83–87.
Panda, T., Kundu, S., & Mazumdar, S. K. (1984). Studies on citric acid production by *Aspergillus niger* using treated Indian cane-molasses. *Journal of Microbiology, 5*, 61–66.
Paul, J. S., Gupta, N., Beliya, E., et al. (2021). Aspects and recent trends in microbial α-amylase: A review. *Applied Biochemistry and Biotechnology, 193*(*8*), *2649–2698*. https://doi.org/10.1007/s12010-021-03546-4
Rana, N., Walia, A., & Gaur, A. (2013). α-amylases from microbial sources and its potential applications in various industries. *National Academy Science Letters, 36*, 9–17. https://doi.org/10.1007/s40009-012-0104-0
Sethi, B. K., Rout, J. R., Das, R. et al. (2013). Lipase production by Aspergillus terreus using mustard seed oil cake as a carbon source. *Annals of Microbiology 63*, 241–252. https://doi.org/10.1007/s13213-012-0467-y
Sethi, B. K., Jana, A., Nanda, P. K., DasMohapatra, P. K., Sahoo, S. L., & Patra, J. K. (2016). Production of α-amylase by aspergillus terreus NCFT 4269.10 using pearl millet and its structural characterization. *Frontiers in Plant Science, 7*, 1–13. https://doi.org/10.3389/fpls.2016.00639
Shirling, E. B., & Gottlieb, D. (1966). Methods for characterization of Streptomyces species. *International Journal of Systemic Bacteriology, 16*(3), 313–340.
Sindhu, R., Binod, P., Madhavan, A., Beevi, U. S., Mathew, A. K., Abraham, A., Pandey, A., & Kumar, V. (2017). Molecular improvements in microbial α-amylases for enhanced stability and catalytic efficiency. *Bioresource Technology, 245*(Part B), 1740–8. https://doi.org/10.1016/j.biortech.2017.04.098

Singh, R., Kumar, M., Mittal, A., & Mehta, P. K. (2016). Microbial enzymes: Industrial progress in 21st century. *3 Biotech, 6*(2), 174. https://doi.org/10.1007/s13205-016-0485-8

Singh, S., Singh, S., Bali, V., Sharma, L., & Mangla, J. (2014). Production of fungal amylases using cheap, readily available agriresidues, for potential application in textile industry. *BioMed Research International, 2014*, 215748. https://doi.org/10.1155/2014/215748

Wang, S., Lin, C., Liu, Y., Shen, Z., Jeyaseelan, J., & Qin, W. (2016). Characterization of a starch-hydrolyzing α-amylase produced by *Aspergillus niger* WLB42 mutated by ethyl methanesulfonate treatment. *International Journal of Biochemistry and Molecular Biology, 7*(1), 1–10.

Chapter 13
Plant–Fungal Interactions

K. Geetha and Vasavi Dathar

Contents

13.1 Introduction

The most intense interactions between microbes and plants take place at the rhizosphere, which is the interface between plant roots and the soil. Soil microbes have a tremendous influence on plant health and productivity (Bloemberg & Lugtenberg, 2001). Interactions between plants and fungi are influenced by cultivars, mutants, and nonhost species. Different stages in the colonization processes may act as control points in different interactions.

K. Geetha
Head, Department of Botany, Telangana Social Welfare Residential Degree College for Women, Palamuru University, Mahabubnagar, Telangana State, India

V. Dathar (✉)
Head, Department of Microbiology, Telangana Social Welfare Residential Degree College for Women, Palamuru University, Mahabubnagar, Telangana State, India

A. C. Shukla (ed.), *Applied Mycology*, Fungal Biology,
https://doi.org/10.1007/978-3-030-90649-8_13

Plant–fungal interactions are of fundamental importance in crop productivity, which may be symbiotic or pathogenic. Plants produce antimicrobial compounds like phytoanticipins and phytoalexins, which act as phyto protectants. Some pathogenic fungi have evolved hydrolytic enzymes that can degrade toxic phyto protectants, thereby allowing the fungus to overcome specific chemical barriers (Tudzynski & Sharon, 2003). Fungi are versatile with respect to their nutritional requirements. Because of this versatility, they interact with plants in different ways. Thus, they play a major role in the natural ecosystems and also in the present agricultural scenario. They decompose and recycle organic materials by interacting with the plants both underground and above ground, that is, with the root and shoot system. They interact with plant roots in the rhizosphere and also above ground and decompose and recycle organic materials. The complex interactions between the fungus and plants lead to diverse outcomes. Several fungi have a combined mode of lifestyles like saprophytic, pathogenic, or symbiotic (Grigoriev, 2013). Techniques like transcriptomics and the omics era have provided an insight into the mechanism of plant–mycorrhizal interactions.

Symbiotic fungi colonize the plants and are beneficial to their hosts by promoting growth, producing secondary metabolites, and enhancing resistance to biotic and abiotic stresses. The symbiotic associations between plants and fungi include endophytic and different types of mycorrhizae. The associations between plants and endophytic fungi are governed by molecular mechanisms and deterministic factors that enhance plant growth, thus providing protection against invading pathogens. In contrast, large numbers of fungi are pathogens that constitute a major threat to crop yield and food security (Shah, 2018).

13.2 Genomics of Plant–Fungal Interactions

Understanding the molecular mechanisms by which fungal pathogens cause disease, subvert plant immune systems, and divert plant developmental and metabolic processes is crucial for development of new antifungal strategies. Identification of key virulence factors of the fungus that allow the pathogen to cause disease or to escape detection by plant's immune system is important. Also, it is important to know the mechanisms by which plants recognize and respond to fungal pathogens. Several fungal and plant genomes are being sequenced and assembled, so that the impact of functional genomics enables us to know how the plant's immune system is able to discriminate between the beneficial and harmful fungi. With the continuous progress being made toward the understanding of plant–fungal interactions, new strategies for crop protection are emerging at an increasingly rapid pace (Shah, 2018).

In nature, plants are generally resistant to many pathogenic fungi and are able to combat infections caused by the pathogenic fungi. Among the plant–fungal associations, symbiotic and neutral associations dominate, whereas parasitic associations are considered to be an exception (Staskawicz, 2001). The plant attracts fungi in the rhizosphere by producing different secretions as a result of its metabolic activity. These secretions are determined by the genetic nature of the plant.

The outcome of the plant–fungal interaction is driven by the receptors and defense proteins expressed by the plant and the specific molecules produced by the fungi. Mutations in the genes of the plant or fungus may cause a change in these interactions where a resistant plant becomes sensitive or susceptible to the fungal infection or vice versa (Stracke et al., 2002; Giraldo & Valent, 2013). Microbes beneficial to plants have evolved strategies that suppress the defenses exhibited by the plants they infect, which allow them to colonize the plants either as epiphytes or endophytes (Zamioudis & Pieterse, 2012). Both symbiotic and pathogenic fungi use mostly similar colonization patterns like development of feeding structure and nutrient sequestration by establishing obligate relationship with the plants (Corradi & Bonfante, 2012). The outcomes of these interactions are contrasting as some plants are rewarded with symbiosis and some others suffer parasitism. In an evolutionary perspective, some fungi have evolved as parasites or pathogens, and, on the other hand, certain fungi have evolved as successful symbionts by their interaction with the host plants.

By decoding the functions of the receptors and signals of both the fungus and plant, we can understand their complex interactions and also their role in inducing immunity in the host plant. Successful colonization of the fungus in the host plant requires all of these contributing factors that also influence the resistance of the plant toward the fungal partner.

13.3 Mycorrhizal Association

Fungi are filamentous in nature and exploit diverse substrates depending on their nutritional strategy. The saprophytic fungi, called as saprobes, thrive in soil, water, and on dead and decaying plant and animal tissues. The other group of fungi, which are the parasitic and mutualistic symbionts, feed on living organisms (Carlile et al., 2001). The mycorrhizal fungi form a heterogeneous group of species that are spread over diverse fungal taxa. The word "mycorrhiza" was introduced by Frank in 1885, which means "fungus roots" (Frank, 1885; Trappe, 2005). Mycorrhiza is the symbiotic association between the roots of a higher plant and a fungus. These fungi can be divided into two major groups: aseptate endophytes such as Glomeromycota or septate Ascomycota and Basidiomycota.

More commonly, mycorrhizal classifications include two broad categories that reflect anatomical aspects (Smith & Read, 2008), referred to as ectomycorrhizae (EMs) where the fungus colonizes intercellular spaces, and endomycorrhizae, in which the fungus develops inside the cells and colonizes the roots. A third category of mycorrhizae is the ectoendomycorrhizae with mixed behavior between ectomycorrhiza and endomycorrhizae (Fig. 13.1). The endomycorrhizae are of three types: orchid, ericoid, and arbuscular mycorrhizae (AMs). The mycorrhizal fungi can be free living, which is a part of their life cycle, and also as symbionts colonizing 90%

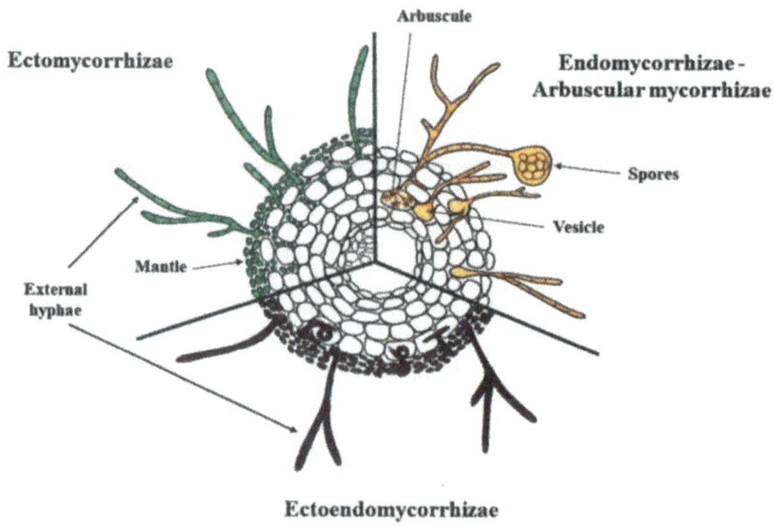

Fig. 13.1 Different types of mycorrhizae. (Courtesy: Ganugi et al., 2019)

of plant species including crop plants, forest trees, and wild grasses. Both the plant and fungi benefit from the relationship where the mycorrhizae improve the nutrient status of the host plant that influences the mineral nutrition, water absorption, growth, and disease resistance. In turn, the host plants help in the growth and reproduction of the fungus (Bonfante & Genre, 2010).

Many land plants have symbiotic association with soil-borne mycorrhizal fungi that provide nutritional benefits to both the partners and make them fit. These benefits are important in solving the food security challenge. Mycorrhizal fungi are found in the environments such as alpine and boreal zones, tropical forests, grasslands, and croplands. The belowground arbuscular mycorrhizal fungal (AMF) communities are affected by biotic factors including animals, plants, fungi, bacteria, and protists, and abiotic factors like water, soil, air, sunlight, temperature, and minerals (Larissa et al., 2019). Mycorrhizal fungi absorb and supply soil nutrients to the plant by the specific activity of the mycelia and thus play a major role in nutrient cycling, but their role in carbon flux is not much defined (Selosse & Roy, 2009).

Mycorrhizae connect plant communities by developing a wide network of the hyphae inside the soil, thus offering a horizontal transfer of nutrients between the plant and fungus (Helgason et al., 1998). Mycorrhizae form special spaces to interact with the host plant through symbiosis (Bonfante, 2001; Harrison, 2005; Parniske, 2008). The ectomycorrhizae form a thick mantle with their hyphae around the epidermal cells of the root tip developing into a Hartig net. The root tip is not colonized by arbuscular mycorrhizae. Hyphae developing from the spore produce a hyphopodium on the root epidermis. Colonization proceeds both intra- and intercellularly, leading to the formation of arbuscules inside inner cortical cells (Fig. 13.1).

13.3.1 Mechanism of Colonization

The study of plant responses illustrates how the mechanisms operate to accommodate the AM fungus inside the plant cell lumen by diverse plants and conserved during evolution. Root colonization is vital to AM fungi (Fig. 13.2); their spores feed germinating hyphae through the catabolism of storage lipids for just a few days (Bonfante & Genre, 2010). The hyphae repeatedly try to explore the soil searching for a host, and when it does not find one, it will arrest its growth and retract the cytoplasm back into the spore. When it finds a host, it re-germinates. Due to the wide host range of the mycorrhizal fungi, this situation is not very frequent in nature, but it shows their endurance because of their strict biotrophy. With the help of root organ cultures, production of nucleic acid extractions on a large scale has been made possible for growing arbuscular mycorrhizal fungi (Becard & Fortin, 1988).

13.3.2 Formation of Hyphopodium

Once a chemical acquaintance has been made between the fungus and the plant, roots and hyphae proliferate and branch in a small volume of the rhizosphere. This pre-symbiotic phase of the AM interaction culminates in a physical encounter between symbionts, when a hyphal tip touches the surface of a root (Fig. 13.2). These fungi start root penetration after carefully selecting a location, and hyphae

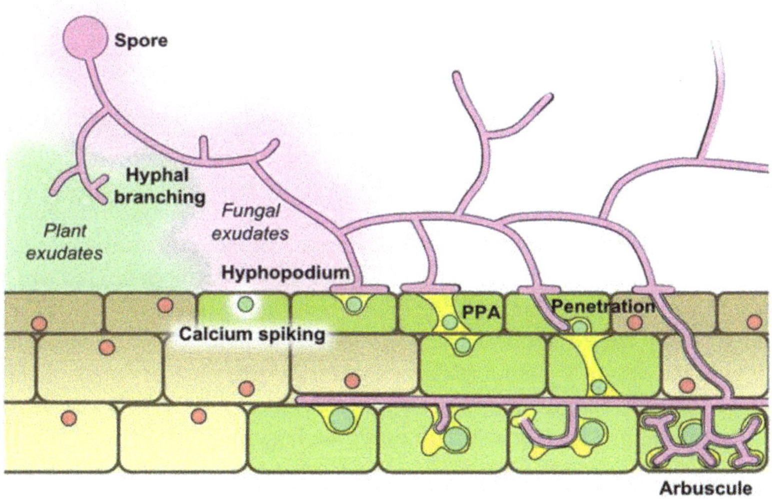

Fig. 13.2 Schematic summary of the root colonization process by AM fungi. (Courtesy: Bonfante & Genre, 2010)

wander along the root surface for several centimeters, forming a long, straight, or slightly curved structure.

This structure swells and flattens on some of the epidermal cell walls and repeatedly branch to form a structure called as hyphopodium (Genre et al., 2005). The arbuscular mycorrhizal fungi concentrate in the young lateral roots of the plants (Kosuta et al., 2003), which are the primary site of colonization of these fungi under laboratory conditions (Chabaud et al., 2002). The hyphopodial wall penetrates into the epidermal cells of the root with many protuberances. But the development of hyphopodium is followed by a pause in fungal growth for about 4–6 hours. Then, the growth of a new tip is initiated and the penetration hypha develops (Genre et al., 2005). The plant cells and tissues now prepare for colonization.

13.3.3 Gene Expression

The gene coding for secretion of a protein responsible for expression of symbiotic pathway, ENOD11, has been reported in the hyphopodium (Chabaud et al., 2002). Many genes of the plant are regulated before the symbiotic association and their expression also changes (Weidmann et al., 2004). During the formation of pre-penetration apparatus, the cell wall remodeling genes and the defense genes become active (Siciliano et al., 2007). The pre-penetration apparatus that is specific to arbuscular mycorrhizal fungi is required for the fungal penetration into the plant (Genre et al., 2005). The endoplasmic reticulum, Golgi bodies, and secretory vesicles are synthesized abundantly, which are concentrated in the pre-penetration apparatus (Genre et al., 2008).

After the completion of the pre-penetration apparatus, the fungus starts growing again. The hyphal tip heads through the epidermal cell wall along the track of PPA. The perifungal membrane may assemble and PPA secretory vesicles may fuse and produce an invagination of the plasma membrane. This marks the formation of a symbiotic interface, and the intracellular compartment allows the growth of arbuscular mycorrhizal fungi without disrupting the plant cell (Bonfante, 2001) (Fig. 13.2).

Fungi evolved much earlier than the plants, and plant–fungal interactions are thought to be started during the evolutionary period of the terrestrial vascular plants (Humphreys et al., 2010; Field et al., 2012). The fungal partners might have helped in the colonization of the plants on the land (Redecker et al., 2000). These associations might have started approximately 400–460 million years ago at the time of evolution of the vascular plants (Remy et al., 1994; Kemen & Jones, 2012). The stability of both the fungus and the plant is provided by beneficial fungi, whereas the pathogenic fungi cause destabilization of the host plant (Jones & Dangl, 2006).

13.3.4 Plant Growth and Development

The symbiotic association between the fungi and plants promotes the growth and development of the plant partner. Here, the fungus improves foraging of the plant, helps to acquire nutrients and water from the soil, and makes the plant in stress tolerant. The plants in turn deliver nutrition to the fungus in the form of carbohydrates (Buscot et al., 2000). All these factors contribute to a stable interaction between the two partners (Fig. 13.3).

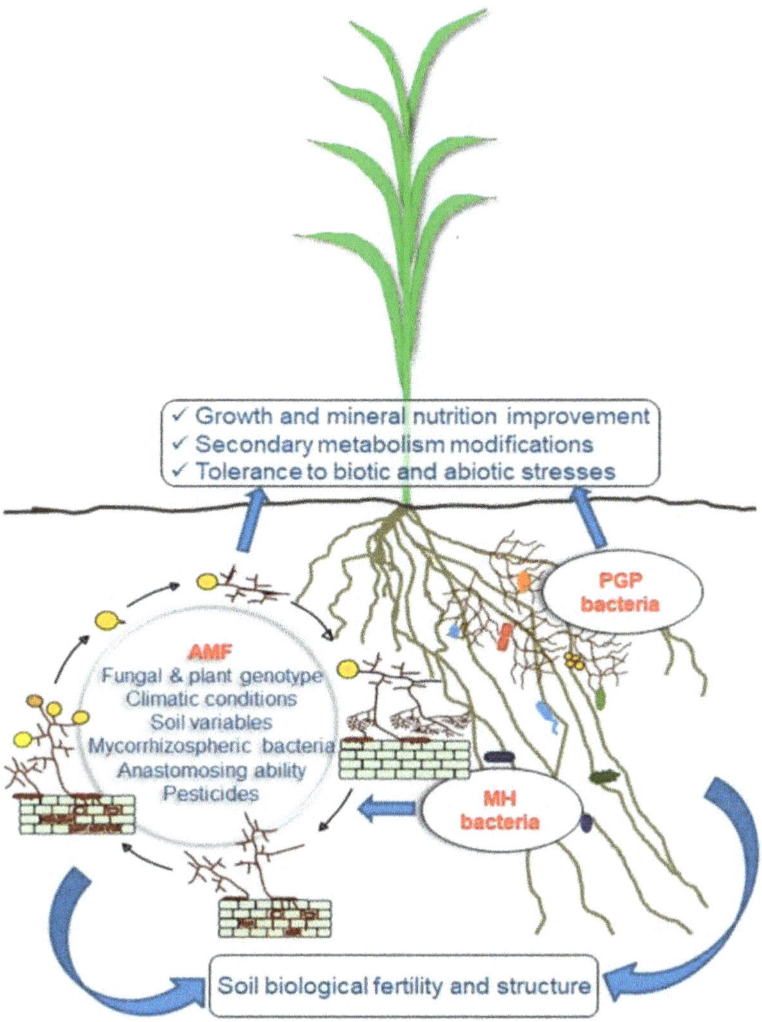

Fig. 13.3 Impact of arbuscular mycorrhizae on plant growth. (Courtesy: Giovannini et al., 2020)

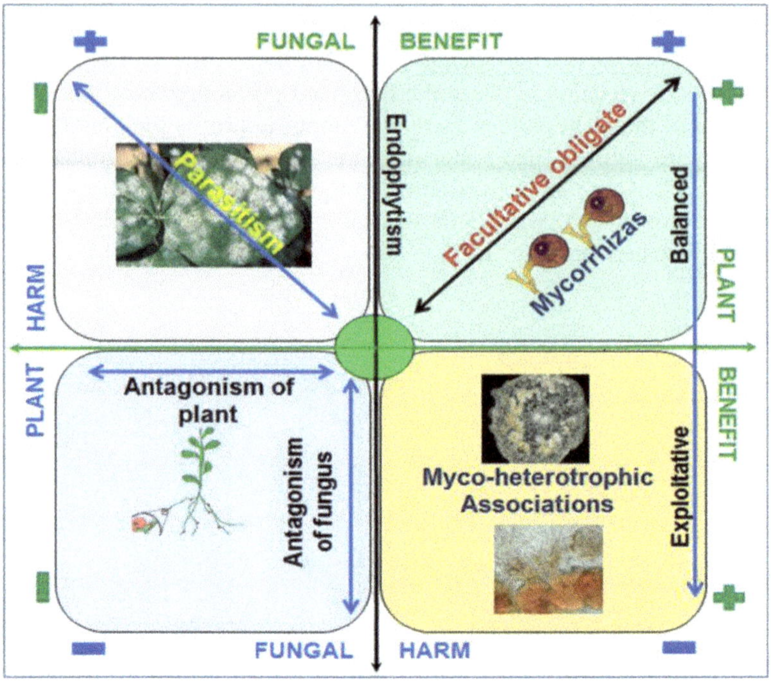

Fig. 13.4 Beneficial and harmful interaction between plant and fungus. (Courtesy: Zeilinger et al., 2016)

The endophytic fungi grow inside the living tissues of plant like the leaves, stems, and roots until the host plant attains senescence. At this stage, the endophyte may become slightly pathogenic (Brundrett, 2004). Because of the abundance of the inhabitant endophytes, most of the plants harbor more than one fungal species and become more resistant to herbivores and pathogens and also become stress tolerant (Strobel & Daisy, 2003). A comparison of different plant–fungal interactions is depicted in Fig. 13.4.

13.4 Fungi as Biofertilizers

In addition to enhancing the nutrient supply to plants, microbes also confer a degree of protection against plant diseases. Various fungi, especially of the genera AMF and *Trichoderma,* produce a range of metabolites against other phytopathogenic fungi (Walsh et al., 2001; Raaijmakers et al., 2002). With development, such microbes could become a realistic alternative to the heavy fungicide regimen used in agriculture at present. A reduction in the use of these chemicals would lead to obvious environmental benefits. The introduction of *Trichoderma* and arbuscular mycorrhizal fungi (AMF) is known to increase the growth of many plant species, crop plants like *Vigna radiata, Saccharum officinarum, Zea mays, Lycopersicum esculentum, Cucumis sativus, Camellia sinensis, Pennisetum glaucum, Cicer arietinum, Raphanus sativus, Sorghum bicolor, Ipomoea batatas, Arachis hypogaea, Oryza sativa, Catharanthus roseus, Mentha piperita, Coleus forskohlii, Andrographis paniculata,* etc. (Selvaraj et al., 2008). Understanding the interaction between consortium of microbial inoculants and plant systems will pave the way to harness more benefits through microbial inoculants for improving plant growth and yield (Raja et al., 2006).

13.5 Disease Production

In contrast to the beneficial effects shown by the symbiotic fungi, the plant–pathogen interaction causes a sequence of events, resulting in the development and maintenance of disease cycle in the host (Daly, 1984). The disease cycle is a chain of interconnected successive events of a pathogen's infection in a host plant. It usually coincides with the lifecycle of the pathogen with a correlation to its host and the environment (Shomrat, 2021).

A disease cycle (Fig. 13.5) in the plants consists of the following phases:

1. The fungi come into contact with the susceptible plant by wind, water, or insect bite or enter through the roots (Travadon et al., 2012).
2. Germination of fungal spores and attachment of the pathogen to the host receptors (Tucker & Talbot, 2001).
3. Penetration of the pathogen into the plant through wounds or natural openings such as stomata or direct penetration with the help of fungal appressoria (Pryce-Jones et al., 1999) or through wounds caused by insects such as Grosmannia clavigera on lodgepole pines (Diguistini et al., 2011) and Ophiostoma ulmi on Dutch elm (D'Arcy, 2000).
4. Establishment of the pathogen by invading into the plant tissues, spread from cell-to-cell causing visible symptoms.
5. Multiplication of the pathogen by reproduction in the host tissues and production of large number of spores.
6. Dispersal of the spores through wind or insects to other susceptible hosts.
7. Survival of the pathogen in a dormant stage under unfavorable conditions (Brown & Ogle, 1997).

Different species evolve together, and this adaptation results in a variety of relationships, such as mutualistic, endosymbiotic, parasitic, competitive, and pathogenic or antagonistic (Faust & Raes, 2012). Many secondary metabolites that are usually bioactive were reported to perform important functions in the fungal and interactions in the ecosystem (Braga et al., 2016) (Table 13.1).

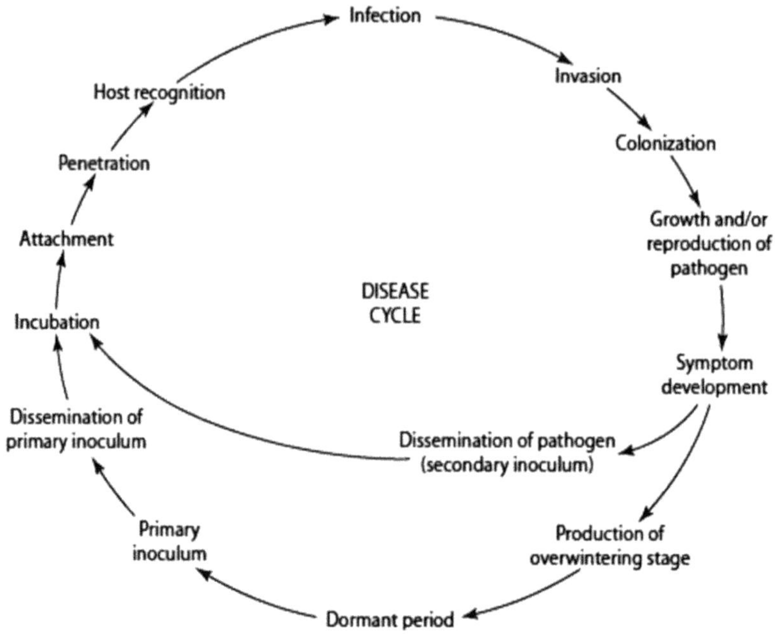

Fig. 13.5 Disease cycle. (Courtesy: Shomrat, 2021)

13.6 Conclusion

Mycorrhizal fungi are important in the fields of environmental change, ecosystem conservation, sustainable agriculture, and development of plants for future needs and food safety. These fungi mobilize P and N, and are an important C sink in the soil, having therefore an important impact on the cycling of these elements (Klironomos et al., 2005). As biofertilizers, they may counteract fertilization excess and thus promote sustainable agriculture. The selection of new crop varieties giving yields on poor soils and in low fertilization conditions should therefore be considered as new aspects, such as their responsiveness to mycorrhizal fungi, which has never consciously been taken into account during plant domestication (Sawers et al., 2008). Arbuscular mycorrhizal symbiosis can be beneficial to crops and agroecosystems in many ways including resistance to pests and improved soil structure (Thomas et al., 2017). The mycorrhizal fungi contribute to the nutritional quality of edible plant organs.

The signaling molecules released by arbuscular and ectomycorrhizal fungi are to be identified, and the potential Myc factor from arbuscular mycorrhizae is under investigation (Bucher et al., 2009), which is based on the potential similarities with the Nod factor. The study of microbes associated with mycorrhizal fungi and their role as a third component of symbiosis is promising. Hence, the mycorrhizal fungal community hidden in the soil can be exploited for the benefit of plants and humans. Newer technologies like CRISPR/Cas9 and proteomics may provide an avenue for accelerating our ability in the discovery of fungal effectors' function (Selin et al., 2016).

Table 13.1 Types and mechanisms of plant–fungal interactions

Fungal partner	Type of interaction	Chemicals or metabolites involved	Effect	References
Moniliophthora roreri and *Trichoderma harzianum*	Phytopathogen–endophyte	Butenolide, harzianolide, sorbicillinol	Compounds produced by the phytopathogen spatially localized in the interaction zone.	Tata et al. (2015)
Trichoderma atroviride and *Arabidopsis* sp.	Endophyte–plant	Indole–acetic acid-related indoles	The fungus colonizes plant roots, thus promoting growth and enhancing systemic disease resistance of the plant.	Salas-Marina et al. (2011)
Stachybotrys elegans and *Rhizoctonia solani*	Mycoparasite–host	Trichothecenes and atranones	*S. elegans* induced alterations in the metabolism and growth of *R. solani*, thereby downregulating the production of antimicrobial compounds by *R. solani*.	Chamoun et al. (2015)
Aspergillus nidulans	Microbial community	Aromatic polyketides	Physical interaction between different microbes leads to the activation of the fungal silent genes coding for secondary metabolite. Alterations in the fungal histone are triggered by the actinomycete.	Schroeckh et al. (2009) and Nutzmann et al. (2011)
Rhizopus species and rice plant	Symbiont–phytopathogen–plant	Rhizoxin, bongkrekic acid, and enacyloxins	In the absence of the endosymbiont, the fungus cannot form spores. The endosymbiont is the causal agent of Rice seedling Blight and produces the phytotoxin rhizoxin. In turn, the growth of the endosymbiont is induced by the fungus.	Partida-Martinez and Hertweck (2005), Partida-Martinez et al. (2007) and Ross et al. (2014)

References

Becard, G., & Fortin, J. A. (1988). Early events of vesicular arbuscular mycorrhiza formation on Ri T-DNA transformed roots. *Plant Phytologist*. https://doi.org/10.1111/j.1469-8137.1988.tb03698.x

Bloemberg, G. V., & Lugtenberg, B. J. J. (2001). Molecular basis of plant growth promotion and biocontrol by rhizobacteria. *Current Opinion in Plant Biology, 4*, 343–350.

Bonfante, P. (2001). At the interface between mycorrhizal fungi and plants: The structural organization of cell wall, plasma membrane and cytoskeleton. In *Mycota, IX Fungal associations* (B. Hock, Ed., pp. 45–91).

Bonfante, P., & Genre, A. (2010). Mechanisms underlying beneficial plant–fungus interactions in mycorrhizal symbiosis. *Nature Communications, 1*, 48.

Braga, R. M., Dourado, M. N., & Araújo, W. L. (2016). Microbial interactions: Ecology in a molecular perspective. *Brazilian Journal of Microbiology, 47*(Suppl 1), 86–98. https://doi.org/10.1016/j.bjm.2016.10.005

Brown, J. F., & Ogle, H. J. (1997). *Plant pathogens and plant diseases*. Rockvale Publication.

Brundrett, M. (2004). Diversity and classification of mycorrhizal associations. *Biological reviews of the Cambridge Philosophical Society, 79*(3), 473–495.

Bucher, M., Wegmüller, S., & Drissner, D. (2009). Chasing the structures of small molecules in arbuscular mycorrhizal signaling. *Current Opinion in Plant Biology, 12*, 500–507.

Buscot, F., Munch, J. C., Charcosset, J. Y., Gardes, M., Nehls, U., & Hampp, R. (2000). Recent advances in exploring physiology and biodiversity of ectomycorrhizas highlight the functioning of these symbioses in ecosystems. *FEMS Microbiology Reviews, 24*(5), 601–614.

Carlile, M. J., Gooday, G. W., & Watkinson, S. C. (2001). *The fungi* (2nd ed.). Cambridge University Press.

Chabaud, M., et al. (2002). Targeted inoculation of *Medicago truncatula* in vitro root cultures reveals MtENOD11 expression during early stages of infection by arbuscular mycorrhizal fungi. *New Phytologist, 156*, 265–273.

Chamoun, R., Aliferis, K. A., & Jabaji, S. (2015). Identification of signatory secondary metabolites during mycoparasitism of *Rhizoctonia solani* by *Stachybotrys elegans*. *Frontiers in Microbiology, 6*, 353.

Corradi, N., & Bonfante, P. (2012). The arbuscular mycorrhizal symbiosis: Origin and evolution of a beneficial plant infection. *PLoSPathogens, 8*(4), e1002600.

Daly, J. M. (1984). The role of recognition in plant disease. *Annual Review of Phytopathology, 22*, 273–273.

D'Arcy, C. J. (2000). Dutch elm disease. *Plant Health Instructor*. https://doi.org/10.1094/PHI-I-2000-0721-02

Diguistini, S., Wang, Y., Liaa, N., et al. (2011). Genome and transcriptome analyses of the mountain pine beetle-fungal symbiont Grosmanniaclavigera, a lodgepole pine pathogen. *Proceedings of the National Academy of Sciences of the United States of America, 108*, 2504–2509.

Faust, K., & Raes, J. (2012). Microbial interactions: From networks to models. *Nature Reviews Microbiology, 10*, 538–550.

Field, K. J., Cameron, D. D., Leake, J. R., Tille, S., Bidartondo, M. I., & Beerling, D. J. (2012). Contrasting arbuscular mycorrhizal responses of vascular and non-vascular plants to a simulated Palaeozoic CO_2 decline. *Nature Communications, 3*, 835.

Frank, A. B. (1885). "Über die auf Wurzelsymbioseberuhende Ernährunggewisser Bäumedurchunterirdische Pilze" [On the nourishing, via root symbiosis, of certain trees by underground fungi]. *Berichte der DeutschenBotanischen Gesellschaft (in German), 3*, 128–145.

Ganugi, P., Masoni, A., Pietramellara, G., & Benedettelli, S. (2019). A review of studies from the last twenty years on plant–arbuscular mycorrhizal fungi associations and their uses for wheat crops. *Agronomy, 9*, 840. https://doi.org/10.3390/agronomy9120840

Genre, A., Chabaud, M., Timmers, T., Bonfante, P., & Barker, D. G. (2005). Arbuscular mycorrhizal fungi elicit a novel intracellular apparatus in *Medicago truncatula* root epidermal cells beforeinfection. *Plant Cell, 17*, 3489–3499.

Genre, A., Chabaud, M., Faccio, A., Barker, D. G., & Bonfante, P. (2008). Prepenetration apparatus assembly precedes and predicts the colonization patterns of arbuscular mycorrhizal fungi within the root cortex of both *Medicago truncatula* and *Daucus carota. Plant Cell, 20*, 1407–1420.

Giovannini, L., Palla, M., Agnolucci, M., Avio, L., Sbrana, C., Turrini, A., & Giovannetti, M. (2020). Arbuscular mycorrhizal fungi and associated microbiota as plant biostimulants: Research strategies for the selection of the best performing inocula. *Agronomy, 10*, 106. https://doi.org/10.3390/agronomy10010106

Giraldo, M. C., & Valent, B. (2013). Filamentous plant pathogen effectors in action. *Nature Reviews Microbiology, 11*(11), 800–814.

Grigoriev, I. (2013). Fungal genomics for energy and environment. In B. Horwitz, P. Mukherjee, M. Mukherjee, et al. (Eds.), *Genomics of soil- and plant-associated fungi: Soil biology* (Vol. 36, pp. 11–27).

Harrison, M. J. (2005). Signaling in the arbuscular mycorrhizal symbiosis. *Annual Review of Microbiology, 59*, 19–42.

Helgason, T., Daniell, T. J., Husband, R., Fitter, A. H., & Young, J. P. W. (1998). Ploughing up the wood-wide web? *Nature, 394*, 431.

Humphreys, C. P., Franks, P. J., Rees, M., Bidartondo, M. I., Leake, J. R., & Beerling, D. J. (2010, November 2). Mutualistic mycorrhiza-like symbiosis in the most ancient group of land plants. *Nature Communications, 1*, 103.

Jones, J. D., & Dangl, J. L. (2006). The plant immune system. *Nature, 444*(7117), 323–329.

Kemen, E., & Jones, J. D. (2012). Obligate biotroph parasitism: Can we link genomes to lifestyles? *Trends in Plant Science, 17*(8), 448–457.

Klironomos, J. N., et al. (2005). Abrupt rise in atmospheric CO2 overestimates community response in a model plant-soil system. *Nature, 433*, 621–624.

Kosuta, S., Chabaud, M., Lougnon, G., Gough, C., Dénarié, J., Barker, D. G., & Bécard, G. (2003). A diffusible factor from arbuscular mycorrhizal fungi induces symbiosis-specific MtENOD11 expression in roots of *Medicago truncatula. Plant Physiology, 131*(3), 952–962.

Larissa, C. V., da Silva, D. K. A., da Silva, I. R., Gonçalves, C. M., de Assis, D. M. A., Oehl, F., & da Silva, G. A. (2019). Ecological aspects of arbuscular mycorrhizal fungal communities in different habitat types of a Brazilian mountainous area. *Ecological Research, 1*, 182–192. https://doi.org/10.1111/1440-1703.1061

Nutzmann, H. W., Reyes-Dominguez, Y., Scherlach, K., et al. (2011). Bacteria-induced natural product formation in the fungus *Aspergillus nidulans* requires Saga/Ada-mediated histone acetylation. *Proceedings of the National Academy of Sciences of the United States of America, 108*, 14282–14287.

Parniske, M. (2008). Arbuscular mycorrhiza: The mother of plant root endosymbioses. *Nature Reviews Microbiology. A comprehensive review on the state of the art research in arbuscular mycorrhizas, 6*, 763–775.

Partida-Martinez, L. P., & Hertweck, C. (2005). Pathogenic fungus harbours endosymbiotic bacteria for toxin production. *Nature, 437*, 884–888.

Partida-Martinez, L. P., Monajembashi, S., Greulich, K. O., & Hertweck, C. (2007). Endosymbiont-dependent host reproduction maintains bacterial-fungal mutualism. *Current Biology, 17*, 773–777.

Pryce-Jones, E., Carver, T. I. M., & Gurr, S. J. (1999). The roles of cellulase enzymes and mechanical force inhost penetration by *Erysiphe graminisf* sp. *hordei. Physiological and Molecular Plant Pathology, 55*, 175–182.

Raaijmakers, J. M., Vlami, M., & de Souza, J. T. (2002). Antibiotic production by bacterial biocontrol agents. *Antonie Van Leeuwenhoek, 81*, 537–547.

Raja, P., Uma, S., Gopal, H., & Govindarajan, K. (2006). Impact of Bioinoculants consortium on rice root exudates, biological nitrogen fixation and plant growth. *Journal of Biological Sciences, 6*, 815–882.

Redecker, D., Kodner, R., & Graham, L. E. (2000). Glomalean fungi from the Ordovician. *Science, 289*, 1920–1921.

Remy, W., Taylor, T. N., Hass, H., & Kerp, H. (1994). Four hundred-million-year-old vesicular arbuscular mycorrhizae. *Proceedings of the National Academy of Sciences of the United States of America, 91*(25), 11841–11843.

Ross, C., Opel, V., Scherlach, K., & Hertweck, C. (2014). Biosynthesis of antifungal and antibacterial polyketides by *Burkholderia gladioli* in co-culture with *Rhizopus* microspores. *Mycoses, 57*, 48–55.

Salas-Marina, M. A., Silva-Flores, M. A., Uresti-Rivera, E. E., et al. (2011). Colonization of *Arabidopsis* roots by *Trichoderma atroviride* promotes growth and enhances systemic disease resistance through jasmonic acid/ethylene and salicylic acid pathways. *European Journal of Plant Pathology, 131*, 15–26.

Sawers, R. J., Gutjahr, C., & Paszkowski, U. (2008). Cereal mycorrhiza: An ancient symbiosis in modern agriculture. *Trends in Plant Science, 13*, 93–97.

Schroeckh, V., Scherlach, K., Nutzmann, & H.W., et al. (2009). Intimate bacterial–fungal interaction triggers biosynthesis of archetypal polyketides in *Aspergillus nidulans*. *Proceedings of the National Academy of Sciences of the United States of America, 106*, 14558–14563.

Selin, C., de Kievit, T. R., Belmonte, M. F., & Dilantha Fernando, W. G. (2016). Elucidating the role of effectors in plant-fungal interactions: Progress and challenges. *Frontiers in Microbiology, 7*, 600. https://doi.org/10.3389/fmicb.2016.00600

Selosse, M. A., & Roy, M. (2009). Green plants that feed on fungi: Facts and questions about mixotrophy. *Trends in Plant Science, 14*, 64–70.

Selvaraj, T., Kumar, S. R., Mathan, C., Nisha, W., & W., & Tesso, M. (2008). Effect of *Glomus mosseae* and plant growth promoting rhizomicroorganisms (PGPR's) on growth, nutrients and content of secondary metabolites in *Begonia malabarica* Lam. Mj. *The International Journal of Science & Technoledge, 2*(03), 516–525.

Shah, D. (2018). Plant–Fungal Interactions. *A special issue of International Journal of Molecular Sciences*. Guest Editor Donald Danforth Plant Science Center, St. Louis, MO 63132, USA (ISSN 1422-0067), section "Molecular Plant Sciences".

Shomrat, A. (2021). *The disease cycle: Stages of disease development*. https://plantlet.org/the-disease-cycle-stages-of-disease-development/

Siciliano, V., et al. (2007). Transcriptome analysis of arbuscular mycorrhizal roots during development of the prepenetration apparatus. *Plant Physiology, 144*, 1455–1466.

Smith, S. E. & Read, D. J. (2008). *Mycorrhizal symbiosis*. Academic Press.

Staskawicz, B. J. (2001). Genetics of plant-pathogen interactions specifying plantdisease resistance. *Plant Physiology, 125*(1), 73–76.

Stracke, S., Kistner, C., Yoshida, S., Mulder, L., Sato, S., Kaneko, T., Tabata, S., Sanda, L. N., Stougaard, J., Szczyglowski, K., & Parniske, M. (2002). A plant receptor-like kinase required for both bacterial and fungal symbiosis. *Nature, 417*(6892), 959–962.

Strobel, G., & Daisy, B. (2003). Bioprospecting for microbial endophytes and their natural products. *Microbiology and Molecular Biology Reviews, 67*(4), 491–502.

Tata, C., Perez, M. L., Campos, M. A., Bayfield, M. N., & Eberlin, & Ifa, D. R. (2015). Imprint desorption electrospray ionization mass spectrometry imaging for monitoring secondary metabolites production during antagonistic interaction of fungi. *Analytical Chemistry, 87*, 12298–12304.

Thirkell, T. J., Charters, M. D., Elliott, A. J., Sait, S. M., & Katie Field, J. (2017). Are mycorrhizal fungi our sustainable saviours? Considerations for achieving food security. *Journal of Ecology, 105*. https://doi.org/10.1111/1365-2745.12788

Trappe, J. M. A. B. (2005). Frank and mycorrhizae: The challenge to evolutionary and ecologic theory. *Mycorrhiza, 15*, 277–281.

Travadon, R., Smith, M. E., Fujiyoshi, P., Douhan, G. W., Rizzo, D. M., & Baumgartner, K. (2012). Inferring dispersal patterns of the generalist root fungus *Armillaria mellea*. *New Phytologist, 193*, 959–969.

Tucker, S. L., & Talbot, N. J. (2001). Surface attachment and pre-penetration stage development by plant pathogenic fungi. *Annual Review of Phytopathology, 39*, 385–417.

Tudzynski, P., & Sharon, A. (2003). Fungal pathogenicity genes. *Applied Mycology and Biotechnology, 3*, 187–212.

Walsh, U. F., Morrissey, J. P., & O'Gara, F. (2001). *Pseudomonas* for biocontrol of phytopathogens: From functional genomics to commercial exploitation. *Current Opinion in Biotechnology, 12*, 289–295.

Weidmann, S., et al. (2004). Fungal elicitation of signal transduction-related plant genes precedes mycorrhiza establishment and requires the dmi3 gene in *Medicago truncatula*. *Molecular Plant-Microbe Interactions, 17*, 1385–1393.

Zamioudis, C., & Pieterse, C. M. (2012). Modulation of host immunity by beneficial microbes. *Molecular Plant-Microbe Interactions, 25*(2), 139–150.

Zeilinger, S., Gupta, V. K., Dahms, T. E., Silva, R. N., Singh, H. B., Upadhyay, R. S., Gomes, E. V., Tsui, C. K., & Nayak, S. C. (2016). Friends or foes? Emerging insights from fungal interactions with plants. *FEMS Microbiology Reviews, 40*(2), 182–207.

Chapter 14
Fungal Biofertilizer: An Alternative for Sustainable Agriculture

Alka Panda

Contents

A. Panda (✉)
Institute of Renewable Energy Technology and Management, Pt Ravishankar Shukla University, Raipur, Chattisgarh, India

A. C. Shukla (ed.), *Applied Mycology*, Fungal Biology,
https://doi.org/10.1007/978-3-030-90649-8_14

14.1 Introduction

Why did the UN World Food Programme (WFP) receive the 2020 Nobel Peace Prize? And what does that have to do with nutrition? The answer to both questions is embedded in renewed concerns globally about food insecurity (Patrick Webb, 2020).

Since the dawn of human civilization, mankind has faced a myriad of pandemics. Presently, the COVID-19 pandemic has not only impacted food trade, food supply chains, and markets, but also people's lives, livelihoods, and nutrition. According to the latest UN estimates, as many as 132 million people went hungry in 2020 as a result of the economic recession triggered by the pandemic. At the same time, 135 million people suffer from acute food insecurity and are in need of urgent humanitarian assistance. Equally urgent is the compounding threat of the pandemic on existing crises – such as conflict, natural disasters, climate change, urban encroachment, loss of biodiversity, pests and animal diseases – that are already stressing our food systems and triggering food insecurity around the globe (HLPE, 2020).

It is in this context that the Nobel Peace Prize committee concluded that tackling hunger and malnutrition, especially in the context of COVID-19, represents a worthy cause, and that the WFP is a worthy champion of that cause (Patrick Webb, 2020).

To urgently minimize COVID-19's damaging effects on food security and nutrition while transforming global food systems to make them more resilient, sustainable, and equitable, FAO calls for immediate action in seven key priority areas, including food security (FAO, 2020).

The CFS Principles for Responsible Investment in Agriculture and Food Systems – also known as RAI – are a set of 10 principles that seek to guide all types of agricultural investment to ensure that it respects a range of environmental, social, and economic goals and supports sustainable food systems (FAO, 2014).

In the midst of such nebulous challenges, the problem of inadequate nutrition is so acute that it is beyond any single type of nutrient source to accept the challenge of appropriate nutrient supply. Integrated use of all the sources such as mineral fertilizers, organic manures, biofertilizers, etc., is the only alternative for improving soil fertility. The use of organic manures and mineral fertilizers is in practice, but the use of biofertilizer in agriculture is not very popular. Hence, there is a need to make its use popular (Singh et al., 2014).

On one hand, it is essential that indigenous agricultural productivity should be enhanced significantly to sustain through the crucial times without degrading the environmental and agricultural ecology further by relying on sustainable agricultural practices, with minimal use of nonrenewable resources (Figueiredo et al., 2017). On the other hand, unless a situation is created for facilitating social and economic access of the rural poor to the developed sources, the organic potential of the country would not become fully available for crop production (Kumar et al., 2010). In the present scenario, exploring biofertilizers is a potent option to look into.

14.1.1 Persisting Effects of Green Revolution

Chemical fertilizers have aided farmers in time and cost-effective production of commercial crops since the 1930s. In India, the green revolution in the 1960s (Somvanshi et al., 2020) led to high productivity of crops through various adapted measures, like adoption of high- yielding variety (HYV) of seeds, which responded well to the highly increased use of inorganic fertilizers and pesticides (Singh, 2000; Brainerd & Menon, 2014). The green revolution, which was beneficial in ensuring food security, has unintended but harmful consequences on agriculture and human health (Singh, 2000; Clasen et al., 2019; Taylor, 2019). Their overuse has been found to harden the soil, cause acidification and dehydration of the soil, decrease fertility, pollute air and water, and release greenhouse gases, thereby bringing hazards to human health and the environment. Extensive use of chemical fertilizers in replenishing NPK depletes essential soil nutrients and minerals, resulting in soil degradation and the loss of equilibrium of a stable soil. This eventually led to poor crop yield, with harvest being more susceptible to pests and diseases. Besides this, chemical fertilizers can cause root burn or fertilizer burn. The nitrate salts present in the nitrogen fertilizers, being water soluble, cause ground and surface water contamination with an accumulative effect. Continuous use of chemical fertilizers is reported to negatively impact the soil inhabitants, beneficial for plant growth. In their larger threat to the environment, animals, and human health, chemical runoff of the excess fertilizer into our water bodies, ponds, streams, ground water, etc., causes eutrophication owing to oxygen depletion that eventually damages the aquatic life. Overapplication of chemical fertilizer to plants may cause the leaves to turn yellow or brown, a condition known as chemical leaf scorch. Leaf scorch can cause the leaves of the plant to wither and may cause the plant to die. Chemical fertilizers destroy soil crumbs, resulting in a highly compacted soil with reduced drainage and air circulation, referred to as soil friability effect. The use of synthetic chemicals jeopardizes soil health, which then leads to fewer or poor quality crops in the long term. Due to the green revolution, India even lost almost 1 lakh varieties of indigenous rice (Prasad, 2016). Globally, agriculture is on an unsustainable track and has a high ecological footprint now (Prasad, 2016).

On one hand, majority of Indian soils are deficient in many macro and micro–nutrients (Fertiliser Association of India, 2011); and on the other hand, with the injudicious application of chemical fertilizers alone, agricultural productivity is declining and environmental quality is deteriorating (Rakshit et al., 2015). Under such circumstances, only bio-based agroeconomy is most viable for agro-industry, farmers, and consumers without compromising on soil and plant health.

14.1.2 Biofertilizer: An Agent of Bio-based Agroeconomy

The solution lies in "biofertilizer," one of the key subsegments of Indian bioeconomy (see Fig. 14.1). Bioeconomy or bio-based economy is a more recent phenomenon. The commonly accepted definition of bioeconomy is the production, utilization, and conservation of biological resources, including related knowledge, science, technology, and innovation, to provide information products, processes, and services across all economic sectors aiming toward a sustainable economy (BIO-ECONOMY INDIA.1615182060_Indian_BioEconomy_Report 2021.pdf).

The benefits of bioeconomy include increased security, economic advantages to farmers, industry, rural communities, and society, environmental benefits at the global, regional, and local levels, and other benefits to society in terms of human health and safety (Kulshreshtha et al., 2011).

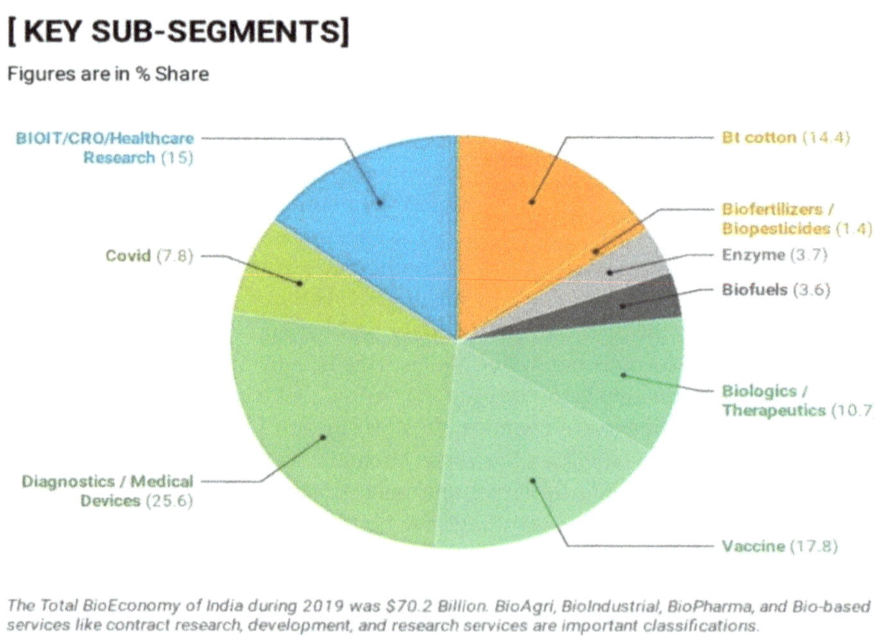

Fig. 14.1 Biofertilizer is one of the key subsegments of bioeconomy. (Source: https://birac.nic.in/webcontent/1615182060_Indian_BioEconomy_Report_2021.pdf

Bioeconomy can be hugely promoted by the extensive use of biofertilizer – a ready-to-use live formulation of soil microorganisms, which on application to seed, root, or soil, mobilizes the availability of nutrients by their biological activity. These are nothing but selected strains of beneficial soil microorganisms cultured in the laboratory and packed in a suitable carrier, which can be used for either seed treatment or soil application. Biofertilizers generate plant nutrients like nitrogen and phosphorous through their activities in the soil or rhizosphere and make them available to plants in a gradual manner. Biofertilizers are gaining momentum recently due to their availability to maintain soil health, minimize environmental pollution, and cut down the use of chemicals in agriculture. In rainfed agriculture, these inputs gain added importance in view of their low cost as most of the farmers are small and marginal and cannot afford expensive chemical fertilizers (Bisen et al., 2015). Biofertilizers are also ideal input for reducing the cost of cultivation and for practicing organic farming. The use of cheap and eco-friendly inputs like biofertilizers is especially important in India where most of the farming will continue to be in the hands of small farmers.

14.1.3 Types of Biofertilizers

Biofertilizers compromise "microbial inoculants" or assemblage of live or latent cells of efficient strains of nitrogen-fixing, phosphate-solubilizing, sulfur-oxidizing, or cellulytic microorganisms used for application of seed, soil, or composting areas with the objective of increasing the numbers of such microorganisms and accelerate certain microbial processes to augment the extent of the availability of nutrients in a form that can be easily assimilated by plants. In large sense, the term may be used to include all organic resources (manure) for plant growth that are rendered in an available form for plant absorption through microorganisms or plant associations or microbial interactions (Rao, 1999). In addition to other biofertilizers available, fungal biofertilizers exert direct or indirect benefits on plant growth and crop yield through biological mechanisms. They can cause stimulation of plant growth by hormone action or antibiosis and by decomposition of organic residues (Pal et al, 2015; Carvajal-Muñoz & Carmona-Garcia, 2012). The following types of biofertilizers are available to the farmers in India:

1. Nitrogen-fixing biofertilizers (N-BF)
   - Symbiotic BF with all legumes – *Rhizobium, Bradyrhizobium*
   - Nonsymbiotic BF for cereals, vegetables, horticultural crops – *Azotobacter*
   - Associative BF for millets, maize, etc. – *Azospirillum* (blue-green algae, *Azolla, Gluconacetobacter diazotrophicus* are also N-BF, but is yet to be included in FCO) (https://geographyandyou.com/biofertilisers-in-indian-agriculture/)

2. Phosphorus biofertilizers (P-BF)
   - Phosphorous-solubilizing biofertilizers (PSB) (*Bacillus, Pseudomonas, Aspergillus, Penicillium, Fusarium, Trichoderma, Mucor, Ovulariopsis, Tritirachium*, and *Candida*)
   - Phosphate-mobilizing biofertilizers – *Mycorrhiza* (*glomus, gigaspora*, etc., for all crops)
3. Potash biofertilizers (K-BF) (microbes like *B. mucilaginosus* and *F. aurantia* for all crops)
4. Zinc solubilizers (Z- BF) (*Bacillus* species are capable of zinc solubilization)
5. Plant growth-promoting biofertilizers (*Pseudomonas*)
6. Enriched compost biofertilizers = cellulolytic fungal cultures (*Chaetomium bostrychodes, C. olivaceum, Humicola fuscoatra, Aspergillus flavus, A. nidulans, A. niger, A. ochraceus, Fusarium solani*, and *F. oxysporum*)

Generally, biofertilizers are available in a solid (using peat, lignite, charcoal, etc., as a carrier) or in a liquid base (using broth involving additives like poly vinyl pyrrolidone, gum, biosurfactants, etc., or by promoting dormant cells/spores). Biofertilizers may be prepared from either single or multiple strains, and experiments are also being conducted for formulating freeze-dried, granular, and polyacrylamide-entrapped inoculants. At present, however, the best production technology and packaging are yet to be obtained (https://geographyandyou.com/biofertilisers-in-indian-agriculture/). Figure 14.2 and Table 14.1 show the general

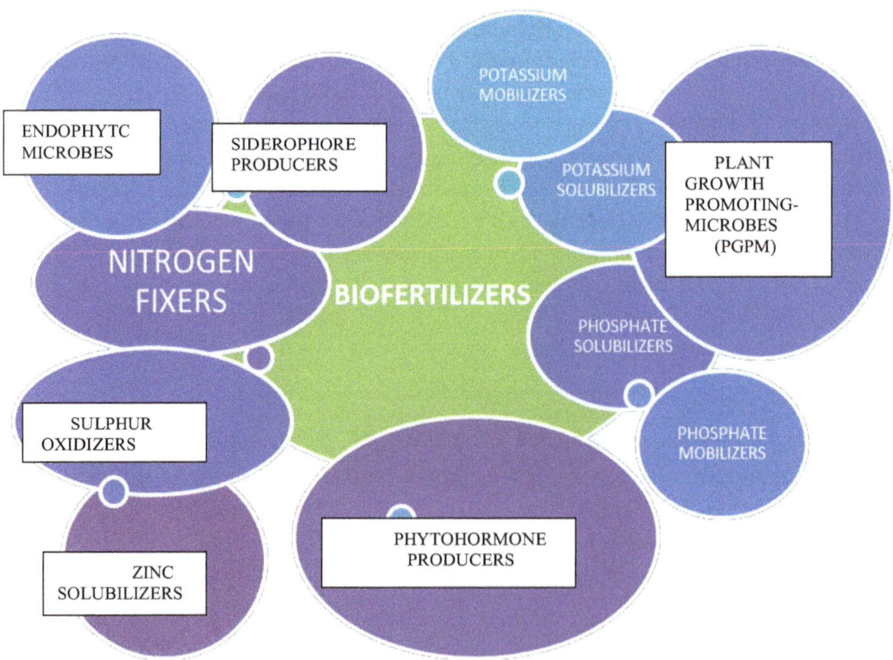

Fig. 14.2 Types of biofertilizers

Table 14.1 Classification of biofertilizers

Types of biofertilizers	Mode of action	Examples	References
N_2-fixing biofertilizers			
1. Free-living; nonsymbiotic	Increase soil nitrogen content by fixing atmospheric N and make it available to the plants.	*Azotobacter, Beijerinckia, Clostridium, Klebsiella, Anabaena, Bacillus polymyxa, Rhodopseudomonas, Rhodospirillum, Chromatium, Nostoc*	Choudhury and Kennedy (2004)
2. Symbiotic		*Rhizobium, Frankia, Xanthomonas* and *Mycobacterium* and *Anabaena azollae*	
3. Associative symbiotic		*Azospirillum* sp., *Enterobacter, Acetobacter diazotrophicus, Herbaspirillum* sp., *Azoarcus* sp., *Alcaligenes*	
P-solubilizing biofertilizers			
1. Bacteria	Solubilize the insoluble forms of P in the soil into soluble forms by secreting organic acids and lowering soil pH to dissolve bound phosphates.	*Bacillus megaterium* var. *phosphaticum, Bacillus subtilis, Bacillus circulans*, and *Pseudomonas striata*	Board (2004)
2. Fungi		*Penicillium* sp., and *Aspergillus awamori*	
P-mobilizing (microphos) biofertilizers			
1. Arbuscular mycorrhiza	Transfer phosphorus from the soil to the root cortex. These are broad spectrum biofertilizers.	*Glomus* sp., *Gigaspora* sp., *Acaulospora* sp., *Scutellospora* sp., *Sclerocystis* sp.	Chang and Yang (2009)
2. Ectomycorrhiza		*Laccaria* sp., *Pisolithus* sp., *Boletus* sp., *Amanita* sp.	
3. Ericoid mycorrhiza		*Pezizella ericae*	
4. Orchid mycorrhiza		*Rhizoctonia solani*	
K-solubilizing biofertilizer			
1. Bacteria	Solubilize potassium (silicates) by producing organic acids that decompose silicates and help in the removal of metal ions and make it available to plants.	*Bacillus mucilaginosus, B. circulanscan, B. edaphicus, Arthrobacter* sp.	Etesami et al. (2017)
2. Fungi		*Aspergillus niger*	
K-mobilizing biofertilizer			
1. Bacteria	They mobilize the inaccessible forms of potassium in the soil.	*Bacillus* sp.	Jha (2017)
2. Fungi		*Aspergillus niger*	

(continued)

Table 14.1 (continued)

Types of biofertilizers	Mode of action	Examples	References
Biofertilizers for micronutrients			
1. Silicate and zinc solubilizers	Solubilize the zinc by proton, chelated ligands, acidification, and by oxidoreductive systems.	*Bacillus* sp., *Bacillus subtilis, Thiobacillus thiooxidans, Saccharomyces* sp., *Mycorrhiza* sp.	Kamran et al. (2017)
2. Sulfur oxidizer	Oxidize sulfur to sulfates that are usable by plants.	*Thiobacillus* sp.	Itelima et al. (2018)
Plant growth–promoting biofertilizers			
1. Plant growth-promoting rhizobacteria	Produce hormones that promote root growth, improve nutrient availability, control plant pathogens, and improve crop yield.	*Pseudomonas fluorescence*	Backer et al. (2018)
2. Plant growth-promoting fungi		*Aspergillus* sp., *Fusarium* sp., *Penicillium* sp., *Piriformospora* sp., *Phoma* sp., *Trichoderma* sp.	

Table 14.2 Differences between the two production technologies of biofertilizers

Carrier-based	Liquid-based
Low shelf-life	Longer shelf-life
Easier to produce	Easier to produce
Temperature sensitive	Temperature tolerant
Less investment	Higher investment for production unit
Low cell counts	High cell counts
Contamination prone	Contamination free
Less effective	More effective
Product may not be 100% sterile	Product can be 100% sterile
Automation difficult	Automation is easier
Low cost	High cost
High dose requirement	Low dose requirement

classification of biofertilizers, and Table 14.2 shows the differences between two different technologies of biofertilizer formulation.

14.1.4 Advantages of Biofertilizer

Biofertilizers are the natural way to get the benefits of synthetic fertilizers without risking the quality of soil health and crop products. Biofertilizers are known to play a number of vital roles in soil fertility, crop productivity, and production in agriculture as they are eco-friendly but cannot replace chemical fertilizers, which are indispensable for getting maximum crop yields. Biofertilizers can increase the crop yield

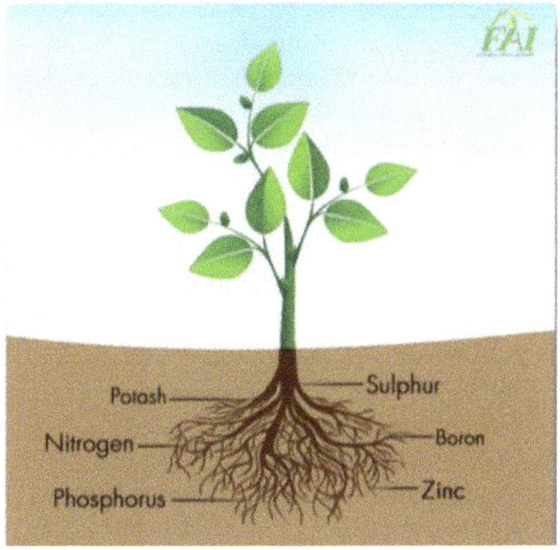

Fig. 14.3 Soil nutrients for plant growth. (Source: FAI)

Fig. 14.4 Food security. (Source: FAI)

by 20–30%. In addition, biofertilizers are cost-effective when compared to synthetic fertilizers. Biofertilizers unleash nutrients (Fig. 14.3) from the soil and provide us food security (Fig. 14.4).

Some of the important functions or roles of biofertilizers in agriculture are

- Phosphate-mobilizing or phosphorus-solubilizing biofertilizers/microorganisms (bacteria, fungi, mycorrhiza, etc.) convert insoluble soil phosphate into soluble forms by secreting several organic acids and under optimum conditions, can solubilize/mobilize about 30–50 kg P_2O_5/ha, and crop yield may increase by 10–20%.

- Mycorrhiza or arbuscular mycorrhiza, when used as biofertilizer, enhance uptake of P, Zn, S, and water, leading to uniform crop growth and increased yield and also enhance resistance to root diseases and improve hardiness of transplant stock (Pal et al., 2014).
- Liberate growth-promoting substances and vitamins and help to maintain soil fertility.
- Act as antagonists and suppress the incidence of soil-borne plant pathogens and, thus, help in the biocontrol of diseases.
- Play an important role in the recycling of plant nutrients.
- Supplement chemical fertilizers for meeting the integrated nutrient demand of the crops.
- Renewable source of nutrients.
- Sustain soil health.
- Supplement chemical fertilizers.
- Replace 25–30% chemical fertilizers.
- Increase the grain yields by 10–40%.
- Decompose plant residues and stabilize C:N ratio of soil.
- Improve texture, structure, and water-holding capacity of soil.
- No adverse effect on plant growth and soil fertility.
- Stimulate plant growth by secreting growth hormones.
- Secrete fungistatic and antibiotic-like substances.
- Solubilize and mobilize nutrients.
- Eco-friendly, non–pollut, and cost-effective.

The microorganisms present in biofertilizers are available in nature. Initially, these organisms are isolated from different sources such as the root nodule for rhizobium, soil for other microbes, etc., and developed in specific media for mass culturing. Hence, the understanding of the biological attributes of beneficial microbes and their mutual interactions in conjunction with the modern agricultural practices would help to develop the most potential biofertilizer for sustainable agro-practice. Potential roles played by biofertilizers for plant growth are illustrated in Fig. 14.5.

14.2 Soil Microbiome: A Repository of Nutrients for Plant–Microbe Holobiont

Years of extensive research have revealed the significance of microorganisms in plant health. The difficulty of culturing transplants of different species in the absence of bacteria and fungi is widely known (Hardoim et al., 2008), which strongly implies the importance of such microorganisms in plant growth. It is readily acknowledged that cooperative microbial symbionts play an important role in their host's life and fitness (Kiers & van der Heijden, 2006). Thus, a plant can be regarded as a holobiont comprising the host plant and its microbiota (Zilber-Rosenberg & Rosenberg,

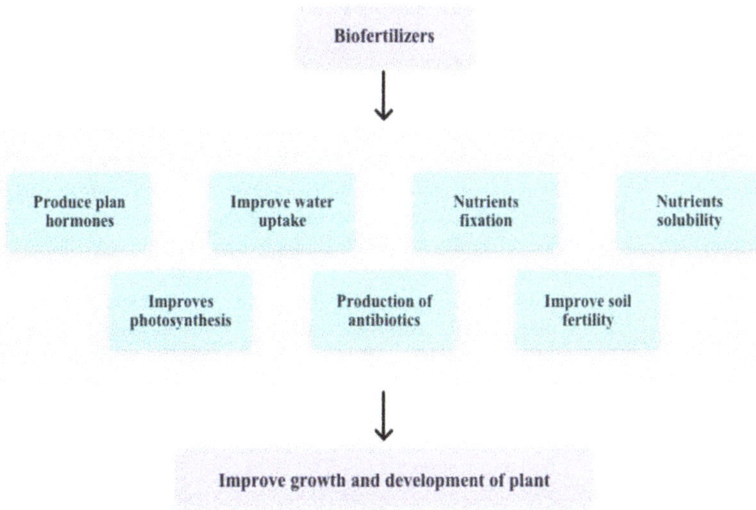

Fig. 14.5 Potential role played by biofertilizers in the growth and development of plants

2008). On the other hand, rhizospheric or root-associated microorganisms, beneficial for plant growth, are strongly influenced by the soil microbiome in the vicinity of the roots.

It is quite evident that the plant–microbiome provides the host plant with a variety of essential functions to adapt to the local environment. However, plant–microbe holobionts are impacted by different factors, generating complex interactive systems (Tosi et al., 2020). One of the most important factors is the soil microbiome, which facilitates plant growth and enhances crop yield. Owing to their rich diversity, complexity of interactions, and numerous metabolic pathways, microbes are an amazing resource for biological activity (Emmert & Handelsman, 1999; Alabouvette et al., 2006; Tejesvi et al., 2007; Mitchell et al., 2008; Raghukumar, 2008) and constitute the most important niches in the soil ecosystem. Such microbiological interactions and activities are responsible for earth geochemical stability, climatic and biogeochemical cycles, and have been exploited for enhanced agricultural, environmental, and forestry management (Barea et al., 2005; Tringe et al., 2005; Hansel et al., 2008; Frey-Klett et al., 2011; Deveau et al., 2018; Odoh et al., 2019a, b). These interactions are also responsible for enhancing plant growth by improving nutrient uptake and developing tolerance for biotic and abiotic stress (Bhardwaj et al., 2014; Clark et al., 2009). Strategic and applied research has demonstrated that certain cooperative microbial species, immersed in a framework of interactions, can be exploited as a low-input biotechnology to help sustainable, environmentally friendly, agrotechnological practices. The global need and demand for enhanced crop production, agroeconomy, and ecological immunity require efficient use of such biological inoculants, composed of beneficial soil microorganisms unlike their chemical counterparts (Sivasakthivelan & Saranraj, 2013). The quantity of nutrient availability rendered by these microorganisms is shown in Table 14.3.

Table 14.3 Benefits rendered by beneficial microbes

Name	Crops suited	Benefits usually seen	Remarks
Rhizobium strains	Legumes like pulses, groundnut, soybean	10–35% yield increase, 50–200 kg N/ha	Fodders give better results. Leaves residual N in the soil.
Azotobacter	Soil treatment for non legume crops including dry land crops	10–15% yield increase – adds 20–25 kg N/ha	Also controls certain diseases.
Azospirillum	Nonlegumes like maize, barley, oats, sorghum, millet, sugarcane, rice, etc.	10–20% yield increase	Fodders give higher/enrich fodder response. Produces growth-promoting substances. It can be applied to legumes as co-inoculant.
Phosphate solubilizers (there are two bacterial and two fungal species in this group)	Soil application for all crops	5–30% yield increase	Can be mixed with rock phosphate.
Blue-green algae and *Azolla*	Rice/wet lands	20–30 kg N/ha, *Azolla* can give biomass up to 40–50 tons and fix 30–100 kg N/ha	Reduces soil alkalinity, can be used for fishes as feed. They have growth-promoting hormonal effects.
Mycorrhizae (VAM)	Many trees, some crops, and some ornamental plants	30–50% yield increase, enhances uptake of P, Zn, S, and water	Usually inoculated to seedlings.

Pathak et al., 2017 stated that fungal–fungal (tropical plant and their foliar endophytes) and bacterial–fungal (arbuscular mycorrhizal fungi and plant growth-promoting rhizobacteria) associations enhance the turnover of soil organic matters apart from facilitating nutrient mobilization for plant growth through nitrogen (N) fixation in leguminous crops (Franco et al., 2011; Nuti & Giovannetti, 2015). The community of mycorrhiza colonizing a plant has been extensively studied over several decades. However, it was only in 2002 that the diversity of fungi colonizing the plant root was shown to be much greater than previously believed (Vandenkoornhuyse et al., 2002), indicating a gap in our knowledge concerning microorganisms colonizing roots. This explains the growing advances in mycobiome, exploring its inherent potentials in sustainable agriculture (Berruti et al., 2016). This review hence focuses on the beneficial soil fungi, which can be developed into potential biofertilizers.

When we focus upon the soil and plant health, we cannot overlook the nuisance created by agrochemicals, including pesticides – one of the deadly human interventions in destabilizing the agroecosystem and environmental ecology. Pesticides are double-edged weapons, which, although targeted on specific targets, inhibit nontarget organisms including the soil mycoflora (Daisley et al., 2022; Lata et al., 2021; Hashimi et al., 2020; Pandey et al., 2005).

Many soil-applied pesticides are intentionally introduced into the soil environment for the control of soil-borne pests and pathogens, resulting in the accumulation of their residues and metabolites in soil at unacceptably high levels (Gamon et al., 2003; Shalaby & Abdou, 2010). On the contrary, the extensive use of pesticides has often resulted in the development and evolution of pesticide resistance in insect pests, plant pathogens, and weeds (Aktar et al., 2009; Pimentel, 2005). Moreover, in the absence of a clear obsolete pesticide management strategy, over the years, significant amounts of obsolete pesticides have been stockpiling in developing countries, which endanger the environment and health of millions of people (Farrera, 2004; Karstensen et al., 2006; Ortiz-Hernandez & Sanchez-Salinas, 2010) and should be managed effectively (Karstensen et al., 2006; Dasgupta et al., 2010; Martinez, 2004).

The most promising opportunity for maximizing benefits and minimizing risks is to invest time, money, and effort into developing a diverse toolbox of crop protection and production strategies that include safe products and practices integrating chemical approaches into an overall and ecologically based framework that will optimize sustainable production, environmental quality, and human health (Damalas, 2009).

When pesticides are used rationally and carefully, in conjunction with other technologies like biopesticides in integrated pest management systems, it is more likely that their use will be justifiable (Cooper & Dobson, 2007). Therefore, biological control agents that are used in the preparation of biopesticides should be tolerant enough to pesticides (Arshad & Aishatul, 2015).

The potentiality of pesticide-tolerant fungi as biofertilizers can pave the way for bio-entrepreneurs. The combination of these potentials when fully harnessed under agricultural scenario will help to sustain agriculture and boost food security globally (Bending et al., 2002; Verma et al., 2014; Tomer et al., 2021). In the present review, we need to understand how agricultural sustainability and economic stability of rural India can largely depend upon fungal biofertilizers.

Focusing on sustainable biofertilizer production and low-cost environmental remediation, this review seeks to identify fungal species being examined individually for their unique capabilities of nutrient mobilization, nutrient solubilization, bioremediation, nitrogen fixation, or biocontrol mechanism order. This will help us to focus on the most promising bioinoculant or assemblage of bioinoculants for promoting soil and plant health.

14.2.1 Soil Mycobiome: Pool of Promising Bioagents for Sustainable Agriculture

The myco-diversity in soil plays a pivotal role affecting soil and plant health as symbionts, decomposers, nutrient mobilizers, agents for bioremediation, antistress, and pest control (Owen et al., 2015; Ajmal et al., 2018). Intensive research on the identity, abundance, distribution, and function of soil mycobiome, and on their various roles in soils has been fundamental for a better understanding of fungal biology, interrelationships, as well as the physiological mechanisms that help in the maintenance of soil health. It is also important to assess the common dynamics that may occur due to the coexistence of fungal populations with other microbes in most cases of host–symbiont associations.

The rapid expansion of the soil mycobiome and their interaction with plants and other microbes pave way for sustainable agriculture to bring increased awareness to the diverse effect of the soil mycobiome on plant and soil health. The objective of the study is to collate a series of reviews and research articles on the role of the soil mycobiome and other associated microbes in sustainable agriculture affecting agro-ecology and agroeconomy of the nation.

14.2.2 Biofertilizer in Agriculture

Biofertilizers include materials derived from living organisms and microbial sources (Rola, 2000; Chen, 2006) with various benefits, such as increased access to nutrients (Chen, 2006; Vessey, 2003; Hart & Trevors, 2005), providing growth-promoting factors for plants, mediating the process of composting, and effective recycling of solid wastes (Gaur & Adholeya, 2004; Das et al., 2007). Biofertilizers are fundamentally microbial inoculants, produced from microbial cultures, that can improve soil fertility and crop productivity such as *mycorrhizae* (Malik et al., 2005; Marin, 2006). *Mycorrhizae* are fungi that form mutualistic relationships with roots of 90% of plants (Rinaldi et al., 2008; Das et al., 2007; Gaur & Adholeya, 2004). *Mycorrhizae* promote absorption of nutrients and water, control plant diseases, and improve soil structure (Rola, 2000; Marin, 2006; Zhao et al., 2003; Chandanie et al., 2006). Plants colonized by *mycorrhizae* grow better than those without them (Yeasmin et al., 2007; Singh et al., 2008) and are beneficial in natural and agricultural systems (Marin, 2006; Adholeya et al., 2005).

Some of the common microorganisms usually used as biofertilizers or in its composition are nitrogen-fixing agents (N-fixer) (Verma et al., 2016; Rana et al., 2019a, b; Odoh et al., 2019a), plant growth-promoting rhizobacteria (PGPRs) (Verma et al., 2016; Yadav et al. 2015, 2018a), potassium (Yadav et al., 2017; Kour et al., 2020), phosphorus solubilizers (Yadav et al., 2017), cyanobacteria, and endo- and ecto-mycorrhizal fungi (López-Bucio et al., 2015; Itelima et al., 2018). Intrinsically, these organisms possess the ability to produce a wide variety of extracellular enzymes for the breakdown of organic matter (Frąc et al., 2018). On the other hand, the bioagent improves plant's vigor and tolerance to abiotic and biotic factors while serving as an eco-friendly and cost-effective alternative (Rouphael et al., 2015; Itelima et al., 2018).

Agroeconomically, biofertilizers are environmental-friendly, cost-effective, and a renewable source of plant nutrients that have gained acceptance over chemical fertilizers (Kour et al., 2020; Verma et al., 2019; Yadav et al., 2020c). These bio-based materials play an important role in maintaining soil fertility, nutritional enrichment, and sustenance of healthy soil for generations (Mishra et al., 2015; Fuentes-Ramirez & Caballero-Mellado, 2005). Some fungal inoculants and their mode of action, as seen in Table 14.4, improve crop yield by enhancing a number of

Table 14.4 Some fungi used as biofertilizers and their mode of action

Biofertilizer agents	Mode of actions/ functions	Microbial species	Plant names
Fusarium spp.	Enrich compost biofertilizers degrade plant residues and maintain C:N balance in the soil	*Fusarium solani* and *F. Oxysporum*	Tomato
Trichoderma spp.	Enhance compost degradation	*Trichoderma harzianum* + *Aspergillus* + *Penicillium*	Potato, corn, tomato, peanut, beans, cotton, soybean
Mycorrhizal fungi	Reduce the impact of environmental stress Improve plant health status while enhancing optimum mineral content	*Glomus intraradices*	Vegetable crops, annuals, perennials, trees and shrubs (eg: onion, garlic, carrot, ornam ental grasses, ferns, aloe, rose bush, fruit trees etc)
Aspergillus spp.	Solubilize unavailable phosphate for plant use through the secretion of organic acid Improve soil quality with subsequent uses	*Aspergillus niger*	
Penicillium spp.	Supply essential minerals (p, Mn, Zn, Fe, Co, Cu, and Mo) to the plants Inducement of abiotic stress resistance in young seedling Protect the plants from fungal pathogens	*P. bilaji, P. italicum, P. albidum, P. frequentans, P. simplicissimum, P. rubrum, P. expansum, P. oxalicum, P. citrinum*	Wheat, soybean, canola
Chaetomium spp.	Enhance compost degradation	*C. bostrychodes, C. oliveaceum*	Tomato, corn, rice, pepper, citrus, durian, birds of paradise, carnation
Glicodium spp.	Reduce the incidence of damping-off issue	*Gliocladium catenulatum*	
Tritirachium spp.	Organic acids secretion for mineral dissolution	*T. album, T. egenum*	

biochemical processes, leading to increase in uptake of nutrients, stimulation of plant growth hormones, antibiosis, and decomposition of organic residues (Odoh et al., 2019a, b).

14.3 Fungal Biofertilizers: Biological Mechanism and Interactions

The vital role of fungal biofertilizer in agrobiology and biotechnological advances has been well documented (Zeilinger-Migsich & Mukherjee, 2014). These fungi, through nutritional versatility, mechanisms, and biochemical processes, interact with other organisms (bacteria, plants, and animals) in their natural habitat. At both intra- and inter-specific interactions, fungal biofertilizers comprise microbial inocula or assemblages of living microorganisms that exert direct or indirect benefits on plant growth and crop yield through different mechanisms (Fuentes-Ramirez & Caballero-Mellado, 2005). These microorganisms are able to fix atmospheric nitrogen or solubilize phosphorus, decompose organic material, or oxidize sulfur in the soil properties (Marin, 2006) that are beneficial to agricultural production in terms of nutrient supply (Malik et al., 2005). One type of biofertilizer is the arbuscular mycorrhizal fungi, which are probably the most abundant fungi in agricultural soil (Khan, 2006; Marin, 2006). The inocula improve crop yield because of increased availability or uptake or absorption of nutrients, stimulation of plant growth by hormone action or antibiosis, and by decomposition of organic residues (Wani & Lee, 2002). Žifčáková et al. (2016) in their report maintained that fungal biofertilizers regulate the balance of carbon and other nutrients in the decomposition process. They play a vital role in maintaining the soil's primary and secondary nutrient requirements – through potassium and phosphate solubilization, nitrogen fixation, and mineralization. They (Table 14.4) also help in the biodegradation of organic matter and pollutants in the soil environment (Adesemoye & Kloepper, 2009; Sinha et al., 2014; Odoh et al., 2019b). Selected fungal species that are used as biofertilizers are mentioned below.

14.3.1 Mycorrhizal Fungi Used as Biofertilizers: Soil–Fungal–Plant (Root) Interactions

Plants that suffer from nutrient scarcity, especially P, N, Zn, Cu, Fe, S, and B, develop mycorrhiza, which includes plants belonging to different groups (Zhu et al., 2008). *Mycorrhizae* uniquely form mutualistic symbiotic relationships with plant roots (Fig. 14.6) of more than 80% of land plants including many important crops and forest tree species (Smith & Zhu, 2001; Gentili & Jumpponen, 2006; Rinaldi et al., 2008) including xerophytes, epiphytes, and hydrophytes (Rai et al., 2013).

Fig. 14.6 Mycorrhizal growth in plant roots

Mycorrhizas enhance uptake of phosphorus (P), nitrogen (N), zinc (Zn), copper (Cu), iron (Fe), sulfur (S), and boron (B).

There are seven types of mycorrhizae: arbutoid mycorrhiza, ectomycorrhiza, endomycorrhiza or arbuscular mycorrhiza, ect-endomycorrhiza, ericoid mycorrhiza, monotropoid mycorrhiza, and orchidoid mycorrhiza (Zhu et al., 2008; Gentili & Jumpponen, 2006; Raina et al., 2000). The two dominant types of mycorrhizae are ectomycorrhizae (ECM) and arbuscular mycorrhizae (AM), which can improve water and nutrient uptake and provide protection from pathogens, but only a few families of plants are able to form functional associations with both AM and ECM fungi (Haskins & Gehring, 2005; Siddiqui & Pichtel, 2008). However, AM fungi are most commonly found in the rhizosphere roots of a wide range of herbaceous and woody plants (Finlay, 2008; Rinaldi et al., 2008). In this review, we focus on ectomycorrhizal fungi and arbuscular mycorrhizal fungi because they are most widespread and are more economically important than others.

14.3.2 Ectomycorrhizae (ECM)

The fungi in this group are characterized by the formation of a thick mantle structure within the intercellular spaces of root cortex. They also form a sheath around the feeder root, thus acting as an interface for nutrient uptake (Bücking & Kafle, 2015). They are surrounded by living cells in the host roots, leading to the development of an extensive network called the Hartig net. This Hartig net acts as both storage and transport organ for phosphorus. Most importantly, ectomycorrhizae are common in members of the families of Pinaceae, Fagaceae, Betulaceae, Salicaceae, and Myrtaceae. Ectomycorrhizal (ECM) fungi form mutualistic symbioses with many tree species (Anderson & Cairney, 2007) without penetrating the living cells in the roots, but only surround them (Gupta et al., 2000; Finlay, 2008; Raina et al., 2000), for example, pine, spruce, larch, hemlock, willow, poplar, oak birch, and eucalyptus (Dahm, 2006; Rinaldi et al., 2008). Most ECM fungi that are associated with forest trees are basidiomycetes, such as *Amanita* sp., *Lactarius* sp., *Pisolithus* sp., and *Rhizopogon* sp., and many of these are edible (Rinaldi et al., 2008; Le et al., 2007). Some ascomycetes also form mycorrhizae such as *Cenococcum* sp., *Elaphomyces* sp., and *Tuber* sp. (Rinaldi et al., 2008; Finlay, 2008). The importance of ECM fungi to trees is in their ability to increase tree growth due to better nutrient acquisition (Gentili & Jumpponen, 2006). ECM fungi help the growth and development of trees because the roots colonized with ectomycorrhiza are able to absorb and accumulate nitrogen, phosphorus, potassium, and calcium more rapidly and over a longer period than nonmycorrhizal roots. ECM fungi help to break down the complex minerals and organic substances in the soil and transfer nutrients to the tree. ECM fungi also appear to increase the tolerance of trees to abiotic stresses like drought, high soil temperatures, soil toxins, and extremes of soil pH. ECM fungi can also protect roots of trees from biotic stress like pathogens (Odoh et al., 2019a, b; Dahm, 2006).

The most commonly widespread ectomycorrhizal product is inoculum of *Pisolithus tinctorius* (Schwartz et al., 2006; Gentili & Jumpponen, 2006). *Pisolithus tinctorius* has a wide host range, and their inoculum can be produced and applied as vegetative mycelium in a peat vermiculite carrier. These fungus inocula are applied to nursery or forestry plantations (Gentili & Jumpponen, 2006). *Piriformospora indica* (Hymenomycetes, Basidiomycota) is another ECM fungus used as a biofertilizer. This taxon can promote plant growth and biomass production and help plant tolerance to herbivory, heat, salt, disease, drought, and increased below and aboveground biomass (Waller et al., 2005; Tejesvi et al., 2007).

14.3.3 Endomycorrhizae

These are a group of fungi that provide biological protection against soil-borne diseases and are associated with most agricultural crops and horticultural soils (Frąc et al., 2018). They occur in most ecosystems of the world and are found in many important crop species such as wheat, maize, rice, grape, soybean, and cotton and also in horticultural species like roses and petunias. The most common among them is arbuscular mycorrhizal fungi (AMF). Bagyaraj and Ashwin (2017), in their work, reported significant increases in crop yield following inoculation with AMF, which aid in stimulating key effects such as root development, improved soil structure, increased nutrient uptake, and mobility of ions. Besides enhancing plant tolerance to stresses, endomycorrhizae improves the general well-being of plants. *Endomycorrhizae* form mutually symbiotic relationships between fungi and plant roots (Ipsilantis & Sylvia, 2007). The plant roots provide substances for the fungi and the fungi transfer nutrients and water to the plantroots (Chen, 2006; Adholeya et al., 2005). Endomycorrhizal fungi are intercellular and penetrate the root cortical cells and form structures called arbuscular vesicles and known as vesicular arbuscular mycorrhiza (VAM), but in some cases no vesicles are formed and they are known as arbuscular mycorrhiza (AM) (Gupta et al., 2000). The agriculturally produced crop plants that form endomycorrhizae of the vesicular–arbuscular mycorrhiza type are now called arbuscular mycorrhizal (AM) fungi (Raja, 2006). AM fungi belong to nine genera: *Acaulospora, Archaeospora, Enterophospora, Gerdemannia, Geosiphon, Gigaspora, Glomus, Paraglomus*, and *Scutellospora* (Finlay, 2008). AM fungi are a widespread group and are found from the arctic to tropics and are present in most agricultural and natural ecosystems. AM fungi help plants to absorb nutrients, especially the less available mineral nutrients such as copper, molybdenum, phosphorus, and zinc (Yeasmin et al., 2007). They increase seedling tolerance to drought, high temperatures, toxic-heavy metals, high or low pH, and even extreme soil acidity (Chen, 2006; Gupta et al., 2000). AM fungi can also affect plant growth indirectly by improving the soil structure, providing antagonist effects against pathogens and altered water relationships (Smith & Zhu, 2001). AM fungi can reduce the severity of soil-borne pathogens and enhance resistance in roots against root rot disease (Akhtar & Siddiqui, 2008a, b; Chen, 2006). This results because of competition for colonization sites or nutrients in the same root tissues and production of fungistatic compounds (Marin, 2006). AM fungi have been found beneficial to the host plants by increasing herbivore tolerance, pollination, soil stability, and heavy metal tolerance (Hart & Trevors, 2005). Mass production of AM fungi has been achieved with several species such as *Acaulospora laevis, Glomus clarum, G. etunicatum, G. intraradices, G. mosseae, Gigaspora ramisporophora*, and *Gigaspora rosea* (Schwartz et al., 2006), but *Glomus intraradices* is the most common inoculum of endomycorrhizae products (Schwartz et al., 2006; Akhtar & Siddiqui, 2008b; Adholeya et al., 2005). Effective management of AM fungi involves increasing population of propagules such as spores, colonized root fragments, and hyphae using host plants and also by adoption of soil management techniques (Smith & Zhu, 2001; Tiwari et al., 2004; Kapoor et al., 2008). AM fungi are phosphate scavengers that expedite the recruitment of soluble phosphate from soil aquifers. Mycorrhizal fungi are efficient in the uptake of specific nutrients, are resistant against soil-borne pathogens, and play a key role in plant growth promotion, plant protection, and soil quality improvements. Also, besides being widely spread in agricultural systems, they have received wide acceptance as organic farming manure, especially in regions where the quest for sustainable crop production is prioritized.

14.3.4 Bacterial–Fungal Interactions (BFIs)

Studies have revealed that bacteria and fungi coexist as dynamic coevolving groups. Interaction between these microbes plays a huge role as they are considered vital players in driving activities such as biochemical cycles and contributing significantly to plant and animal pathogens (Deveau et al., 2018). BFI by-products are exploited for enhanced agricultural, environmental, and forestry management (Frey-Klett et al., 2011; Odoh et al., 2019a, b). In characterizing BFI, understanding of microbiomes using molecular tools, chemical and microbial ecology, genomics, and biophysics is essential (Thompson et al., 2017; Bergelson et al., 2019). Freyklett et al. (2011) in their work suggested a shift from disordered polymicrobial communities to a highly specific symbiotic association of fungal hyphae and bacterial cells, following their complexities. Arbuscular mycorrhizal fungi (AMF) and bacteria (plant growth-promoting rhizobacteria [PGPR]) association has been reported to promote crop growth (Pathak et al., 2017). According to Philippot et al. (2013), *Bacillus* sp., *Pseudomonas* sp. (PGPR), and AMF interaction proffers strong viability and causes significant improvement in field study when used singly or in combined application (Pathak et al., 2017). Although this association has positive influence on crop yield, it also enhances soil nutritional status and soil microbial biodata. Studies have shown that PGPR and AMF are key bioinoculants with potential to stop plant's dependence on agrochemicals, thus aiding in sustainable agricultural practices (Pathak et al., 2017; Franco et al., 2011). PGPR boosts plant growth directly (growth-promoting hormones) and indirectly (synthesis of antimicrobial substances Zheng et al., 2018). As aid to the process of mycorrhization, mycorrhizal-helping bacteria (MHB) and PGPR symbiotically interact with mycorrhizal fungi, and mycorrhizal roots for nutrient uptake. Studies have revealed that rhizospheric AMF and PGPR elicit systemic host immune responses for plant resistance (Singh, 2018), just as experimental evidence has proven that co-inoculants of AMF and PGPR give synergistic advantage to crop growth especially in nutrient-limited agricultural soil (Gouda et al., 2018).

14.4 Production, Formulation, and Application of Mycorrhizal Fungi as Fungal Biofertilizers

AM fungi are obligate symbiotic microorganisms since they cannot be grown without the plant host on synthetic media (Hart & Trevors, 2005). As AM fungal inocula must be produced in association with the host plant, therefore there are many constraints to large-scale commercial production. Mass production by pot culture either in the greenhouse or in growth chambers is the most commonly used production method (Marin, 2006; Gentili & Jumpponen, 2006; Bagyaraj & Ashwin, 2017; Raja, 2006; Kapoor et al., 2008). AM fungal inocula have to be prepared by multiplication of the selected fungi in roots of susceptible host plants growing in the sterilized soil or substrates, for example, perlite, vermiculite, peat, sand, or a mixture of them (Naqvi & Mukerji, 2000). The inocula of AM fungi can be applied as spores, or fragments of colonized roots. The spores and hyphae can be isolated from the soil rhizosphere and mixed with carrier substrates (Gentili & Jumpponen, 2006). Spore inocula are the most resistant and can survive unfavorable environmental conditions for a long period, but they colonize new root systems more slowly than other preparations. Therefore, both types of inocula, for example, spores and fragments of colonized roots, should be combined in commercial products (Marin, 2006).

Root-based bulk inoculum production technology utilizes mass-produced seedlings grown in sterilized soil infected with selected AM fungi using spores from fruiting bodies from cultivated plants. This technology results in seedlings with infected root systems, and the roots and adhering soil are chopped up and used as the starter inoculum for scale-up production. The inocula are produced in bulk by infecting fresh seedling of selected plants (Gentili & Jumpponen, 2006). The root inocula are kept in polythene bags and used for pelleting seeds or in the preparation of granules for seed bed inoculation (Singh & Tilak, 2002). The other methods such as soil-free aeroponic, nutrient film, and root organ culture system have been used for the production of AM, but these methods are costly and preclude commercial mass production (Gentili & Jumpponen, 2006). It may be possible to mass produce plants in tissue culture in sterile agar media and induce mycorrhizal associations using spores from fruiting bodies of selected mycorrhizal fungi. The dried root tissues and fungal mycelia could then be developed into mycorrhizal seeding products.

Some steps are essential for development of a commercial fungal biofertilizers. They include selection, large-scale production, carrier selection and preparation, mixing and curing, maintenance of appropriate numbers of inocula, and strong quality control (Malik et al., 2005). The criteria for selecting AM fungi will depend on details of the local environment, soil conditions, and host plants. The AM fungi must (1) colonize roots rapidly after inoculation, (2) absorb phosphate from the soil, (3) transfer phosphorus to the plant, (4) increase plant growth, (5) persist in soil and re-establish mycorrhizal symbiosis during the following seasons, and (6) form propagules that remain viable during and after inoculum production (Adholeya, 2006).

The success of a formulation depends on whether it (1) is economically viable to produce, (2) does not alter the viability and function of the inoculum, and (3) is easy to carry and enhance dispersal during application. The inoculum formulation may comprise one or more AM fungi and other organisms that together enhance the ability of the inoculum to form mycorrhizal associations with the target plant.

The formulations are available in the form of powder, tablets/pellets or granules, gel beads, and balls (Adholeya et al., 2005). There are many ways to apply the AM inocula (Adholeya et al., 2005; Schwartz et al., 2006), including scattering by hand, in-furrow application, seed coating, root dipping, and seedling inoculation. The efficacy of the application of AM inocula depends on the product, environmental condition, delivery method, and other variables. The success of AM fungi inoculation depends on crop species, size and effectiveness of indigenous AM fungi populations, fertility of the soil, and cultural practices (Adholeya et al., 2005). The production of commercial mycorrhizal inoculum has evolved considerably, with the availability of various types of microbial cultures and inoculants in the market today that are rapidly increasing because of the technological advances (Raja, 2006). A simple flow chart illustrating biofertilizer's production is shown in Fig. 14.7.

There are more than 30 companies worldwide marketing mycorrhiza products (Tables 14.5 and 14.6) comprising one or multiple mycorrhizal fungal inocula. These products are plant growth promoters and are to be used in horticulture, agriculture, restoration, and forestry (Schwartz et al., 2006).

Other fungi used as biofertilizers (Table 14.4) are discussed below.

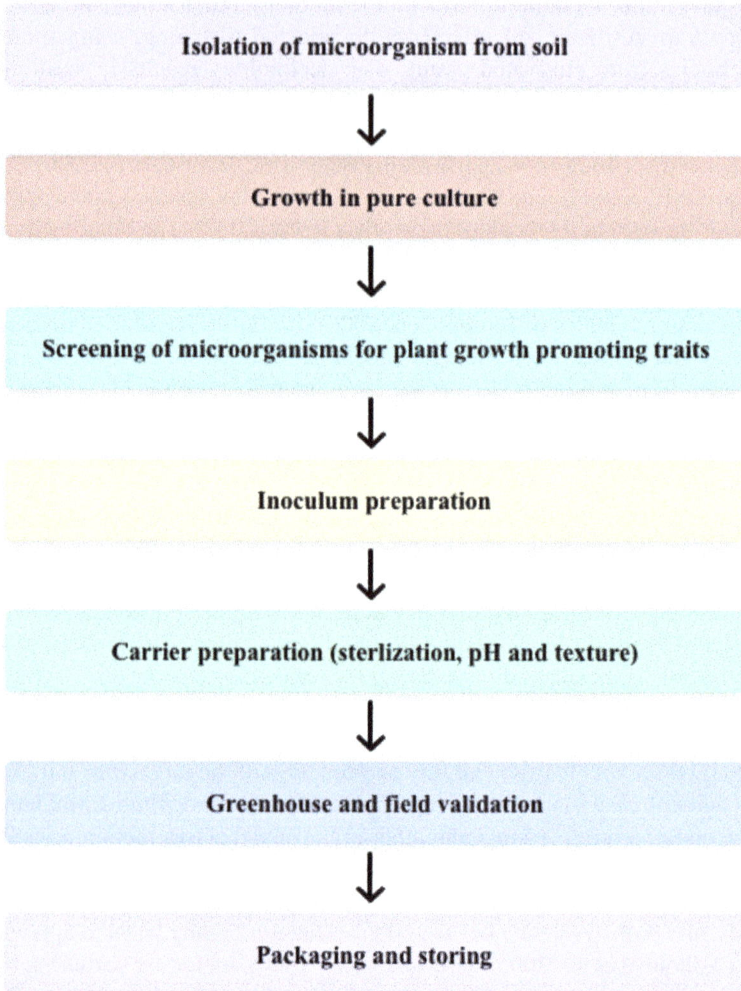

Fig. 14.7 Simplified flow chart for the production of biofertilizer

14.4.1 Penicillium *spp.*

Penicillium spp. are phosphorus-solubilizing agents that improve phosphorus absorption in plants and stimulate plant growth (Wakelin et al., 2004; Pradhan & Sukla, 2005). They act as biofertilizers by enriching compost and soil quality (Sharma et al., 2013). Some of the important species in this genus are *P. bilaji, P. italicum, P. simplicissimum, P. oxalicum, P. frequentans*, and P. *rubrum* (Yadav et al., 2018b). They also act as zinc solubilizers (Anitha et al., 2015; Ghosh et al., 2019).

Table 14.5 Contribution of arbuscular mycorrhizal fungi to plant growth promotion and soil nutrients

Mycorrhizal fungi	Plants	Effect on plant	Effect on soil	References
Glomus versiforme *Glomus mosseae*	Tomato	Promotes growth and yield under water stress and more efficient conditions	Increases phosphorus concentration in the soil	Maaloum et al. (2020)
Glomus etunicatum	*Maize*	Improves chlorophyll content and nutrient uptake in maize	Increases soil quality	Xu et al. (2019)
Acaulospora lacunosa	Strawberry	Enhances nutrient uptake in strawberry	Increases soil nutrient for horticultural crops productivity	Chiomento et al. (2019)
Rhizophagus irregularis	Wheat	Improves tolerance to stress, enhances plant growth, and increases seed yield	Increases soil nutrient needed for wheat production	Tohidi et al. (2018)
R. irregularis	Maize	Enhances tolerance to salt stress and improves growth parameters	Reduces the concentration of salt in the soil for better plant development	Krishnamoorthy et al. (2016)
G. mosseae and *G. geosporus*	Strawberry	Enhances growth and improves its tolerance to water stress	Increases soil nutrient to enhance its colonization on the plant root system	Boyer et al. (2015)
Rhizophagus irregularis	Tomato	Protects plants against pathogens (*Sclerotinia sclerotiorum*) and improves nutrient uptake in plants	Increases soil micronutrient and triggers the defense of the plant against pathogens	Mora-Romero et al. (2015)
Glomus deserticola	Snapdragon	Increases the total dry matter, chlorophyll content, and improves Snapdragon tolerance to water stress	Increases soil nutrients needed for plant growth promotion	Tognon et al. (2016)
Glomus spp. and *Mortierella* spp.	Seashore mallow	Increases shoot and root weight under salt stress	Increases soil nutrient and enhances its absorption by plants	Mota et al. (2019)
Glomus versiforme	*Mentha arvensis* L.	Increases dry weight and improves nutrient uptake in salt stress conditions	Increases soil nutrient and enhances its absorption by the plant to enhance its tolerance to salinity	Bharti et al. (2016)

Table 14.6 Fungal biofertilizers available in India

Biofertilizer	Active ingredient	Mode of action	Dose	Manufacturer
A. Phosphate-solubilizing microorganisms				
Grotop PSB Powder	Phosphate-solubilizing microorganisms (Bacillus sp.), powder 107–109 cfu g^{-1} and liquid 109 cfu ml^{-1}	Solubilize unavailable organic and inorganic forms of phosphorus (80%).	Seed: 5–10 g kg^{-1} seed Soil: 0.5–1 kg $acre^{-1}$ along with 40–50 kg manure Foliar: 3 ml l^{-1} water	MD Biocoals Pvt. Ltd., Haryana
Mani Dharma's biopromoter	*Bacillus megaterium* + *Aspergillus niger*	Biopromoter facilitates root formation and plant growth. It improves soil quality with subsequent uses. It should not be mixed with antibacterial agents and inorganic fertilizers.	Foliar: 200 g/200 ml^{-1} of "rice kanji" or 5% jaggery water	Mani Dharma Biotech Private Limited, Tamil Nadu
Multiplex Nalapak	Homogenous mixture of *Azotobacter* + *Azospirillum* + phosphate solubilizer + potash mobilizer	It produces amino acids, vitamins, and growth-promoting substances like IAA, GA, and cytokines, which helps in better growth and development of crop plants. Improves physical, chemical, and biological properties of the soil.	Soil: 500 ml 5 kg^{-1} $acre^{-1}$ along with 100 kg well decomposed FYM Foliar: 10 g l^{-1} water	Multiplex Bio-Tech Pvt. Ltd., Karnataka
Ambiphos	Phosphate-solubilizing microorganism (*Aspergillus niger*)	PSM secrete organic acids, which dissolve unavailable phosphate into soluble form and make it available to the plants.	Foliar: 3–5 ml l^{-1} water	Ambika Biotech & Agro Services, Madhya Pradesh,

(continued)

Table 14.6 (continued)

Biofertilizer	Active ingredient	Mode of action	Dose	Manufacturer
Biophos	*Bacillus megaterium* var. *phosphaticum*	PSM secrete organic acids, which dissolve unavailable phosphate into soluble form and make it available to the plants.	Seed: 5–10 ml kg^{-1} of seed Seedling: 125–250 ml in 25–50 liter of water as coating, sett treatment 125–250 ml in 60–80 l ha^{-1} for 30 min Soil: 500–625 ml ha^{-1} mixing with 250–375 kg FYM	Biotech International Limited, Delhi
BioP-P	Phosphate-solubilizing microorganism (2 × 108 CFU g^{-1})	PSM secrete organic acids, which dissolve unavailable phosphate into soluble form and make it available to the plants.	Foliar: 3–5 ml l^{-1} water	Sundaram Overseas Cooperation, Gujarat
PSM	Phosphate-solubilizing microorganisms	PSM secrete organic acids, which dissolve unavailable phosphate into soluble form and make it available to the plants.	Soil: 0.5–1 kg $acre^{-1}$ along with 40–50 kg FYM	Shree Biocare India, Shree Biocare Solution Pvt Ltd, Gujarat
PSM	Phosphate-solubilizing microorganisms	PSM secrete organic acids, which dissolve unavailable phosphate into soluble form and make it available to the plants.	Soil: 0.5–1 kg ha^{-1} a	KRIBHCO, UP

(continued)

Table 14.6 (continued)

Biofertilizer	Active ingredient	Mode of action	Dose	Manufacturer
B. Enriched compost				
Multiplex Sagar (compost poly culture)	Homogenous mixture of *Azospirillum* + *Trichoderma*+ *Pleurotus*	Nitrogen-fixing potential, biopesticidal activity, enhances compost degradation.	**Soil:** 1 kg acre^{-1} along with 1000 kg well-decomposed FYM **Foliar:** 100 ml 10 l^{-1} water	Multiplex Bio-Tech Pvt. Ltd., Karnataka
Enriched compost culture	*Trichoderma harzianum*+ *Aspergillus* + *Penicillium*	Enhances compost degradation F.	**Foliar:** 100 ml 10 l^{-1} water	Organic Biotech Pvt Limited, Maharashtra
Biomanure culture	*Trichoderma harzianum* + *Aspergillus*	Enhances compost degradation F.	**Foliar:** 100 ml 10 l^{-1} water	Uno Natural and Greens Private Limited, Tamil Nadu
LignoBiocompost culture	*Trichoderma resei, Phanerochaete chrysosporium*, and *Aspergillus awamori*	Enhances compost degradation.	**Foliar:** 75–100 ml 10 l^{-1} water	Peak Chemical Industries Limited, West Bengal
C. Phosphate-mobilizing biofertilizer				
ManiDharma VAM	Vesicular arbuscular endomycorrhiza (VAM)	Solubilize phosphate and supplies P, Zn, Mn, Fe, Cu, Co, and Mo to the plants. Increases the plant vigor by inducing drought resistance in young seedlings. Protects the plants from the fungal pathogens.	**Soil:** 200 g/m2 or 2–5 g seedling^{-1}; 50–200 g trees^{-1}; 3–5 kg acre^{-1} (2–3 cm depth).	Mani Dharma Biotech Pvt Ltd, Tamil Nadu
Ecorhiza-VAM/ Nurserrhiza-VAM	Arbuscular mycorrhiza	Improved uptake of nutrients, root development, and growth in plant.	**Soil:** 200 g/m2 or 2–5 g seedling^{-1}; 50–200 g trees^{-1}; 3–5 kg acre^{-1} (2–3 cm depth)	Mani Dharma Biotech Pvt Ltd, Tamil Nadu
Ecorhiza-VAM/ Nurserrhiza-VAM	Arbuscular mycorrhiza	Improved uptake of nutrients, root development, and growth in plant.	**Soil:** 3–5 kg acre^{-1} with the 200–250 kg FYM, one tablet plant^{-1} in 2–4 inches deep hole near the plant root	TERI, New Delhi

(continued)

Table 14.6 (continued)

Biofertilizer	Active ingredient	Mode of action	Dose	Manufacturer
Root care	Mycorrhiza (*Glomus intraradices*)	Improved plant health, uptake of nutrients and reduction of environmental stress.	Soil: 5 kg $acre^{-1}$	Ambica Biotech, MP
Mycorrhizae	105 propagules/kg with carrier material (talc powder/ vermiculite)	Mobilize major nutrients like phosphorus and potassium and certain micronutrients like zinc, calcium, etc.	Soil: 5–10 kg of VAM ha^{-1} is recommended along with the 1 q of FYM; 25 g $plant^{-1}$	Dr. Rajan Laboratories, Tamil Nadu
JOSH Super/ JOSH Plus	Mycorrhizal (*Glomus intraradices*)	Root development in plant enhances growth.	Soil: 5 kg $acre^{-1}$ for all crops, 60 infective propagules g^{-1}	Cadila Pharmaceuticals Limited, Gujarat
Shubhodaya	Vesicular arbuscular mycorrhiza (with three species: 1. isolated from desert and adoptable to harsh water-strained condition; 2. adoptable to water-lodging conditions; and 3. adaptable to general and acidic/ basic soil conditions)	Improved uptake of nutrients.	Soil: 5–10 kg $acre^{-1}$	Cosme Biotech, Goa
TARI VAM	Vesicular arbuscular mycorrhiza	Mobilize nutrients like, phosphorus and zinc.	Soil: 5–10 kg $acre^{-1}$	TARI Biotech, Tamil Nadu
BioCarry	Vesicular arbuscular mycorrhiza (VAM)	Helps in efficient mobilization and uptake of fertilizers and other nutrients by plants.	Soil: 5 kg $acre^{-1}$	Sundaram Overseas Operation, Gujarat
Symbion VAM	Solid formulations of arbuscular mycorrhiza	Improved uptake of nutrients, root development in plant and growth s.	Soil: 5 kg $acre^{-1}$	T. Stanes and Company Limited, Tamil Nadu

(continued)

Table 14.6 (continued)

Biofertilizer	Active ingredient	Mode of action	Dose	Manufacturer
CAMBAY's VARDHAK	Arbuscular mycorrhiza (powder and tablets)	Helps in efficient mobilization of nutrients.	Soil: 5 kg $acre^{-1}$; one tablet $plant^{-1}$	Neesa Agritech Private Limited
Mycorrhiza–VAM	Arbuscular mycorrhiza	Mobilize nutrients like P, Zn, Cu, and B.	Soil:10 kg $acre^{-1}$	KCP Sugar and Industries Corporation Ltd
Mycorrhiza–AM Biofertilizer	Arbuscular mycorrhiza	Helps in efficient mobilization of nutrients and improved plant growth.	Soil: 5 kg $acre^{-1}$	Majestic Agronomics Pvt. Ltd., HP
Colonizer	Arbuscular mycorrhiza (powder)	Colonizes living plants roots. Improves phosphorus uptake and imparts stress and disease resistance.	Soil: 2–3 kg $acre^{-1}$	Krishidhan Seeds Pvt. Ltd, Maharashtra

Penicillium simplicissimum and *Penicillium* sp. in combination with *Aspergillus awamori* are used as phosphate-mobilizing biofertilizers. Mishra et al. (2013) in their finding suggested that a combination of *Penicillium* and *Trichoderma* increases the development of wheat and soybean under nursery condition. *Penicillium bilaiae* was formulated as a commercial product named Jumpstart® (wettable powder) and has been found to enhance accumulation of phosphorus (P) and increase dry matter and seed yield in canola (*Brassica napus*) (Wakelin et al., 2004; Sane & Mehta, 2015; Chandrashekrappa & Basalingappa, 2018; Whitelaw et al., 1999). *Penicillium radicum*, isolated from the rhizosphere of wheat roots, has shown a good promise in plant growth promotion (Whitelaw et al., 1999), while *P. italicum* from the rhizosphere soil has the ability to solubilize tricalcium phosphate (TCP) and promote efficient soybean growth and production (Ram et al., 2015). In general, *Penicillium* spp. serves as phosphate solubilizer and general plant growth stimulator (Pal et al., 2006).

14.4.2 Aspergillus *spp.*

Aspergillus spp. helps in unleashing the soil phosphate complexes through the synthesis of an organic acid, which dissolves the soil phosphate for plant uptake (Zarafi & Dauda, 2019). A good example is *Aspergillus niger*, which facilitates plant growth through root formation (biopromoter). It improves soil quality and can be used to produce zinc-solubilizing biofertilizers. An example of the biofertilizer made from *Aspergillus niger* includes "Ambiphos." It secretes organic acids, which aid in dissolving unavailable phosphate into a soluble form and makes it available for plants use. In addition, "Ligno Biocom" post culture is also a good biofertilizer made from *Aspergillus awamori*, which serves as an enhancer for compost degradation (Pal et al., 2006). Most importantly, these fungi genera (*A. flavus, A. tubingensis, A. awamori, A. terreus, A. fumigates, A. niger*, and *A. melleus*) are able to solubilize inorganic phosphate through the production of citric acid, gluconic acid, glycolic acid, oxalic acids, and succinic acid (Barroso et al., 2006; Akintokun et al., 2007). *Aspergillus fumigatus*, which most often isolated from compost, has been reported to be a potassium-releasing fungus (Barroso et al., 2006; Akintokun et al., 2007; Sharma et al., 2013). *Aspergillus* spp. are also considered as microphos biofertilizers because of their ability to release phosphate from the bounded and insoluble state (Jehangir et al., 2017).

14.4.3 **Chaetomium** *spp.*

Chaetomium species are found mostly in soil and organic compost. It is a mycofungicide agent used mostly for protective and curative purposes. According to Dhingra et al. (2003), *Ch. bostrychodes* and *Ch. olivaceum* have been used severally as biofertilizers for compost enrichment. Most interesting is that the product of *Chaetomium* species has in recent time been used as fungal biofertilizers. A good example is seen in Ketomium, a bioproduct formulated from *Ch. globosum* and *Ch. cupreum*, which serves as plant growth stimulants (Jehangir et al., 2017). As a unique strategy, *Chaetomium* spp. have developed the ability to enable the suppression of bacterial and other fungal growth through competition, mycoparasitism, and antibiosis (Marwah et al., 2007). *Chaetomium globosum* and *Ch. cupreum* most importantly control root rot disease in black pepper, citrus, soybean stem, and strawberry with tendency of reducing sugar beet damping-off diseases (Dhingra et al., 2003; Zarafi & Dauda, 2019). Tomato, corn, rice, pepper, citrus, durian, birds of paradise, and carnation treated with Ketomium® have a greater plant growth and high yields than nontreated plants (Soytong et al., 2001).

14.4.4 **Gliocladium** *spp.*

Gliocladium species are common soil saprobes. They are parasites to many plant pathogens such as *Sporidesmium sclerotiorum* and *Fusarium* spp. They attack the fungal host by direct hyphal contact and forms pseudo appressoria. Engineered strain of *Gliocladium catenulatum* (strain JI446) has also been used as a wettable powder applied in soils, roots, and foliage to reduce the incidence of damping-off disease in the greenhouse (Viterbo et al., 2007; Youssef & Eissa, 2014). *Gliocladium virens* has been used as a biological control agent against a wide range of soil-borne pathogens such as *Pythium* and *Rhizoctonia* under greenhouse and field conditions. They also produce antibiotic metabolites such as gliotoxin, which has antibacterial, antifungal, antiviral, and antitumor activities. A good example is *Gliocladium catenulatum*-JI446, a commercialized bioagent that has been used to reduce the impact of damping-off disease caused by *Pythium ultimum* and *Rhizoctonia solani* (Kaewchai et al., 2009; Nissipaul et al., 2017).

14.4.5 Trichoderma *spp.*

Trichoderma is a fungal genus commonly found in most habitats. They are predominantly found in root and soil ecosystems and considered as ubiquitous saprobes. *Trichoderma* spp. can be easily isolated from decaying wood soil and organic materials (Rosa et al., 2012). Studies have shown a number of successfully reduced plant diseases through their highly effective antagonistic and mycoparasitic activity (Kour et al., 2019, Sharma et al., 2019; Verma et al., 2017). Some *Trichoderma* strains interact with their plant's host, increasing directly their growth potential and imposing resistance against diseases and abiotic stresses (Rosa et al., 2012). Some of the fungal disease-causing agents acted upon by *Trichoderma* (*Trichoderma harzianum* T-22) include *Rhizoctonia* spp., *Pythium* spp., *Botrytis cinerea, Pythium, Rhizoctonia,* and *Fusarium* spp. As an agent with mycofungicides properties, *Trichoderma* spp. has been engineered overtime to successfully support crops such as potato, corn, tomato, peanut, beans, cotton, and soybean (Sneha et al., 2018). They are considered an excellent competitor within the rhizosphere because they exercise resistance against soil unfavorable conditions and high efficiency in harnessing soil nutrients while also being hostile to phytopathogens (Chandrashekrappa & Basalingappa, 2018).

Members of the fungal genus *Trichoderma* are found to establish beneficial interactions with plants either antagonizing phytopathogens or directly influencing morphogenesis (Benítez et al., 2004; Harman et al., 2004; Harman, 2006; Hermosa et al., 2012). Furthermore, *Trichoderma* induces plant defense responses and improves crop performance under different stress conditions (Rawat et al., 2013; Hashem et al., 2014).

Mastouri et al. (2010) demonstrated that, when tomato seeds were treated with *Trichoderma harzianum* strain T-22 (T22), a range of biotic and abiotic stresses were alleviated. This is because after the colonization of roots *Trichoderma* spp. are able to chemically communicate with the plant and, sometimes, they act as endophytic symbionts. They are thus capable of correctly altering the expression of various plant genes and, as a consequence, plant physiology (Harman et al., 2012).

Recently, the use of the T22 strain of *T. harzianum* was also investigated as a new approach for controlling viruses, considering that chemical treatments have no effect on these pathogens (Vitti et al., 2015).

Trichoderma species improve mineral uptake, release minerals from soil and organic matter, enhance plant hormone production, induce systematic resistance mechanisms, and induce root systems in hydroponics (Ali et al., 2021). For these reasons, *Trichoderma* species are known as plant growth-promoting fungi. *Trichoderma* species have therefore successfully been used as biofungicides and biofertilizers in greenhouse and field plant production (Harman et al., 2012; Harman, 2006; Hermosa et al., 2012). These studies emphasize that *Trichoderma* can be widely applied as important sources of antibiotics, enzymes, decomposers, biopesticides, and plant growth promoters in agricultural practices. There are many *Trichoderma* products as fungal biofertilizers available in the market (some of them are shown in Table 14.6).

14.4.6 Formulation of Fungal Biofertilizers

Fungal biofertilizer is formulated either in liquid or solid state. Often, these bioproducts are packaged as dry biomass (dusts, granules, and briquettes) and suspension (water-based and emulsions). Dust particle size ranges from 5 to 20 mm and contains <10% of an organism by weight. Granules are discrete masses, 5–10 mm^3 in size. Pellets are >10mm^3, and briquettes have large blocks up to several cubic centimeters. The granules are discrete masses contain an inert carrier like charcoal, lignite, clay minerals (vermiculite, bentonite), starch polymers, dry fertilizers, and ground plant residues. In some cases, the choice of carrier depends on absorption, hardness, bulk density, and product disintegration rate in water. During formulation, these microbial bioagents get coated with various materials to slow and control the rate of their release. In the solid-based state, however, their shelf-life is about 6 months as they are not tolerant to UV rays and temperatures greater than 30 °C. Usually, the population density of these microbes ranging from 108 CFU/ml reduces with time (106 CFU/mL) at fourth month, reaching almost nil at the end of sixth month. Again, the improper sterilization of carrier material and their handling such as mixing the organism with carrier and packing serve as source of contaminations. Because of this, the inoculant packet could not hold desired biofertilizer organism for longer time and has not become effective and popular among the farmers in some climes. Furthermore, the quality of the biofertilizers gets deteriorated with longer and unsafe storage conditions. These have in many ways contributed to the limitation associated with biofertilizer application and in some cases where it could not be able to give viable results in the field. Owing to these challenges, the liquid formulation of organisms with a count of 109 CFU/ml (*Azospirillum* and *Phosphobacteria*) was developed to avoid drawbacks of biofertilizers and to increase the quality and shelf-life of bioinoculants. Here, preservatives are added with the microbial cultures and packed in bottles with improved shelf-life of about 2 years. According to Pal et al. (2015), the application of 1 ml of liquid biofertilizers is equivalent to the application of 1 kg of 5-month-old carrier-based biofertilizers. Increasing research has often suggested amendment of carrier materials to increase effectiveness of the biofertilizers (Itelima et al., 2018). These carriers are the delivery vehicle of live biofertilizers from the factory to the field that upon application

Table 14.7 Classification of carrier materials for the production of biofertilizer

Categories of carrier material	Carrier materials
Natural materials	Peat, lignite, coal, clay, and organic soil
Inert materials	Talc, vermiculite, perlite kaolin, bentonite, silicate, rock phosphate, calcium sulfate, and zeolite
Synthetic polymers	Polyacrylamide, polystyrene, and polyurethane
Natural polymers	Xanthan gum, carrageenan, agar agar, and agarose
Organic materials	Charcoal, biochar, composts, farmyard manure, sawdust, maize straw, vermicompost, cow dung, corn cob, and wheat husk
Agro-industry by-product	Sludge ash, jaggery

turns into organic, inorganic, and/or synthetic materials in the environment (Mahanty et al., 2016). Bashan et al. (2014) in their work reported that carriers have tremendous importance in delivering the right number of viable cells in good physiological condition. It also provides a short-term protective niche to the biofertilizers in soil, either physically via the provision of a protective surface of pore space or nutritionally via the provision of a specific substrate (Arora et al., 2010); hence, a good carrier should be nontoxic, of high moisture absorption capability, easy to process, and free of lump-forming materials, easy to sterilize, available in adequate amounts, and cost-effective with good pH buffering capacity (Mahanty et al., 2016). Different carrier materials used for formulating biofertilizers are shown in Table 14.7.

14.4.7 Mode of Application of Biofertilizers

Given below, are several ways that biofertilizers can be applied:

- Seed inoculation with powder formulations
- Dry biofertilizers mixed with the seeds in the seed hopper
- Sprinkle method (a small amount of water mixed with seeds before peat powder is added and mixed)
- Slurry method (the biofertilizer is suspended in water then added to the seeds and mixed)
- Seed pelleting
- Biofertilizers and adhesive are applied as slurry to seeds and coated with ground material like lime
- Peat suspension in water sprayed into the furrow during sowing

14.5 Role of Fungal Biofertilizer in Sustainable Agro-Practices

The use of fungal biofertilizers in sustainable agriculture has been gaining prominence in recent times, not only because it allows for effective utilization of mineral elements such as nitrogen and phosphorus, but also enhances mineral acquisition, transportation, as well as acting as an efficient biocontrol agent. The use of fungi as agents to improve soil fertility, water uptake, nutrient availability, induce environmental stress tolerance, biocontrol, as well as in limiting the use of agrochemicals has long been identified (Odoh et al., 2019a; Figueiredo et al., 2017). In a nutshell, the potentiality of fungal biofertilizers can be attributed to its unique functional niche in the agroecosystem, which are as follows:

- The fungal mycorrhizal symbiosis is arguably the most important symbiosis because of its agricultural importance (Yadav et al., 2019). Fundamentally, these plant species form mycorrhizal associations by acting as a critical linkage between soil and plant roots. This association is usually characterized by the movement of fungal-acquired nutrients to plants and the movement of plant-produced carbon to fungus (Johri et al., 2015). The fungal mycelium that extends from the root surfaces into the soil matrix captures nutrients from soil solution while the minuscule diameter of the fungal hyphae increases the surface area for absorption, thus increasing its ability for nutrient acquisition and utilization. Interestingly, these occur when mycorrhizal fungi colonize plant root systems; just as seen in the endo (arbuscular), ecto, ectendo, arbutoid, ericoid, monotropoid, and orchidaceous mycorrhizae. Pal et al. (2014) noted that when mycorrhizae are used as biofertilizers they enhance uptake of phosphate, zinc, and water, leading to improved hardiness and uniform growth of transplanted stock. The arbuscular mycorrhizal fungi also improve soil structure and control plant diseases, thus implying that plants colonized by these microbes grow better than those without them (Tripathi et al., 2017). Arbuscular mycorrhizal fungi (AMF), a group of soil fungi that live in symbiosis with most agricultural crops, have the ability to mobilize mineral nutrients from soil for their plant hosts.
- Fungal agents act as inhibitors or antagonizers of phytopathogens (biocontrollers), promote plant growth (biostimulants), activate mineral nutrients, and fix atmospheric nitrogen (biofertilizers) as part of sustainable agricultural practices (Odoh et al., 2019a).
- Many fungal bioinoculants such as endophytic fungi (*Penicillium, Aspergillus, Piriformospora*, and *Curvularia*) solubilize and mineralize phosphorus (Rai et al., 2013). However, for phosphate solubilization, endophytes are the major contributors. Here, they colonize plants without inducing symptoms of disease, making them a more aggressive colonizer of plant root than non-endophytic microbes. These endophytes convert insoluble soil phosphate into soluble forms through the secretion of organic acids with concomitant changes in the texture, structure, and water-holding capacity of the soil.

- *Trichoderma* biofertilizers (Zhang et al., 2018) were shown to increase soil antifungal compounds that suppressed pathogenic fungi and allowed for improved grassland biomass. This suggests that *Trichoderma* biofertilizers could be an important tool for sustainable management of soil and plant productivity.

14.5.1 Fungal Biofertilizers as an Alternative to Chemical Fertilizers

Fungal biofertilizers are a renewable source of plant rhizospheric nutrients that could serve as a close alternative to chemical fertilizers (Kannaiyan, 2002). The use of fungal biofertilizers can help to curb the excessive use of synthetic fertilizers, which when used indiscriminately causes detrimental effects to the environment and human health (Odoh et al., 2019b; López-Bucio et al., 2015; Yadav et al., 2020c). Fungal biofertilizers include plant growth-stimulating fungi (*Trichoderma*), mycorrhizal fungi (ectomycorrhiza and arbuscular mycorrhizae), enzyme-producing fungi for compost production, and P/K-solubilizing fungi. Mycorrhiza fungal biofertilizers, also known as vesicular arbuscular mycorrhiza (VAM), help in retaining moisture around the root region and increase resistivity towards different root and soil pathogens unlike the chemical fertilizers (Sadhana, 2014).

14.5.2 Fungal Biofertilizers Enhance Mineral Acquisition and Transport

The primary goal of AMF inoculation is to increase and enhance mineral acquisition and transport. Clark (1997) in his work observed increased plant acquisition of mineral nutrients upon AMF addition. During seasonal variation (dry season), soil structure and porosity determine soil water retention ability. As a result, this agent "AMF" influences hormonal flow and changes information from plant roots to shoots, and affects stomatal responses when soil water potential is lowered (Odoh et al., 2019a, b). In the work of Jagnaseni et al. (2016), they maintained that mycorrhizal associations increase nutrient uptake during water stress through hydraulic conductivity in the roots. They occur as the hyphal sheath, Hartig net, and extraradical mycelium. These fungal mantles represent a significant apoplastic barrier creating a closed interfacial apoplast, thus enabling nutrient transfer (Bücking & Kafle, 2015). Upon application, the differential expression of plant and fungal uptake transporters in the mycorrhizal interface plays a role in the development of a strong concentration gradient across the mycorrhizal interface. High-affinity phosphorus (P) and nitrogen (N) transporters of ECM and AM fungi are expressed in the ericoid mycorrhiza (ERM), but downregulated in the intraradical mycelium (IRM). This favors the active absorption of nutrients by the ERM, but reduces nutrients reabsorption process in IRM (Bücking & Kafle, 2015). Consequently, AM and ECM fungi store significant amounts of P as poly P where they are assumed to play an important role in the transport of P and N from the ERM to the IRM (Bücking & Kafle, 2015), unlike in hydrolysis of poly P where IRM would release Pi and Arg into the fungal cytoplasm to facilitate the efflux catabolic product of the urea cycle into the mycorrhizal interface. In general, AM and ECM fungi regulate nutrient transport in host cells via accumulation of poly P while inducing carbon supply in host plant, a process that also triggers poly P hydrolysis (Bücking & Kafle, 2015).

14.5.3 Fungal Biofertilizers Enhanced Nitrogen Fixation

Soil nitrogen is considered a valuable nutrient for plant growth. It is also one of the most essential micronutrients for food production. They exist in the form of ammonium, nitrate, and amino acids. Bücking and Kafle (2015) in their research reported that the ERM of AM and ECM fungi takes up inorganic N sources (ammonium NH^{4+}or nitrate NO^{3-} from the soil). Through protein transporter named AMT1, nitrogen is taken up by plants as ammonium. During glutamine synthetase formation, this ammonium in combination with glutamate forms glutamine in the extraradical mycelium. Interestingly, NH^{4+} has been described as the preferred N source of mycorrhizal fungi because its uptake is energetically more efficient than the uptake of NO^{3-}. In *Hebeloma cylindrosporum*, AMT1 and AMT2 are expressed as NH4+ transporters and regulated by exogenous NH4+ supply. The expression of both transporters is upregulated under low NH^{4+} supply conditions, but downregulated in response to an exogenous supply of $NH^{4+\cdot}$ More importantly, low-affinity NH^{4+} transporter (AMT3) also enables the fungus to maintain a basal level of N uptake and assimilation at high exogenous supply conditions (Bücking & Kafle, 2015).

14.5.4 Fungal Biofertilizers Enhanced Phosphorus Content

Phosphorus (P) is an extremely immobile element present in the soil. It is very vital for plant growth. One of the major roles of vascular–arbuscular (VA) fungi is the supply of phosphorus to plant roots via phosphate transporters in the hyphal membrane (Rastegari et al., 2020). This enables the fungi to provide phosphorus as poly P pool to the plants. The mycorrhizal hyphae present outside the root of the host plant known as the networks of filamentous, extraradical hyphae of AM fungi help in the uptake of freely available phosphates. Generally, the fungal hyphae extend beyond the host root to enable greater soil volume contact for phosphate acquisition. This fungus (AM fungi) hydrolyzes organic phosphates to soluble forms for plant use. This uptake is done using the phosphate transporter of the Pht1 family. According to Vergara et al. (2019), this fungus has two separate uptake systems for phosphorus: (a) a high-affinity system that works against an electrochemical potential gradient. This takes up the Pi from the soil via proton co-transport and (b) a low-affinity system that facilitates the diffusion of Pi across the fungal plasma membrane. The expression of these transporters is often regulated by responses to externally available P concentration and P demand of the fungus (Johri et al., 2015).

14.5.5 Fungal Biofertilizers as an Efficient Biocontrol and Bioremedial Agent

Biological control agents are referred to as organisms that suppress plant diseases. In other words, they are considered as the utilization of introduced or resident living organisms such as fungi to reduce pressure and effects exerted by plant pathogens or suppressing reproduction of their kinds (Yadav et al., 2020a). Various fungal biofertilizers possess this ability, for example, *Trichoderma harzianum* – a species with biocontrol potential against *Botrytis cinerea, Fusarium, Pythium*, and *Rhizoctonia* (Redda et al., 2018); *Chaetomium globosum* and *C. cupreum*, having biocontrol activity against root rot disease caused by *Fusarium, Phytophthora*, and *Pythium* (Aggarwal, 2015). Besides inducing biocontrol, mycorrhizal associations also protect plants against heavy metal toxicity (copper, zinc, iron, manganese, cadmium, nickel, etc.). Ectomycorrhizal fungi protect trees by accumulating and immobilizing these toxic metals in the mycorrhizal mantle. Also, this association enhances detoxification mechanisms through extracellular heavy metal chelation (e.g., glycoprotein glomalin), binding of heavy metals to rhizodermal cell walls, and distorting heavy metal uptake mechanism. In addition, fungal biofertilizers also act as biocontrol agents (BCA) by competing for nutrients and space or by producing metabolites that impede spore germination (fungistasis), kill the cells (antibiosis), or modify the rhizosphere (acidifying the soil) so as to prevent pathogens multiplication (Khokhar et al., 2012). In mycoparasitism, biocontrol may result from a direct interaction between the pathogen itself and the BCA as they rely on the recognition, binding, and enzymatic disruption of the host fungus cell wall (Abdallah et al., 2018).

14.6 Fungal Biofertilizers as an Agent of Pesticide Tolerance

The most concerning agroecological issue is the unintended detrimental effects of pesticides due to its extensive use for decades for crop protection and yield. The versatile biological mechanism and interaction of fungi with soil and plant microbiome assessed under pesticide application can pave the way to isolate and develop fungal strains with high tolerance for pesticides, making it most compatible in the integrated agricultural management. Since the application of pesticides is unavoidable in conventional agricultural practice, to develop effective biofertilizers, with both pesticide tolerance and plant growth-promoting (PGP) traits are of paramount significance. A few such examples are discussed below.

14.6.1 *Compatibility of* Aspergillus niger *Isolates with Pesticides*

One such example is *Aspergillus niger Van Tieghem*, which is a good phosphate solubilizer (Khan & Anwer, 2007), siderophore producer (Khan & Anwer, 2007), antagonistic to many pathogens (Khan & Anwer, 2007, 2008, 2018; Khan et al. 2006a, b), and does not produce mycotoxins and ribotoxin (Campbell, 1994). Interestingly, the fungus is reported to decrease aflatoxin contamination. It is xerotolerant and grows at a wide range of temperatures (10–50 °C), pH (2.0–11.0), and osmolarity (from nearly pure water up to 34% salt), and can also improve its thermostability (Zhang et al., 2007). The fungus is extremely resistant to herbicides, fungicides, and pesticides at very high concentrations (Braud et al., 2006). In view of the effectiveness of *A. niger* for their metal tolerance (Ahmad et al., 2006), biosorption potential, and pesticides tolerance, it was considered desirable to identify more efficient soil isolates among the diverse *A. niger* aggregate collected from different crop fields of several districts in Uttar Pradesh (North India). The compatibility test of *A. niger* with different groups of pesticides varied with isolates (Geetha et al., 2016). Some strains of *A. niger* have shown to degrade endosulfan pesticides (Bhalerao & Puranik, 2007) and some of them have produced organophosphate-degrading enzymes (organophosphorus hydrolase) that split P-S bonds of organophosphate (Liu et al., 2001, 2004). *A. niger* has also been reported to degrade carbamates by hydrolyzing several N-methylcarbamate insecticides (Qing et al., 2006) and pyrethroids pesticides through a novel pyrethroid hydrolase from their cell extracts (Liang et al., 2005).

14.6.2 *Compatibility of* Trichoderma *Isolates with Pesticides*

Similarly, the compatibility of fungal biocontrol agents (*Trichoderma harzianum*, *Trichoderma virens*, and *Pochonia chlamydosporia*) was assessed with different concentration of six pesticides, viz., carbendazim, mancozeb, metalaxyl, captan, thiram, and nemacur commonly used by farmers in India for the control of soil-borne plant pathogens (Mohiddin & Khan, 2013). Their efficacy in pesticide tolerance broadened the use of these strains as biopesticides under integrated disease management for managing soil-borne plant pathogens.

Currently, *Trichoderma* spp. share almost 70% of fungal biocontrol agents (BCAs) market. *Trichoderma harzianum* is extensively used for its dual advantages of controlling several soil and seed-borne phytopathogens in addition to its capacity to act as an effective biofertilizer (Keswani et al., 2013). Competence of *Trichoderma* spp. to other fungal biocontrol agents owes to its relatively faster metabolic rates; secretion of antimicrobial secondary metabolites and physiological conformation are some of the key factors that contribute to its vigorous antagonistic effect in field conditions. On the other hand, *Trichoderma* spp. has also been exploited for commercial production of enzymes like cellulases, hemicellulases, proteases, and â-1, 3-glucanase. This article (Keswani et al., 2013) advertises the commercial potential of *Trichoderma* spp. discussing its status in various aspects, viz., pest control, plant growth promotion, bioremediation, and production processes.

14.6.3 Compatibility of ECM with Pesticides: A Novel Path

Endeavors pertaining to application of intriguing ectomycorrhizal association for on-site pesticide remediation are acquiring huge consideration from global scientific community (Assad et al., 2020). These ectomycorrhizal (ECM) fungi represent a functional group of enormous importance, that interconnects plants through below-ground colossal hyphal networks and serve as potential indicators of changes in environmental quality. Owing to their specialized lifestyle and mycorrhizal remediation potential, ECM fungi remediate pesticides through enzymatic conversion of noxious pesticides into innocuous residues. As an innovative emerging discipline, ectomycorrhizal biotechnology engrosses application of subsist ECM fungal strains or their prominent fungal enzymes, via contemporary biotechnological approaches, for sustainable remediation and recuperation of gravely disturbed environs. This technology has persuasive potential to arise as an effectual substitute of conventional remediation system. Ectomycorrhizal ecology, mycorrhizoremediation of pesticides, fungal enzymes entailed in remediation, role of ectomycorrhizal biotechnology in pesticide remediation, and plant–pollutant–mycorrhiza interactions with reference to climate change are the novel paths to be explored.

14.6.4 Plant Growth-Promoting Microorganism (PGPM): Prospects and Constraints

Bioremediation and microbial degradation are being used for the destruction of chemicals in soil and ground water. These pollutants are removed from the soil by indigenous microorganisms, including bacteria for their extraordinary characteristic to use a wide range of xenobiotics as sole energy and carbon source (Siddique et al., 2003).

In other words, indigenous microorganism is also known as plant growth-promoting microorganism (PGPM), as represented by the microbial cells of bacteria, actinomycetes, and fungi. PGPM inoculants are presently commercialized as an innovative and efficient solution for plant growth enhancement through direct and indirect mechanisms.

Presently, the PGPM available in the market as the only single or dual microbial consortium is not effective for all the crops and exhibit the inability to degrade or tolerate pesticides. They have been sourced from other agroclimatic region, and that is why they lose their potency to function and survive in the indigenous soil, with different edaphic and climatic conditions. This poses a huge problem with PGPM as biofertilizers, currently available in the market for farmer's application. In order to overcome these problems, the need of alternative practices like the effective pesticide-degrading plant growth-promoting microbial consortia are required for improving plant growth of vegetables and degrade or minimize the pesticides in soils. These consortia can degrade and utilize the pesticides for co-metabolism or as a sole source of carbon and enhance plant growth. Being indigenous in nature, they may easily grow and help in pesticide bioremediation in crop fields. Therefore, the future prospects lie in screening indigenous fungal strains for the development of efficient indigenous pesticide-tolerant plant growth-promoting microbial consortia from rhizospheric and non-rhizospheric soil.

Similar studies were undertaken with an intention to isolate and characterize the plant growth-promoting properties and pesticide (chlorpyrifos) tolerance ability of microbial strains for sustainable vegetable productions (Verma et al.,2016a, b). Fifty microbial isolates from vegetable field were successively tested on the basis of morphological, biochemical, and plant growth-promoting properties. Ultimately, three strains were found as effective pesticide-tolerant plant growth-promoting microbial consortia for vegetable production under sustainable agriculture.

According to results of microbial populations, fungal populations were recorded comparatively higher than actinomycetes and bacteria. Therefore, fungi are found to have higher resistance to pesticides (Ambethgar, 2009).

A similar study was conducted to screen for pesticide-resistant isolates of plant growth-promoting microorganisms such as *Azotobacter* spp., *Azospirillum* spp., *Rhizobium* spp., and phosphate solubilizers, the activity of which would not be affected by the presence of pesticides owing to their pesticide resistance, and to check their efficacy as biofertilizers. Screening was carried out using rhizospheric soil samples from fields treated with pesticides (Prabhu & Raut, 2011). Plant studies and viability studies proved that the isolates, mainly the consortium, not only proved to be effective in promoting plant growth in the presence of 1000 ppm of six commonly used pesticides, but also proliferated in pesticide-contaminated soils and improved the soil quality. These studies indicate that the isolates can be used as potential biofertilizers in pesticide-contaminated fields, thus bringing about enhanced plant growth even in the presence of pesticides.

Mishra and Sundari (2015) investigated the effects of direct inoculation of selected consortia on plants to study its effects on supporting plant growth in the presence of root pathogen *Sclerotium rolfsii* and organophosphate pesticides Malathion (ML) and Methyl Parathion (MP). Candidate plant growth-promoting microbial (PGPM) isolates chosen for the study are two bacterial isolates (PGPM2, a diazotrophic bacterium, PGPM9, a fluorescent Pseudomonad) and one fungal species (T103, a biocontrol fungus), originating from native agricultural fields of western U.P., India. Host plants inoculated with individual species showed a distributed growth enhancement pattern, while isolate T103 improved root biomass, isolate PGPM9 enriched photosynthetic pigment content and isolate PGPM2 expanded root and shoot lengths.

The results affirm the hypothesis that synergistic action of carefully selected PGPMs can escalate the benefits of plant growth promotion even in presence of pathogen and pesticide; hence, this consortium may be a valuable option for sustained plant growth in modern agriculture systems.

14.6.5 Developing Nation Through Developing Effective Biofertilizers

Shen et al. (2019) screened rice endophytic biofertilizers for fungicide tolerance and plant growth-promoting characteristics. Out of 17 different strains obtained from rice seedling roots, except for one, all strains could tolerate two or more fungicides. Rice roots inoculated with the endophytic bacteria conferred rice growth-promoting ability and should be considered a potential bacterial biofertilizer.

However, to develop effective biofertilizers, PGP traits (Mahanty et al., 2017) along with pesticide tolerance should be confirmed before their application because applying pesticides is inevitable in most agriculture practices (Ahemad & Khan, 2012). The hypothesis of such study should be based upon the application of biofertilizers with pesticide-tolerant ability to the soil, where they would be likely to retain their functions and would be capable of promoting the growth of the crops treated with pesticides.

Conclusively, all such studies suggest that it is important to examine the impact of pesticides during the process of developing biofertilizers with pesticide tolerance and plant growth-promoting characteristics. The rising demand for environmentally friendly, organic, and sustainable agricultural practices is driving the application on the use of beneficial biological products. Sustainability has become an integral component of the agriculture system. The use of fungi in agriculture sector is potentially useful for improved plant health and growth, water uptake, nutrient availability, stress tolerance, and biocontrol. Fungal species served as very important biological tools in sustainable agricultural ecosystem with the process of mycoremediation, mycocontrol–mycoherbicides, mycoinsecticides, and as mycorrhiza fungi (Thakur, 2020) and have potential role for attaining sustainable agricultural systems. Examples of fungi used as mycoremediators are *Pleurotus ostreatus, Rhizopus arrhizus, Phanerochaete chrysosporium, P. sordid, Trametes hirsute, T. versicolor, Lentinus edodes*, and *L. tigrinus.*

14.7 Biofertilizer Production: Indian Scenario

- Since the time the first commercial production of biofertilizer was initiated in 1956 at the Indian Agricultural Research Institute, New Delhi, and Agricultural College and Research Institute, Coimbatore, India has shown slow yet a steady growth in its production.

- The Ministry of Agriculture under the Ninth Plan initiated the real effort to popularize and promote the input with the setting up of the National Project on Development and Use of Biofertilizers (NPDB) during 1983–84.
- During 1965–1990, around 30 biofertilizer production laboratories were set up in the country to meet the demand, and a lot of schemes formulated to popularize their use in different legume crops. The 1990s saw a dramatic surge in biofertilizer industries with adding of new biofertilizers such as *Azotobacter*, *Azospirillum*, PSBs, and VAM (Adholeya, 2012) added to the list and total production jump was observed ten times in between 1989 to 2000 (Bhattacharya & Dwivedi, 2004).
- Starting from a few tons, production and consumption increased gradually and reached a moderate figure of 28,000 MT by 2009–10 (Fertiliser Association of India, 2011).
- Research in the field of microbial fertilizers is in progress in different research institutes of India. Several institutes of ICAR are producing commercial microbial fertilizers. However, further researches and encouragement from the government are needed to isolate and characterize more salt-tolerant microbial strains using modern biotechnological tools.
- The government should campaign for the utilization of these efficient and economic fertilizers because instead of considerable efforts by scientists, majority of the farmers in India are unaware of the use and fruitfulness of these microbial fertilizers.
- The share of bioinputs in agriculture is abysmally low. The market share is not even 1% in India. Relatively speaking there is more rigor in estimation of biofertilizer market in India because of the presence of some large producers in production of biofertilizers and comparatively this is more organized than other green inputs market (Rakshit & Bhadoria, 2002).
- Based on the gross cropped area in India and recommended doses of biofertilizers, potential demand is segregated into different categories of biofertilizer, such as *Rhizobium, Azotobacters, Azospirillum*, BGA, and phosphate solubilizer, where demands differ widely.
- Contrary to the world biofertilizer scenario, Indian biofertilizer industry is not restricted to *Rhizobium*. Presently phosphorous-solubilizing bacteria accounts for about 55% compared to other nitrogenous biofertilizers, which account for 45% (NCOF Annual Report, 2007–08).
- Although actual production and distribution of biofertilizers are below the targeted level, but there has been a positive trend in India, with respect to production.
- The growth rate in installed biofertilizer capacity is comparatively more stable than the growth rate in total production, consumption, and distribution.
- This is an indication that there is not only a need but also a role for market development for green agriculture inputs in India.
- In spite of the impressive growth rate of more than 200% in production capacity and around 300% growth rate experienced in production and consumption of biofertilizers in India, in the recent past yet it is only around 1.5% of the estimated demand potential for biofertilizers in the country.

- The region-wise distribution of biofertilizers is more dispersed in relation to chemical fertilizers across the country. As per the latest data available on biofertilizers, South Indian state Andhra Pradesh has overtaken another South Indian state Tamil Nadu in biofertilizer production to reach the top. The other major producers of biofertilizers are Karnataka, Kerala, Maharashtra, and Madhya Pradesh. The highest number of units is located in the state of Maharashtra.
- A good degree of large industrial chemical fertilizer units like IFFCO, KRIBHCO, TCL, MFL, NFL, and RCF are also involved in biofertilizer production as their corporate social responsibility along with chemicals-based fertilizers, insecticides, and pesticides. Among the PSUs, KRIBHCO produced highest amount followed by IFFCO, MFL, NFL, and RCF.
- Due to intensive and continued assistance from Government of India in this regard, overall production of biofertilizers in the country is continuing with a positive growth rate. However, the market for biofertilizers is still not well developed and the biofertilizer industry has not anticipated the growth.
- The current strategy of marketing biofertilizer products in India is through niche markets. The scope for a biofertilizer is often perceived to be limited.
- Biofertilizers are often perceived to be more expensive than the chemical fertilizers. This is more so since the farmers and small holders received fertilizers heavily subsided by the government, enabling a few farmers, who appreciate the benefits of use of certain biofertilizers.
- Another perception on biofertilizer is its slow effect on the crops as compared to chemical fertilizers.
- Special care (storage, mixing with powders, etc.) is also needed to handle microbial inocula, so that these remain effective for extended use. These inoculants too favors certain environment. Concerning microbial inoculants, some users realized their potential.
- There was difference of opinion on the effectiveness of microbial inoculants available in the market. Some felt that the performance of these products is often disappointing, unreliable, and not as claimed by the manufacturer. Some products, however, do give good results. All these perceptions contribute to influencing the user on the use of microbial inoculants and biofertilizers.
- The way to forward the produce is that it may satisfy the users in terms of versatility, ease of use, and cost. The use of biofertilizers has still not spread uniformly, although there has been a steady rise in their use by certain group of farmers.

14.7.1 Quality Control

- Quality of biofertilizer is one of the most important factors resulting in their success or failure and acceptance or rejection by end user, the farmers. Efforts to promote the use of these substances in the past have been hampered by poor and uneven quality. A survey of farmers shows that poor quality of biofertilizers is largely responsible for the poor acceptance by users.

- Basically, the quality means the number of selected microorganisms in the active form per gram or milliliter of biofertilizer. Specifications of biofertilizer differ from country to country and may contain parameters like the microbial density at the time of its manufacturing, microbial density at the time of expiry, expiry period, permissible contamination, pH, moisture, microbial strain, and the carrier.
- Quality has to be controlled at various stages of production (during mother culture stage, carrier selection, broth culture stage, mixing of broth and culture, packing and storage). Inoculant shall be packed in 50–75 microns 1000 density polythene packets and should be marked prominently the following:

  1. Name of the product
  2. Name and address of manufacturer
  3. Name of the carrier
  4. Batch number
  5. Date of manufacturing
  6. Date of expiry
  7. Net mass
  8. Crop for which intended
  9. Storage instruction (15–30 ± 0.2 °C)

Countries like India have regulations for inoculant quality, but neither it is regulated properly nor are the existing regulations well enforced. Bureau of India Standards (BIS) has published necessary specifications/standards for different biofertilizers. But these specifications are purely voluntary in nature and are being regulated on firms and producers, who have opted for BIS certification and putting for BIS certification and putting ISI mark on their products. Gaps exist between India and some of the leading countries in all areas of biofertilizer production and application technology. These gaps are important, particularly in the field of strains, techniques used for sterilization, fermentation, and carriers. The effects of these are aggravated further by poor quality control maintained by most of the producers. Under such circumstances, a significant increase in the level of penetration and demand of biofertilizers are possible only if these gaps are reduced and strict quality control is maintained. While promotional efforts are important, but the success of such efforts will depend on the availability of biofertilizers of high and consistent quality. A system by which the quality is monitored by the central and state-level authorities must be devised and enforced.

14.7.2 Challenges and Prospects of Fungal Biofertilizer Usage

Despite significant improvement and progress in biofertilizer technology over the years, the progress in the field of its production technology is below satisfaction. Further, there do exist obstacles for biofertilizer use. These are not only technical problems, but also socio-economic and human resource obstacles. The technical problems can be addressed through a comprehensive program of basic and applied research up to a certain extent. Overcoming the socioeconomic and human resource obstacles would require an emphasis on education, training, and promotion of private enterprise. The difficulties to expand the use of biofertilizer by farmers in India are as follows:

1. Difficult handling of biofertilizer
2. Problems with distribution
3. Low quality of biofertilizer
4. Cannot be stored for longer period
5. Lack of demonstration and low visual effect of biofertilizer
6. Low in public relation and technology transfer
7. Low knowledge of farmer on sustainable agriculture and environmental effect of biofertilizer

One of the most important limitations of biofertilizers is seen in their nutrient content when compared to inorganic fertilizers as a number of works have reported deficiency symptoms associated with plants grown in biofertilizer-treated soils (Itelima et al., 2018). However, this challenge is curbed by the addition of substances such as bone meal (rich in phosphorus), wood ash (rich in potassium), or other substances of natural origin such as phosphate rock to enrich the fertilizer (Itelima et al., 2018). Also, the use of nutrient-rich organic materials such as palm wastes (rich in potassium) and wood ash (rich in potassium also) in making biofertilizers serves as a good remedy. Mahimairaja et al. (2008) stated that the addition of phosphorus to wastes makes the biofertilizers more balanced and reduces nitrogen losses. Notably also are the limitations associated with the storage conditions of biofertilizers; this goes a long way in affecting its efficacy. Biofertilizers should be stored in cold storage conditions away from heat and the polythene used in packaging should be of low-density grade and thickness of about 50–75 microns (Mishra & Dadhich, 2010). As good as they are (biofertilizers), sometimes they do bring about the expected outcome owing to exposure to high temperatures and other harsh environmental conditions.

Other constraints limiting the use of biofertilizers technology may include human resource, unavailability of suitable strains, and unavailability of suitable carrier. Chen (2006) has identified some bottlenecks for effective biofertilizer inoculation:

- Unavailability of suitable strains: This is one of the major constraints in the production of biofertilizers because only suitable strains have the ability to survive both in the broth and in the inoculant carrier.
- Unavailability of suitable carrier: If suitable carrier material is not available, it becomes difficult to maintain the shelf-life of biofertilizer. In terms of suitability of carrier, the order is peat, lignite, charcoal, farmyard manure, soil, and rice bran.
- Lack of awareness among farmers: Farmers are mostly not aware of biofertilizers and their usefulness in increasing crop yields. They are unaware of the damages caused on the ecosystem by continuous application of inorganic fertilizer and increasing bioaccumulation of toxic materials in plants.

- Skill and technical knowhow: It is another problem. This is because the unskilled and inadequate staff farmers are not given proper instructions about the application. This however affects in general terms the success of biofertilizers.
- Environmental constraints: Soil characteristics like salinity, acidity, drought, water logging, and a number of climatic conditions affect the use of biofertilizers. Table 14.8 throws light on various constraints in the production and commercialization of biofertilizers.

Table 14.8 Constraints/limitations in the production and commercialization of biofertilizers

Constraints/limitations		Recommendations
Technical constraints	Unavailability of suitable carrier material	Identification and selection of appropriate economic carrier to maintain shelf-life and effectiveness of biofertilizers
		Skilled staff should be hired and manpower should be trained via proper training
	Lack of skilled staff in production	Frequent monitoring of the biofertilizer production units for quality assurance
Marketing constraints	Lack of regulation and units standards for biofertilizer	Necessary legislation for monitoring biofertilizers should be done by the government
	Limited transportation and storage facilities	
	Poor and incomplete labeling of biofertilizer products	Proper labeling of biofertilizers should be done (giving genus name, viable count, and expiry date, etc.)
	Lack of promotion network and publicity among the end users.	Wide publicity of biofertilizers should be done through scientific training, fairs, exhibitions, or media
Biological constraints	Unavailability of appropriate strain. Tendency of strain to mutate during fermentation	Continuous efforts for identification of strains and screening for their efficiency across the type of soil, agro-climatic conditions, and farming situations are recommended
	Nonavailability of right inoculant at the right place at the right time	Standardization of biofertilizer dose in a particular crop and soil.
		Understanding of strain effectiveness should be strengthened through extensive research in this field
Field-level constraints	Soil conditions like acidity, presence of salts and toxic elements in the soil, and extreme climatic conditions may make the results of biofertilizers inconsistent	It is needed to strengthen the research and technologies to reduce effects and to counteract stated soil and environmental conditions
	Inadequate awareness among the farming community about bioinoculants	A strong training and awareness program may be initiated to aware and motivate farmers
Financial constraints	Nonavailability of sufficient fund and equipment from government and private bodies	Use of high-tech equipment is required for quality products
		Government should provide funding and loans for development of production units

14.8 Conclusions

- Fungi as biofertilizers in agriculture sector are one of the emerging areas for growth and enhancement of crop production for sustainable agriculture. This review envisioned the future beneficial role of plant growth-promoting fungal communities in plant growth promotion and soil fertility. These PGP fungi can be used as biofertilizers, bioinoculants, and biocontrol agents in place of chemical fertilizers and pesticides in an environmentally and eco-friendly manner.
- Use of beneficial microbes as biofertilizers and biopesticides in agriculture sector are one of the methods that can pave way for the next agricultural green revolution. Beneficial fungi, associated with plants, are playing an essential role in the development and growth of plants through various mechanisms including solubilization of different insoluble and unavailable nutrients and production of plant growth regulators. Fungi also help in alleviating biotic stresses like plant pest and pathogen and abiotic stresses like drought, salinity, temperature, heavy metals, etc. Fungi are playing significant role in bioremediation of heavy metals and other pesticides also.
- There are many commercial fungal biofertilizers available worldwide. Using fungal biofertilizers offers more environmentally friendly alternatives than chemical fertilizers. However, there are some limitations in using these products. The success can be affected by environmental conditions, while application difficulties, limited shelf-life, and slow action as compared to chemical products may discourage farmers to use the biofertilizers. A lot of research done in the past few decades has enabled these fungi to emerge as a potential biofertilizer, a cheap and environment-friendly alternative to expensive and harmful chemical fertilizers. This aspect of a conventional to alternative route to more food grain production in a sustainable manner especially gains significance for developing countries. Among the fungal biofertilizers, mycorrhizal biofertilizer products greatly appeal to the agro-industry mainly due to its versatility and use of environment-friendly technology. The AM biofertilizer technology can be called poor man's technology.
- Microbial mixed inoculum (biofertilizers and biopesticides) or consortium, used as an alternate to synthetic fertilizers and pesticides to increase the soil fertility and control disease and pests in agriculture is gaining prominence. Biofertilizers and biopesticides are environment-friendly products and can be used in integrated nutrient management (INM) and integrated pest management (IPM) techniques.
- White biotechnology refers to the use of living cells and/or their enzymes to create industrial products that are more easily degradable, require less energy, and create less waste during production. The fungal white biotechnology includes biodiversity of fungi from different habitats, including extreme environments (high temperature, low temperature, salinity, and pH) and associated with plants (epiphytic, endophytic, and rhizospheric) and their industrial applications in diverse sectors. Endophytes have generally greater beneficial effects (FAO,

2014) and are less likely to encounter an extreme change of the soil environment because of the insulation of plants. Thus, to achieve better yield and protection of crop, the use of endophytic biofertilizers can be preferred over other microbial species. Endophytes can be explored as effective biofertilizers as they have an advantage of living within a plant's tissues and therefore more readily exert a direct beneficial effect (Vandenkoornhuyse et al., 2002).

- Future research in rhizosphere biology will rely on the development of molecular and biotechnological approaches to increase our knowledge of rhizosphere biology and to achieve an integrated management of soil microbial populations.
- In the agricultural sector, vegetables form the most vital part of Indian agriculture and nutritional security due to their short life cycle, high yield, nutritional richness, economic viability, and also provide employment opportunities to rural nonfarming and farming sector. India has diverse climatic conditions with different weather phenomena, which help in growing a wide range of vegetable and fruit production. India ranks second in worldwide vegetable and fruit production. In our continual combat with pests along with other agro-ecological and environmental challenges, while meeting the nation's demand to feed the ever-increasing human population, the use of pesticide is unavoidable. The inefficient farm management practices and improper application of agrochemicals and pesticides have caused degradation of soil fertility and health apart from affecting different biotic and abiotic factors (Setboonsarng & Gilman, 2006).
- The need of the hour is to develop effective biofertilizers, with both pesticide tolerance and plant growth-promoting (PGP) traits, isolated indigenously from the crop field soil. Sustainable agricultural practices are essential for providing better food and fiber for a healthy nation. Only effective microbes are suitable alternatives to chemical fertilizers and pesticides for enhancing plant production under sustainable agriculture. These microbes also have the ability to survive under pesticide threat and adverse environmental conditions because indigenous microbes can change their physiological and genetic character according to environmental conditions.
- The interest in biofertilizers is increasing and so is the potential for their use in sustainable agriculture. However, many of the products that are currently available worldwide are often of very poor quality, resulting in the loss of confidence from farmers. The formulation of an inoculant is a crucial multistep process that should result in one or several strains of microorganisms included in a suitable carrier, providing a safe environment to protect them from the often harsh conditions during storage and ensuring survival and establishment after introduction into soils. One of the key issues in formulation, development, and production is the quality control of the products at each stage of the process. Quality control of products is of great importance. This is always a challenge with microbial products because the mass production of living organisms tends to select those best suited to mass production rather than those found more effective in the field. Research in the field, supported by advanced technology, will enhance biofertilizer use in the country as well as profitability for the small and marginal farmers.

- Potentiality of pesticide-tolerant fungi as biofertilizers can pave the way for bio-entrepreneurship. Agricultural sustainability and economic stability of rural India depend upon scientific capacity building amongst the farmers. The combination of the biological potentials of plant–holobiont and soil mycobiome, when fully harnessed under agricultural scenario, will help to boost food security globally. The biofertilizers are important to ensure a healthy future for the generations to come. Long-term use of biofertilizers is economical, eco-friendly, more efficient, productive, and accessible to marginal and small farmers over chemical fertilizers.

In a nutshell, the future is envisioned to be most promising for rural India by following the key points mentioned below:

- Shifting the focus from chemical fertilizers to biological fertilizers will lead to the proliferation of biofertilizer industries and incentivizing the ecological benefits will encourage the farming community for application of biofertilizers.
- Biofertilizers would play a key role in productivity and sustainability of soil and protect the environment as eco-friendly and cost-effective inputs for the farmers.
- Strain improvement using genomic and biotechnological tools for better nutrient fixation/solubilization and plant growth-promoting attribute will further lead to gaining confidence in biofertilizers by stakeholders.
- Selection of effective and competitive multifunctional biofertilizers for a variety of crops.
- Quality control system for the production of inoculants and their application in the field.
- Study of microbial persistence of biofertilizers in soil under stressful environments conditions.
- Establishment of "Bio-fertilizer Act" and strict regulation for quality control in markets and application.

References

Abdallah, M. F., Ameye, M., De Saeger, S., Audenaert, K., & Haesaert, G. (2018). Biological control of mycotoxigenic fungi and their toxins: An update for the pre-harvest approach. In *Mycotoxins: Impact and management strategies*. Intechopen.

Adesemoye, A. O., & Kloepper, J. W. (2009). Plant-microbe's interactions in enhanced fertilizer use efficiency. *Applied Microbiology and Biotechnology, 1*, 1–12.

Adholeya, A. (2006). Potential of arbuscular mycorrhizae in organic farming system. In Handbook of Microbial Biofertilizers:223

Adholeya, A. (2012). Mycorrhiza: A friendly association with plants towards sustainable food production. *Journal of Mycology and Plant Pathology, 42*, 183–188.

Adholeya, A., Tiwari, P., & Singh, R. (2005). Largescale inoculum production of arbuscular mycorrhizal fungi on root organs and inoculation strategies. In S. Declerck, D.-G. Strullu, & A. Fortin (Eds.), *Soil biology, volume 4 in vitro culture of Mycorrhizae* (pp. 315–338). Springer.

Aggarwal, R. (2015). *Chaetomium globosum*: A potential biocontrol agent and its mechanism of action. *Indian Phytopathology, 68*, 8–24.

Ahemad, M., & Khan, M. S. (2012). Effect of fungicides on plant growth promoting activities of phosphate solubilizing *Pseudomonas putida* isolated from mustard (*Brassica compestris*) rhizosphere. *Chemosphere, 86*, 945–950.

Ahmad, I., Ansari, M. I., & Aqil, F. (2006). Biosorption of Ni, Cr and Cd by metal tolerant *Aspergillus niger* and *Penicillium* sp. Using single and multi-metal solution. *Indian Journal of Experimental Biology, 44*, 73–76.

Ajmal, M., Ali, H. I., Saeed, R., Akhtar, A., Tahir, M., Mehboob, M. Z., & Ayub, A. (2018). Biofertilizer as an alternative for chemical fertilizers. *The Journal ofAgricultural Science, 7*, 1–7.

Akhtar, S. M., & Siddiqui, Z. A. (2008a). Arbuscular mycorrhizal fungi as potential bioprotectants against plant pathogens. In Z. A. Siddiqui, M. S. Akhtar, & K. Futai (Eds.), *Mycorrhizae: Sustainable agriculture and forestry* (pp. 61–97). Springer.

Akhtar, S. M., & Siddiqui, Z. A. (2008b). Biocontrol of a root-rot disease complex of chickpea by Glomus intraradices, Rhizobium sp. and Pseudomonas straita. *Crop Protection, 27*, 410–417.

Akintokun, A. K., Akande, G. A., Akintokun, P. O., Popoola, T. O. S., & Babalola, A. O. (2007). Solubilization on insoluble phosphate by organic acid-producing fungi isolated from Nigerian soil. *International Journal of Soil Science, 2*, 301–307.

Aktar, W., Sengupta, D., & Chowdhury, A. (2009). Impact of pesticide use in agriculture; their benefits and hazards. *Interdisciplinary Toxicology, 2*(1), 1–12.

Alabouvette, C., Olivain, C., & Steinberg, C. (2006). Biological control of plant diseases: The European situation. *European Journal of Plant Pathology, 114*, 329–341.

Ali, A. F., Suhail, F. M., Salim, H. A., & Abed, A. H. (2021). *Azotobacter chroococcum, Azospirillum brasilense, Trichoderma harzianum* and Tri-calcium phosphate on hydroponically grown barley grains, (Vol.735,No.1,p.012056). IOP Publishing.

Ambethgar, V. (2009). Potential of entomopathogenic fungi in insecticide resistance management (IRM): A review. *Journal of Biopesticides, 2*(2), 177–193.

Anderson, C. I., & Cairney, W. G. J. (2007). Ectomycorrhizal fungi: Exploring the mycelial frontier. *FEMS Microbilogy Review, 31*, 388–406.

Anitha, S., Devi, S. N. P., & Kumari, K. S. (2015). Isolation and identification of zinc solubilizing fungal isolates from agricultural fields. *Indian Journal ofAgricultural Sciences, 85*, 1638–1642.

Arora, N. K., Khare, E., & Maheshwari, D. K. (2010). Plant growth promoting rhizobacteria: Constraints in bioformulation, commercialization and future strategies. In *Plant growth and health promoting bacteria* (pp. 97–11). Springer.

Arshad, A. M., & Aishatul, B. (2015). *Aspergillus niger*-a novel heavy metal bio-absorbent and pesticide tolerant fungus. *Research Journal of Chemistry and Environment, 19*(2), 57–66.

Assad, R., Reshi, Z. A., Rashid, I., Shouche, Y., & Dhotre, D. (2020). Role of ectomycorrhizal biotechnology in pesticide remediation. In R. Bhat, K. Hakeem, & N. Saud Al-Saud (Eds.), *Bioremediation and biotechnology, Vol 3*. Springer. https://doi.org/10.1007/978-3-030-46075-4_14

Backer, R., Rokem, J. S., Ilangumaran, G., Lamont, J., Praslickova, D., Ricci, E., Subramanian, S., & Smith, D. L. (2018). Plant growth-promoting rhizobacteria: Context, mechanisms of action, and roadmap to commercialization of biostimulants for sustainable agriculture. *Frontiers in Plant Science, 9*, 1473.

Bagyaraj, D. J., & Ashwin, R. (2017). Soil biodiversity: Role in sustainable horticulture. *Biodiversity in Horticultural Crops, 5*, 1–18.

Barea, J. M., Pozo, M. J., & Azcón, R. (2005). Concepción Azcón-Aguilar, Microbial cooperation in the rhizosphere. *Journal of Experimental Botany, 56*(417), 1761–1778. https://doi.org/10.1093/jxb/eri197

Barroso, C. B., Pereira, G. T., & Nahas, E. (2006). Solubilization of $CAHPO_4$ and $ALPO_4$ by *Aspergillus niger* in culture media with different carbon and nitrogen sources. *Brazillian Journal of Microbiology, 37*, 434–438.

Bashan, Y., de Bashan, L. E., Prabhu, S. R., & Hernandez, J. P. (2014). Advances in plant growth-promoting bacterial inoculant technology: Formulations and practical perspectives. *Plant and Soil, 378*, 1–33.

Bending, G. D., Friloux, M., & Walker, A. (2002). Degradation of contrasting pesticides by white rot fungi and its relationship with ligninolytic potential. *FEMS Microbiology Letters, 212*(1), 59–63. https://doi.org/10.1111/j.1574-6968.2002.tb11245.x
Benítez, T., Rincón, A. M., Limón, M. C., & Codón, A. C. (2004). Biocontrol mechanisms of *Trichoderma* strains. *International Microbiology, 7*, 249–260.
Bergelson, J., Mittelstrass, J., & Horton, M. W. (2019). Characterizing both bacteria and fungi improves understanding of the *Arabidopsis* root microbiome. *Scientific Reports, 9*, 24.
Berruti, A., Lumini, E., Balestrini, R., & Bianciotto, V. (2016). *Arbuscular mycorrhizal* fungi as natural bio-fertilizers: Let's benefit from past successes. *Frontiers in Microbiology, 6*, 1559.
Bhalerao, T. S., & Puranik, P. R. (2007). Biodegradation of organochlorine pesticide, endosulfan, by a fungal soil isolate. *Aspergillus niger, International Biodeterioration and Biodegradation, 59*, 315–321.
Bhardwaj, D., Ansari, M. W., Sahoo, R. K., & Tuteja, N. (2014). Biofertilizers function as key player in sus-tainable agriculture by improving soil fertility, plant tolerance and crop productivity. *Microbial Cell Factories, 13*, 1–10.
Bharti, N., Pandey, S. S., Barnawal, D., Patel, V. K., & Kalra, A. (2016). Plant growth promoting rhizobacteria *Dietzia natronolimnaea* modulates the expression of stress responsive genes providing protection of wheat from salinity stress. *Scientific Reports, 6*, 1–16.
Bhattacharya, P., & Dwivedi. (2004). *Proceedings of national conference on quality control of biofertilizers*. National Bio-fertilizer Development Centre.
BIO-ECONOMY INDIA.1615182060_Indian_BioEconomy_Report 2021.pdf
Bisen, K., Keswani, C., Mishra, S., Saxena, A., Rakshit, A., & Singh, H. B. (2015). Unrealized potential of seed biopriming for versatile agriculture. In *Nutrient use efficiency: From basics to advances* (pp. 193–206). Springer.
Board, N. (2004). *The complete technology book on bio-fertilizer and organic farming*. National Institute of Industrial Research.
Boyer, L. R., Brain, P., Xu, X.-M., & Jeffries, P. (2015). Inoculation of drought-stressed strawberry with a mixed inoculum of two arbuscular mycorrhizal fungi: Effects on population dynamics of fungal species in roots and consequential plant tolerance to water deficiency. *Mycorrhiza, 25*, 215–227.
Brainerd, E., & Menon, N. (2014). Seasonal effects of water quality: The hidden costs of the Green Revolution to infant and child health in India. *Journal of Development Economics, 107*, 49–64. https://doi.org/10.1016/j.jdeveco.2013.11.004
Braud, A., Karine, J. E., Vieille, E., Triller, A., & Lebeau, T. (2006). Changes in extractability of Cr and Pb in a polycontaminated soil after bioaugmentation with microbial producers of biosurfactants, organic acids and siderophores. *Water, Air, & Soil Pollution: Focus, 6*, 261–279.
Bücking, H., & Kafle, A. (2015). Role of arbuscular mycorrhizal fungi in the nitrogen uptake of plants: Current knowledge and research gaps. *Agronomy, 5*, 587–612.
Campbell, C. K. (1994). Forms of Aspergillosis. In K. A. Powell, A. Renwick, & J. F. Peberdy (Eds.), *The genus Aspergillus* (pp. 313–319). Plenum Press.
Carvajal-Muñoz, J. S., & Carmona-Garcia, C. E. (2012). Benefits and limitations of biofertilization in agricultural practices. *Livestock Research for Rural Development, 24*, Article 43. http://www.lrrd.org/lrrd24/3/carv24043.htm
Chandanie, W. A., Kubota, M., & Hyakumachi, M. (2006). Interactions between plant growth promoting fungi and arbuscular mycorrhizal fungus *Glomus mosseae* and induction of systemic resistance to anthracnose disease in cucumber. *Plant and Soil, 286*, 209–217.
Chandrashekrappa, G. K., & Basalingappa, K. M. (2018). Biofertilizer for crop production and soil fertility. *Academia Journal of Agricultural Research, 6*, 299–306.
Chang, C. H., & Yang, S. S. (2009). Thermo-tolerant phosphate-solubilizing microbes for multi-functional biofertilizer preparation. *Bioresource Technology, 100*, 1648–1658.
Chen, J. H. (2006, 16–20 October). *The combined use of chemical and organic fertilizers and/or biofertilizer for crop growth and soil fertility*. International workshop on sustained management of the soil-rhizosphere system for efficient crop production and fertilizer Use (pp. 1–11). Land Development Department, Bangkok 10900 Thailand.

Chiomento, J. L. T., Stürmer, S. L., Carrenho, R., da Costa, R. C., Scheffer-Basso, S. M., Antunes, L. E. C., Nienow, A. A., & Calvete, E. O. (2019). Composition of arbuscular mycorrhizal fungi communities signals generalist species in soils cultivated with strawberry. *Scientia Horticulturae, 253*, 286–294.

Choudhury, A., & Kennedy, I. (2004). Prospects and potentials for systems of biological nitrogen fixation in sustainable rice production. *Biology and Fertility of Soils, 39*, 219–227.

Clark, R. B. (1997). Arbuscular mycorrhizal adaptation, spore germination, root colonization and host plant growth and mineral acquisition at low pH. *Plant and Soil, 192*, 15–22.

Clark, D. P., Dunlap, P. V., & Madigan, M. T. (2009). In T. D. Brock & M. T. Madigan (Eds.), *Biology of microorganisms* (Vol. 485). Pearson.

Clasen, B., Murussi, C., & Storck, T. (2019). Pesticide contamination in Southern Brazil. In L. M. Gómez-Oliván (Ed.), *Pollution of Water Bodies in Latin America* (pp. 43–54). Springer. https://doi.org/10.1007/978-3-030-27296-8_3

Cooper, J., & Dobson, H. (2007). The benefits of pesticides to mankind and the environment. *Crop Protection, 26*(9), 1337–1348.

Dahm, H. (2006). Role of mycorrhizae in the forestry. In M. K. Rai (Ed.), *Handbook of microbial biofertilizers* (pp. 241–270). Food Products Press.

Daisley, B. A., Chernyshova, A. M., Thompson, G. J., & Allen-Vercoe, E. (2022). Deteriorating microbiomes in agriculture-the unintended effects of pesticides on microbial life. *Microbiome Research Reports,1*(1), 6.

Damalas, C. A. (2009). Understanding benefits and risks of pesticide use. *Scientific Research and Essay, 4*(10), 945–949.

Das, A., Prasad, R., Srivastava, A., Giang, H. P., Bhatnagar, K., & Varma, A. (2007). Fungal siderophores: Structure, functions and regulation. In A. Varma & S. B. Chincholkar (Eds.), *Soil biology volume 12 microbial Siderophores* (pp. 1–42). Springer.

Dasgupta, S., Meisner, C., & Wheeler, D. (2010). Stockpiles of obsolete pesticides and cleanup priorities, a methodology and application for Tunisia. *Journal of Environmental Management, 91*(4), 824–830.

Deveau, A., Bonito, G., Uehling, J., Paoletti, M., Becker, M., et al. (2018). Bacterial-fungal interactions: Ecology, mechanisms and challenges. *FEMS Microbiology Reviews, 42*, 335–352.

Dhingra, O. D., Mizubuti, E. S. G., & Santana, F. M. (2003). *Chaetomium globosum* for reducing primary inoculum of *Diaporthe phaseolorum* sp. *meridionalis* in soil-surface soybean stable in field condition. *Biological Control, 26*, 302–310.

El Maaloum, S., Elabed, A., Alaoui-Talibi, Z. E., Meddich, A., Filali-Maltouf, A., Douira, A., Ibnsouda-Koraichi, S., Amir, S., & El Modafar, C. (2020). Effect of arbuscular mycorrhizal fungi and phosphate-solubilizing bacteria consortia associated with phosphor compost on phosphorus solubilization and growth of tomato seedlings (*Solanum lycopersicum* L.). *Communications in Soil Science and Plant Analysis, 51*, 622–634.

Emmert, E. A. B., & Handelsman, J. (1999). Biocontrol of plant disease: A (Gram-) positive perspective. *Microbiology Letters, 171*, 1–9.

Etesami, H., Emami, S., & Alikhani, H. A. (2017). Potassium solubilizing bacteria (KSB): Mechanisms, promotion of plant growth, and future prospects a review. *Journal of Soil Science and Plant Nutrition, 17*, 897–911.

Ali,A.F., Suhail, F. M., Salim, H. A., & Hussein, A. A. (2021). Influence of *Azotobacter chroococcum, Azospirillum brasilense, Trichoderma harzianum* and Tri-calcium phosphate on hydroponically grown barley grains. *IOP Conference Series: Earth and Environmental Science, 735*(2021), 012056.

FAO. (2014). *The state of world fisheries and aquaculture 2014. Opportunity and challenge*. FAO.

FAO (Food and Agriculture Organization of the United Nations), WHO (World Health Organization). (2020). *COVID-19 and Food Safety: Guidance for food businesses: Interim guidance* [Online]. http://www.fao.org/3/ca8660en/CA8660EN.pdf. Accessed on 17 Apr 2020.

Farrera, P. R. (2004). Acerca de los plaguicidas y su uso en la agricultura, Revista Digital Ceniap Hoy, Número 6, septiembrediciembre, Maracay, Venezuela.

Fertiliser Association of India. (2011). *Fertiliser Statistics 2009–10 and earlier issues*. The Fertilizer Association of India.
Figueiredo, G. G., Lopes, V., Fendrich, R., & Szilagyi-Zecchin, V. (2017). Interaction between beneficial bacteria and sugarcane. In D. P. Singh, H. B. Singh, & R. Prabha (Eds.), *Plant-microbe interactions in agro-ecological perspectives* (pp. 1–27). Springer.
Finlay, R. D. (2008). Ecological aspects of mycorrhizal symbiosis: With special emphasis on the functional diversity of interactions involving the extraradical mycelium. *Journal of Experimental Botany, 59*, 1115–1128.
Frąc, M., Hannula, S. E., Bełka, M., & Jędryczka, M. (2018). Fungal biodiversity and their role in soil health. *Frontiers in Microbiology, 9*, 707.
Franco, J. A., Bañón, S., Vicente, M. J., Miralles, J., & Martínez-Sánchez, J. J. (2011). Root development in horticultural plants grown under abiotic stress conditions-a review. *The Journal of Horticultural Science and Biotechnology, 86*, 543–556.
Frey-Klett, P., Burlinson, P., Deveau, A., Barret, M., Tarkka, M., & Sarniguet, A. (2011). Bacterial-fungal interactions: Hyphens between agricultural, clinical, environmental, and food microbiologists. *Microbiology and Molecular Biology Reviews, 75*, 583–609.
Fuentes-Ramirez, L. E., & Caballero-Mellado, J. (2005). Bacterial biofertilizers. In A. Siddiqui (Ed.), *PGPR: Biocontrol and biofertilization* (pp. 143–172). Springer.
Gamon, M. E., Saez, E., Gil, J., & Boluda, R. (2003). Direct and indirect exogenous contamination by pesticides of rice-farming soils in a Mediterranean wetland. *Archives of Environmental Contamination and Toxicology, 44*, 141–151.
Gaur, A., & Adholeya, A. (2004). Prospects of arbuscular mycorrhizal fungi in phytoremediation of heavy metal contaminated soils. *Current Science, 86*, 528–534.
Geetha, S., Kaparapu, & Jyothi. (2016). Effect of pesticides on *aspergillus niger* from agricultural soil. *World Journal of Pharmacy and Pharmaceutical Sciences, 5*(5), 731–739. https://doi.org/10.20959/wjpps20165-6591
Gentili, F., & Jumpponen, A. (2006). Potential and possible uses of bacterial and fungal biofertilizers. In M. K. Rai (Ed.), *Handbook of microbial biofertilizers* (pp. 1–28). Food Products Press.
Ghosh, S., Pareek, N., Rawerkar, K. P., Chandra, R., Pachauri, S. P., & Kaushik, S. (2019). Prospective zinc solubilizing microrganisms for enhanced growth and nutrition in Maize(Zeamays L.) *International Journal of Current Microbiology and Applied Sciences, 8*(8), 2771–2727.
Gouda, S., Kerry, R. G., Das, G., Paramithiotis, S., Shin, H. S., & Patra, J. K. (2018). Revitalization of plant growth promoting rhizobacteria for sustainable development in agriculture. *Microbiological Research, 206*, 131–140.
Gupta, V., Satyanarayana, T., & Garg, S. (2000). General aspects of mycorrhiza. In K. G. Mukerji, J. Singh, & B. P. Chamola (Eds.), *Mycorrhizal biology* (pp. 27–44). Kluwer Academic/Plenum Plublishers.
Hansel, C. M., Fendorf, S., Jardine, P. M., & Francis, C. A. (2008). Changes in bacterial and archaeal commu-nity structure and functional diversity along a geochemically variable soil profile. *Applied and Environmental Microbiology, 74*, 1620–1633.
Hardoim, P. R., van Overbeek, L. S., & Elsas, J. D. V. (2008). Properties of bacterial endophytes and their proposed role in plant growth. *Trends in Microbiology, 16*, 463–471.
Harman, G. E. (2006). Overview of mechanisms and uses of *Trichoderma* spp. *Phytopathology, 96*, 190–194. https://doi.org/10.1094/PHYTO-96-0190
Harman, G. E., Howell, C. R., Viterbo, A., Chet, I., & Lorito, M. (2004). *Trichoderma* species-opportunistic, avirulent plant symbionts. *Nature Reviews Microbiology, 2*, 43–56. https://doi.org/10.1038/nrmicro797
Harman, G. E., Herrera-Estrella, A. H., Horwitz, B. A., & Lorito, M. (2012). *Trichoderma* – From basic biology to biotechnology. *Microbiology, 158*, 1–2. https://doi.org/10.1099/mic.0.056424-0.
Hart, M. M., & Trevors, J. T. (2005). Microbe management: Application of mycorrhyzal fungi in sustainable agriculture. *Frontiers in Ecology and the Environment, 3*, 533–539.

Hashem, A., Allah, E. F., Alqarawib, A. A., Al Huqaila, A., & Egamberdieva, D. (2014). Alleviation of abiotic salt stress in *Ochradenus baccatus* (Del.) by *Trichoderma hamatum* (Bonord.) Bainier. *Journal of Plant Interactions, 9*, 857–868. https://doi.org/10.1080/17429145.2014.983568

Hashimi, M. H., Hashimi, R., & Ryan, Q. (2020). Toxic effects of pesticides on humans,plants,animals,pollinators and beneficial organisms. *APRJ, 5*(4), 37–47.

Haskins, K. E., & Gehring, C. A. (2005). Evidence for mutualist limitation: The impacts of conspecific density on the mycorrhizal inoculum potential of woodland soils. *Oecologia, 145*, 123–131.

Hermosa, R., Viterbo, A., Chet, I., & Monte, E. (2012). Plant-beneficial effects of *Trichoderma* and of its genes. *Microbiology, 158*, 17–25. https://doi.org/10.1099/mic.0.052274-0

HLPE. (2020). *Impacts of COVID-19 on food security and nutrition: Developing effective policy responses to address the hunger and malnutrition pandemic*. Rome. https://doi.org/10.4060/cb1000en, https://geographyandyou.com/biofertilisers-in-indian-agriculture/

Ipsilantis, I., & Sylvia, D. M. (2007). A bundance of fungi and bacteria in a nutrient-impacted Florida wetland. *Applied Soil Ecology, 35*, 272–280.

Itelima, J., Bang, W., Onyimba, I., & Oj, E. (2018). A review: Biofertilizer; a key player in enhancing soil fertility and crop productivity. *Journal of Microbiology and Biotechnology Reports, 2*, 22–28.

Jagnaseni, B., Aveek, S., Babita, S., & Siraj, D. (2016). Mycorrhiza: The oldest association between plant and fungi. *Resonance, 21*, 1093–1104.

Jehangir, I. A., Mir, M. A., Bhat, M. A., & Ahangar, M. A. (2017). Biofertilizers an approach to sustainability in agriculture: A review. *Indian Journal of Pure & Applied Biosciences, 5*, 327–334.

Jha, Y. (2017). Potassium mobilizing bacteria: Enhance potassium intake in paddy to regulates membrane permeability and accumulate carbohydrates under salinity stress. *Brazilian Journal of Biological Sciences, 4*, 333–344.

Johri, A. K., Oelmüller, R., Dua, M., Yadav, V., Kumar, M., Tuteja, N., & Stroud, R. M. (2015). Fungal association and utilization of phosphate by plants: Success, limitations, and future prospects. *Frontiers in Microbiology, 6*, 984.

Kaewchai, S., Soytong, K., & Hyde, K. D. (2009). Mycofungicides and fungal biofertilizers. *Fungal Diversity, 38*, 25–50.

Kamran, S., Shahid, I., Baig, D. N., Rizwan, M., Malik, K. A., & Mehnaz, S. (2017). Contribution of zinc solubilizing bacteria in growth promotion and zinc content of wheat. *Frontiers in Microbiology, 8*, 2593.

Kannaiyan, S. (2002). Biofertilizers for sustainable crop production. In S. Kannaiyan (Ed.), *Biotechnology of biofertilizers* (pp. 9–50). *Springer*.

Kapoor, R., Sharma, D., & Bhatnagar, A. K. (2008). Arbuscular mycorrhizae in micropropagation systems and their potential applications. *Scientia Horticulturae, 116*, 227–239.

Karstensen, K. H., Nguyen, K. K., Le, B. T., Pham, H. V., Nguyen, D. T., Doan, T. T., Nguyen, H. H., Tao, M. Q., Luong, D. H., & Doan, H. T. (2006). Environmentally sound destruction of obsolete pesticides in developing countries using cement kilns. *Environmental Science & Policy, 9*(6), 577–586.

Keswani, C., Singh, S., Singh, & Harikesh. (2013). A superstar in biocontrol enterprise: *Trichoderma* spp. *Biotech Today An International Journal of Biological Sciences, 3*(2), 27.

Khan, A. G. (2006). Mycorrhizoremediation-an enhanced form of phytoremediation. *Journal of Zhejiang University Science B, 7*, 503–514.

Khan, M. R., & Anwer, M. A. (2007). Molecular and biochemical characterization of soil isolates of *Aspergillus niger* aggregate and an assessment of their antagonism against *Rhizoctonia solani*. *Phytopathologia Mediterranea, 46*, 304–315.

Khan, M. R., & Anwer, M. A. (2018). DNA and some laboratory tests of nematode suppressing efficient soil isolates of *Aspergillus niger*. *Indian Phytopath, 61*(2), 212–225.

Khan, M. R., Anwer, M. A., Khan, S. M., & Khan, M. M. (2006a). An evaluation of isolates of *Aspergillus niger* against *Rhizoctonia solani*. *Tests of Agrochemicals and Cultivars, 27*, 31–32.
Khan, M. R., Anwer, M. A., Khan, S. M., & Mohiddin, F. A. (2006b). A characterization of some isolates of *Aspergillus niger* and evaluation of their effects against Meloidogyne incognita. *Tests of Agrochemicals and Cultivars, 27*, 33–34.
Khokhar, M., Renu, G., & Sharma, R. (2012). Biological control of plant pathogens using biotechnological aspects: A review. *Epidemiology, 1*, 277–289.
Kiers, E. T., & van der Heijden, M. G. A. (2006). Mutualistic stability in the arbuscular mycorrhizal symbiosis: Exploring hypotheses of evolutionary cooperation. *Ecology, 87*, 1627–1636.
Kour, D., Rana, K. L., Yadav, N., Yadav, A. N., Singh, J., Rastegari, A. A., & Saxena, A. K. (2019). Agriculturally and industrially important fungi: Current developments and potential biotechnological applications. In A. N. Yadav, S. Singh, S. Mishra, & A. Gupta (Eds.), *Recent advancement in white biotechnology through fungi, volume 2: Perspective for value-added products and environments* (pp. 1–64). Springer. https://doi.org/10.1007/978-3-030-14846-1_1
Kour, D., Rana, K. L., Yadav, A. N., Yadav, N., Kumar, M., Kumar, V., Vyas, P., Dhaliwal, H. S., & Saxena, A. K. (2020). Microbial biofertilizers: Bioresources and eco-friendly technologies for agricultural and environmental sustainability. *Biocatalysis and Agricultural Biotechnology, 23*, 101487. https://doi.org/10.1016/j.bcab.2019.101487
Krishnamoorthy, R., Kim, K., Subramanian, P., Senthilkumar, M., Anandham, R., & Sa, T. (2016). Arbuscular mycorrhizal fungi and associated bacteria isolated from salt-affected soil enhances the tolerance of maize to salinity in coastal reclamation soil. *Agriculture, Ecosystems and Environment, 231*, 233–239.
Kulshreshtha, S., McConkey, B. G., Liu, T. T., Dyer, J. A., Vergé, X. P. C., & Desjardins, R. L. (2011). Biobased economy – Sustainable use of agricultural resources. *Environmental Impact of Biofuels*. https://doi.org/10.5772/19989
Kumar, V., Chandra, A., & Singh, G. (2010). Efficacy of fly-ash based bio-fertilizers vs perfected chemical fertilizers in wheat (*Triticum aestivum*). *International Journal of Scientific & Technology Research, 2*, 31–35.
Lata, R., Thapa, K., Kanojia, N., Sharma, N., Singh, S., Grewal, A. S., Srivastav, A. L., & Kaushal, J. (2021). An extensive review on the consequences of chemical pesticides on human health and environment. *Journal of Cleaner Production, 283*, 124657.
Le, H. T., Stubbe, D., Verbeken, A., Nuytinck, J., Lumyong, S., & Desjardin, D. E. (2007). *Lactarius* in Northern Thailand: 2, *Lactarius* subgenus *Plinthogali*. *Fungal Diversity, 27*, 61–94.
Liang, W. Q., Wang, Z. Y., Li, H., Wu, P. C., Hu, J. M., Li, N. L., Cao, X., & Liu, Y. H. (2005). Purification and characterization of a novel pyrethroid hydrolase from *Aspergillus niger* ZD11. *Journal of Agricultural and Food Chemistry, 53*(19), 7415–7420.
Liu, Y. H., Chung, Y. C., & Xiong, Y. (2001). Purification and characterization of a dimethoate degrading enzyme of *Aspergillus niger* ZHY256, isolated from sewage. *Applied and Environmental Microbiology, 67*, 3746–3749.
Liu, Y. H., Liu, H., Chen, Z. H., Lian, J., Huang, X., & Chung, Y. C. (2004). Purification and characterization of a novel organophosphorus pesticide hydrolase from *Penicillium lilacinum* BP303. *Enzyme and Microbial Technology, 34*, 297–303.
López-Bucio, J., Pelagio-Flores, R., & Herrera-Estrell, A. (2015). *Trichoderma* as biostimulant: Exploiting the multilevel properties of a plant beneficial fungus. *Scientia Horticulturae, 196*, 109–123.
Mahanty, T., Bhattacharjee, S., Goswami, M., Bhattacharyya, P., Das, B., Ghosh, A., & Tribedi, P. (2016). Biofertilizers: A potential approach for sustainable agriculture development. *Environmental Science and Pollution Research, 24*(4), 3315–3335.
Mahimairaja, S., Dooraisamy, P., Lakshmanan, A., Rajannah, G., Udayasoorian, C., & Natarajan, S. (2008). Composting technology and organic waste utilization. *Journal of Sciences, 1*, 332–560.
Malik, K. A., Hafeez, F. Y., Mirza, M. S., Hameed, S., Rasul, G., & Bilal, R. (2005). Rhizospheric plant fungal diversity – Microbe interactions for sustainable agriculture. In Y.-P. Wang, M. Lin,

Z.-X. Tian, C. Elmerich, & W. E. Newton (Eds.), *Biological nitrogen fixation, sustainable agriculture and the environment* (pp. 257–260). Springer.

Marin, M. (2006). Arbuscular mycorrhizal inoculation in nursery practice. In M. K. Rai (Ed.), *Handbook of Microbial Biofertilizers* (pp. 289–324).

Martinez, J. (2004). Practical guideline on environmentally sound management of obsolete pesticides in the Latin America and Caribbean Countries, Basel Convention Coordinating Centre for Latin America and the Caribbean, Montevideo, Uruguay, 66.

Marwah, R. G., Fatope, M. O., Deadman, M. L., Al- Maqbali, Y. M., & Husband, J. (2007). Musanahol: A new aureonitol-related metabolite from a *Chaetomium* sp. *Tetrahedron, 63*, 8174–8180.

Mastouri, F., Björkman, T., & Harman, G. E. (2010). Seed treatment with *Trichoderma harzianum* alleviates biotic, abiotic, and physiological stresses in germinating seeds and seedlings. *Phytopathology, 100*, 1213–1221. https://doi.org/10.1094/PHYTO-03-10-0091

Mishra, B. K., & Dadhich, S. K. (2010). Methodology of nitrogen biofertilizer production. *International Journal of Advanced Research and Development, 1*, 3–6.

Mishra, N., & Sundari, S. K. (2015). Native PGPM consortium: A beneficial solution to support plant growth in the presence of phytopathogens and residual organophosphate pesticides. *Journal of Bioprocessing and Biotechniques, 5*(2). https://doi.org/10.4172/2155-9821.1000202

Mishra, D. J., Rajivir, S., Mishrra, U. K., & Kumar, S. S. (2013). Role of biofertilizers in organic agriculture: A review. *Research Journal of Recent Sciences, 2*, 39–41.

Mishra, S., Singh, A., Keswani, C., Saxena, A., Sarma, B. K., & Singh, H. B. (2015). Harnessing plant-microbe interactions for enhanced protection against phytopathogens. In *Plant microbes symbiosis: Applied facets* (pp. 111–125). Springer.

Mitchell, A. M., Strobel, G. A., Hess, W. M., Vargas, P. N., & Ezra, D. (2008). *Muscodor crispans*, a novel endophyte from *Ananas ananassoides* in the Bolivian Amazon. *Fungal Diversity, 31*, 37–43.

Mohiddin, F. A., & Khan, M. R. (2013). Tolerance of fungal and bacterial biocontrol agents to six pesticides commonly used in the control of soil borne plant pathogens. *African Journal of Agricultural Research, 8*(43), 5331–5334.

Mora-Romero, G., Cervantes-Gámez, R., Galindo-Flores, H., González-Ortíz, M., Félix-Gastélum, R., Maldonado-Mendoza, I. E., Pérez, R. S., León-Félix, J., Martínez-Valenzuela, M., & Lopez-Meyer, M. (2015). Mycorrhiza-induced protection against pathogens is both genotype-specific and graft-transmissible. *Symbiosis, 66*, 55–64.

Mota, R. M. A., Fernández, A. D., Jr., Trejo, J. B., de la Cruz, & Elizondo, Y. (2019). Inoculación de hongos solubilizadores de fósforo y micorrizas arbusculares en plantas de jitomate. *Revista Mexicana de Ciencias Agrícolas, 10*, 1747–1757.

Naqvi, N. S., & Mukerji, K. G. (2000). Mycorrhizal technology in plant micropropagation system. In K. G. Mukerji, B. P. Chamola, & J. Singh (Eds.), *Micorrhizal biology* (pp. 217–233). Klumer Academic/Plenum Publishers.

Nissipaul, M., Sodimalla, T., Subhashreddy, R., & Suman, B. (2017). Novel biofilm biofertilizers for nutrient management and Fusarium wilt control in chickpea. *International Journal of Current Microbiology and Applied Sciences, 6*, 1846–1852.

Nuti, M., & Giovannetti, G. (2015). Borderline products between bio-fertilizers/bio-effectors and plant protectants: The role of microbial consortia. *Journal of Agriculture, Science and Technology, 5*, 305–315.

Odoh, C. K., Eze, C. N., Akpi, U. K., & Unah, V. U. (2019a). Plant growth promoting rhizobacteria (PGPR): A novel agent for sustainable food production. *American Journal of Agricultural and Biological Sciences, 14*, 35–54.

Odoh, C. K., Zabbey, N., Sam, K., & Eze, C. N. (2019b). Status, progress and challenges of phytoremediation – An African scenario. *Journal of Environmental Management, 237*, 365–378.

Ortiz-Hernandez, M. L., & Sanchez-Salinas, E. (2010). Biodegradation of the organophosphate pesticide tetrachlorvinphos by bacteria isolated from agricultural soils in Mexico. *Revista Internacional de Contaminacion Ambiental, 26*(1), 27–38.

Owen, D., Williams, P., Griffith, G., & Withers, P. (2015). Use of commercial bio-inoculants to increase agricultural production through improved phosphorus acquisition. *Applied Soil Ecology, 86*, 41–54.

Pal, S., Singh, H. B., & Rakshit, A. (2014). The arbuscular mycorrhizal symbiosis, an underground world wide web. In D. P. Singh & H. B. Singh (Eds.), *Microbial communities for sustainable soil health and ecosystem productivity* (pp. 219–253). Studium Press LLC.

Pal, S., Singh, H. B., Farooqui, A., & Rakshit, A. (2015). Fungal biofertilizers in Indian agriculture: Perception, demand and promotion. *Journal of Eco-friendly Agriculture, 10*, (2)101–113.

Pandey, A. K., Alka, N., & Belapurkar, U. (2005). Screening of potential phosphate solubilizing fungi inhabiting pesticide contaminated crop fields soil at Jabalpur. *Journal of Basic and Applied Mycology, 4*(I & II), 1–5.

Pathak, D., Lone, R., & Koul, K. K. (2017). Arbuscular mycorrhizal fungi (AMF) and plant growth- promoting rhizobacteria (PGPR) association in potato (Solanum tuberosum L.) A brief review. In V. Kumar, M. Kumar, S. Sharma, & R. Prasad (Eds.), *Probiotics and plant health* (pp. 401–420). Springer.

Patrick Webb. (2020). The 2020 Nobel Peace Prize rewards the persistent vision of a world without hunger, famine, or malnutrition. *The American Journal of Clinical Nutrition, 113*(2) February 2021, 290–293. https://doi.org/10.1093/ajcn/nqaa355

Philippot, L., Raaijmakers, J. M., Lemanceau, P., & van der Putten, W. H. (2013). Going back to the roots: The microbial ecology of the rhizosphere. *Nature Reviews Microbiology, 11*, 789–799.

Pimentel, D. (2005). Environmental and economic costs of the application of pesticides primarily in the United States. *Environment, Development and Sustainability, 7*, 229–252.

Prabhu, A. M., & Raut, S. V. (2011). Isolation and characterization of pesticide resistant strains of *Azotobacter* spp., *Azospirullum* spp., *Rhizobium* spp. and phosphate solubilizing microorganisms and their application as biofertilizers. *Trends in Biosciences, 4*(2), 194–197.

Pradhan, N., & Sukla, L. B. (2005). Solubilization of inorganic phosphates by fungi isolated from agriculture soil. *African Journal of Biotechnology, 5*, 850–854.

Prasad, S. C. (2016). Innovating at the margins: The System of Rice Intensification in India and transformative social innovation. *Ecology and Society, 21*, 7. https://doi.org/10.5751/ES-08718-210407

Qing, Z., Yang, L., & Huan, L. Y. (2006). Purification and characterization of a novel carbaryl hydrolase from *Aspergillus niger* PY168. *FEMS Microbiology Letters, 228*(1), 39–44.

Raghukumar, C. (2008). Marine fungal biotechnology: An ecological perspective. *Fungal Diversity, 31*, 19–35.

Rai, A., Rai, S., & Rakshit, A. (2013). Mycorrhiza-mediated phosphorus use efficiency in plants. *Environmental and Experimental Biology, 11*, 107–117.

Raina, S., Chamola, B. P., & Mukerji, K. G. (2000). Evolution of Mycorrhiza. In K. G. Mukerji, J. Singh, & B. P. Chamola (Eds.), *Mycorrhizal biology* (pp. 1–25). Kluwer Academic/Plenum Plublishers.

Raja, P. (2006). Status of endomycorrhizal (AMF) biofertilizer in the global market. In M. K. Rai (Ed.), *Handbook of microbial biofertilizers* (pp. 395–416). Food Products Press.

Rakshit, A., & Bhadoria, P. B. S. (2002). Biotechnology for sustainable agriculture. *India Grains,* June 2002, 7–9.

Rakshit, A., Sen, A., & Singh (H.B. (2015). *Nutrient use efficiency: From basics to advances* (p. 417). Springer.

Ram, H., Malik, S. S., Dhaliwal, S. S., Kumar, B., & Singh, Y. (2015). Growth and productivity of wheat affected by phosphorus-solubilizing fungi and phosphorus levels. *Plant, Soil and Environment, 61*, 122–126.

Rana, K. L., Kour, D., Sheikh, I., Dhiman, A., Yadav, N., Yadav, A. N., Rastegari, A. A., Singh, K., & Saxena, A. K. (2019a). Endophytic fungi: Biodiversity, ecological significance and potential industrial applications. In A. N. Yadav, S. Mishra, S. Singh, & A. Gupta (Eds.), *Recent advancement in white biotechnology through Fungi, diversity and enzymes perspectives, vol 1* (pp. 1–62). Springer.

Rana, K. L., Kour, D., & Yadav, A. N. (2019b). Endophytic microbiomes: Biodiversity, ecological significance and biotechnological applications. *Research Journal of Biotechnology, 14*, 142–162.

Rao, D. L. N. (1999). Biofertilizer situation in India and future thrusts. *Indian Council of Agricultural Research (ICAR)*, 1–3.

Rastegari, A. A., Yadav, A. N., & Yadav, N. (2020). *Trends of microbial biotechnology for sustainable agriculture and biomedicine systems: Perspectives for human health*. Elsevier.

Rawat, L., Singh, Y., Shukla, N., & Kumar, J. (2013). Salinity tolerant *Trichoderma harzianum* reinforces NaCl tolerance and reduces population dynamics of *Fusarium oxysporum* f. sp. ciceri in chickpea (*Cicer arietinum* L.) under salt stress conditions. *Archives of Phytopathology and Plant Protection, 46*, 1442–1467. https://doi.org/10.1080/03235408.2013.769316

Redda, E. T., Ma, J., Mei, J., Li, M., & Wu, B. (2018). Antagonistic potential of different isolates of *Trichoderma* against *Fusarium oxysporum, Rhizoctonia solani*, and *Botrytis cinerea*. *European Journal of Experimental Biology, 8*, 12.

Rinaldi, A. C., Comandini, O., & Kuyper, T. W. (2008). Ectomycorrhizal fungal diversity: Separating the wheat from the chaff. *Fungal Diversity, 33*, 1–45.

Rola, C. A. (2000). *Economic perspective for agricultural biotechnology research planning*. Philippine institute for development studies, Discussion paper No. 2000–10, 28.

Rosa, H., Ada, V., Ilan, C., & Enrique, M. (2012). Plant-beneficial effects of Trichoderma and of its genes. *Journal of Microbiology, 158*, 17–25.

Rouphael, Y., Franken, P., Schneider, C., Schwarz, D., Giovannetti, M., Agnolucci, M., et al. (2015). Arbuscular mycorrhizal fungi act as biostimulants in horticultural crops. *Scientia Horticulturae, 196*, 91–108.

Sadhana, B. (2014). Arbuscular mycorrhizal fungi (AMF) as a biofertilizer–a review. *International Journal of Current Microbiology and Applied Sciences, 3*, 384–400.

Sane, S. A., & Mehta, S. K. (2015). Isolation and evaluation of rock phosphate solubilizing fungi as potential bio-fertilizer. *Journal of Fertilizers & Pesticides, 6*, 156.

Schwartz, M. W., Hoeksema, J. D., Gehring, C. A., Johnson, N. C., Klironomos, J. N., Abbott, L. K., & Pringle, A. (2006). The promise and the potential consequences of the global transport of mycorrhizal fungal inoculum. *Ecology Letters, 9*, 501–515.

Setboonsarng, S., & Gilman, J. (2006). *Alternative agriculture in Thailand and Japan*. [internet] http://www.solutionssite.org/artman/publish/article_15.shtml

Shalaby, S. E. M., & Abdou, G. Y. (2010). The influence of soil microorganisms and bio- or -organic fertilizers on dissipation of some pesticides in soil and potato tube. *Journal of Plant Protection Research, 50*(1), 86–92.

Sharma, S. B., Sayyed, R. Z., & Trivedi, M. H. (2013). Phosphate solubilizing microbes: Sustainable approach for managing phosphorus deficiency in agricultural soils. *Springer Plus, 2*, 587.

Sharma, S., Kour, D., Rana, K. L., Dhiman, A., Thakur, S., Thakur, P., Thakur, S., Thakur, N., Sudheer, S., Yadav, N., Yadav, A. N., Rastegari, A. A., & Singh, K. (2019). *Trichoderma*: Biodiversity, ecological significances, and industrial applications. In A. N. Yadav, S. Mishra, S. Singh, & A. Gupta (Eds.), *Recent advancement in white biotechnology through Fungi. Diversity and enzymes perspectives, vol 1* (pp. 85–120). Springer. https://doi.org/10.1007/978-3-030-10480-1_3

Shen, F.-T., Yen, J.-H., Liao, C.-S., et al. (2019). Screening of Rice endophytic biofertilizers with fungicide tolerance and plant growth-promoting characteristics. *Sustainability, 11*, 1133. https://doi.org/10.3390/su11041133

Siddique, T., Okeke, B. C., Arshad, M., & Frankenberger, W. T., Jr. (2003). Biodegradation kinetics of endosulfan by *Fusarium vetricosum* and a *Pandoraea* species. *Journal of Agricultural and Food Chemistry, 51*, 8015–8019.

Siddiqui, Z. A., & Pichtel, J. (2008). Mycorrhixae: An overview. In Z. A. Siddiqui, M. S. Akhtar, & K. Futai (Eds.), *Mycorrhizae: Sustainable agriculture and forestry* (pp. 1–36). Springer.

Singh, R. (2000). Environmental consequences of agricultural development: A case study from the Green Revolution state of Haryana, India. *Agriculture, Ecosystems & Environment, 82*, 97–103. https://doi.org/10.1016/S0167-8809(00)00219-X

Singh, I. (2018). Plant growth promoting rhizobacteria (PGPR) and their various mechanisms for plant growth enhancement in stressful conditions: A review. *European Journal of Biological Research, 8*, 191–213.

Singh, G., & Tilak, K. V. B. R. (2002). Vesicular arbuscular mycorrhizal as bioinoculant. In S. Kannaiyan (Ed.), *Biotechnology of biofertilizers* (pp. 312–322). Kluwer Academic Publishers.

Singh, S., Pandey, A., & Palni, L. M. S. (2008). Screening of arbuscular mycorrhizal fungal consortia developed from the rhizospheres of natural and cultivated tea plants for growth promotion in tea (Camellia sinensis (L.) O.Kuntze). *Pedobiologia, 52*, 119–125.

Singh, S., Singh, B. K., Yadav, S. M., & Gupta, A. K. (2014). Potential of biofertilizers in crop production in Indian Agriculture. *American Journal of Plant Nutrition and Fertilization Technology, 4*, 33–40.

Sinha, R. K., Valani, D., Chauhan, K., & Agarwal, S. (2014). Embarking on a second green revolution for sustainable agriculture by vermiculture biotechnology using earthworms: Reviving the dreams of Sir Charles Darwin. *Journal of Agricultural Safety and Health, 1*, 50–64.

Sivasakthivelan, P., & Saranraj, P. (2013). *Azospirillum* and its formulations: A review. *International Journal of Microbiology Research, 4*, 275–287.

Smith, S. E., & Zhu, Y. G. (2001). Application of arbuscular mycorrhizal fungi: Potentials and challenges. In B. P. Stephen & K. D. Hyde (Eds.), *Bio-exploitation of filamentous fungi* (*Fungal Diversity Research Series, 6*) (pp. 291–308).

Sneha, S., Anitha, B., Sahair, R. A., Raghu, N., Gopenath, T. S., Chandrashekrappa, G. K., & Basalingappa, K. M. (2018). Biofertilizer for crop production and soil fertility. *Academia Journal of Agricultural Research, 6*, 299–306.

Somvanshi, P. S., Pandiaraj, T., & Singh, R. P. (2020). An unexplored story of successful green revolution of India and steps towards ever green revolution. *Journal of Pharmacognosy and Phytochemistry, 9*, 1270–1273. Available online at https://www.phytojournal.com/archives/2020/vol9issue1/PartU/9-1-256-412.pdf

Soytong, K., Kanokmadhakul, S., Kukongviriyapa, V., & Isobe, M. (2001). Application of *Chaetomium species* (Ketomium®) as a new broad spectrum biological fungicide for plant disease control: A review article. *Fungal Diversity, 7*, 1–15.

Sumita, P., Singh, H. B., Alvina, F., & Adholeya, A. (2006). Potential of Arbuscular mycorrhizae in organic farming system. In M. K. Rai (Ed.), *Handbook of microbial biofertilizers* (pp. 223–239). Food Products Press.

Taylor, M. (2019). Hybrid realities: Making a new Green Revolution for rice in south India. *The Journal of Peasant Studies, 47*, 483–502. https://doi.org/10.1080/03066150.2019.1568246

Tejesvi, M. V., Kini, K. R., Prakash, H. S., Subbiah, V., & Shetty, H. S. (2007). Genetic diversity and antifungal activity of species of *Pestalotiopsis* isolated as endophytes from medicinal plants. *Fungal Diversity, 24*, 37–54.

Thakur, M. (2020). Fungi as a biological tool for sustainable agriculture. In A. Yadav, S. Mishra, D. Kour, N. Yadav, & A. Kumar (Eds.), *Agriculturally important fungi for sustainable agriculture* (Fungal Biology). Springer. https://doi.org/10.1007/978-3-030-45971-0_11

Thompson, L. R., Sanders, J. G., McDonald, D., et al. (2017). A communal catalogue reveals Earth's multiscale microbial diversity. *Nature, 551*, 457–463.

Tiwari, P., Adholeya, A., Prakash, A., & Arora, D.K. (2004). Commercialization of arbuscular mycorrhizal biofertilizers. *Fungal biotechnology in agricultural, food, and environmental applications.21*,195–203.

Tognon, G. B., Sanmartín, C., Alcolea, V., Cuquel, F. L., & Goicoechea, N. (2016). Mycorrhizal inoculation and/or selenium application affect post-harvest performance of snapdragon flowers. *Plant Growth Regulation, 78*, 389–400.

Tohidi, M. H., Heydarian, A., Donath, T., & Sohrabi, M. (2018). Study of effect of arbuscular mycorrhiza (*Glomus intraradices*) fungus on wheat under nickel stress. *Agronomy Research, 16*, 1660–1667.

Tomer, A., Singh, R., Singh, S. K., Dwivedi, S. A., Reddy, C. U., Keloth, M. R. A., & Rachel, R. (2021). Role of fungi in bioremediation and environmental sustainability. *Mycoremediation and Environmental Sustainability*. https://doi.org/10.1007/978-3-030-54422-5_8,187-200

Tosi, M., Mitter, E. K., Gaiero, J., & Dunfield, K. (2020). It takes three to tango: The importance of microbes, host plant, and soil management to elucidate manipulation strategies for the plant microbiome. *Canadian Journal of Microbiology, 66*(7), 413–433.

Tringe, S. G., von Mering, C., Kobayashi, A., Salamov, A. A., Chen, K., Chang, H. W., Podar, M., Short, J. M., Mathur, E. J., Detter, J. C., Bork, P., Hugenholtz, P., & Rubin, E. M. (2005). Comparative metagenomics of microbial communities. *Science, 308*, 554–557.

Tripathi, S., Mishra, S. K., & Varma, A. (2017). *Mycorrhizal fungi as control agents against plant pathogens. Ecorestoration* (pp. 161–178). *Springer*.

Vandenkoornhuyse, P., Baldauf, S. L., Leyval, C., Straczek, J., & Young, J. P. W. (2002). Extensive fungal diversity in plant roots. *Science, 295*, 2051.

Vergara, C., Araujo, K., Emanuelle, C. S., Sônia, R., Nivaldo, S. J., Orivaldo, J. S., Marcus, V. L., & Zilli, J. É. (2019). Plant-mycorrhizal fungi interaction and response to inoculation with different growth promoting fungi. *Pesquisa Agropecuária Brasileira, 54*, e25140.

Verma, J. P., Jaiswal, D. K., & Sagar, R. (2014). Pesticide Relevance and their microbial degradation: A-state-of-art. *Reviews in Environmental Science and Bio/Technology, 13*, 429–466.

Verma, J. P., Jaiswal, D. K., & Maurya, P. K. (2016a). Screening of bacterial strains for developing effective pesticide tolerant plant growth-promoting microbial consortia from rhizosphere soils of vegetable fields of eastern Uttar Pradesh. *Energy, Ecology and Environment, 1*(6), 408–418.

Verma, P., Yadav, A. N., Khannam, K. S., Kumar, S., Saxena, A. K., & Suman, A. (2016). Molecular diversity and multifarious plant growth promoting attributes of *Bacilli* associated with wheat (*Triticum aestivum* L.) rhizosphere from six diverse agro-ecological zones of India. *Journal of Basic Microbiology, 56*, 44–58.

Verma, P., Yadav, A. N., Kumar, V., Singh, D. P., & Saxena, A. K. (2017). Beneficial plant-microbes interactions: Biodiversity of microbes from diverse extreme environments and its impact for crop improvement. In *Microbial interactions and agro-ecological impacts, vol 2* (pp. 543–580). Springer. https://doi.org/10.1007/978-981-10-6593-4_22

Verma, P., Yadav, A. N., Khannam, K. S., Mishra, S., Kumar, S., Saxena, A. K., & Suman, A. (2019). Appraisal of diversity and functional attributes of thermotolerant wheat associated bacteria from the peninsular zone of India. *Saudi Journal of Biological Sciences, 26*, 1882–1895. https://doi.org/10.1016/j.sjbs.2016.01.042

Vessey, J. K. (2003). Plant growth promoting rhizobacteria as biofertilizers. *Plant and Soil, 255*, 571–586.

Viterbo, A., Inbar, J., Hadar, Y., & Chet, I. (2007). Plant disease biocontrol and induced resistance via fungal mycoparasites. In C. P. Kubicek & I. S. Druzhinina (Eds.), *Environmental and microbial relationships, The Mycota IV* (2nd ed., pp. 127–146). Springer.

Vitti, A., La Monaca, E., Sofo, A., Scopa, A., Cuypers, A., & Nuzzaci, M. (2015). Beneficial effects of *Trichoderma harzianum* T-22 in tomato seedlings infected by *Cucumber mosaic virus* (CMV). *BioControl, 60*, 135–147. https://doi.org/10.1007/s10526-014-9626-3

Wakelin, S. A., Werren, P. R., & Ryder, H. M. (2004). Phosphate solubilization by Penicillium spp. closely associated with wheat root. *Biology and Fertility of Soils, 40*, 36–43.

Waller, F., Achatz, B., Baltruschat, H., Fodor, J., Becker, K., Fischer, M., Heier, T., Huckelhoven, R., Neumann, C., Von Wettstein, D., Franken, P., & Kogel, K. H. (2005). The endophytic fungus *Piriformospora indica* reprograms barley to saltstress tolerance, disease resistance, and higher yield. *Proceedings of the National Academy of Sciences of the United States of America, 102*, 13386–13391.

Wani, S. P., & Lee, K. K. (2002). Biofertilizers for sustaining cereal crops production. In S. Kannaiyan (Ed.), *Biotechnology of biofertilizers* (pp. 50–64). Narosa Publishing House.

Whitelaw, M. A., Hardena, T. J., & Helyar, K. R. (1999). Phosphate solubilisation in solution culture by the soil fungus *Penicillium radicum*. *Soil Biology and Biochemistry, 3*, 655–665.

Xu, H., Shao, H., & Lu, Y. (2019). Arbuscular mycorrhiza fungi and related soil microbial activity drive carbon mineralization in the maize rhizosphere. *Ecotoxicology and Environmental Safety, 182*, 109476.
Yadav, A. N., Sachan, S. G., Verma, P., & Saxena, A. K. (2015). Prospecting cold deserts of north western Himalayas for microbial diversity and plant growth promoting attributes. *Journal of Bioscience and Bioengineering, 119*, 683–693.
Yadav, A. N., Kumar, R., Kumar, S., Kumar, V., Sugitha, T., Singh, B., Chauhan, V., Dhaliwal, H. S., & Saxena, A. K. (2017). Beneficial microbiomes: Biodiversity and potential biotechnological applications for sustainable agriculture and human health. *Journal of Applied Biology & Biotechnology, 5*, 45–57.
Yadav, A. N., Kumar, V., Prasad, R., Saxena, A. K., & Dhaliwal, H. S. (2018a). Microbiome in crops: Diversity, distribution and potential role in crops improvements. In R. Prasad, S. S. Gill, & N. Tuteja (Eds.), *Crop improvement through microbial biotechnology* (pp. 305–332). Elsevier.
Yadav, A. N., Verma, P., Kumar, V., Sangwan, P., Mishra, S., Panjiar, N., Gupta, V. K., & Saxena, A. K. (2018b). Biodiversity of the genus penicillium in different habitats. In V. K. Gupta & S. Rodriguez-Couto (Eds.), *New and future developments in microbial biotechnology and bioengineering. Penicillium system properties and applications* (pp. 3–18). Elsevier. https://doi.org/10.1016/B978-0-444-63501-3.00001-6
Yadav, A. N., Singh, S., Mishra, S., & Gupta, A. (2019). *Recent advancement in white biotechnology through Fungi. Volume 3, perspective for sustainable environments*. Springer.
Yadav, A. N., Rastegari, A. A., Yadav, N., & Kour, D. (2020a). *Advances in plant microbiome and sustainable agriculture: Diversity and biotechnological applications*. Springer.
Yadav, A. N., Singh, J., Rastegari, A. A., & Yadav, N. (2020c). *Plant microbiomes for sustainable agriculture*. Springer.
Yeasmin, T., Zaman, P., Rahman, A., Absar, N., & Khanum, N. S. (2007). Arbuscular mycorrhizal fungus inoculum production in rice plants. *African Journal of Agricultural Research, 2*, 463–467.
Youssef, M. A., & Eissa, F. M. (2014). Biofertilizers and their role in management of plant parasitic nematodes: A review. *Biotechnology and Pharmaceutical Research, 5*, 1–6.
Zarafi, A. B., & Dauda, W. P. (2019). Exploring the importance of fungi in agricultural biotechnology. *International Journal of Agricultural Sciences and Veterinary, 7*, 1–12.
Zeilinger-Migsich, S., & Mukherjee, P. K. (2014). Fungus-fungus interactions. *The Open Mycology Journal, 8*, 27.
Zhang, W., Mullaney, E. J., & Lei, X. G. (2007). Adopting selected hydrogen bonding and ionic interactions from *Aspergillus fumigatus* phytase structure improves the thermostability of *Aspergillus niger* PhyA phytase. *Applied and Environmental Microbiology, 73*, 3069–3076.
Zhang, F., Huo, Y., Cobb, B., Luo, G., Zhou, J., Yang, G., Wilson, W. T., & Zhang, Y. (2018). *Trichoderma* biofertilizer links to altered soil chemistry, altered microbial communities and improved grassland biomass. *Frontiers in Microbiology, 9*, 848.
Zhao, Z. W., Wang, G. H., & Yang, L. (2003). Biodiversity of arbuscular mycorrhizal fungi in tropical rainforests of *Xishuangbanna*, southwest China. *Fungal Diversity, 13*, 233–242.
Zheng, W., Zeng, S., Bais, H., LaManna, J. M., Hussey, D. S., Jacobson, D. L., & Jin,Y. (2018). Plant growth promoting rhizobacteria (PGPR) reduce evaporation and increase soil water retention. *Water Resources Research, 54(5),3673–3687*
Zhu, G. S., Yu, Z. N., Gui, Y., & Liu, Z. Y. (2008). A novel technique for isolating orchid mycorrhizal fungi. *Fungal Diversity, 33*, 123–137.
Žifčáková, L., Vetrovský, T., Howe, A., & Baldrian, P. (2016). Microbial activity in forest soil reflects the changes in ecosystem properties between summer and winter. *Environmental Microbiology, 18*, 288–301.
Zilber-Rosenberg, I., & Rosenberg, E. (2008). Role of microorganisms in the evolution of animals and plants: The hologenome theory of evolution. *FEMS Microbiology Reviews, 32*, 723–735.

Chapter 15
Efficacy of Plant Products as Biofungicides for Postharvest Decay of Root, Tuber, and Bulb Crops: An Opportunity for Bioentrepreneurship

Ashirbad Mohapatra, Akhtari Khatoon, and Kunja Bihari Satapathy

Contents

15.1 Introduction

There is a worldwide swing to the use of eco-friendly methods for protecting the crops from pests and diseases as the use of potentially harmful chemicals is viewed with several disadvantages. Plants have been known for their medicinal and antimicrobial properties since ancient times, thereby indicating that the plant kingdom is a vast storehouse of chemicals that can check many plant pathogens. Recently, in different parts of the world, attention has been paid toward exploitation of plants as novel chemotherapeutants in the plant protection in view of their long-term effect on crop disease management, low cost, and safety to ecosystem. Biofungicide

A. Mohapatra · A. Khatoon
Department of Botany, Sri Jayadev College of Education and Technology, Bhubaneswar, Odisha, India

K. B. Satapathy (✉)
Department of Botany, School of Applied Sciences, Centurion University of Technology and Management, Bhubaneswar, Odisha, India

A. C. Shukla (ed.), *Applied Mycology*, Fungal Biology,
https://doi.org/10.1007/978-3-030-90649-8_15

means fungicides of biological origin which may be microorganism such as bacteria, fungi, and animal or plant-based product like secondary metabolite.

Vegetables are an important source of human nutrition and are a rich and comparatively cheaper source of vitamins and minerals. They are important for the maintenance of good health and are also beneficial in protecting against some degenerative diseases. The vegetables grouped under root, tuber, and bulb crops are most important as compared to other vegetables being nonperishable in nature. After harvest, they can be stored and consumed throughout the year. Some important vegetables included in this group are potato (*Solanum tuberosum* L.), sweet potato [*Ipomoea batatas* (L.) Lam.], radish (*Raphanus sativus* L.), carrot (*Daucus carota* L.), beet (*Beta vulgaris* L.), colocasia [*Colocasia esculenta* (L.) Schott.], yam (*Dioscorea alata* L.), Elephant's foot yam [*Amorphophallus paeoniifolius* (Dennst.) Nicolson], onion (*Allium cepa* L.), and garlic (*Allium sativus* L.). After harvest, the produce is stored in locally prepared bins, in open store houses, or in bamboo baskets for their consumption and marketing throughout the year.

The issue of postharvest disease management of these crops has become an important concern due to inadequate agricultural storage practices for produce preservation from microbes-induced spoilages (Kana et al., 2012). These crops are drought tolerant and provide a wide harvesting window which makes it act as a famine reserve. However, postharvest deteriorations caused by microbial invasion of the tubers are the most important causes of loss in its production and contribute hugely to the unsuccessful long-term storage of the root tubers. The pathogenic organisms including fungi gain entry into tubers probably through the area where the tubers are separated from stems at harvest or the root tips which break during harvesting, or other natural openings and cracks on the tuber's surfaces sustained during harvesting, transit, or storage (Okigbo, 2009). Improper handling leaves off bruises on the produce during harvesting, transit, or storage which may become contaminated from adherent soil or air-borne propagules which later colonies such abrasions to cause rots (Okigbo & Nmeka, 2005). In Odisha, the postharvest decay of these vegetables during storage poses serious problems and sustains heavy losses particularly for the warm and humid climate. According to various workers, rotting usually starts from the field and progresses in storage (Okigbo & Ikediugwu, 2000). Unhealthy store conditions lead to the absorption of moisture by produce in storage as a result of defects in the storage facility, thus encouraging the development of hotspots and molds.

Antimicrobial chemicals such as benzimidazoles, aromatic hydrocarbons, and sterol biosynthesis inhibitors are often used in the control of plant diseases in agriculture. However, there is a series of problems against the effective use of these chemicals in areas where the fungi have developed resistance (Brent & Hollomon, 1998). In order to overcome this problem, higher concentrations of the chemicals had been used, but this increased the risk of high-level toxic residues in the products which were not beneficial for human beings and the environment. Thus, there has been a growing interest on the research of possible use of plant extracts for control of pest and diseases in agriculture which was less harmful to human health and the environment (Costa et al., 2000; Duru et al., 2003).

To avoid use of synthetic fungicides, it is imperative to find alternative sources from naturally occurring compounds that are easily biodegradable and of low mammalian toxicity for a safe control of fungal pathogens (Adongo et al., 2012). Plants are the source of natural pesticides that make excellent leads for new pesticide developments (Shanmugavalli et al., 2009). Recent studies on the use of plant extract have opened a new approach to control plant diseases. These plant extracts have been reported to be safe, nontoxic to man, but effective against plant pathogens (Shivpuri et al., 1997). Plant diversity serves humankind as renewable natural resources for a variety of biologically active chemicals. Active constituents of the medicinal and aromatic plants have been found to be less phytotoxic, more systemic, and easily biodegradable (Fawcett & Spencer, 1970). Plants are the sources of natural pesticides that make excellent leads for new pesticide development (Gangadevi et al., 2008; Satish et al., 2008; Shanmugavalli et al., 2009; Swarna Latha & Neelakanta Reddy, 2009).

In India, Dastur (1916) first studied the control measure for ripe rot of plantains. After a long period of gap, studies in this field gained rapid momentum, and within two decades, much knowledge was gained on fungi causing postharvest rot of fruits and vegetables from several parts of the country, particularly from Pune, Allahabad, Lucknow, Delhi, Kolkata, Bhagalpur, and Bangalore. Unfortunately, little work has been done so far in this field in Odisha (Das, 2012). In continuation of the previous work, the present topic was selected in view of the limited studies on postharvest fungal rots of root, tuber, and bulb vegetables. The present article is, therefore, based upon the postharvest decay of root, tuber, and bulb vegetables and their control by biofungicides. It is expected that the present investigations shall throw some light on the etiology and control of postharvest storage decay of vegetables such as potato, sweet potato, colocasia, yam, amorphophallus, beet, carrot, radish, onion, and garlic caused by various species of fungi and their management by application of botanicals to avoid serious economic losses in the markets and store houses.

15.2 Methodology

15.2.1 Pathological Studies

(i) *Collection of Rotten Samples of Root, Tubers, and Bulb Crops*

A total of 1000 samples of rotten potatoes, sweet potatoes, radish, carrot, beet, colocasia, yam, amorphophallus, onion, and garlic were collected from various markets and store houses of Odisha during 2014–2016. Rotten plant samples were collected in separate polythene bags with labels in respect of (1) date of collection, (2) place of collection, (3) extent of damage in sample lots, (4) viable symptoms (smell, color, and mycelial growth), and (5) name of collector. All the samples were brought to the laboratory and surface sterilized with 0.1% mercuric chloride solution for 2–3 min. Some specimens from each sample were examined for the presence of fungal infections by taking a small portion of the rotten tissues on a clean glass slides, macerated with lactophenol-cotton blue, and examined under a microscope for the occurrence of fungal mycelium and fruiting bodies, if any. Specimens showing the signs of fungal infections were immediately processed for phytopathological analysis, after recording the visual symptoms of rotting.

(ii) *Isolation of Fungi from Rotten Samples*

Sterile moist chambers were prepared out of Petri dishes, 100 mm in diameter, by placing triple-layer moist filter paper over which a platform was prepared by placing two pieces of clean glass slides over a pair of equal-sized glass rods. It was sterilized at 15 PSI for 20 min in an autoclave.

Small pieces of tissues including both rotten and healthy areas were surface sterilized with 0.1% mercuric chloride solution for 2–3 min, washed thoroughly three times with sterile distilled water, and transferred aseptically into the sterile moist chamber at the rate of 5 pieces per chamber. Similar sterilized tissues were also transferred into petriplates containing potato dextrose agar (PDA) at the rate of 5 pieces per petri plate.

All the materials were incubated at room temperature (28 °C ± 1 °C) for four to five days. At the end of incubation, the sterile chambers as well as the petriplates were examined for the presence of fungi. Fungi of all kinds appearing on the incubated materials were transferred aseptically into PDA slants. Each fungus as obtained in PDA slants was subsequently purified by single spore/hyphal tip isolation technique and maintained by periodic subculturing in PDA slants for further studies. The frequency of incidence of each fungus on different samples of rotten samples was estimated by the formula:

$$\text{Frequency of incidence} = X / Y \times 100$$

Where X = the number of pieces showing the presence of different fungi and
Y = total number of pieces investigated. The frequency of incidence was expressed in terms of percentage.

(iii) *Identification of Fungi*

The colony characteristics of each fungal isolates on PDA were recorded. The details of morphology of each isolate were studied under varying magnifications (×100 to ×1000) of a microscope. The characteristics of mycelium and fructifications were recorded. The dimensions of each structure were measured with the help of an Ocular micrometer. Each fungal isolate was taxonomically identified with the help of available literatures and confirmed by the microbiology Department, OUAT, Bhubaneswar, and Odisha. The fungal isolate which was found to be new to their molecular identification was carried out by sequencing of 26S rRNA gene followed by submission of sequence to NCBI GenBank. Genomic DNA was extracted from these fungal isolates (Sambrook et al., 1989) followed by amplification of 26S rRNA gene using universal primers DF (5′-ACCCGCTGAACTTAAGC-3′) and DR (5′-GGTCCGTGTTTCAAGACGG-3′) and sequencing at Xcelris Genomics, India.

Raw sequences were analyzed by Bio-Edit software (v7.0.5.3) and identification of the isolates by BLASTN in NCBI database (www.ncbi.nlm.nih.gov/nucleotide). The phylogenetic tree was constructed using the UPGMA distance algorithm (Wheeler & Kececioglu, 2007) using Mega software v5.1. The resultant tree topologies were evaluated by boot-strap analysis of muscle datasets based on 500 resamplings. The final sequences were submitted to NCBI GenBank to obtain the accession number of each isolate.

15.2.2 In Vitro *Evaluation of the Antifungal Efficacy of Plant Extracts*

(i) *Selection of the Test Fungal Isolates*

A total of ten fungal species were found to be associated with postharvest decay of potato, sweet potato, radish, carrot, beet, taro, yam, amorphophallus, onion, and garlic during the present investigations. All the fungal species were purified, identified, and maintained in potato dextrose agar slants by periodically subculturing to use in the present studies.

(ii) *Selection of Plants and Plant Parts*

On the basis of local availability and medicinal values, the leaves of the 16 plant species such as *Abutilon indicum* (L.) Sweet (Malvaceae), *Ageratum conyzoides* L. (Asteraceae), *Alstonia scholaris* (L.) R.Br. (Apocynaceae), *Artocarpus heterophyllus* Lam. (Moraceae), *Averrhoa carambola* L. (Averrhoaceae), *Cassia fistula* L. (Caesalpiniaceae), *Centella asiatica* Urban. (Apiaceae), *Dillenia indica* L. (Dilleniaceae), *Eucalyptus globulus* Labill (Myrtaceae), *Haldinia cordifolia* (Roxb.) Rids (Rubiaceae), *Justicia adhatoda* L. (Acanthaceae), *Lawsoniainermis* L. (Lythraceae), *Murraya paniculata* Jack. (Rutaceae), *Pithecellobium dulce* (Roxb.) Benth. (Mimosaceae), *Pongamia pinnata* (L.) Panigrahi (Fabaceae), and *Tamarindus indica* L. (Caesalpiniaceae) were selected for the present investigations. Some of these plants have been reported to exhibit antifungal properties.

(iii) *Collection, Processing, and Extraction of Plant Material*

Leaves of 16 angiosperm plants belonging to diverse plant families were collected from different parts of Odisha. Each material was collected in separate polythene bags and brought to the laboratory of Department of Botany, S.J. College of Education and Technology, Bhubaneswar, Odisha (India), for further studies.

Three solvent systems were used for leaf compound extraction, viz., aqueous, petroleum ether, and methanol. For the preparation of aqueous extract (cold water extract), fresh plant leaves were washed thoroughly with sterile water, chopped into pieces, and the extracts were obtained by macerating with 100 ml of sterile distilled water for each 20 g of fresh leaf material. The extracts thus obtained were separately filtered through a three-fold sterilized muslin cloth into sterilized conical flasks.

Then the extracts were refiltered with Whatman No 1 filter paper using a power-operated suction filter pump. These extracts formed the standard plant extract solution (20%) and were stored at 5 °C in the refrigerator for further use. For dried leaf extraction, reflux method of solvent extraction was used for purification of different organic constituent. Solvent series for successive separation was as follows: petroleum ether and methanol. First, the soxhelt extraction unit was cleaned and rinsed with petroleum ether and then 30 g of air-dried and coarse-powdered plant material were kept in it. Then, this unit was kept on water bath and exhaustively extracted with petroleum ether for 30–36 h. During extraction, the round bottom flask (RBF) which contains the extracts should be with porcelain chips for well siphonation. To confirm whether the extraction is complete or not, the extract from siphonating tube of soxhelt was taken in a watch glass. After petroleum ether extraction, the extracted plant material was then air dried, repacked in soxhelt apparatus, extracted with methanol for such time hour, and maintained again. The crude extract and the fraction obtained at every step were filtered and distilled to evaporate the solvent from the extract. The liquid extracts were concentrated separately by using a rotary vacuume evaporator, and the resulting dried extract was preserved in desiccators until further use. The diluted plant extracts were prepared by dissolving purified petroleum ether and methanolic extract in dimethyl sulfoxide (DMSO) @ 20 mg plant extract in 1 ml DMSO.

(iv) *Evaluation of Antifungal Efficacy of Plant Extracts*

For the evaluation of antifungal effect of plant extracts on inhibition of mycelia growth of ten isolated fungal species, namely, *Apergillus flavus*, *Apergillus niger*, *Fusarium oxysporum*, *Geotrichum candidum*, *Junghuhnia* sp. AK15 (NCBI Gene Bank accession number KT 946990), *Mucor* sp. BOT15 (NCBI Gene Bank accession number KT 946994), *Neoscytallidium* sp. AM15 (NCBI Gene Bank accession number- KT 946992), *Penicillium* sp. KBS15 (NCBI Gene Bank accession number KT 946991), *Penicillium* sp. UU15 (NCBI Gene Bank accession number KT 946993), and *Rhizopus oryzae,* poison food technique was adopted. For this purpose, Potato Dextrose Agar (PDA) medium was prepared and sterilized at 15 PSI for 20 min. After sterilization, 1 ml of each diluted plant extract solution (20 mg extract/1 ml DMSO) was mixed with 19 ml of PDA (Potato Dextrose Agar) and poured into each petriplate @ 1 mg/ml, mixed thoroughly, and allowed to solidify. PDA medium without the plant extracts but substituted with 1 ml distilled water was served as control.

The mycelial disc of 5 mm in diameter was taken from the periphery of 2- to 3-day-old colonies of the desired fungi with the help of a sterilized cork-borer. These discs were transferred aseptically inside the laminar flow with the provision of ultraviolet light and inoculated at the center of each petriplate containing PDA with plant extracts. Three replications were maintained for each fungus × plant extract combination. A separate set of petri plates containing only PDA was inoculated by each test fungus to serve as control. All the test petriplates were incubated at room temperature (28 ± 1 °C) and examined at regular intervals. Growth of the isolates was considered as optimum for recording the observations when it reached around 8 cm in any of the control sets. *Rhizopus oryzae* was incubated for 1 day, *Neoscytallidium* sp. AM15 for 3 days, *Junghuhnia* sp. AK15 and *Mucor* sp. BOT15 for 5 days, *Aspergillus flavus* for 7 days, *Aspergillus niger* for 8 days, *Penicillium* sp. UU15 and *Fusarium oxysporum* for 10 days *Geotrichum candidum* for 11 days, and *Penicillium* sp. KBS15 for 15 days depending upon their complete growth (8 cm) on the petriplate. At the end of incubation period, the diameter of the colony in each petriplate was measured in two planes, one at the right angle to the other. Further, the data of each test plant extract was compared with that of the control in order to estimate the inhibition (positive) or stimulation (negative) effects of the plant extract tested. The formula used for this purpose was:

$$\%\text{inhibition} / \text{Stimulation} = X - Y / X \times 100$$

Where X = colony diameter in control and Y = colony diameter in treated plated.

For the sake of convenience of description of the results, it was classified by the method of Rath and Mohanthy (1978) with slight modifications as given below:

% of inhibition	Rating
0 or negative	No inhibition
1–10	Incipient inhibition
11–30	Mild inhibition
31–70	Moderate inhibition
71–100	Severe inhibition

While screening the data on the basis of the above-noted scale, it was noted that most of the effective inhibition of the colony diameter was in the range of 31–70%. Only in a few instances, it exceeded 70%.

(v) *Comparison of Plant Extracts and Estimation of Inhibition of Mycelial Growth by Selected Plant Extracts*

For this purpose, three plants were selected that show the mean maximum inhibition of mycelial growth of ten test fungal species, namely, *Aspergillus flavus*, *Aspergillus niger*, *Fusarium oxysporum*, *Geotrichum candidum*, *Junghuhnia* sp. AK15, *Mucor* sp. BOT15, *Neoscytallidium* sp. AM15, *Penicillium* sp. KBS15, *Penicillium* sp. UU15, and *Rhizopus oryzae*. The selected plant species were *Ageratum conyzoides*, *Alstonia scholaris,* and *Averrhoa carambola*.

(vi) *Comparison of the Efficacy of Synthetic Fungicides and Selected Plant Extracts*

A separate experiment was conducted in order to compare the relative efficacy of four synthetic fungicides, viz., Dhanustin, Blitox-50, Indofil-M-45, and Mancozeb, with the petroleum ether and methanolic leaf extracts of *Ageratum conyzoides*, *Alstonia scholaris,* and *Averrhoa carambola*. The details of concentrations are given in Table 15.1:

Table 15.1 Comparison of plants and fungicides used along with their concentration

Treatment	Solvent system	Concentration mg/ml	Chemical constituent
Blitox-50		0.1	Copper oxychloride
Dhanustin		0.1	
Indofil-M-45		0.1	Carbamate
Mancozeb		0.1	Coordination product of zinc and magnesium ethylene bis dithiocarbamate
Ageratum conyzoides	*Aqueous*	4	Natural plant product
	Pet-ether	1	
	Methanol	1	
Alstonia scholaris	*Aqueous*	4	Natural plant product
	Pet-ether	1	
	Methanol	1	
Averrhoa carambola	*Aqueous*	4	Natural plant product
	Pet-ether	1	
	Methanol	1	
Control	–	–	No fungicide or plant product

Incorporation of the test fungicides/natural plant extracts was carried out as per the method described earlier. The required quantity of fungicides and plant extracts was incorporated after sterilization of the medium. The petriplates containing PDA and fungicides/plant extracts were inoculated by 10 test fungal isolates in three replicates and incubated for 1st to 15th days depending upon their respective complete growth as described earlier. At the end of incubation period, the colony diameter of each fungus-fungicide/plant extract combination was measured and transferred to percentage of mycelial inhibition as per the methods described earlier.

(vii) *Formulation of Herbal Fungicides and Recommendation for Use*

A combination of petroleum ether leaf extracts of *Ageratum conyzoides* (10 mg) and *Alstonia scholaris* (10 mg); *Ageratum conyzoides* (10 mg) and *Averrhoa carambola* (10 mg); and *Alstonia scholaris* (10 mg) and *Averrhoa carambola* (10 mg) was tested against ten fungi tested. The concentration of the formulated combination was 20 mg/ml DMSO (10 mg extract of each plant).

15.3 Results and Discussion

The data on the incidence and taxonomic identity of fungi in root, tuber, and bulb crops revealed that eight genera comprising of ten species of fungi were responsible for storage decay of ten root, tuber, and bulb vegetables collected from different market places and storehouses of Odisha in frequencies ranging from 9.74% to 50.3% of the cases studied. *Aspergillus*, *Geotrichum*, *Penicillium,* and *Rhizopus* were found to be the dominant fungi causing maximum loss of those vegetables during storage due to rotting. It was also noted that these fungi are generally the soil inhabitants which come in contact with the tubers and fleshy roots of these vegetables during the crop stand as surface microflora. After harvest and during transport, transit, and storage, they infected the tubers and roots through wounds arising out of harvest, transportation, and storage handlings or by injuries caused by insects and other microorganisms when the environmental factors are favorable for their growth and disease development, incurring heavy economic losses to both growers and consumers. Experiments conducted on the pathogenicity tests indicated that fungi isolated from rotten samples of potato, colocasia, amorphophallus, sweet potato, yam, carrot, radish, beet, onion, and garlic were pathogenic to their respective hosts in an artificial inoculation test.

The data on the pathogenicity test of six fungi on potato tubers indicated that *Aspergillus niger* was the most pathogenic one and caused the complete rotting of tuber within 10 days of incubation. The pathogenicity test data of colocasia rotting caused by five fungi indicated that *Rhizopus oryzae* was the most pathogenic to corms than other isolates, and complete rotting of the corm within 10 days of inoculation was recorded. Amorphophallus rotting was caused by three fungi, and the most virulent was *Geotrichum candidum* which caused the rotting within 30 days of inoculation. Artificial rotting of this host took a long time. Among the two fungi, *Rhizopus oryzae* was the most virulent one causing the complete rotting of sweet potato within 15 days. Similarly, among the five fungi causing storage rot in yam, *Neoscytallidium* sp. AM15 was the most virulent to cause complete rotting within 20 days. Rotting in carrot, radish, and beet root was caused by three fungi, and among them, the most virulent fungus for carrot was *Rhizopus oryzae,* while for radish and beet was *Aspergillus niger,* but both the fungi cause complete rotting of their respective host within 10 days of inoculation. Rotting of onion and garlic was found to be caused by two and three fungi, respectively. Among them, *Aspergillus niger,* for onion, and *Penicillium* sp. UU15, for garlic, were found to be more virulent than others. They cause the complete rotting of the respective host within 15 days of inoculation.

Studies on the inhibition of ten fungal species causing storage rot of root, tuber, and bulb vegetables, by aqueous, petroleum ether, and methanolic leaf extract of sixteen angiospermic plants revealed that petroleum ether and methanolic leaf extracts were more efficient than aqueous extracts (Tables 15.2, 15.3, and 15.4 and Figs. 15.1, 15.2, and 15.3). Petroleum ether and methanolic extracts of all the sixteen test plants were severe to moderately efficient against the ten fungi tested, while the aqueous leaf extract was mild to incipiently efficient to inhibit the mycelial growth. Out of the sixteen plants tested, extracts of *Ageratum conyzoides*, *Alstonia scholaris,* and *Averrhoa carambola* were comparatively more efficient than extracts of other plants tested (Table 15.5 and Fig. 15.4).

Comparative efficacy of plant extracts (petroleum ether and methanolic extract of *Ageratum conyzoides*, *Alstonia scholaris,* and *Averrhoa carambola*) with four commercial fungicides (Blitox-50, Dhanustin, Indofil, and Mancozeb) revealed that plant extracts were more efficient to inhibit the mycelial growth of ten fungi tested (Table 15.6 and Fig. 15.5). Mishra and Dixit (1976) reported that the leaf extracts of

Table 15.2 Screening of aqueous, petroleum ether, and methanolic extracts of *Ageratum conyzoides* L. against ten test fungi

Test fungi	% of Inhibition of mycelial growth			
	Aqueous extract	Petroleum ether extract	Methanolic extract	Mean
Aspergillus flavus	25.13 ± 0.56	62.90 ± 0.17	56.95 ± 0.68	48.33
Aspergillus niger	21.25 ± 0.73	62.54 ± 0.17	46.51 ± 0.63	43.43
Fusarium oxysporum	15.68 ±0.29	59.80 ± 0.76	57.41 ± 1.05	44.3
Geotrichum candidum	16.42 ± 0.63	89.03 ± 0.56	79.08 ± 0.65	61.51
Junghuhnia sp. AK15	33.12 ± 0.51	98.13 ± 0.38	100	77.08
Mucor sp. BOT15	42.77 ± 0.73	99.9 ± 0.08	88.48 ± 1.19	77.05
Neoscytallidium sp. AM15	20.8 ± 0.80	86.86 ± 0.26	87.87 ± 0.67	65.17
Penicillium sp. KBS15	26.81 ± 0.66	70.29 ± 0.80	58.72 ± 1.04	51.94
Penicillium sp. UU15	14.79 ± 0.40	58.35 ± 0.36	59.5 ± 0.88	44.21
Rhizopus oryzae	11.78 ± 0.35	88.97 ± 1	88.88 ± 0.59	63.21
Mean	22.85	77.68	72.34	

Results expressed as mean ± SEM of three determinations

Table 15.3 Screening of aqueous, petroleum ether, and methanolic extracts of *Alstonia scholaris* (L.) R.Br. against ten test fungi

Test fungi	% of Inhibition of mycelial growth			
	Aqueous extract	Petroleum ether extract	Methanolic extract	Mean
Aspergillus flavus	16.98 ± 0.62	54.26 ± 0.82	45.66 ± 1.44	38.97
Aspergillus niger	10.88 ± 0.48	34.32 ± 0.36	54.72 ± 0.98	33.3
Fusarium oxysporum	5.58 ± 0.54	58.55 ± 0.22	57.33 ± 1.51	40.48
Geotrichum candidum	18.44 ± 1.11	73.41 ±1.87	80.14 ± 1.03	57.33
Junghuhnia sp. AK15	25.46 ± 0.65	100	99.46 ± 0.23	74.97
Mucor sp. BOT15	15.97 ± 0.99	100	74.48 ± 0.83	63.48
Neoscytallidium sp. AM15	8.99 ± 0.53	84.28 ± 0.75	87.25 ± 1.03	60.17
Penicillium sp. KBS15	10.5 ± 0.23	51.66 ± 0.98	59.91 ± 0.53	40.69
Penicillium sp. UU15	14.14 ± 0.65	63.1 ± 1.15	79.04 ± 1.05	52.09
Rhizopus oryzae	21.18 ± 0.93	86.29 ± 0.68	89.14 ± 0.65	65.54
Mean	14.81	70.58	72.71	

Results expressed as mean ± SEM of three determinations

Clematis gouriana were superior to Blitox-50, Dithane M-45, Dithane Z-78, and Ziram, while Renu et al. (1985) reported that essential oil from leaf of *Aegle marmelos* possessed greater fungitoxicity than that of six commercial fungicides. El-Shami et al. (1985) evaluated garlic extracts along with five different fungicides *in vitro* against *Fusarium oxysporum* sp. and found that the garlic extract inhibited spore germination and mycelia growth of the fungus in a similar manner to five different fungicides. Bandara et al. (1989) reported that the petroleum and ethyl acetate

Table 15.4 Screening of aqueous, petroleum ether, and methanolic extracts of *Averrhoa carambola* L. against ten test fungi

Test fungi	% of Inhibition of mycelial growth			
	Aqueous extract	Petroleum ether extract	Methanolic extract	Mean
Aspergillus flavus	17.49 ± 0.38	57.39 ± 0.46	54.57 ± 0.86	43.15
Aspergillus niger	20.66 ± 0.64	27.8 ± 0.12	45.47 ± 0.75	31.31
Fusarium oxysporum	14.44 ± 0.72	73.5 ± 0.62	58.57 ± 0.44	48.83
Geotrichum candidum	14.03 ± 1.22	77.75 ± 0.65	70.46 ± 0.84	54.08
Junghuhnia sp. AK15	19.19 ± 0.45	98.93 ± 0.42	99.66 ± 0.27	72.59
Mucor sp. BOT15	9.3 ± 0.53	84.81 ± 0.53	79.83 ± 0.47	57.98
Neoscytallidium sp. AM15	22.86 ± 0.69	88.93 ± 0.67	90.21 ± 0.74	67.33
Penicillium sp. KBS15	15.93 ± 0.26	75.20 ± 0.73	73.08 ± 0.73	54.74
Penicillium sp. UU15	26.08 ± 0.58	67.38 ± 1.01	67.5 ± 0.7	53.65
Rhizopus oryzae	22.63 ± 0.65	87.94 ± 0.62	89.7 ± 1.56	66.75
Mean	18.26	73.96	72.9	

Results expressed as mean ± SEM of three determinations

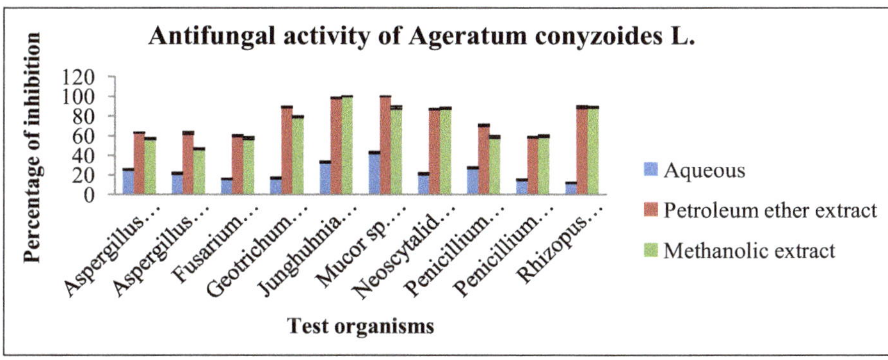

Fig. 15.1 Antifungal activity of *Ageratum conyzoides* L. against ten tested fungi

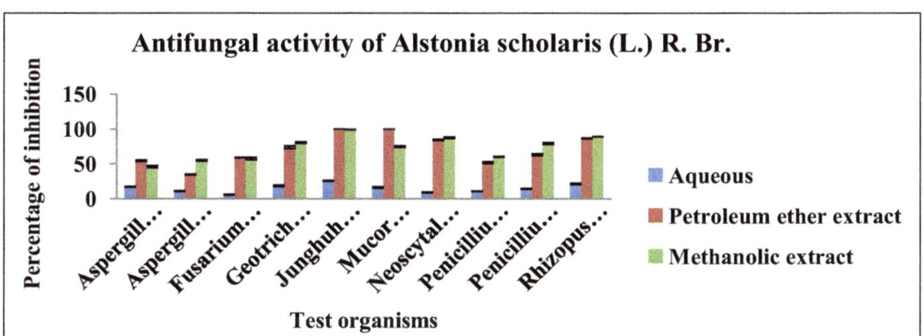

Fig. 15.2 Antifungal activity of *Alstonia scholaris* (L.) R.Br. against ten test fungi

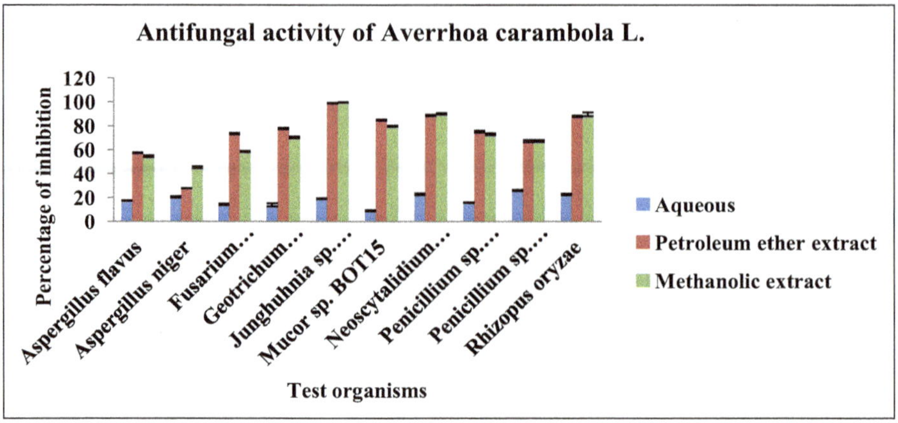

Fig. 15.3 Antifungal activity of *Averrhoa carambola* L. against ten tested fungi

extracts of *Butea monosperma* were fungitoxic to *Cladosporium cladosporioides.* During the present investigations, similar study on comparing the effect of four commercial fungicides with three promising plant extracts could reveal that the activity of a particular plant extract or fungicide effective against a particular test fungus was quite ineffective against others in inhibiting the mycelial growth. For inhibiting the mycelial growth of *Aspergillus flavus*, *Fusarium oxysporum,* and *Penicillium* sp. KBS15, Dhanustin and Mancozeb were found to be superior over the three test plants, but the activity of plant extracts was found to be more efficient than Blitox-50 and Indofil. For growth retardation of *Aspergillus niger*, the plant extracts were found to be more efficient than the test commercial fungicides, except Dhanustin. Against *Geotrichum candidum*, the fungitoxic potential of *Ageratum conyzoides* was found to be equal with Mancozeb, while the effect of the rest of the plant extracts was more than rest of the three commercial fungicides. However, the antifungal activity of plant extracts was found to be superior over fungicides against *Mucor* sp. BOT15. Against *Junghuhnia* sp. AK15, all the test plant's extract and commercial fungicides showed significant inhibitory effect, except Blitox-50. The effect of Dhanustin, which completely inhibited the mycelial growth of *Aspergillus flavus* and *Penicillium* sp. UU15, confirmed its superiority over all the test fungicides and three plant species. For the complete inhibition of mycelial growth in the case of *Geotrichum candidum*, *Junghuhnia* sp. AK15, *Mucor* sp. BOT15, *Neoscytallidium* sp. AM15, *Penicillium* sp. KBS15, *Penicillium* sp. UU15, and *Rhizopus oryzae,* all the three test plant species were significantly superior over all the four commercial fungicides.

Studies on the antifungal properties and potentials of plants and plant products have now been extended in several directions, namely identification of the potential microbiocidal products *in vitro*; their isolation, identification, and characterization in the laboratory; comparison of plant products and extracts with commercial fungicides; and testing the efficacy of plant products in the field, *in vivo*. All these activities put together shall enable us not only locate and use the present herbal

Table 15.5 Comparative efficacy of antifungal activity of sixteen angiospermic plants

Growth inhibition of test fungal species (%)											
Test plants	Af	An	Fo	Gc	Ja	Mb	Na	Pk	Pu	Ro	Mean
Abutilon indicum	37.99 ± 19.24	26.4 ± 8.97	28.5 ± 4.74	43.75 ± 16.8	76.72 ± 28.59	51.86 ± 24.83	58.53 ± 24.42	28.59 ± 14.63	41.62 ± 11.63	52.2 ± 33.95	**44.61**
Ageratum conyzoides	48.33 ± 16.58	43.43 ± 16.99	44.3 ± 20.26	61.51 ± 32.14	77.08 ± 31.09	77.05 ± 24.68	65.17 ± 31.38	51.94 ± 18.38	44.21 ± 20.8	63.21 ± 36.36	**57.62**
Alstonia scholaris	38.97 ± 15.93	33.3 ± 17.91	40.48 ± 24.68	57.33 ± 27.63	74.97 ± 35.01	63.48 ± 35.17	60.17 ± 36.21	40.69 ± 21.61	52.09 ± 27.61	65.54 ± 31.38	**52.7**
Artocarpus heterophyllus	42.53 ± 25.42	31.03 ± 12.2	40.84 ± 17.07	61.77 ± 25.62	73.44 ± 34.5	63.85 ± 32.76	52.54 ± 28.29	41.33 ± 18.05	38.91 ± 23.46	64.42 ± 37.39	**51.06**
Averrhoa carambola	43.15 ± 18.18	31.31 ± 10.42	48.83 ± 25.07	54.08 ± 28.47	72.59 ± 37.76	57.98 ± 34.48	67.33 ± 31.45	54.74 ± 27.45	53.65 ± 19.49	66.75 ± 31.21	**55.04**
Cassia fistula	34.27 ± 20.27	35.14 ± 11.34	41.31 ± 14.25	54.07 ± 28.91	68.98 ± 33.4	63.01 ± 33.39	60.8 ± 36.34	35.94 ± 16.12	46.89 ± 22.98	62.94 ± 32,.26	**50.33**
Centella asiatica	38.56 ± 18.0	27 ± 13.78	22.08 ± 12.46	47.99 ± 19.95	68.76 ± 33.07	50.48 ± 23.06	57.16 ± 24.61	24.37 ± 15.74	40.96 ± 16.1	58.95 ± 24.92	**43.63**
Dillenia indica	34.76 ± 19.98	27.5 ± 18.7	33.62 ± 17.93	51.63 ± 25.85	69.07 ± 34.09	50.39 ± 19.7	63.01 ± 27.96	44.51 ± 22.13	51.51 ± 25.18	60.82 ± 26.06	**48.68**
Eucalyptus globulus	40.95 ± 16.88	37.19 ± 15.11	32.52 ± 19.36	57.6 ± 27.68	72.15 ± 32.4	63.85 ± 28.9	56.06 ± 27.8	39.75 ± 20.06	46.82 ± 22.4	55.63 ± 24.76	**50.25**
Haldina cordifolia	36.71 ± 17.23	32.09 ± 13.78	49.14 ± 23.26	53.07 ± 25.58	70.16 ± 30.99	59.31 ± 35.64	65.75 ± 30.3	46.64 ± 20.42	43.01 ± 24.2	55.2 ± 33.05	**51.1**
Justicia adhatoda	33.55 ± 17.24	31.08 ± 12.08	37.97 ± 23.38	56.01 ± 30.01	70.88 ± 33.16	59.78 ± 29.62	43 ± 25.08	29.95 ± 11.8	35.63 ± 15.95	59.67 ± 31.89	**45.75**
Lawsoniainermis	33.47 ± 21.8	37.68 ± 17.9	28.7 ± 18.71	55.62 ± 26.17	68.69 ± 34.41	52.08 ± 24.06	62.96 ± 32.67	36.06 ± 21.95	48.04 ± 23.28	74.28 ± 36.36	**49.75**
Murraya paniculata	40.49 ± 22.33	26.81 ± 15.82	32.51 ± 12.89	48.84 ± 22.56	66.98 ± 38.56	52.66 ± 26.47	56.41 ± 25.05	27.56 ± 16.92	40.01 ± 22.7	48.68 ± 20.57	**44.09**
Pithecellobium dulce	36.74 ± 17.06	22.13 ± 14.93	27.41 ± 15.72	49.81 ± 26.88	68.93 ± 32.57	39.7 ± 22.15	57.77 ± 29.79	26.96 ± 19.58	45.71 ± 23.17	53.32 ± 24.93	**42.84**

(continued)

Table 15.5 (continued)

Growth inhibition of test fungal species (%)											
Test plants	Af	An	Fo	Gc	Ja	Mb	Na	Pk	Pu	Ro	Mean
Pongamia pinnata	37.26 ± 18.25	35.5 ± 18.5	36.52 ± 15.68	52.18 ± 28.56	74.77 ± 31.27	64.74 ± 28.64	54.92 ± 26.67	49.39 ± 20.67	40.35 ± 22.88	56.95 ± 28.60	**50.25**
Tamarindus indica	19.31 ± 14.32	16.02 ± 11.33	33.06 ± 18.13	54.5 ± 27.95	68.88 ± 36.15	53.51 ± 24.5	52.4 ± 23.92	35.41 ± 19.01	33.32 ± 17.66	65.51 ± 32.01	**43.19**
Mean	**37.31**	**30.85**	**36.11**	**53.73**	**71.44**	**57.73**	**58.37**	**38.36**	**43.92**	**60.25**	

Af Aspergillus flavus, *An Aspergillus niger*, *Fo Fusarium oxysporum*, *Gc Geotrichum candidum*, *Ja Junghuhnia* sp. AK15, *Mb Mucor* sp. BOT15, *Na Neoscytallidium* sp. AM15, *Pk Penicillium* sp. KBS15, *Pu Penicillium* sp. UU15, *Ro Rhizopus oryzae*.

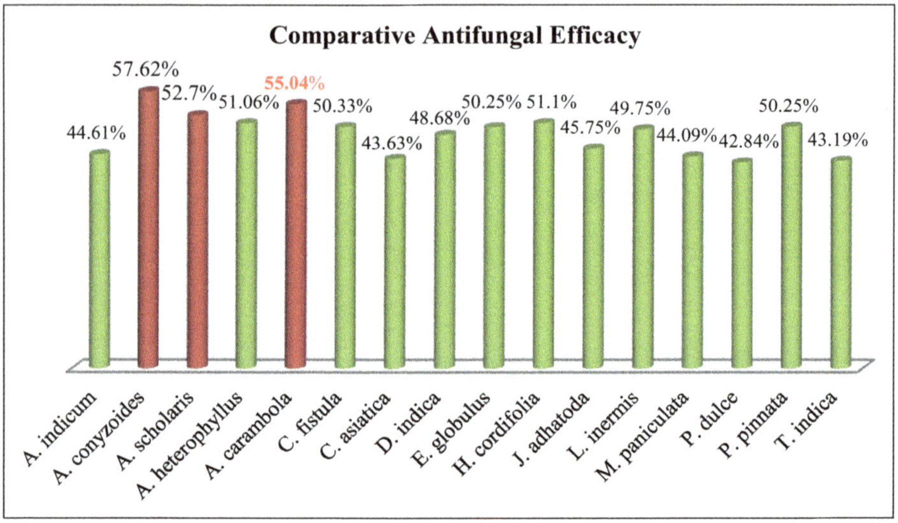

Fig. 15.4 Comparative efficacy of antifungal activity of sixteen medicinal plants

recommendations against postharvest management of root, tuber, and bulb crops, but also the application of biofungicides would certainly be an endeavor for sustainable agricultural practices. As the plant kingdom is a rich reservoir of useful materials, the future of plant products in plant diseases control seems to be very bright.

It is suggested that the extracts of effective experimental plants of the present study may be utilized as postharvest dips for the control of storage decay of root, tuber, and bulb crops in order to avoid the side effects caused due to use of commercial fungicides. The findings of the present investigation will be of immense help to the agrochemical industries to prepare and biofungicides for sustainable agricultural practices.

Table 15.6 Comparative efficacy of three selected plants and four commercial fungicides on mycelia inhibition of ten fungal species

Percentage of inhibition of mycelial growth								
Test fungi	Ag	As	Av	Blitox-50	Dhanustin (0.1 mg/ml)	Indofil (0.1 mg/ml)	Mancozeb (0.1 mg/ml)	Mean (0.1 mg/ml)
A. flavus	48.33	38.97	43.15	21.74	100	19.63	82.5	**54.99**
A. niger	43.43	33.3	31.31	17.66	81.41	22.75	18.11	**39.10**
F. oxysporum	44.3	40.48	48.83	0	86.7	29.33	79.53	**62.81**
G. candidum	61.51	57.33	54.08	17.09	31.41	31.41	61.05	**53.12**
***J.* sp. AK15**	77.08	74.97	72.59	17.79	76.24	72.5	80.43	**78.024**
***M.* sp. BOT15**	77.05	63.48	57.98	24.8	11.83	33.17	30.75	**51.73**
***N.* sp. AM15**	65.17	60.17	67.33	58.33	67.13	77.28	73.2	**76.69**
***P.* sp. KBS15**	51.94	40.69	54.74	12.5	68.58	19.25	72.57	**51.78**
***P.* sp. UU15**	44.21	52.09	53.65	56.25	100	100	100	**78.57**
R. oryzae	63.21	65.54	66.75	100	6.48	31.75	37.16	**56.04**
Mean	**57.62**	**52.70**	**55.04**	**22.61**	**62.98**	**43.7**	**63.53**	

Ag Ageratum conyzoides, *As Alstonia scholaris*, *Av Averrhoa carambola*

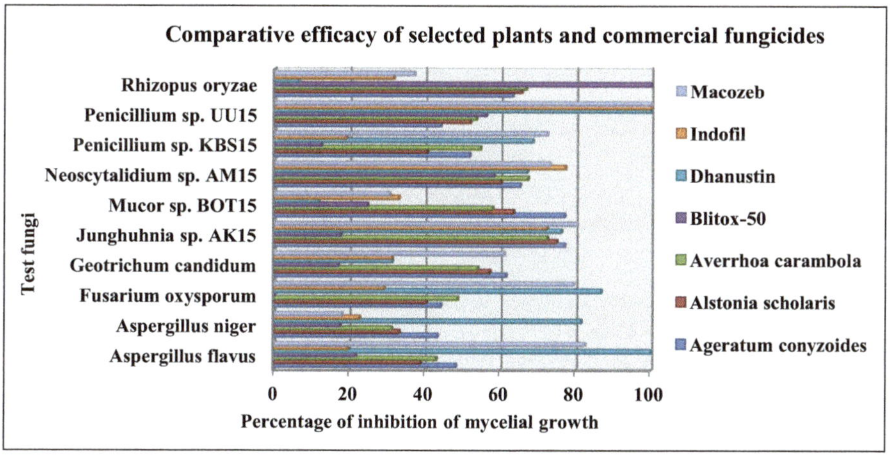

Fig. 15.5 Comparative efficacy of three selected plants and four commercial fungicides on mycelial inhibition of ten test fungal species

15.4 Conclusion

Indian economy is dependent upon agriculture, and agriculture has major problems of fungal diseases. Fungi can cause serious damage in agriculture, resulting in critical losses of yield, quality, and profit. Fortunately, agriculture has made tremendous progress during the last century, and part of this progress has been the development of modern means of plant disease control. In particular, the introduction of chemicals as disease control agents has contributed to a substantial increase in crop production, to a smoothing of annual undulations in crop yields, and, ultimately, to today's high level of food security. Despite the fact that agrochemicals control diseases of crops, yet the constant and regular use of chemicals has resulted in detrimental effects to the environment and human health since 0.1% of agrochemicals used in crop protection reach the target pest leaving the remaining 99.5% to enter the environment to cause hazards to nontarget organisms including human. Broad-spectrum activity of agrochemicals has increased the risk of residue toxicity, prices, and pathogen resistances. This has urged agriculturists to look for available alternatives. Use of naturally occurring beneficial microorganism may be safe and alternative approach to compensate multiple effects of fertilizers and pesticides. In nature, microorganisms grow in various associations ranging from antibiosis, commensalism, parasitism, and symbiosis. This association can be exploited in biological control of plant pathogen. Advantages of biofungicides are that they are biodegradable in nature, have no effect on nontarget species, are cheaper than agrochemicals, are less toxic, and take less time to develop. The present study was made to evaluate the fungi toxic potential of plant-based bioformulation and to further develop a bioentrepreneurship protocol of test plants for formulation of biofungicide to be used against storage pathogens associated with decay of root, tuber, and bulb crops in Odisha.

References

Adongo, J. O., Njue, A. W., Omolo, J. O., Cheplogoi, P. K., & Otaye, D. O. (2012). *In-vitro*inhibition of *Botrytis cinerea* – Causative agent for grey mold by crude extracts of Basidiomycetes fungi. *Science Journal of Biochemistry*, 173–175.

Bandara, B. M. R., Fernando, H. I. S., Hewage, C. M., Karunarathe, V., Adikaram, N. K. B., & Wijsundara, D. S. A. (1989). Antifungal activity of some medicinal plants of Srilanka. *Journal of the National Science Council of Srilanka, 17*(1), 1–3.

Brent, K. J., & Hollomon, D. W. (1998). Fungicide resistance: The assessment of risk. FRAC, Global Crop Protection Federation, Brussels. *Monograph, 2*, 1–48.

Costa, T. R., Fernandes, F. L. F., Santos, S. C., Oliveria, C. M. A., Liao, L. M., Ferri, P. H., Paulo, J. R., Ferreira, H. D., Sales, B. H. N., & Silva, M. R. R. (2000). Antifungal activity of volatile constituents of *Eugenia dysenterica*leaf oil. *Journal of Ethnopharmacology, 72*(1–2), 111–117.

Das, S. (2012). Antimicrobial and Antioxidant activities of green and ripe fruits of *Averrhoa carambola* Linn. and *Zizyphusmauritiana*Lam. *Asian Journal of Pharmaceutical and Clinical Research, 5*(3), 102–105.

Dastur, J. F. (1916). Spraying for ripe rot of plantain fruits. *Agricultural Journal of India, 11*, 142–149.

Duru, M. E., Cakir, A., Kordali, S., Zengin, H., Harmandar, M., Izumi, S., & Hirata, T. (2003). Antifungal activities of the leaves of three *Pistacia* species grown in Turkey. *Fitoterapia, 74*(1–2), 170–176.

El-Shami, M. A., Fadl, F. A., Tawfick, K. A., Sirry, A. R., & Zayat, M. M. (1985). Antifungal property of garlic clove juice compared with fungicidal treatments against *Fusarium* wilt of watermelon. *Egyptian Journal of Phytopathlology, 17*(1), 55–62.

Fawcett, C. H., & Spancer, D. M. (1970). Plant Chemotherapy with natural products. *Annual Review of Phytopathology, 8*, 403–418.

Gangadevi, V., Yogeswari, S., Kamalraj, S., Rani, G., & Muthumary, J. (2008). The antibacterial activity of *Acalypha indica* L. *Indian Journal of Science and Technology, 1*(6), 1–5.

Kana, H. A., Aliyu, I. A., & Chammang, H. B. (2012). Review on neglected and underutilized root and tuber crops as food security in achieving the millennium development goals in Nigeria. *Journal of Agriculture and Veterinary Science, 4*, 27–33.

Mishra, S. B., & Dixit, S. N. (1976). Comparision of activity of the active principle of *Clematis gouriana*with commercial fungicides. *Indian Phytopathology, 29*, 448.

Okigbo, R. N. (2009). Variation in phytochemical properties of selected fungicidal aqueous extract of some plant leaves in Kogi State, Nigeria. *American-Eurasian Journal of Sustainable Agriculture, 3*(3), 407–409.

Okigbo, R. N., & Ikediugwu, F. E. O. (2000). Studies on biological control of post-harvest rot in yams (*Dioscorea*spp.) using *Trichoderma viride*. *Journal of Phytopathology, 148*, 351–335.

Okigbo, R. N., & Nmeka, I. A. (2005). Control of yam tuber rot with leaf extract of *Xylopia aethiopica* and *Zinguber officinale*. *African Journal of Biotechnology, 4*(8), 804–807.

Rath, G. C., & Mohanty, G. N. (1978). Control of Macrophomina rot of stored garlic. *Science and Culture, 44*, 467–468.

Renu, H. V., Dubey, N. K., & Dixit, S. N. (1985). Mycotoxic properties of the essential oil *Aegle marmelos*. Corr. *BeitrageZurBiologie der pflanzen, 60*, 325–361.

Sambrook, J., Fritschi, E. F., & Maniatis, T. (1989). *Molecular cloning: a laboratory manual*. Cold Spring Harbor Laboratory Press.

Satish, S., Raghavendra, M. P., Mohana, D. C., & Raveesha, K. A. (2008). Antifungal activity of a known medicinal plant *Mimusopselengi* L. against grain moulds. *Journal of Agricultural Technology, 4*(1), 151–165.

Shanmugavalli, N., Umashankar, V., & Raheem. (2009). Anitmicrobial activity of *Vanilla planifolia*. *Indian Journal of Science and Technology, 2*(3), 37–40.

Shivpuri, A., Sharma, O. P., & Thamaria, S. (1997). Fungitoxic properties of plant extracts against pathogenic fungi. *Journal of Mycology and Plant Pathology, 27*(1), 29–31.

Swarna, L. L., & Neelakanta, R. P. (2009). Antimicrobial, antidiarrhoeal and analysis of phytochemical constituents of *Sphaeranthusamaranthoides*. *Indian Journal of Science and Technology, 2*(3), 45–48.

Wheeler, T. J., & Kececioglu, J. D. (2007). Multiple alignment by ligning alignments. *Bioinformatics, 23*(1), 559–568.

Chapter 16
Plant Pathogenic Fungi and Their Phytotoxins as Bioherbicides

Saurabh Gupta, Sushma Dubey, and Jaya Singh

Contents

16.1 Introduction

Legume crops are the second most important group of food plants and the major source of protein in the predominately vegetarian diet of the people of India. We have the distinction of being the world's single largest producer of legumes/pulses, having an area approximately of 20–24 million hectares under legumes. Madhya Pradesh is contributing a major percentage to the total legume production in the country. Pests especially weeds have increasingly become a major threat to sustainability of legume crops. Weeds are ubiquitous and continually changing pests in agriculture. They claim their own share of soil fertility and productivity at the cost of crop yield. They impose severe allelopathic effects on crops. Conventional

S. Gupta (✉) · S. Dubey
Department of Biotechnology, Kalinga University, Raipur, Chhattisgarh, India

J. Singh
Biodiversity Conservation and Rural Biotechnology Centre (BCRBC), Jabalpur, MP, India

A. C. Shukla (ed.), *Applied Mycology*, Fungal Biology,
https://doi.org/10.1007/978-3-030-90649-8_16

methods of weed control have failed due to one or other reasons, while microbial management either directly or via their secondary metabolites has received significant momentum as evidenced by commercialization of several products such as mycoherbicides or biorationals.

Weeds are ubiquitous and are considered to be unwanted plants in agriculture and other settings. More specifically, the term is often used to describe native or nonnative plants that grow and reproduce aggressively. About 1800 of the weeds cause serious economic losses in crop production, and about 300 weed species are serious in cultivated crops throughout the world (Burnside, 1979; Holm, 1969; Holm et al., 1977).

Losses in crop yield due to weed infestation are very heavy, and reduction of 20–40% is not uncommon. The range of losses has been reported to be 30–100% in India, 20–40% in USSR, 8–24% in the USA, and 6–50% in the UK (Anonymous., 1979). A variety of microbes are pathogenic to plants. Recently, interest has developed in exploiting these pathogens and their phytotoxin as bioherbicides (Duke et al., 1991).

Some traditional synthetic bioherbicides have had no environmental impact, and some researchers believe that natural products are less toxic or at least more biodegradable. Traditional screening methods for the discovery of bioherbicides have also reached a point of diminishing returns (Wright et al., 1991). Fungal phytotoxic (CFCF) natural products may have novel mechanisms of action that are yet to discover potential new tools to combat bioherbicide resistance. The potential value of bioproducts as a source of compounds for development is widely underappreciated. Types of species of *Alternaria petroselini* (FCLW#23) are known to produce many phytotoxins, but only some have been proven to play roles in pathogenesis (Scheffer, 1992; Otani & Kohmoto, 1992; Montemurro & Visconti, 1992; Kohmoto et al., 1993). Several *Alternaria* sp. have a multitoxin system, and they produce more than one toxin which is important in disease causation.

All the compounds are now low-molecular-weight cyclodepsipeptides (Montemurro & Visconti, 1992; Ayer & Pena-Rodriguez, 1987; Ayer et al., 1987; Bains & Tewari, 1987; Buchwaldt & Jensen, 1991; Buiatti et al., 1987). Destruxin B ($C_{30}H_{51}$, N_5O_7, MW = 593) is the major phytotoxin produced by *Alternaria brassicae*. Two other phytotoxins, homodestruxin B ($C_{31}H_{53}N_{507}$, MW = 607) and destruxin B2 ($C_{29}H_{49}Np_7$, MW = 579) are produced in much the same amounts. Desmethyldestruxin B ($C_{29}H_{49}N_{507}$, MW = 579) is also produced in trace quantities because of its phytotoxic activity. All these phytotoxins except dihydrotentoxin have phytotoxic activity.

Fungal phytotoxin is also a source of new bioherbicides (Duke, 1990, 1991; Lydon & Duke, 1989). This review will first address strategies for herbicide discovery. Then, host-specific from *Alternaria petroselini* (FCLW#23) fungi will be discussed with a review of the available literature as well as recent findings from our laboratory. Finally, the development of bioherbicides from phytotoxins of *Alternaria petroselini* (FCLW#23) will be discussed.

Conventional techniques of screening for resistance to pathogens suffer from several drawbacks such as scarce efficiency of selection, high time, and space requirements. Environmental factors like temperature, Ph, effect of light, and C:N ratio may modify the host–pathogen interaction, often making some disease resistances very difficult to identify. In particular, partial resistances determined by small individual effects of different genes are not easily detected. Uniform inoculation and incubation of large numbers of plants is logistically problematic and may result in high frequencies of escapes. A further limitation of traditional techniques of screening for disease resistance sometimes consists in the lack of standardized infection conditions which cause discrepancies between greenhouse and field disease responses.

16.2 Phytotoxins and Culture Filtrates

The use of (CFCF) phytotoxins or crude extracts of pathogen for the individuation of disease-resistant plants has drawn considerable attention. Produced by many plant pathogens, they may represent practical and appropriate agents for the selection of disease resistance, if *in vitro* response to toxin and *in vivo* reaction to disease at the whole plant are well correlated. Toxins act at cellular level, allowing a uniform exposure of very large populations to the selection pressure. When added to *in vitro* culture, the multiple-step regimen based on gradual increase of the inhibitory level of toxins may often result in more appropriate resistant and vigorously growing cultures (Jones, 1990). This procedure may reduce the risk of physiological habituation of the host tissue to the selective agent.

In vitro selection involves extensive preliminary work, and it is currently restricted to toxigenic pathogens. The poor knowledge of biochemical and pathogenetic events occurring in infected plants is undoubtedly the major obstacle in its development.

When *in vitro* selection with pathogen or toxin is ineffective, artificial inoculation of regenerated weed in the greenhouse or other in vitro indicators become of particular relevance and may be used as an alternative (Megnegneau & Branchard, 1991; Storti et al., 1992; Buiatti et al., 1987). Although the screening on regenerated weed at the whole plant level is considerably laborious and requires a large space, increased resistance to specific pathogens, obtained by means of this technique, has been reported.

Some work has been done toward developing bioassay methods for symptom development in seedlings and intact plants, detached and attached leaves, detached branch bearing fruits of rapeseed, and effect on pollen germination (Ayer & Pena-Rodriguez, 1987; Bains & Tewari, 1987; Buchwaldt & Green, 1992). These bioassay methods are less sensitive than the ones based on the effects of other toxins on some physiological or biochemical processes, or on isolated cells and protoplasts (Gardner et al., 1986; Yoder, 1981; Yoder et al., 1977). Since many aspects of toxin research are dependent on the suitability of bioassay methods, there is a need to develop a sensitive, quantitative, reproducible, and easy-to-perform bioassay method.

16.3 Considerations in the Use and Development of Microbial Herbicides by *Alternaria petroselini* (FCLW#23)

There have been basically four types of strategies used to develop bioherbicides. The first type is random screening. Synthesized biochemicals are screened against leguminous weed species for phytotoxic activity. Compounds related to those with activity are then studied for structure/activity relationships to optimize bioherbicidal activity.

A secondary strategy is to design bioherbicides to attack with leguminous weeds a certain molecular site of action and to optimize this activity by the study of structure–activity relationships. This is sometimes called the biorational approach. Metribuzin resulted from this process (Wright et al., 1991). With the third strategy, herbicides may be designed similar to, but beyond the scope of, those patented by competitors. This, however, is unlikely to result in novel compounds or important advances in technology.

The end of the strategy is to isolate natural products from biological sources of fungi and screen for bioherbicidal activity (Tachibana & Kaneko, 1986; Mullner et al., 1993). Although random screening has to date been most rewarding as a source of herbicides, this strategy has approached the point of diminishing returns (Wright et al., 1991).

Sites attacked by natural products may have few and sometimes structurally complex but effective inhibitors. Since these sites may have relatively few effective inhibitors, the odds of discovery of a bioherbicide that attacks such a site by random screening may be very low. The known molecular sites of action of microbial phytotoxins differ in almost every case from the sites of action of commercial bioherbicides (Devine et al., 1993; Duke et al., 1991).

Bioproducts from plant pathogens often are more selective than synthetic compounds, perhaps due to their isolation from host-specific weed hosts. This can be a desirable property, as avoidance of injury to crop plants is a goal of synthetic herbicide development. However, if the compound is too selective killing only one or a few weed species, it may not be a viable herbicide candidate.

16.4 Potential Problems Associated with Bioherbicides

Herbicide discovery can be a much more complicated process with microbially derived herbicides than with synthetic herbicides. First, the microbe must be isolated from its source. A method of growing the fungus in large quantities must be devised. Often, microbes are not stable in culture, and many mutate and lose their virulence. The toxic compound must be isolated and purified prior to toxicity testing. Obtaining sufficient quantities of the toxin may be difficult.

Because of the small quantities obtained, microbioassays have been developed for screening natural phytotoxins. A leguminous weed bioassay has been developed, as well as bioassays involving radicle growth from small-seed plants, leaf discs, and growth and development of intact, small plants (Abbas et al., 1993a,

Table 16.1 Effect of different concentrations of cell-free culture filtrate of *Alternaria petroselini* (FCLW#23) on petriplate bioassay studies of leguminous weeds

Seed germination inhibited (in %)							
Concentration of CFCF (%)	24 hpt	48 hpt	72 hpt	96 hpt	120 hpt	144 hpt	168 hpt
Control	0.0	0.0	0.0	0.0	0.0	0.0	0.0
5	0.0	0.0	0.0	0.0	0.0	0.0	0.0
10	0.0	0.0	0.0	0.0	10	25	30
20	0.0	0.0	0.0	10	25	51	65
25	0.0	5	20	35	40	68	75
50	45	65	88	100	100	100	100
75	60	75	90	100	100	100	100
100	80	95	100	100	100	100	100

Culture medium used = modified Richard's broth, Temperature = 28 °C, pH = 5

Table 16.2 Effect of different concentrations of cell-free culture filtrate of *Alternaria petroselini* (FCLW#23) on whole plant bioassay studies of leguminous weeds

Whole plant inhibited (in %)							
Phytotoxic damage (in %)	24 hpt	48 hpt	72 hpt	96 hpt	120 hpt	144 hpt	168 hpt
Control	0.0	0.0	0.0	0.0	0.0	0.0	0.0
5	0.0	0.0	0.0	0.0	0.0	0.0	0.0
10	0.0	0.0	0.0	0.0	0.0	0.0	10
20	0.0	0.0	0.0	10	30	45	50
25	0.0	0.0	10	35	50	80	90
50	30	65	80	90	100	100	100
75	50	75	85	100	100	100	100
100	60	85	95	100	100	100	100

Culture medium used = modified Richard's broth, Temperature = 28 °C, pH = 5

1995; Tanaka et al., 1993). These assays may not be as accurate in predicting the potential of herbicides as greenhouse screening test methods because they may not reflect the effect on intact terrestrial plants in soil (Tables 16.1 and 16.2).

Following isolation of the phytotoxin, determination of its structure is necessary. Because of the complex nature of some of these compounds, this can be difficult. Despite the host specificity of the pathogen, it is possible that the same compound may have been recovered from a different organism. Ayer et al. (1989) showed that, in an herbicide discovery effort based on microbial sources, 72% of compounds, whose structures were determined, were already known. They had already excluded many known compounds through a phytotoxin profile database.

Once a compound is identified, another difficulty is producing it at an economically feasible price. Such compounds may be too complex to lend themselves to chemical synthesis. The cost of production by the microbes themselves may prove prohibitive. It is possible that these compounds may be used as leads to develop structural analogs which are simpler to synthesize (Ito et al., 1974).

Characteristics of microbial products may make their use as herbicides unsuitable due to factors such as extremely short half-lives or rapid degradation in certain environmental conditions. Although the phytotoxins may have activity when introduced by the plant pathogen, they may not be taken up by plant roots or leaf cuticles when applied as herbicides. They also may not be translocated to the site of action. The activity in leaf disc bioassays often does not predict the activity of an herbicide in whole plants (Nandihalli et al., 1992) (Figs. 16.1, 16.2, 16.3 and 16.4).

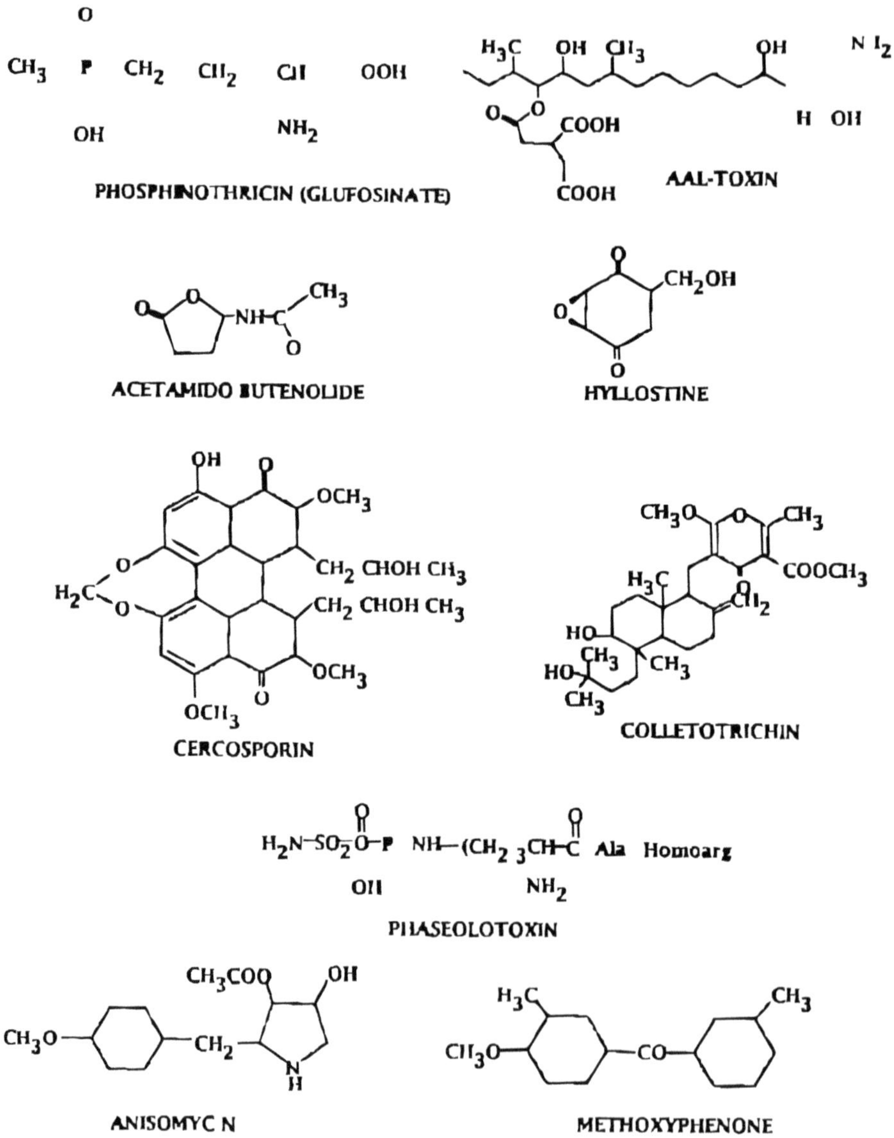

Cont...

Fig. 16.1 Structures of phytotoxin maintained in the test

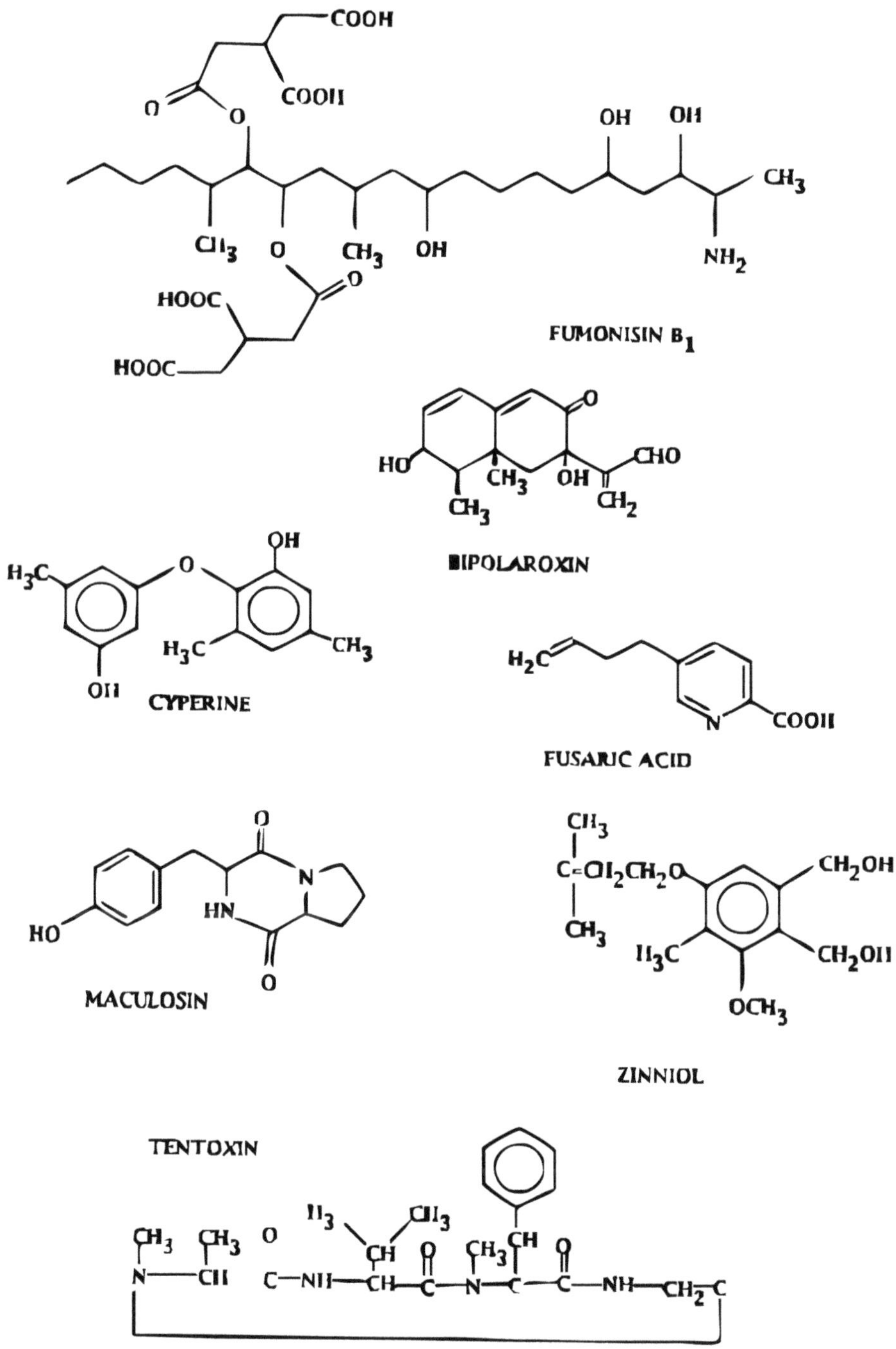

Fig. 16.1 (continued)

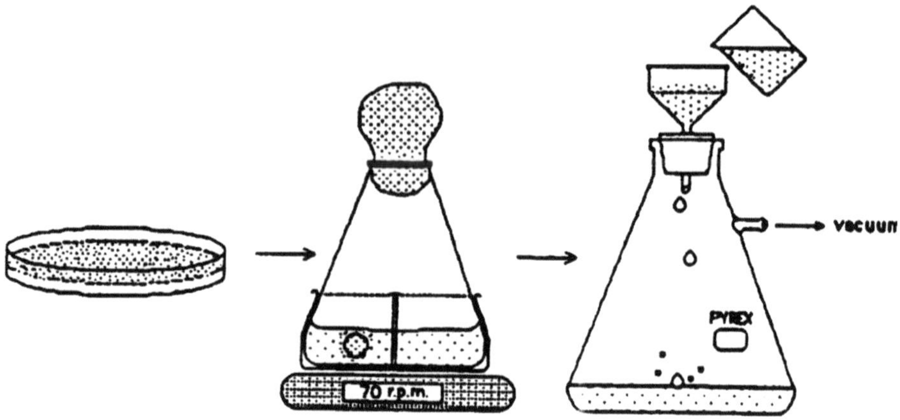

Fig. 16.2 Flow diagram of production of fungal cell-free cultural filtrate (CFCF)

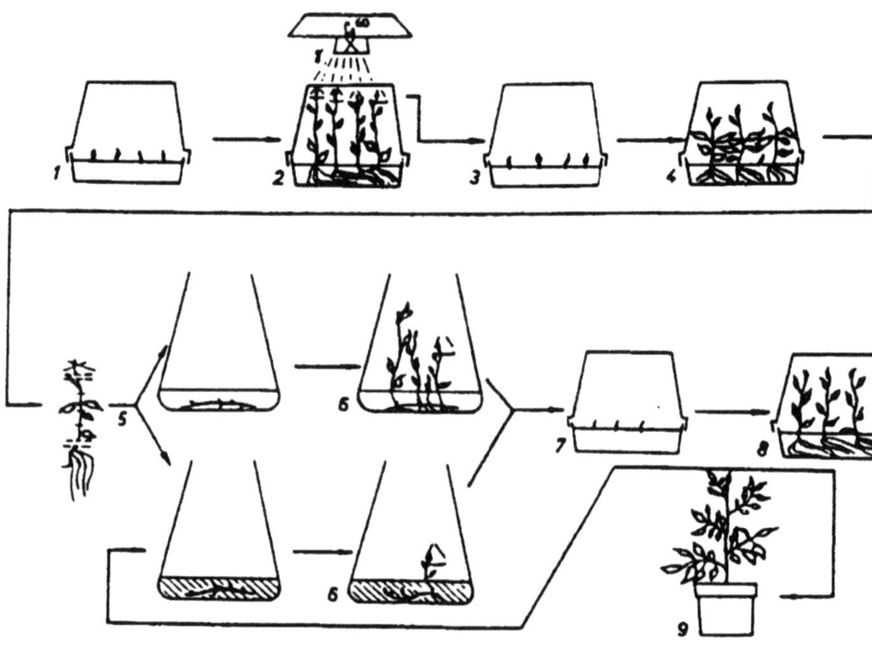

Fig. 16.3 Flow diagram of *in vitro* selection by bioassay

16.5 Use of Live Plant Pathogens *Alternaria petroselini* (FCLW#23) or Their Phytotoxins as Bioherbicides

In many instances, the plant pathogen causes the same effect on weeds as the phytotoxin it produces. It would be possible to use either the pathogen or the phytotoxin as an herbicide. Plant pathogens, however, have limitations that often make their use impractical with current technology (Julien, 1992; Zomer et al., 1993).

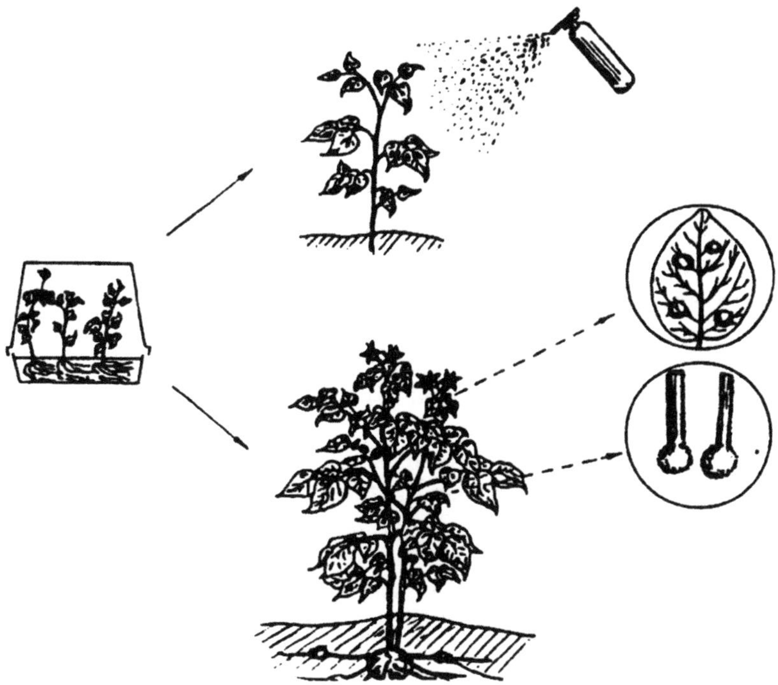

Fig. 16.4 Check of resistance of fungus by different procedures of *in vivo* artificial inoculation

Most plant pathogens that attack weeds are too host-specific, attacking only one or a very few weed species. As a multitude of weed species may affect the same crop, control of only one weed species would be of little interest to the farmer. However, toxins often have broader spectrums than the pathogens that produce them.

For example, AAL-toxin has been isolated from strains of *Alternaria petroselini* that are pathogenic only to certain cultivars of leguminous crops. However, Aflatoxin itself has a much broader range of activity and is toxic to Johnson grass [*Sorghum halepense* (L.) Pers.], black nightshade (*Solanum nigrum* L.), jimsonweed (*Datura stramonium* L.), prickly sida (*Sida spinosa* L.), hemp sesbania (*Sesbania exaltata* Rydb. ex A.W. Hill), and other formulations.

Living organisms also have the disadvantage of requiring special storage conditions such as temperature and humidity. They often have limited periods of viability and may have to be used within one season.

16.6 Host-Specific Phytotoxins

Although most known phytotoxins affect a variety of plant species, host-specific phytotoxins affect only one, or a very few, species (Scheffer & Livingston, 1984). Many are isolated from a pathogen of the affected species, and all known host-specific phytotoxins are from fungal pathogens.

Most known host-specific phytotoxins have been isolated from crop pathogens, and over twenty are known to exist. However, these toxins have not, in most cases, been tested for phytotoxicity to weeds.

AAL-toxin, produced by *Alternaria petroselini* (FCLW#23) lycopersici, is a hydroxylated long-chain alkylamine with a tricarboxylic acid moiety attached. AAL-toxin was initially reported to be host specific to only certain cultivars of leguminous crops, those with the Ase/Ase genotype. Heterozygous or Ase/Ase tomatoes are not affected.

We tested AAL-toxin on a variety of leguminous weed species. We found that AAL-toxin is phytotoxic to a number of weeds [4]. Therefore, AAL-toxin can no longer be considered to be truly host-specific (Tanaka et al., 1993; Duke et al., 1991; Abbas et al., 1993b). It is also likely that, with further testing, other so-called host-specific phytotoxins will be discovered to have a broader host range and may find applications in weed control.

To date, only one truly host-specific phytotoxin has been isolated from a weed pathogen, the cyclic dipeptide maculosin (Stierle et al., 1989; Strobel et al., 1990). Maculosin is derived from a pathovar of *Alternaria petroselini* (FCLW#23) and is host-specific for leguminous weeds. Maculosin is nontoxic to all other weed and crop species tested, including monocots and dicots.

Host-specific phytotoxins will probably be of less use as herbicides than those with · broader ranges. Most crops have a combination of problem weed species, and, in most cases, it would be prohibitively expensive to develop and use a different bioherbicide for each weed species. However, relative host selectivity may be an advantage in some situations, and resistant crop plants might be developed or selected.

16.7 Development Considerations

The use of bioproducts as bioherbicides depends on the ability to produce them at a cost that makes the process profitable. If an herbicide is not cost-effective, it should not be produced despite its effectiveness, since these phytotoxins are produced by microbes, and fermentation is one of the methods that could be used for production. Fermentation may also not be feasible in some instances because of instability in the producing strain. However, recent developments in biotechnology have improved the yield and quality of the fermentation process; so costs may decrease with time as technology advances.

The other option for natural products is to produce them by chemical synthesis. Glufosinate, the other commercially viable herbicide derived from microbial sources, is synthetically produced. Some compounds are too complex for economical synthesis. However, active analogs may be produced that are simple to synthesize, as in the case of methoxyphenone which was modified from anisomycin (Yamada et al., 1974; Ito et al., 1974).

16.8 Conclusion

Plant pathogens remain a largely untapped reservoir of potential bioherbicides. Preliminary studies have identified a host-specific phytotoxins isolated from *A. petroselini* (FCLW#23) that deserve further investigation.

Host-specific phytotoxins are less numerous and sometimes have broader spectra than reported when tested on leguminous weed. This may allow the application of some of these toxins or their analogs in leguminous weed management.

The study of phytotoxins produced by weed-specific pathogens is relatively new, and weed pathogens may be where the greatest chance of developing commercial herbicides lies. Because they are derived from weeds, such phytotoxins are more likely to be toxic to arid weeds less likely to be damaging to crops. However, natural products may seem to be more environment-friendly.

References

Abbas, H. K., Mirocha, C. J., Kommedahl, T., Vesonder, R. F., & Golinski, P. (1989). Production of trichothecene and non-trichothecene mycotoxins by Fusarium species isolated from maize in Minnesota. *Mycopathologia, 108*, 55–58.

Abbas, H. K., Gelderblom, W. C. A., Cawood, M. F., & Shier, W. T. (1993a). Biological activities of fumonisins, mycotoxins from *Fusarium moniliforme*, in jimsonweed (*Datura stramonium* L.) and mammalian cell cultures. *Toxicon, 31*, 345–353.

Abbas, H. K., Vesonder, R. F., Boyette, C. D., & Peterson, S. W. (1993b). Phytotoxicity of AAL-toxin and other compounds produced by Aftemariaalternata to jimsonweed (*Datura stramonium*). *Canadian Journal of Botany, 71*, 155–160.

Abbas, H. K., Tanaka, T., & Duke, S. O. (1995). Pathogenicity of Aftemaria a/temata and Fusarium moniliforme and phytotoxicity of AAL-toxin and fumonisin B, on tomato cultivars. *Journal of Phytopathology, 143*, 329–334.

Anonymous. (1979). A look at world pesticide markets. *Farm Chemistry, 142*(9), 61–68.

Ayer, W. A., & Pena-Rodriguez, L. M. (1987). Metabolites produced by Altemaria brassicae, the blackspot pathogen of canola. Part 1, the phytotoxic components. *Journal of Natural Products, 50*, 400–407.

Ayer, W. A., Bains, P. S., Pena-Rodriguez, L. M., & Tewari, J. P. (1987). Production of host-specific phytotoxin by Altemariabrassicae. In *Proceeding of the 7th lnternational rapeseed congress*, Poznan, Poland, 1256–1261.

Ayer, S. W., Issac, B. G., Krupa, D. M., Crosby, K. E., Letendre, L. J., & Stonard, R. J. (1989). Herbicidal compounds from microorganisms. *Pesticide Science, 27*, 221–223.

Bains, P. S., & Tewari, J. P. (1987). Purification chemical characterization and hostspecificity of the toxin produced by Altemariabrassicae. *Physiological and Molecular Plant Pathology, 30*, 259–271.

Buchwaldt, L., & Green, H. (1992). Phytotoxicity of destruxin Band its possible role in the pathogenesis of Altemariabrassicae. *Plant Pathology, 41*, 55–63.

Buchwaldt, L., & Jensen, J. S. (1991). HPLC pudflcation of destruxins produced by Altemariabrassicae in culture and in leaves of *Brassica napus*. *Phytochemistry, 30*, 2311–2316.

Buiatti, M., Marcheschi, G., Venturo, R., Bettini, P., Bogani, P., Morpurgo, R., Nacmias, B., & Pellegrini, M. G. (1987). In vitro response to *Fusarium* elicitor and toxic substances in crosses between resistant and susceptible carnation cultivars. *Plant Breeding, 98*, 346–348.

Burnside, O. C. (1979). Weeds. In W. B. Ennis (Ed.), *Introduction to crop protection*. American Society of Agronomy.
Devine, M. D., Duke, S. O., & Fedtke, C. (1993). *Physiology of herbicide action* (p. 441). *Englewood Cliffs*.
Duke, S. O. (1990). Natural pesticides from plants. In J. Janick & J. E. Simon (Eds.), *Advances in new crops* (pp. 511–517). Timber Press.
Duke, S. O. (1991). Plant terpenoids as pesticides. In R. F. Keeler & A. T. Tu (Eds.), *Handbook of natural toxins, toxicology of plant and fungal compounds* (Vol. 6, pp. 269–296). Dekker.
Duke, S. O., Abbas, H. K., Boyette, C. D., & Gohbara, M. (1991). Microbial compounds with the potential of herbicidal use. *Brighton crop protection conference, weeds – 1991, 1*, 155–164.
Gardner, J. M., Kono, Y., & Chandler, J. L. (1986). Bioassay and host selectivity of *Alternaria citri* toxin affecting rough lemon and mandarins. *Physiological and Molecular Plant Pathology, 29*, 293–304.
Holm, L. G. (1969). Weed problems in developing countries. *Weed Science, 17*, 113–118.
Holm, L. G., Plucknett, D. L., Pancho, J., & Herberger, J. P. (1977). The world's worst weeds. In *Distribution and biology*. University Press of Hawaii.
Ito, K., Futatsuya, F., Hibi, K., Ishida, S., Yamada, O., & Munakata, K. (1974). Herbicidal activity of 3,3'-dimethyl-4-methoxybenzophenone (NK-049) in paddy field. I. Herbicidal characteristics of NK-049 on weeds. *Weed Science (Japan), 18*, 10–15.
Jones, P. W. (1990). In vitro selection for disease resistance. In P. J. Dix (Ed.), *Plant cell line selection: Procedures and applications* (pp. 113–149). VCH.
Julien, M. H. (Ed.). (1992). *Biological control of weeds* (p. 186). C.A.B. International.
Kohmoto, K., Itoh, Y., Shimomura, N., Kondoh, Y., Otani, H., Kodama, M., Nishimura, S., & Nakatsuka, S. (1993). Isolation and biological activities of two host-specific toxins from the tangerine pathotype of *Altemaria altemata. Phytopathology, 83*, 495–502.
Lydon, J., & Duke, S. O. (1989). The potential of pesticides from plants. In L. E. Craker & J. E. Simon (Eds.), *Herbs, spices, and medicinal plants-recent advances in botany, horticulture and pharmacology* (Vol. 4, pp. 1–41). Oryx Press.
Megnegneau, B., & Branchard, M. (1991). Effects of fungal culture filtrates on tissue from susceptible and resistant genotypes of muskmelon to *Fusarium oxysporum* f. sp. *melonis. Plant Science, 79*, 105–110.
Montemurro, N., & Visconti, A. (1992). *Altemaria* metabolites-chemical and biological data. In *Altemaria: Biology, plant diseases, and metabolites* (Topics in secondary metabolism) (pp. 449–557).
Mullner, H., Eckes, P., & Donn, G. (1993). Engineering crop resistance to the naturally occurring glutamine synthetase inhibitor phospinothricin (glufosinate). In S. O. Duke, J. J. Menn, & J. R. Plimmer (Eds.), *Pest control with enhanced environmental safety* (American chemical society symposium series, 524) (pp. 135–141). ACS.
Nandihalli, U. B., Duke, M. V., & Duke, S. O. (1992). Relationships between molecular properties and biological activities of 0-phenyl pyrrolidino- and piperidino-carbamate herbicides. *Journal of Agricultural and Food Chemistry, 40*, 1993–2000.
Otani, H., & Kohmoto, K. (1992). Host specific toxins of *Altemaria* species. In J. Chelkowski & A. Visconti (Eds.), *Alternaria – Biology, plant diseases, and metabolites* (pp. 123–156). Elsevier Science Publishers, B.V.
Scheffer, R. P. (1992). Ecological and evolutionary roles of toxins from *Alternaria* species pathogenic to plants. In J. Chelkowski & A. Visconti (Eds.), *Altemaria: Biology, plant diseases, and metabolites, topics in secondary metabolism* (pp. 101–122). Elsevier Science Publishers.
Scheffer, R. P., & Livingston, R. S. (1984). Host-selective toxins and their role in plant diseases. *Science, 112*, 17–21.
Stierle, A., Cardellina, J. H., & Strobe, G. (1989). Maculosin, a host-specific phytotoxin for spotted knapweed from *Altemaria altemata. Proceedings of the National Academy of Sciences of the United States of America, 85*, 8008–8011.

Storti, E., Latil, C., Salli, S., Bettini, P., Bogani, P., Pellegrini, M. G., Simeti, C., Molnar, A., & Buiatti, M. (1992). The in vitro physiological phenotype of tomato resistance to *Fusarium oxysporum* f. sp. lycopersici. *Theoretical and Applied Genetics, 84*(1–2), 123–128.

Strobel, G., Stierle, A., Park, S. H., & Cardellina, J. (1990). *Maculosin: A hostspecific phytotoxin from Altemaria alternata on spotted knapweed* (American Chemical Society Symposium Series, 439) (pp. 53–62).

Tachibana, T., & Kaneko, K. (1986). Development of a new herbicide, bialaphos. *Journal of Pesticide Science, 11*, 297–304.

Tanaka, T., Abbas, H. K., & Duke, S. O. (1993). Structure-dependent phytotoxicity of fumonisins and related compounds in a duckweed bioassay. *Phytochemistry, 33*, 779–785.

Wright, K., Pillmoor, J. B., & Briggs, G. C. (1991). Rationality in herbicide design. In N. R. Baker & M. P. Percival (Eds.), *Topics in photosynthesis, herbicides* (Vol. 19, pp. 337–368). Elsevier.

Yamada, O., Kurozumi, A., Futatsuya, F., Ito, K., Ishida, S., & Munakata, K. (1974). Studies on plant growth-regulating activities of anisomycin and toyocamycin. *Agricultural and Biological Chemistry, 36*, 2013–2015.

Yoder, O. C. (1981). Assays. In R. D. Durbin (Ed.), *Toxins in plant diseases* (pp. 45–78). Academic.

Yoder, O. C., Payne, G. A., Gregory, P., & Gracen, V. E. (1977). Bioassays for detection and quantification of *Helminthosporium maydis* race T toxin: A comparison. *Physiological Plant Pathology, 10*, 237–245.

Zomer, P. S., Evans, S. L., & Savage, S. D. (1993). Perspectives on providing a realistic technical foundation for the commercialization of bioherbicides. In S. O. Duke, J. J. Menn, & J. R. Plimmer (Eds.), *Pest control with enhanced environmental safety* (American chemical society symposium series, 524) (pp. 79–86). ACS.

Chapter 17
Mycotoxins: A Concealed Threat in Agri-Food Sector

Manisha Shukla and M. Mishra

Contents

17.1 Introduction

Like all organisms, fungi manufacture byproducts of metabolism as they grow. The bulk of those low mass secondary metabolites are harmless to humans; however, some will deleteriously have an effect on human health, and these secondary metabolites are named as mycotoxins. These are toxic and have a major impact if they enter the food chain. Mycotoxins have attracted worldwide attention because of their impact on human health, immense economic losses, and domestic and foreign

M. Shukla (✉)
P. G. Department of Plant Science, Avvaiyar Govt. College for Women, Karaikal, Puducherry, India

M. Mishra
E.G.S. Pillay Engineering College, Nagapattinam, Tamil Nadu, India

A. C. Shukla (ed.), *Applied Mycology*, Fungal Biology,
https://doi.org/10.1007/978-3-030-90649-8_17

trade. It ought to even be remembered that some fungal metabolites are helpful to humans (e.g., antibiotics produced by fungi), whereas there are approximately 300–400 or so recognized mycotoxins, and solely ten are ordinarily ascertained in sickness, that are together called mycotoxicoses. Plant toxin contamination in foods and fodder has been changing into a world concern day by day. In line with Food and Agricultural Organization (FAO) reports, it's calculable that plant toxin affects nearly 25th of the world's crop annually and is inflicting immense agricultural product and industrial losses in billions of dollars (Alshannaq & Yu, 2017).

The best-known mycotoxins are the Aflatoxin family and the ergot alkaloids. This chapter addresses the kind of mycotoxins and role of plant's toxin-detoxifying agents.

There are four major aflatoxins (B1, B2, G1, and G2), but B1 is the most toxic. Most mycotoxins are chemical-stable and survive the food process. Several hundred different mycotoxins have been identified, but the most commonly observed mycotoxins that present a concern to human health and livestock include aflatoxins, ochratoxin A, patulin, fumonisins, zearalenone, and nivalenol/deoxynivalenol.

17.2 Characteristics of Mycotoxins (Cinar and Onbaşı 2018; Etzel 1999)

Mycotoxins are secondary toxic metabolites with a wide variety of chemical structures synthesized by fungi (mold) (Turnera et al., 2009). These secondary metabolites do not seem to be essential for vegetative growth and so could have very little or no primary function inside the organism.

Secondary metabolites are made once the organism enters the stationary phase, once the initial phase of ascension has declined. The metabolites made during this phase are typically related to differentiation and sporulation and might have profound biological activities that in some instances are exploited economically.

Mycotoxins are thought to be a sort of "chemical defence system" to guard mold from insects, microorganisms, nematodes, grazing animals, and humans (Etzel, 1999). Improper storage, transport, and selling may also cause mold growth and synthesis of mycotoxins (Etzel, 1999).

Factors that have an effect on mold growth and mycotoxin production are temperature, relative humidity, fungicides and/or fertilizers, interaction between the colonizing toxigenic fungal species, type of substrate and nutritional factors, geographical location, genetic necessities, and bug infestation (Tola & Kebede, 2016; Deligöz & Smith, 2005) (Fig. 17.1). A type of mold could produce one mycotoxin, a mycotoxin that will be synthesized by several molds.

Mycotoxins cause totally different degrees of toxicity consistent with exposure time, phytotoxin quantity, physiological condition, and sensitivity of the organism in humans and animals. In the majority of animal feed and products like wheat bran, noug cake, pea hulls, maize grain, milk and meat, and conjointly human food like cereal, fruit and vegetables, spice, etc., appearance of mycotoxins was discovered (Tola & Kebede, 2016). These foods create serious health risks in humans and every animal species.

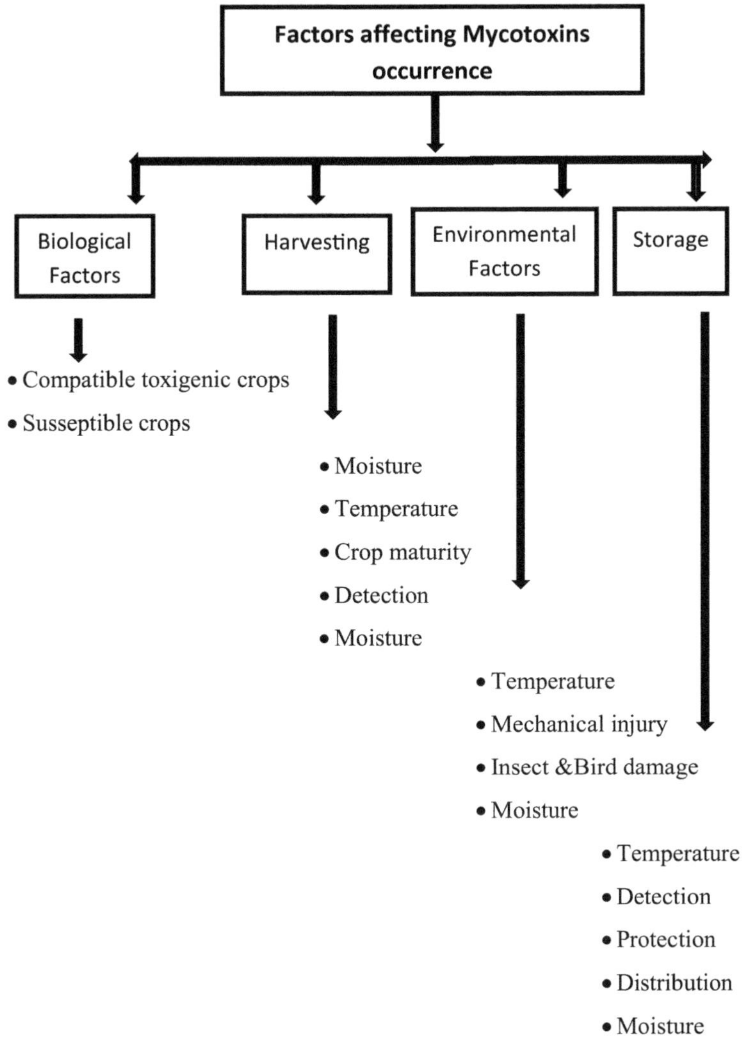

Fig. 17.1 Factors Affecting Mycotoxin occurrence in the agri-food sector

17.2.1 Difference Between Primary and Secondary Metabolites

A number of distinct variations are apparent between primary and secondary metabolites. First, they have been shown to possess an enormous variety of biosynthetic origins and structures that are not in general found among the primary metabolites.

Second, their incidence tends to be restricted to a tiny low variety of organisms and so will vary between isolated strains of constant species.

Finally, their production is characterized by the generation of groups of closely connected compounds which can have terribly totally different biological properties. Vital samples of secondary metabolites embrace medically vital compounds like antibiotics, statins, cyclosporins, and ergot alkaloids. Agriculturally vital secondary metabolites embrace strobilirubin, associate antifungal compound, and plant hormones like gibberellin.

Secondary metabolites of molds particularly belonging to genera *Aspergillus, Penicillium,* and *Fusarium* are called Mycotoxins. Out of 300 identified secondary metabolites, only around 30 have toxic properties which are of some concern.

17.3 Toxicogenic (Mycotoxin-Producing) Fungi Can Be Distinguished into Two Groups

17.3.1 Preharvest

Fungi (such as *Fusarium*) which invade their substrate and produce mycotoxin on the growing plants before harvesting this is the category of field (preharvest) toxins. Aflatoxins and *Fusarium* toxins are included in this group.

17.3.2 Postharvest

These are fungi which produce toxins after harvesting and during crop storage and transportation. These toxins are named storage (or postharvest) toxins, and Ochratoxin A belongs to this group.

Mycotoxins were "discovered" following a sharp and fatal outbreak that occurred in 1960 on Turkey farms in Great Britain (Asao et al., 1963). This acute case led to the identification of aflatoxins and consecutively the link between molds, their toxins, and mycotoxicosis.

It is well known that mycotoxins may also cause chronic diseases in animals and may have completely different effects (hepatotoxicity, genotoxicity, nephrotoxicity, neurotoxicity, nephrotoxicity, immunotoxicity, etc.). It ought to be noted that toxicity might vary significantly at intervals in a structural group of mycotoxins and that the danger might not continuously result from the toxin itself but from its metabolites.

Toxic effects of mycotoxin once present at very low levels don't seem to be investigated till now.

Mycotoxin's area unit is usually present at the same time in staple. Moreover, complete feed is formed of various raw materials. Thus, animals are usually not exposed to at least one mycotoxin but to many toxins at the same time. Once mycotoxins are present at the same time, interactive effects are classified as additive, antagonistic, or synergistic.

Multiple contaminants may additionally occur attributable to the power of a given mold species to provide many varieties of mycotoxins in one type of food ingredient, or else, many forms of mycotoxins are also found within the same food or feed containing completely different contaminated ingredients or raw materials.

Mycotoxins may occur in conjugated type, either soluble (masked mycotoxins) or incorporated into/associated with/attached to macromolecules (bound mycotoxins). These conjugated mycotoxins will emerge when metabolized by living plants, fungi, and mammals or when food is processed (Berthiller et al., 2009). Masked mycotoxins show lower toxicity than the initial, nonconjugated, toxic parent compounds (Gareis, 1994). The presence of mycotoxins in feed may create a risk to consumers if these toxins and/or their metabolites are excreted and accumulated in animal products like milk and meat. The presence of mycotoxins in milk could be a public health concern and should be regularly monitored. A great majority of mycotoxins found in feeds do not cause issues because either they are not excreted into milk like the bulk of Fusarium toxins, or they are excreted as a less toxic metabolite like ochratoxinα.

The mycotoxins of major concern as feed contaminants that are potentially removable from feed are mainly aflatoxins, ochratoxin A, and Fusarium toxins (trichothecenes like deoxynivalenol, diacetoxyscirpenol, nivalenol, T2-toxin/HT2-toxin, zearalenone, and fumonisins) (Figs. 17.2 and 17.3). One among the methods

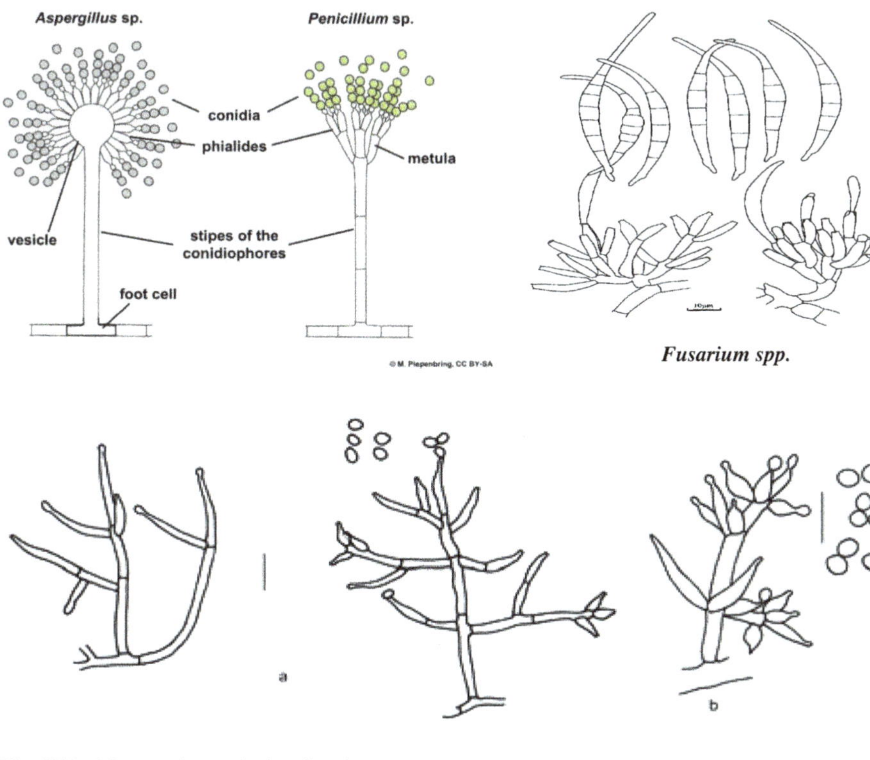

Fig. 17.2 Mycotoxin-producing fungi

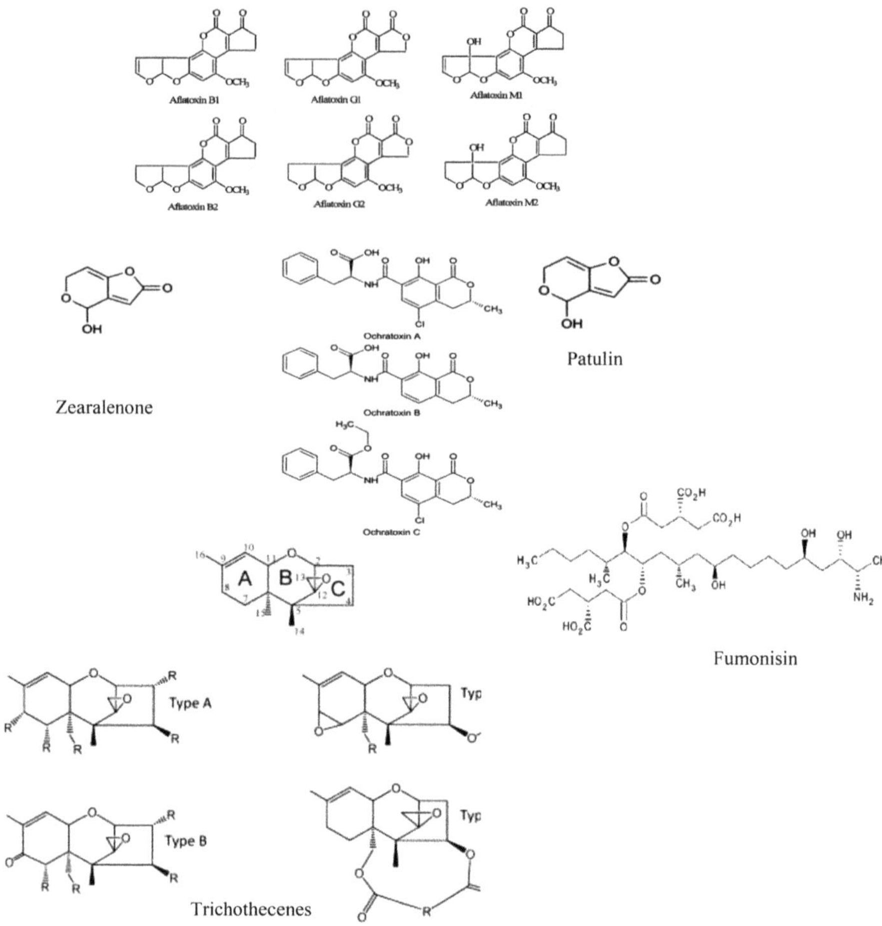

Fig. 17.3 Chemical structure of mycotoxins

for reducing the exposure to mycotoxins is to decrease their bioavailability together with numerous mycotoxin-adsorbing agents in the compound feed, which results in a reduction of mycotoxin uptake as well as distribution to the blood and target organs. Another strategy is the degradation of mycotoxins into nontoxic metabolites by using biotransforming agents like bacteria/fungi or enzymes.

These fungi frequently contaminate and produce toxins while growing on crops like corn and peanuts that are later on consumed by animals and humans. Consumption of meat and milk from cows that are exposed to the mycotoxins also can result in exposure of humans to the toxins, whereas acute aflatoxicosis may be a comparatively rare phenomenon, and eaten aflatoxins are notorious for their carcinogenicity as a result of chronic exposure. The mutagenic nature of aflatoxins is believed to be due to the DNA-damaging properties of aflatoxin metabolic derivatives. The foremost common type of unwellness associated with dietary exposure to aflatoxins is liver cancer; but it should also be involved in different varieties of cancer. Because genus *Aspergillus* species are present within the surroundings, it's not possible to prevent foodstuff contamination with fungi. However, this contamination is reduced by the utilization of rigorous production, storage, and watching procedures.

Ergotism (also referred to as St Anthony's fire) may be an unwellness related to the uptake of cereals contaminated with the fungus ergot. This fungus, which infects the flowers of grasses and cereals, produces a variety of alkaloids, of which alkaloid is the best known. Alkaloid is said to be the psychoactive drug LSD, and uptake of cereals and cereal-derived merchandise like rye bread that are contaminated with the fungus may end up in serious symptoms of unwellness, like convulsions and gangrene. It's been instructed that mycotoxin may additionally exert a harmful result through inhalation, instead of by uptake. Most homes, workplace buildings, and factories provide several niches appropriate for the expansion of a myriad of filamentous fungi, together with genus *Aspergillus, Claviceps*, and *Stachybotrys* spp., and mycotoxins made by these species may contribute to the development referred to as "sick building syndrome." This syndrome has been related to a cluster of nonspecific symptoms (usually together with fatigue, minor metabolism issues, and headache) that are solely practiced inside a specific building. Poor air quality and ventilation, cleansing chemicals, and microbial contamination are the characteristic options of sick building syndrome. The role of fungi within the syndrome is a supply of considerable conjecture.

Mycotoxins are little and quite stable molecules that are very tough to get rid of or eradicate and that enter the feed chain while keeping their unhealthful properties.

Toxigenic molds could develop underneath all climates on any solid or liquid supports as soon as biological process substances and wet (water activity Aw over 0.6) are present, hence the wide range of contaminated foodstuff substrates (Table 17.1). These toxins are found as natural contaminants in several foodstuffs of plant origin, particularly cereals, however, additionally fruits, hazelnuts, almonds, seeds, fodder, and foods consisting of, or factory-made from, these products and supposed for human or animal consumption (Binder & Binder, 1998).

Mycotoxigenic molds are uncontrollable due to their global presence in nature. Prevention of mycotoxin synthesis in all stages of food processing is an essential point for public health and economic reasons.

17.4 Mycotoxin-Detoxifying Agents

The Commission regulation (EC) No 386/2009 of May 12, 2006 outlines a new functional group of feed additives as "substances for reduction of the contamination of feed by mycotoxins: substances that may suppress or cut back the absorption promote the excretion of mycotoxins or modify their modes of action" (Kavin, 2018).

Table 17.1 Name of some important mycotoxin-producing fungi, susceptible foods, and mycotoxin effects on humans and animals

Mycotoxins	Genus/species	Major food	Toxic effects and diseases
Aflatoxin	*Aspergillus flavus* *A. parasiticus* *A. nomius* *Penicillium*	Cereals, feeds, oilseeds and pulp, coconut	Carcinogenic, hepatotoxicity, teratogenicity, decreasing immune systems, affecting the structure of DNA, hepatitis, bleeding, kidney lesions
Fumonisin	*Fusarium verticillioides* *F. culmorum*	Cereals, corn	Encephalomalacia, pulmonary edema, carcinogenic, neurotoxicity, liver damage, heart failure, esophageal cancer in humans
OchratoxinOTA	*Aspergillus* *Penicillium* *A. ochraceus* *P. nordicum* *P. verrucosum*	Cereals, herbs, oil seeds, figs, beef jerky, fruits, and wine	Kidney and liver damage, loss of appetite, nausea, vomiting, suppression of immune system, carcinogenic
Patulin	*Aspergillus terreus* *A. clavatus* *Penicillium* *Penicillium carneum* *P. clavigerum* *P. griseofulvum*	Silage, wheat, feeds, apples, grapes, peaches, pears, apricots, olives, cereals	Neural syndromes, brain hemorrhage, skin lesions, skin cancer, lung, mutagenicity, antibacterial effect
Trichothecenes (T2, DON, DAS, HT2)	*Fusarium* *Cephalosporium* *Trichoderma* *Fusarium oxysporum*	Cereals, feeds, silage, legumes, fruits, and vegetables	Immune suppression, cytotoxic, skin necrosis, hemorrhage, anemia, granulocytopenia, oral epithelial lesions, GIS lesions, hematopoietic, alimentary toxic aleukia (ATA), hypotension, coagulopathy
Zearalenone	*Fusarium* *F. graminearum* *F. culmorum*	Cereals, corn, silage, timothy grass, fodder	Carcinogenic, hormonal imbalance estrogenic effect, reproductive problems, teratogenic

Table adapted from AycanCinarElif (2018) OnbustMycotoxins

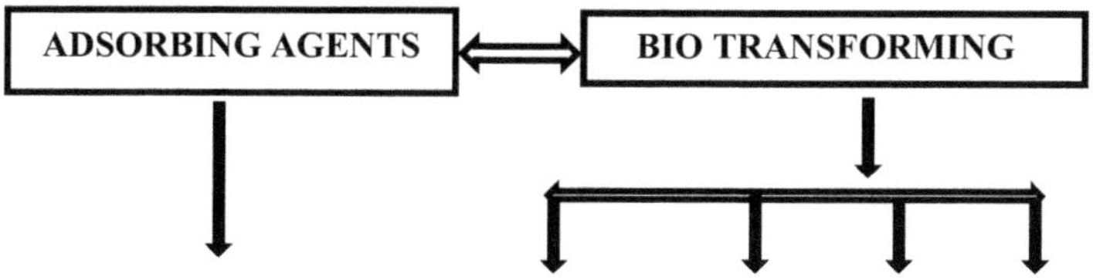

- **AluminosilicatesBacteria Fungi Yeast Enzymes**
- **Tectosilicates**
- **Bentonites**
- **Montmorillonite**
- **Zeolites**
- **Hydrated Na Caaluminosilicate**
- **Activated carbon (AC)**
- **Yeast cell walls**
- **Micronized fibres**
- **Bacteria**
- **Polymers**
- **Polyvinylpyrrolidone**

Depending on their mode of action, these feed additives could act by reducing the bioavailability of the mycotoxins or by degrading them or remodeling them into less toxic metabolites.

Therefore, we will outline at least two main categories:

- *Adsorbing agents* One of the methods for reducing the exposure to mycotoxins is to decrease their bioavailability together with varied mycotoxin-adsorbing agents within the compound feed, which ends up in a reduction of mycotoxin uptake, moreover as distribution to the blood and target organs. Adsorbing agents are known as binding agents, adsorbents, or binders.
- *Biotransforming agents* Another strategy is the degradation of mycotoxins into nontoxic metabolites by mistreatment of biotransforming agents like bacteria/fungi or enzymes. Substances that don't directly act with mycotoxins, i.e., inhibitor agents or immunostimulatory agents, are reviewed; however, they don't seem to be thought of sensu stricto as mycotoxin-detoxifying agents. However, we tend to be aware that such compounds could also be terribly economical for reducing the toxicity of mycotoxins.

17.4.1 Mycotoxin-Adsorbing Agents

Mycotoxin-adsorbing agents are large giant mass compounds that ought to be able to bind the mycotoxins in contaminated feed while not dissociating within the large number of animals. In this way, the toxin-adsorbing agent advances through the animal and is eliminated via the excreta. This prevents or minimizes exposure of animals to mycotoxins.

Mycotoxin-adsorbing agents are silica-based inorganic compounds or carbon-based organic polymers. The inorganic adsorbing agents presently on the market embrace natural clay products similarly as artificial polymers.

- *Aluminosilicates*: Silicate minerals are the most important category of mycotoxin sequestering agents, and most studies on the alleviation of mycotoxicosis by the utilization of adsorbing agents have centered on aluminosilicates. Inside this group, there is a pair of vital taxonomic categories: phyllosilicate taxonomic category and also the tectosilicate subclass.
- *Phyllosilicates*: They embrace bentonites, montmorillonites, smectites, kaolinites, and illites.
- *Tectosilicates*: They embrace zeolites.
- *Bentonites*: They are originally created from the weathering of volcanic ash in place (Ramos et al., 1996). They belong to the phyllosilicate group and are adsorbing agents with a superimposed crystalline microstructure and variable composition. Bentonites are typically impure clay consisting principally of montmorillonite. Because of their montmorillonite content, bentonites swell and form thixotropic gels (Diaz & Smith, 2005).
- *Montmorillonite*: It may be a superimposed salt that absorbs organic substances either on its external surfaces or inside its interlaminar areas (Ramos et al., 1996). Changed montmorillonite nanocomposite (MMN) may be a new sportive additive. Developed with nanomodification techniques, MMN features a sizable area, higher body, and stronger ion exchange activities beside a lot of active sites that build its nanoparticle impact straightforward to exert, and as a result, its surface assimilation effectivity is greatly increased.
- *Zeolites*: They are crystalline hydrous aluminosilicates of alkali and alkaline-earth cations characterized by an infinite three-dimensional structure. Zeolites are a bunch of silicates consisting of interlocking tetrahedrons of SiO4 and AlO4 (Kabak & Dobson, 2006; Ramos & Hernandez, 1997). Zeolites have large pores that give area for large cations like Na, potassium, and calcium, and they're characterized by their ability to lose and absorb water and exchange constituent cations while not harming the crystalline structure (Diaz & Smith, 2005; Papaioannou et al., 2002a, b). Clinoptilolite may be a natural mineral whose main application is the surface assimilation of heavy metals from liquid solutions (Papaioannou et al., 2002a, b).
- *Hydrated Na Ca aluminosilicate* (HSCAS): It is probably the foremost studied mycotoxin-sequestering agent among the mineral clays (Diaz & Smith, 2005; Kabak & Dobson, 2006; Kleiner et al., 2001). It's a naturally occurring and

heat-processed Ca montmorillonite that's ordinarily used as an anticaking additive in animal feed (Galvano et al., 2001).

- *Activated carbon (AC)*: It may be a nonsoluble powder shaped by transmutation of many organic compounds and made by activation processes aimed toward developing an extremely porous structure (Kleiner et al., 2001). C is thought jointly of the foremost effective and nontoxic groups of sorbents and has been shown to be a tenacious sorb agent of a good sort of medication and cytotoxic agents. It has been ordinarily used as a medical treatment for severe intoxications since the nineteenth century (Wang et al., 2008). The sequestrant properties of AC rely on several factors together with pore size, area, structure of the phytotoxin, and doses. Superactivated carbon differs from AC, wherein the particle size is reduced, thereby increasing area. The precise surface area of AC so varies from 500 m^2/g to 3500 m^2/g for superactivated charcoals (Ramos & Hernandez, 1997).
- *Yeast cell walls*, derived from the baker's yeast, are used as a dietary mycotoxin-adsorbing agent. Yeast cell walls consist virtually entirely of proteins and carbohydrates. The sugar fraction consists primarily of aldohexose, mannose, and N-acetylglucosamine. Glucans and mannans, the 2 main sugars, are found in equal concentrations in baker's yeast. Yeast mannan chains of varied sizes are exposed on the external surface and are coupled to cell membrane proteins (Huwig et al., 2001). The cell walls harboring polysaccharides, proteins, and lipids exhibit various completely different and straightforward accessible surface assimilation centers.
- *Micronized fibers*: They are obtained from completely different plant materials like cereals (wheat, barley, oat), pea hulls, apple, bamboo, etc. They're deep-rooted mainly of polysaccharide, hemicelluloses, and polymer and may be obtained in ultrafine (<100 μ) or less fine (>100 μ) fractions (Evans & Dawson, 2007).
- *Bacteria*: It has been recommended that cell membrane peptidoglycans and polysaccharides are the 2 most significant parts answerable for binding by carboxylic acid microorganism. Carboxylic acid microorganisms (LAB) are a group of gram-positive, acid-tolerant, typically nonsporulating microorganisms that have common metabolic and physiological characteristics. These microorganisms, sometimes found in rotten plants and drink products, manufacture carboxylic acid because of the major metabolic end-product of sugar fermentation. The most strains that comprise the work are *Eubacteria, Leuconostoc, Pediococcus, Lactococcus,* and true bacteria similarly as a lot of peripheral *Aerococcus, Carnobacterium, Enterococcus, Oenococcus,* and *Sporolactobacillus*. Strains of carboxylic acid microorganisms like *Eubacteria rhamnosus* strain GG and *Eubacteria rhamnosus* strain LC-705 are accustomed to take away mycotoxins. True bacteria thermophilus NG40Z and C5 have additionally been tested for their ability to detoxify mycotoxins (Aoudia et al., 1963).
- *Polymers*: Cholestyramine is an insoluble, quaternary ammonium ion exchange resin that powerfully binds anionic compounds. It's been used as medication in human for gripping digestive juice acids within the canal so as to scale back

sterol (Diaz & Smith, 2005). Polyvinylpyrrolidone may be an extremely polar amphiprotic chemical compound (Underhill et al., 1995).

17.4.2 Mycotoxin-Biotransforming Agents

Some studies have shown that some microorganisms have the power to degrade mycotoxins. Biotransforming agents embody bacterium, yeasts, fungi, and enzymes. The concept is for every plant toxin, or category of mycotoxins, to use enzymes that specifically degrade the poisonous substance into a nontoxic compound. Such enzymes are delineated in bacterium, yeast, or fungi. Refined enzymes have additionally been tested for this purpose.

(i) *Bacteria*

1. Gram-positive, anaerobic bacterium. The bacterial strain BBSH 797, a microbial feed additive, was isolated from rumen fluid. It's a gram-positive, nonspore-forming irregular rod living strictly anaerobic. It is 0.2 ± 0.4 and 1 ± 1.5 mm and occurs separately and in long chains up to 100 mm (Celik et al., 2000). Analysis of the 16s rRNA and therefore the guanine/cytosine content beside the specialized nutritionary demands and different physiological characteristics indicates a new species of the genus *Eubacterium* (Fuchs et al., 2002).
2. Gram-positive, aerobic bacterium *Nocardia asteroides* may be a species of *Nocardia* that may be a genus of weakly staining gram-positive, catalase-positive, rod-shaped bacterium. It forms partly acid-fast beaded branching filaments. *Nocardia asteroides* are pathogenic: They'll cause nocardiosis, a severe respiratory organ infection in immunocompromised hosts. *Nocardia* are found worldwide in soil that's rich with organic matter (Binder et al., 2007).
3. *Corynebacterium m*ay be a genus of gram-positive, enzyme positive, nonspore-forming, nonmotile, rod-formed bacterium that is straight or slightly curving. Their size falls between 2.6 μm long and 0.5 μm in diameter. They're defined by high G:C content, with close phylogenetic relationship to *Arthrobacter, mycobacterium, Nocardia,* and *actinomycete*. They are cosmopolitan in nature and are mostly innocuous. Some are helpful in industrial settings like *C. glutamicum*, whereas others cause human sickness like *C. diphtheria*, the infectious agent liable for contagious disease. *Eubacteria rubrum,* named during this means due to the brilliant red color of its colonies, may be a nonacid fast bacterium capable of manufacturing massive quantities of lipids (Binder et al., 2007).
4. Mycobacteria are aerobic and nonmotile bacteria that are characteristically acid-alcohol fast. Mycobacteria are sometimes classified as gram-positive due to their lack of Associate in Nursing outer semipermeable membrane. *Mycobacterium* strain DSM 44556T was isolated with fluoranthene because

of the single carbon supply from the soil of a former coal gas plant, polluted with polycyclic aromatic hydrocarbons. The physiological properties, carboxylic acid pattern, and therefore the 16S ribosomal polymer gene sequence indicated membership to the genus *Mycobacterium*; however, they were totally different from all types of strains of mycobacterium species. This strain delineated a new species that the name *Mycobacterium fluoranthenivorans sp. nov* was proposed for it (Binder et al., 2007).

5. *Rhodococcus erythropolis* is an aerobic, gram-positive, nonmotile, catalase-positive eubacterium that forms rods to extensively branched vegetative mycelium. It's an expedient infectious agent in immunocompromised patients.
6. *Curtobacterium* sp. strain 114-2 belongs to the genus of *Curtobacterium*. It is bacterium of the order actinomycetales. They're gram-positive soil organisms (Wu et al., 2009).
7. Gram-negative, aerobic bacterium *Flavobacterium aurantiacum* (NRRL B-184) may be a species of Flavobacterium that may be a genus of gram-negative, nonmotile, rod-shaped bacterium consisting of about 10 recognized species. Flavobacteria are found in soil and water in a variety of environments. Many species are famous to cause sickness in seafood (Binder et al., 2007).
8. *Pseudomonas fluorescens* may be a common gram-negative, rod-shaped microorganism. It belongs to the genus *Pseudomonas bacteria*. *P. fluorescens* has multiple flagella. It has a very versatile metabolism and may be found within the soil and in water. It's an obligate organism; however, certain strains are capable of exploiting nitrate rather than oxygen as a final electron acceptor throughout metabolism (Ueno et al., 1983).
9. Alcaligenes may be a genus of gram-negative, aerobic, rod-shaped bacterium. Alcaligenes species have been used for the economic production of nonstandard amino acids (Ueno et al., 1983). *Flavobacterium, Pseudomonas, Alcaligenes*, and *Bacilli* are often employed in a combination.

(ii) *Fungi*

1. *Aspergillus* species are highly aerobic and are found in most oxygen-rich environments, wherever they normally grow as molds on the surface of a substrate, as a result of the high oxygen tension. Commonly, fungi grow on carbon-rich substrates like monosaccharides and polysaccharides. *Aspergillus niger* may be a fungus and one of the foremost common species of the genus *Aspergillus*. It's a typical contamination of food. It's omnipresent in soil and is often reported from indoor environments.
2. *Aspergillus flavus* may be a common mold within the surroundings and may cause storage issues in stored grains. *A. flavus* is especially common on corn and peanuts, moreover as water damaged carpets. It can even be a human infectious agent, related to aspergillosis of the lungs and generally causing membrane, otomycosis, and naso-orbital infections. Several strains

produce important quantities of aflatoxin. *A. flavus* spores are allergenic (Megharaj et al., 1997).

3. *Aspergillus candidus* belongs to the genus *Aspergillus* and may be found in warm soils, grain, and within the secondary decay of vegetation (Binder et al., 2007). *Aspergillus parasiticus* may be a mold renowned to provide aflatoxin and generally found on black olives.
4. *Eurotium herbariorum* contains a moderately rapid growth rate with downy to powdery colonies. The *Eurotium* is usually found in tropical and semi-tropic zones. *Eurotium* may be a xerophilic fungus and is isolated primarily from soil, plants, stored grains, and house dirt (Megharaj et al., 1997).
5. *Rhizopus* is a genus of molds that has cosmopolitan thread-like fungi found in soil, decaying fruit and vegetables, animal excreta, and old bread. Mold grows well on general plant media. Mold species might overgrow and inhibit other fungi. Some structures are visible to the naked eye, i.e., sporangia seem macroscopically as black dots within the thick of white, cottony mycelium (Megharaj et al., 1997).
6. *Penicillium raistricki* belongs to *Penicillium* fungus genus that may be a genus of *Ascomycetes* fungi of major importance within the environment, food, and drug production. The mycelium generally consists of an extremely branched network of cell organs, septate, typically colorless hyphae (Binder et al., 2007).
7. *Rhinocladiella atrovirens* colonies are restricted, velvety, or lanose, usually slightly mucoid at the center. Conidiogenous cells are cylindrical, intercalary or free, 9–19 × 1.6–2.2 μm; rough rachis up to 15 μm long, with packed, flat, or butt-shaped, unpigmented conidial denticles. Rhinocladiella contains six to eight species, with two species of medical interest, *R. atrovirens* and *R. aquaspersa* (Nakazato et al., 1990).

(iii) *Yeast*

1. *Trichosporon mycotoxinivorans* belongs to the genus *Trichosporon,* which is characterized by the production of arthroconidia. Thirty-five species are delineated within the genus until now. This genus is monophyletic, on the idea of 18S and 26S rDNA sequences. *T. mycotoxinivorans* may be a yeast strain isolated from the hindgut of the lower termite (Mastotermitidae). The name of this recently isolated strain refers to a vital characteristic of *T. mycotoxinivorans* to detoxify mycotoxins like OTA and ZEA (Blackwell et al., 1999).
2. *Phaffia rhodozyma* and *Xanthophyllomyces dendrorhous* isolates: *Phaffia rhodozyma* was isolated within the 1960s. A minimum of 2 species appear to exist, together with the anamorph *Phaffia rhodozyma* and also the teleomorph *Xanthophyllomyces dendrorhous.* The yeast has attracted considerable biotechnological interest as a result of its ability to synthesize the antioxidant astaxanthin as its major pigment. This property has stimulated research on the biology of the yeast, moreover as development of the yeast as an industrial microorganism for astaxanthin production by fermentation (Molnar et al., 2004).

(iv) *Enzymes*

1. Protease A are enzymes which break down proteins (proteolysis), by reaction of the amide bonds linking amino acids together within the peptide chain. Proteases work best in acidic conditions. Peptidase A is obtained through fermentation processes from selected fungus genus *A. niger* strains (Peteri et al., 2007).
2. Pancreatin may be a mixture of many exocrine gland enzymes made by the gland cells of the pancreas. It's composed of enzyme, amylase, enzyme, and peptidase (Peteri et al., 2007).
3. Carboxypeptidase A typically refers to the pancreatic exopeptidase which hydrolyzes amide bonds of C-terminal residues with aromatic or open-chain facet chains. Most scientists within the field currently refer to this catalyst as CPA1 (Abrunhosa et al., 2006).
4. Epoxidases are enzymes that are able to biotransform epoxide groups into diene groups (Abrunhosa et al., 2006).
5. Lactone hydrolases are enzymes that catalyze the hydrolysis of lactone rings (intramolecular cyclic esters) to provide a hydroxyl group and a carboxyl group (Schatzmayr et al., 2006).

17.5 Impact of Mycotoxins on Livestock and Human

Mycotoxins are toxicant fungal products that are made once fungi grow in human and animal foods. A large range of food and drinkable things is contaminated with mycotoxins. Exposure to mycotoxins causes illness in humans and animals. Recent studies using biomarkers of exposure, internal dose, and adverse effects have shown that mycotoxins are underappreciated as a cause of illness. Some mycotoxins cross the placenta and are present within the craniate at birth, and others are excreted in milk. Some mycotoxins cause pathological processes (cancer) in humans and animals, thus causing kidney and medicine diseases. Animals are exposed to mycotoxins through consumption, inhalation, and/or skin contact. Several mycotoxins are potent in low doses, that is, quite little amounts of the compound will represent vital health effects (Fig. 17.4). Since filamentous fungi are common and expedient organisms, mycotoxins are widespread (Bennett & Krska, 2009). They are found in foods and feeds across the world and are of accelerating concern as potential contaminants within the indoor surroundings. Many of them are involved as warfare agents. Mycotoxins are to blame for cancers in addition to many alternative disorders moving the epithelial duct, urogenital, vascular, kidney, and nervous systems. Some mycotoxins are immunocompromising, thereby reducing resistance to communicable disease. It's estimated that two fifth of the world's crops, as well as several basic foods, are contaminated by mycotoxin-producing fungi (Eaton & Gallagher, 1994). The degree of biological effect caused by mycotoxins varies counting on the kind and supply, exposure route and dose, species status, and underlying subclinical

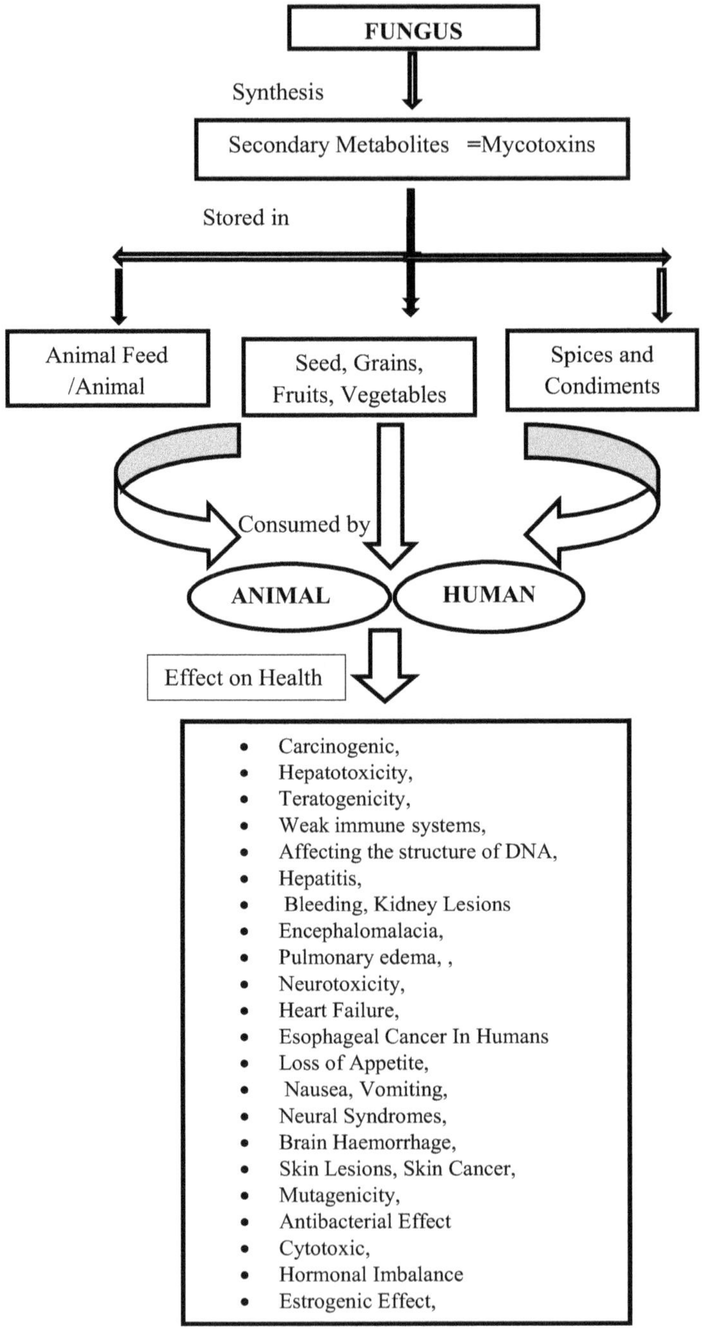

Fig. 17.4 Impact of fungal mycotoxin

conditions of the topic. For example, however, aflatoxin B1 (AFB1) is one in all the foremost potent liver carcinogens in humans (Fletcher & Blaney, 2016). Many countries have set limits for human and animal exposure to those mycotoxins known to cause adverse health outcomes. These laws vary by country, mycotoxin type, species, and food supply. International food laws are expected to be enforced within the close to future; however, adherence to such laws is difficult, particularly in countries where food provides are scarce. Higher quantities of mycotoxins are required in food and feed to produce adverse health symptoms in human and animals. It damages the system. These mycotoxins are restricted by limiting the consumption of contaminated food and feeds. Accessibility of mycotoxins information in numerous commodities is important for the adoption of safety against mycotoxins, and pesticides mycotoxin contamination is managed by spreading data regarding mycotoxins, predicting and preventing contamination, police investigation mycotoxins in grains, effective use of contaminated grain, and breeding crops for fungous resistance (Kam et al., 2007). Mycotoxin plays a big risk and food safety issue to the health and well-being of human and animals. Therefore, though the mycotoxin concentrations might considerably scale back throughout food process by physical treatment as cleansing and edge or by change of state, frying, baking, cooking, and extrusion, i.e., thermal process (Mitchell et al., 2014), nevertheless, it can't be fully eliminated. It's concluded that mycotoxins represent a big downside for the animal feed trade and on-going risk to the safety of the feed supply. Thus, because of the increasing public pressure, it's vital to produce an additional safe and eco-friendly way to manage plant toxin contamination. Use of microorganism antagonists as bacterium, yeast, and fungi is appropriate to cut back the chemical use and facilitate as biomanagement agents. Recent approach for safeguarding animals from harmful effects of mycotoxin contaminated feed is to use substances as mycotoxin binders to cut back the plant toxin contamination. These binders in addition to the diet scale back the absorption of mycotoxins from the alimentary canal and their future distribution to blood and target organs, so preventing or reducing mycotoxicosis in stock.

17.6 Conclusion

Mycotoxin is a well-known food safety risk that could be a threat to human and livestock health and has high economic significance within the food industry. Recently, the food industry has become tuned in to the new term for modified mycotoxins introduced by Rychlik et al. (masked mycotoxin) (Takahashi-Ando et al., 2002).

Prevention is the most vital and effective method in reducing fungal growth and mycotoxin production to ensure food safety, good agricultural practices (GAP) in the field, management measures of harvest and storage, physical processes (cleaning, milling, etc.), implementation of biotechnological application, biological management through the utilization of controlled atmosphere throughout storage, detoxification/degradation, and fermentation techniques are some preventive measures to regulate mycotoxin prevalence. Physical ways (dehulling, washing, sorting, and cleansing of visible moldy seed) reduce completely different mycotoxin species in foods despite grain genre (Rychlik et al., 2014; AycanCinar Elif, 2018). Scudamore and Patel (2000) and Patel (2011) ascertained a discount of T-2 (62%) and HT-2 (53%) and DON (50%) in wheat seeds after cleaning. Scudamore and Patel also reported a 32% reduction in fumonisin levels in corn in an industrial enterprise (Scudamore & Patel, 2008). Moreover, milling is an important effect in the reduction of *Fusarium mycotoxins* in grains particularly wet milling of maize, which has shown to result in the degradation of mycotoxins (Scudamore & Patel, 2000).

Preharvesting is taken into account initially and one among the foremost necessary stages to forestall mold growth and mycotoxin synthesis. Many methods are out there for the production of healthy products and to cut back the mold formation at preharvesting, together with selection of plants per the soil structure and production capability, use of plant, that is immune to fungi and insects, irrigation time, create fertilization, and use of pesticides to forestall insect injury (Park, 2002).

Harvesting at the suitable time periods (low wetness and full maturity) is important for reducing the chance of a mycotoxin contamination since overmaturity creates sensitivity to mold growth. To boot, appropriate harvest instrumentation and procedures ought to be used, and crops ought to be dried once maturity to each grain wetness reaches safe levels (Barkai & Paster, 2008; World Health Organization, 2018).

17.7 Summary

Mycotoxins are substances produced by fungi and are poisonous to humans and other animals if consumed in sufficient quantities. Toxicity resulting from consumption of mold-contaminated foods has been reported to occur since ancient times. Mycotoxins produced (notably and aflatoxins) have caused many epidemics in the human populations of various countries throughout history and in the current age. They can severely impact our food and must be managed through programs of active surveillance and management.

Toxicity most commonly arises after ingestion; however, dermal and inhalation exposure routes can also result in adverse effects. Different mycotoxins and disrupt and critically damage organs. compromise the immune system, or cause cancer. The degree of biological effect caused by mycotoxins varies depending on the type and source, exposure route and dose, species susceptibility, and underlying subclinical conditions of the subject. However, mycotoxin contamination can be managed by disseminating knowledge about mycotoxins, predicting and preventing contamination, detecting mycotoxins in grains, effective use of contaminated grain, and breeding crops for fungal resistance. But, the efficacy of the preharvest and postharvest methods for mycotoxin reduction varies, and none of them confers absolute protection; therefore, the use of multiple strategies is desirable to minimize mycotoxin exposure.

References

Abrunhosa, L., Santos, L., & Venancio, A. (2006). Degradation of ochratoxin A by proteases and by a crude enzyme of *Aspergillus niger*. *Food Biotechnology, 20*, 231–242.

Alshannaq, A., & Yu, J. H. (2017). Occurrence, toxicity, and analysis of major mycotoxins in food. *International Journal of Environmental Research and Public Health, 14*(6), 632.

Aoudia, N., Callu, P., Grosjean, F., & Wogan, G. N. (1963). Aflatoxins B and G. *Journal of the American Chemical Society, 85*, 1706–1707.

Asao, T., Buchi, G., Adbel-Kader, M.M., Chang, S.B., Wick, E.L., Wogan, G.N., (1963). Aflatoxins B and G. *Journal of American Chemical Society, 85*, 1706–1707.

Barkai, G. R., & Paster, N. (2008). Moldy fruits and vegetables as a source of mycotoxins: Part 1. *World Mycotoxin Journal, 1*(2), 147–159.

Bennett, J. W., & Krska, R. (2009). Formation, determination and significance of masked and other conjugated mycotoxins. *Analytical and Bioanalytical Chemistry, 395*(5), 1243–1252.

Berthiller, F., Schuhmacher, R., Adam, G., Krska, R., (2009). Formation, determination and significance of masked and other conjugated mycotoxins Analytical and Bioanalytical Chemistry.

Binder, E. M., & Binder, J. (1998). Microorganism, method for obtaining same and feed additive. *PAT*, EP1042449.

Binder, E. M., Tan, L. M., Chin, L. J., Handl, J., & Richard, J. (2007). Worldwide occurrence of mycotoxins in commodities, feeds and feed ingredients. *Animal Feed Science and Technology, 137*, 265282.

Blackwell, B. A., Gilliam, J. T., Savard, E. M., Miller, J. D., & Duvick, J. P. (1999). Oxidative deamination of hydrolyzed fumonisin B1 (AP1) by cultures of Exophialaspinifera. *Natural Toxins, 7*, 3138.

Celik, I., Oguz, H., Demet, O., Donmez, H. H., Boydak, M., & Sur, E. (2000). Efficacy of polyvinylpolypyrrolidone in reducing the immunotoxicity of aflatoxin in growing broilers. *British Poultry Science, 41*, 430–439.

Cinar, A., & Onbaşı, E. (2018). *On bust mycotoxins: The hidden danger in foods*. https://doi.org/10.5772/intechopen.89001

Deligöz, E., & Smith, T. K. (2005). *Mycotoxin sequestering agents: Practical tools for the neutralisation of mycotoxins* (pp. 323–340). The Mycotoxin Blue Book, Utah University Publications.

Eaton, D. L., & Gallagher, E. P. (1994). Mechanisms of aflatoxin carcinogenesis. *Annual Review of Pharmacology and Toxicology, 34*(1), 135–172.

Etzel, R. A. (1999). Indoor mold and children's health. *Environmental Health Perspectives, 107*(Suppl 3), 463.

Evans, J., & Dawson, K. (2007). The ability of Mycosorb to bind toxins present in endophyte-infected tall fescue. In *Science and Technology in the feed industry.* North American Biosciences Center, Alltech, Inc.

Fletcher, M. T., & Blaney, B. J. (2016). Mycotoxins. In *Encyclopaedia of food grains* (2nd ed., pp. 290–296). Elsevier.

Fuchs, E., Binder, E. M., Heidler, D., & Krska, R. (2002). Structural characterization of metabolites after the microbial degradation of type A trichothecenes by the bacterial strain BBSH 797. *Food Additives and Contaminants, 19*, 379–386.

Galvano, F., Piva, A., Ritieni, A., & Galvano, G. (2001). Dietary strategies to counteract the effects of mycotoxins: A review. *Journal of Food Protection, 64*, 120–131.

Gareis, M. (1994). Maskiertemykotoxine. *UbersTierernahr, 22*, 104–113.

Huwig, A., Freimund, S., Kappeli, O., & Dutler, H. (2001). Mycotoxin detoxication of animal feed by different adsorbents. *Toxicology Letters, 122*, 179–188.

Kabak, B., & Dobson, A. D. W. (2006). Strategies to prevent mycotoxin contamination of food and animal feed: A review. *Critical Reviews in Food Science and Nutrition, 1*(46), 593–619.

Kam, P., Bianchini, A., & Bullerman, L. B. (2007). Inhibition of mold growth by sourdough bread cultures. *Rurals, 2*.

Kavin, K. (2018). Fungi biology & application. In *Fungal infections of humans* (3rd ed., pp. 251–270). Willey Blackwell.
Kleiner, I., Flegar-Mestric, Z., Zadro, R., Breljac, D., StanovicJanda, S., Stojkovic, R., Marusic, M., Radacic, M., & Boranic, M. (2001). The effect of the zeolite clinoptilolite on serum chemistry and hematopoiesis in mice. *Food and Chemical Toxicology, 39*, 717–727.
Megharaj, M., Garthwaite, I., & Thiele, J. H. (1997). Total biodegradation of the oestrogenic mycotoxin zearalenone by a bacterial culture. *Letters in Applied Microbiology, 24*, 329–333.
Mitchell, N. J., Marroquín-Cardona, A. G., Romoser, A., Phillips, T. D., & Hayes, A. W. (2014). *Mycotoxins. Reference module in biomedical sciences*. Elsevier.
Molnar, O., Schatzmayr, G., Fuchs, E., & Prillinger, H. (2004). *Trichosporon mycotoxinivorans* sp. *nov*., A new yeast species useful in biological detoxification of various mycotoxins. *Systematic and Applied Microbiology, 27*, 661–671.
Nakazato, M., Morozumi, S., Saito, K., Fujinuma, K., Nishima, T., & Kasai, N. (1990). Interconversion of aflatoxin B1 and aflatoxicol by several fungi. *Applied and Environmental Microbiology, 56*, 1465–1470.
Papaioannou, D. S., Kyriakis, C. S., Papasteriadis, A., & Roumbies, N. (2002a). A field study on the effect of in-feed inclusion of a natural zeolite (clinoptilolite) on health status and performance of sows/gilts and their litters. *Research in Veterinary Science, 72*, 51–59.
Papaioannou, D. S., Kyriakys, S. C., Papasteriadis, A., Roumbies, N., Yannakopoulos, A., & Alexopoulos, C. (2002b). Effect of in-feed inclusion of a natural zeolite (clinoptilolite) on certain vitamin, macro and trace element concentrations in the blood, liver, and kidney tissus of sows. *Research in Veterinary Science, 72*, 61–68.
Park, D. L. (2002). Effect of processing on aflatoxin. In J. W. DeVries, M. W. Trucksess, & L. S. Jackson (Eds.), *Mycotoxins and food safety* (pp. 173–179). Springer.
Peteri, Z., Téren, J., Vágvölgyi, C., & Varga, J. (2007). Ochratoxin degradation and adsorption caused by astaxanthin-producing yeasts. *Food Microbiology, 24*, 205–210.
Ramos, A. J., & Hernandez, E. (1997). Prevention of aflatoxicosis in farm animals by means of hydrated sodium calcium aluminosilicate addition to feedstuffs: A review. *Animal Feed Science and Technology, 65*, 197–206.
Ramos, A. J., Fink-Gremmels, J., & Hernandez, E. (1996). Prevention of toxic effects of mycotoxins by means of non-nutritive adsorbent compounds. *Journal of Food Protection, 59*, 631–641.
Rychlik, M., Humpf, H. U., Marko, D., Dänicke, S., Mally, A., Berthiller, F., Klaffke, H., & Lorenz, N. (2014). Proposal of a comprehensive definition of modified and other forms of mycotoxins including "masked" mycotoxins. *Mycotoxin Research, 30*(4), 197–205.
Schatzmayr, G., Zehner, F., TÃubel, M., Schatzmayr, D., Klimitsch, A., Loibner, A. P., & Binder, E. M. (2006). Microbiologicals for deactivating mycotoxins. *Molecular Nutrition & Food Research, 50*, 543–551.
Scudamore, K. A., & Patel, S. (2000). Survey for aflatoxins, ochratoxin A, zearalenone and fumonisins in maize imported into the United Kingdom. *Food Additives and Contaminants, 17*(5), 407–416.
Scudamore, K. A., & Patel, S. (2008). The fate of deoxynivalenol and fumonisins in wheat and maize during commercial breakfast cereal production. *World Mycotoxin Journal, 1*, 437–448.
Takahashi-Ando, N., Kimura, M., Kakeya, H., Osada, H., & Yamaguchi, I. (2002). A novel lactonohydrolase responsible for the detoxification of zearalenone: Enzyme purification and gene cloning. *Biochemical Journal, 365*, 1–6.
Tola, M., & Kebede, B. (2016). Occurrence, importance and control of mycotoxins: A review. *Food Science and Technology, 2*, 1191103.
Turnera, N. W., Subrahmanyam, S., & Piletskyb, S. A. (2009). Analytical methods for determination of mycotoxin: A review. *AnayticaChimicaActa, 632*, 168–180.
Ueno, Y., Nakayama, K., Ishii, K., Tashiro, F., Minoda, Y., Omori, T., & Komagata, K. (1983). Metabolism of T-2 toxin in *Curtobacterium* sp. Strain 114-2. *Applied and Environmental Microbiology, 46*, 120–127.

Underhill, K. L., Rotter, B. A., Thompson, B. K., Prelusky, D. B., & Trenholm, H. L. (1995). Effectiveness of cholestyramine in the detoxification of zearalenone as determined in mice. *Bulletin of Environmental Contamination and Toxicology, 54*, 128–134.

Wang, P., Afriyie-Gyawu, E., Tang, Y., Johnson, N., Xu, L., Tang, L., Huebner, H., Ankrah, N. A., Ofori-Adjei, D., Ellis, R. W., Jolly, P., Williams, J., Wang, J. S., & Phillips, T. D. (2008). NovaSil clay intervention in Ghanaians at high risk for aflatoxicosis: II. Reduction in biomarkers of aflatoxin exposure in blood and urine. *Food Additives and Contaminants, 25*, 622–634.

World Health Organization. (2018). *Mycotoxins*. Courtesy: https://www.who.int/news-room/fact-sheets/detail/mycotoxins

Wu, Q. K., Jezkova, A., Yuan, Z., Pavlikova, L., Dohnal, V., & Kuca, K. (2009). Biological degradation of aflatoxins. *Drug Metabolism Reviews, 41*, 1–7.

Chapter 18
Efficacy of Phytohormones on Mycotoxin Treated Maize Seeds (*Zea mays* L.)

Gajendra Prasad, Nitu Kumari, and Khwaja Salahuddin

Contents

18.1 Introduction

Phytohormones are also known as plant growth regulators that influence several biological activities. These are important for growth and development of plants (Sembedner & Parthier, 1993). Phytohormones function as important chemical messengers as they modulate many cellular processes in plants and also coordinate different signaling pathways during exposure to mycotoxins (Vob et al., 2014).

G. Prasad · N. Kumari
Seed Pathology and Mycotoxin Laboratory, University Department of Botany, L. N. Mithila University, Darbhanga, India

K. Salahuddin (✉)
Department of Botany, M.R.M College, L. N. Mithila University, Darbhanga, India

A. C. Shukla (ed.), *Applied Mycology*, Fungal Biology,
https://doi.org/10.1007/978-3-030-90649-8_18

Maize is one of the most important staple food cereals grown throughout the world. It is commonly and directly consumed as processed food, indirectly as additive in most of food products and used as animal fodder (Lutz, 1994). Maize may be contaminated with storage fungi, producing mycotoxin that can be detrimental to health of human beings and animals. It is susceptible to a number of ear and kernel rots that can cause damage in humid areas.

Mycotoxins are naturally occurring toxins by certain fungi. These are found in food and can cause a variety of health effects, posing a serious health threat to both humans and live stocks. These fungi grow rapidly on a variety of natural substrates, and consumption of fungi-contaminated food can be detrimental (Kpodo et al., 1996). The crops that are frequently affected by *Aspergillus* spp. include cereals (corns, wheat, rice, and millet), oilseeds (pea nut, soya bean, sunflower), and spices (coriander, chilly, pepper, black pepper, turmeric, and ginger). The production of mycotoxin depends on high moisture content (20–25%) and high relative humidity (70–90%) (Shah et al., 2010).

The fungi-producing mycotoxins fall broadly into two groups that invade before harvest commonly called field fungi, and those that occur only after harvest called storage fungi. The four major aflatoxins are called B_1, B_2, G_1, and G_2 based on their fluorescence under UV light. Aflatoxin B_1 is the most potent, natural, highly toxic, and carcinogenic secondary metabolite produced by *Aspergillus flavus* on agricultural crops (Lentopoloys et al., 2003). It is evident that aflatoxin is known for phytotoxicity with respect to seed germination and inhibition of root and hypocotyl elongation (Daskek & Llwellyn, 1983). AFB_1 was found at the highest concentration in contaminated food and feed. The different concentrations of AFB_1 inhibited chlorophyll, carotenoids, protein, and lipid contents and reduced the growth and germination of *Zea mays* L. and *Vicia faba* seeds (El-Naghy et al., 1999). It is properly documented that mycotoxin produced by specific filamentous fungi also causes significant reduction in crop yield and economic losses (Bhatnagar & Garcia, 2001). Aflatoxin content may vary with the season and storage time. It was observed that toxin contamination was higher during the rainy season and increased with increasing storage time (Ahmad, 1993). Aflatoxins are found in food chain in different pathways and make important detrimental damages in the metabolism of organisms (Hela et al., 2000). The seeds have been shown to be naturally contaminated with various levels of aflatoxin in the field (Ayalew, 2010) and during storage (Bilgrami & Sinha, 1992). Aflatoxins cause economic loss at all levels of food and feed production, processing, and distribution.

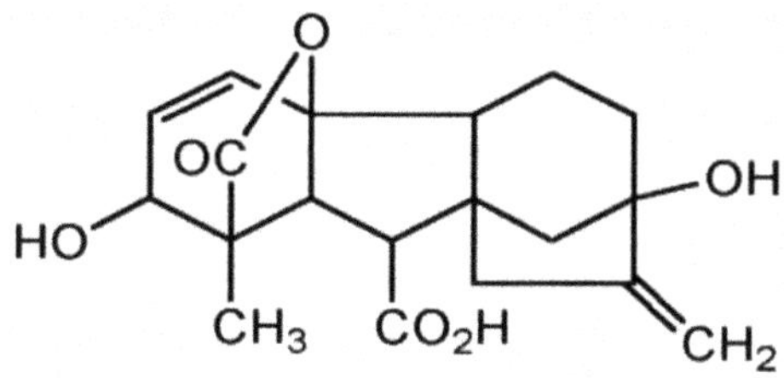

Chemical structure of Gibberellin (GA_3)

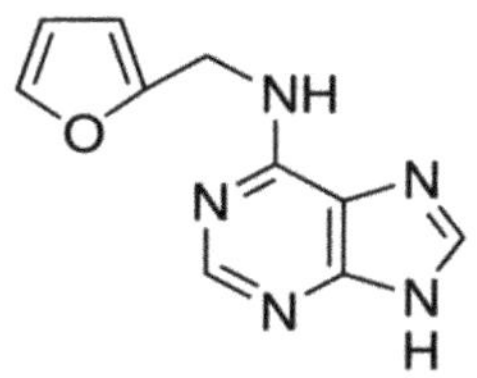

Chemical structure of Kinetin (Kn)

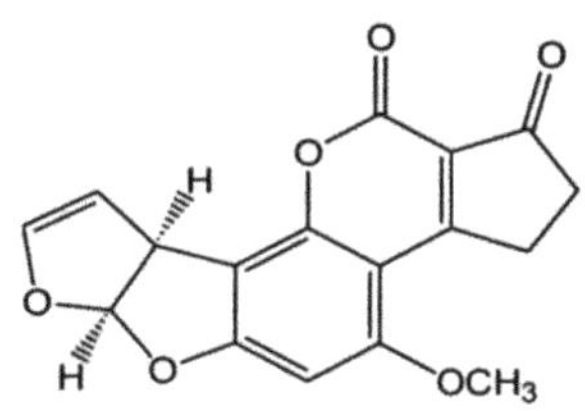

Chemical structure of Aflatoxin (AFB_1)

This article discusses hormonal action on maize seed, the action of mycotoxin, and finally the reversal effect with the application of phytohormones leading to the production of crops in the future.

18.2 Material and Methods

18.2.1 Collection of Healthy Seeds

The healthy seeds of maize (Madhuri-01) were obtained from Dayal Traders Manures and seed storage house, Darbhanga, India. The seeds were evaluated physically and were surface sterilized with 0.1% $HgCl_2$ for 2 min and then washed with sterilized distilled water.

18.2.2 Collection of Aflatoxin B1 and Phytohormones

The stock solution of AFB_1 was obtained from Sigma, USA. Pure phytohormones such as GA_3 and Kn were obtained from the local scientific stores, Darbhanga, India, and stored before analysis.

18.2.3 Preparation of Stock Solution

The stock solution of AFB_1 and phytohormones were prepared separately in ethanol from, which the dilution (i.e., 0.1, 0.25, 0.5, 1.0, and 2.0 ppm) was made in sterilized distilled water. Solutions of aflatoxin B_1 and phytohormones (2 ppm) were also mixed in different ratios (1:1, 1:2, 2:1, 1:3, 3:1 v/v) in order to record the combined effects.

The seeds were soaked initially in distilled water for 1 h and subsequently in different combinations of AFB_1 and phytohormones for 20 h. For each treatment, 100 seeds were taken in triplicates. The soaked seeds were then placed on moist blotter paper in sterilized petriplates and were kept for germination under automatic regulated seed germinator at 28 ± 2 °C (McLean et al., 1995; Sinha, 1990).

18.2.4 Seed Germination Index (GI)

It was calculated after 5 days of incubation according to the formula as given below:

$$GI = \frac{\text{Number of germinated seeds}}{\text{Number of seeds observed}} \times 100$$

The seedling growth (root and shoot lengths) was determined on the 7th day by measuring the lengths of radicle and plumule.

The data were analyzed statistically, i.e., t-test for seed germination and F-test for seedling growth. Statistical calculations were carried out using the ANOVA test (Dospekov, 1984).

18.2.5 Quantitative Estimation of Starch

About 200 mg freshly grounded sample was thoroughly shaken in 80% warm ethanol. After 5 min, the supernatant was decanted and centrifuged. Starch was estimated from the residue by adding 5 ml distilled water and mixing it thoroughly with 6.5 ml of 52% perchloric acid (prepared by adding 270 ml of 70% perchloric acid into 100 ml of distilled water). This was followed by constant stirring for 10 min. After 15 min, again 20 ml of distilled water was added and centrifuged and then the supernatant was decanted. The extraction was repeated thrice and the supernatant was mixed together. The volume was raised to 100 ml by adding extra distilled water (stock solution). One milliliter of this stock solution was mixed with 10 ml at 0.1% anthrone reagent (760 ml conc. H_2SO_4 was diluted up to 1 l and 1 g anthrone dissolved) and heated at 100 °C for 12 min. The bluish-green solution was cooled at room temperature. Optical density was recorded at 630 nm under spectrophotometer. Readings were compared with the standard curve of starch prepared through similar procedure.

18.2.5.1 Assay of α-Amylase

It was assayed according to the method of Bernfeld (Bernfeld, 1955). One milliliter of starch solution was taken in experimental and control tubes. Approximately 0.5 ml of enzyme source (homogenate, i.e., 100 mg powder/ml acetate buffer, pH 4.8) was added in experimental and control tubes, but in the latter, the homogenate was previously dipped in boiling water bath for 5 min before being added. Both experimental and control tubes were incubated for 30 min at 37 °C. After incubation, both were dipped in boiling water bath for 5 min and subsequently cooled in running tap water. Three milliliter of KI solution (prepared by adding 25.4 mg I_2 and 400 mg KI in 100 ml of distilled water) was added in both tubes and mixed with the help of cyclomixer. The supernatant was used as crude extract for the extraction of enzyme. Colorimetric reading was taken at 620 nm, and difference of control and experimental reading gave the activity of α-amylase which was expressed in the value of optical density. Data recorded at each stage were the average of the three replicates.

18.2.6 Quantitative Estimation of Protein

It was done by the method of Lowry in both control and treated seeds (Lowry et al., 1951). Approximately 100 mg of seed flour were crushed in 100 ml of acetate buffer (pH 4.8) and centrifuged. The test solution (2 ml) was taken from the supernatant. To this, 10 ml of alkaline reagent (prepared by mixing 50 ml of 2% Na_2CO_3 solution in 0.1 N NaOH solution and 1 ml 0.5% $CuSO_4$ in 1% Na-K tartrate solution) was mixed thoroughly and was allowed to stand at room temperature for 10 min. One milliliter of diluted Folin–Ciocalteu reagent (1:3 in distilled water) was added. After 10 min, the extraction was added at 600 nm against the blank prepared by albumin.

18.3 Results and Discussion

The effects of phytohormones (GA_3 and Kn) and AFB_1on individual and combined ratios were observed on the seed germination of maize (Table 18.1; Figs. 18.1 and 18.2). A significant fall in seed germination at all concentrations of AFB_1 was noticed, but the maximum inhibition (79%) in all parameters was observed at the highest concentration (2 ppm) of AFB_1. The seed germination decreased by prolonged soaking intervals than the control treatment. When aflatoxin was mixed with GA_3, then inhibition was reversed up to 20.40%. The same case was noticed when aflatoxin was mixed with Kn. The toxin functioned as anti-gibberellin by inhibiting DNA synthesis (Cavusoglu & Solusoglu, 2015). This is due to the fact that GA_3 is capable of breaking the dormancy and inducing seed germination and seedling growth.

The effects of phytohormones (GA_3 and Kn) and AFB_1 on the length of roots were observed (Table 18.2). The root length showed positive response to GA_3 and Kn and grew up to 12.11 cm and 11.76 cm, respectively, but when AFB_1 was applied separately, then the root length grew up to 3.42 cm showing 65% inhibition. But when AFB_1 and GA_3 were treated, the root length was reversed up to 8.92 cm, and similar response (8.65 cm) showed if AFB_1 and Kn were applied.

The effects of phytohormones (GA_3 and Kn) and AFB_1 on the length of shoots were observed (Table 18.3). The shoot length showed positive response to GA_3 and Kn and grew up to 8.89 cm and 8.76 cm, respectively, but when AFB_1 was applied separately, then the shoot length grew up only 2.25 cm showing 63% inhibition. AFB_1 restricted the growth of plant by inhibiting seed germination (Reddy et al., 2010). But when AFB_1 and GA_3 were treated, then the shoot length was reversed up to 4.94 cm, and similar response (4.83 cm) showed if AFB_1 and Kn were applied.

The effects of phytohormones (GA_3 and Kn) and AFB_1 on the content of starch were observed (Table 18.4). Starch is accumulated during seed development, and germination is characterized by its degradation. The dissolution of insoluble starch to soluble maltose-dextrins occurs in the presence of α-amylase which catalyzes the hydrolysis of glucosidic linkage (Prasad et al., 2018a). A fluctuation in the starch content was observed in maize seeds due to the treatment of phytohormones and

Table 18.1 Impact of hormones with respect to seed germination on AFB_1-treated maize seeds

Observations	Seed germination (Mean value ± SE)	Difference with control	Percentage Inhibition
Control	98 ± 0.47	–	–
GA_3 (2 ppm)	100 ± 1	2	2
Kn (2 ppm)	100 ± 1	2	2
AFB_1 (2 ppm)	21 ± 0.57	13.50	79
AFB_1 + GA_3 (2 ppm v/v)	78 ± 0.47	22.27	20.40
AFB_1 + Kn (2 ppm v/v	79 ± 0.74	27.02	19.38

Fig. 18.1 Combined effect of AFB_1 and GA_3 (2 ppm v/v) on maize seedling growth

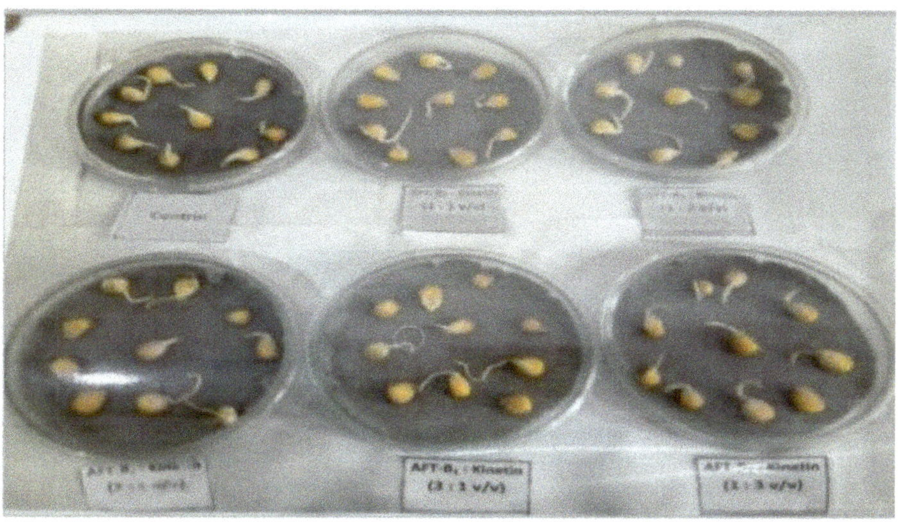

Fig. 18.2 Combined effect of AFB_1 and Kn (2 ppm v/v) on maize seed germination

mycotoxin. The maximum amount of starch (62.41 mg/100 mg) was recorded when maize seeds were treated with AFB_1 at 2 ppm concentration. The amount was decreased considerably up to 12.25 mg/100 mg and 13.12 mg/100 mg when AFB_1 was mixed with GA_3 and Kn, respectively.

The effects of phytohormones (GA_3 and Kn) and AFB_1 on the content of α-amylase were observed in different time duration (Table 18.5). The enzyme activity was completely lost in control set as well as GA_3 and Kn treatment in maize seeds after 90-min interval. However, enzyme activity was also lost in maize seeds due to treatment of GA_3 and Kn individually after 60-min interval.(Sakdeo 2016). The maximum inhibition as a result of AFB_1 treatment was observed. With increase

Table 18.2 Impact of hormones with respect to root length on AFB_1-treated maize seeds

Observations	Root length (cm)	Difference with control	Percentage Inhibition
Control	9.78 ± 0.12	–	–
GA_3 (2 ppm)	12.11 ± 1.00	2.33	23.82
Kn (2 ppm)	11.76 ± 0.15	0.35	20.24
AFB_1 (2 ppm)	3.42 ± 0.09	8.69	65.03
AFB_1 + GA_3 (2 ppm v/v)	8.92 ± 0.07	3.19	8.79
AFB_1 + Kn (2 ppm v/v	8.65 ± 0.04	3.46	11.55

Table 18.3 Impact of hormones with respect to shoot length on AFB_1-treated maize seeds

Observations	Shoot length (cm)	Difference with control	Percentage inhibition
Control	6.15 ± 0.24	–	–
GA_3 (2 ppm)	8.89 ± 0.08	2.74	44.50
Kn (2 ppm)	8.76 ± 0.09	2.61	42.00
AFB_1 (2 ppm)	2.25 ± 0.08	3.40	63.00
AFB_1 + GA_3 (2 ppm v/v)	4.94 ± 0.12	0.12	19.00
AFB_1 + Kn (2 ppm v/v	4.83 ± 0.12	1.32	21.00

Table 18.4 Impact of hormones on starch content on AFB_1-treated maize seeds

Observations	Starch content (mg/100 mg)
Control	4.75 ± 0.13
GA_3 (2 ppm)	0.93 ± 0.17
Kn (2 ppm)	1.14 ± 0.24
AFB_1 (2 ppm)	62.41 ± 0.23
AFB_1 + GA_3 (2 ppm v/v)	12.25 ± 0.18
AFB_1 + Kn (2 ppm v/v)	13.12 ± 0.10

Table 18.5 Impact of hormones on α-amylase content on AFB_1-treated maize seeds

	α-amylase activity			
Observations	0 min	30 min	60 min	90 min
Control	0.73 ± 0.09	0.34 ± 0.09	0.14 ± 0.08	–
GA_3 (2 ppm)	0.56 ± 0.12	0.24 ± 0.08	–	–
Kn (2 ppm)	0.62 ± 0.07	0.26 ± 0.12	–	–
AFB_1 (2 ppm)	0.97 ± 0.07	0.61 ± 0.07	0.30 ± 0.07	0.12 ± 0.04
AFB_1 + GA_3 (2 ppm v/v)	0.65 ± 0.09	0.31 ± 0.10	0.13 ± 0.04	0.09 ± 0.57
AFB_1 + Kn (2 ppm v/v)	0.73 ± 0.17	0.34 ± 0.08	0.21 ± 0.07	0.08 ± 0.04

in time incubation, the α-amylase activity gradually decreased in both germinated and nongerminated seeds.

The effects of phytohormones (GA_3 and Kn) and AFB_1 on the content of protein were observed in different time duration (Table 18.6). A highly significant fall in the level of protein in maize seedlings was observed during the treatment of different

Table 18.6 Impact of AFB_1 on protein content of maize seeds

Concentration of AFB1 (ppm)	Protein (mg/100 mg)	Difference with control	Percentage Inhibition
0.00	8.43 ± 0.10	–	–
0.10	7.14 ± 0.02	1.29	15.30
0.25	5.88 ± 0.02	2.55	30.24
0.50	4.65 ± 0.03	3.78	44.88
1.00	3.19 ± 0.05	5.24	62.15
2.00	2.23 ± 0.08	6.20	73.44

Table 18.7 Combined effects of AFB1 and GA3 on protein content of maize seeds

	Protein content (mg/100 mg)		
Concentration of AFB_1: GA_3 (2 ppm v/v)	Amount (Mean value ± SE)	Difference with control	Percentage inhibition
0:0	8.35 ± 0.04	–	–
1:1	4.86 ± 0.06	3.49	41.79
1:2	5.59 ± 0.05	2.76	33.05
2:1	3.87 ± 0.02	4.48	53.65
1:3	7.12 ± 0.05	1.23	14.73
3:1	3.19 ± 0.01	5.16	61.79

Table 18.8 Combined effects of AFB1 and Kn on protein content of maize seeds

	Protein content (mg/100 mg)		
Concentration of AFB_1: Kn (2 ppm v/v)	Amount (Mean value ± SE)	Difference with control	Percentage inhibition
0:0	8.45 ± 0.04	–	–
1:1	4.76 ± 0.06	3.69	43.66
1:2	5.56 ± 0.05	2.89	34.20
2:1	3.85 ± 0.02	4.60	54.43
1:3	7.10 ± 0.05	1.35	15.97
3:1	3.17 ± 0.01	5.28	62.48

concentration of AFB_1. The percentage inhibition was found to be the maximum at (2 ppm) concentration of AFB_1. Protein is inhibited due to nonavailability of mRNA.

The biochemical effect of AFB_1 in combination with GA_3 on protein content was noticed when treated in various combination ratios (Table 18.7). The maximum and minimum inhibitions were observed when concentrations of AFB_1 and GA_3 are in the ratio of 3:1 and 1:3, respectively (Prasad et al. 2018b).

The biochemical effect of AFB_1 in combination with Kn on protein content was noticed when treated in various combination ratios (Table 18.8). The maximum and minimum inhibitions were observed when concentrations of AFB_1 and GA_3 are in the ratio of 3:1 and 1:3, respectively (Dilip et al., 2017; Pratiwi et al., 2015).

18.4 Conclusion

Maize is an important agricultural crop and has suffered from various fungal diseases resulting in the loss of its productivity. Phytohormones have played pivotal roles in increasing the length of root, shoot, and seed germination. Definitely aflatoxin has caused much damage to the yield of maize crop if applied separately. The reversal effect was noticed if phytohormones and aflatoxin were applied in combination and has gained much prominence with respect to the yield in such infected maize crops. If this method is applied by farmers, then it will be financially fruitful due to high yields as well as beneficial for the consumers with respect to the protein quality intake.

References

Ahmad, S. K. (1993). Mycoflora changes and aflatoxin production in stored black gram seeds. *The Journal of Stored Products Research, 29*, 33–36.

Ayalew, A. (2010). Mycotoxins and surface and internal fungi of maize in Ethiopia. *African Journal of Food, Agriculture, Nutrition and Development, 10*, 4109–4123.

Bernfeld, P. (1955). Amylase α and β-methods in enzymology. *Scientific Research, 1*, 149–158.

Bhatnagar, D., & Garcia, S. (2001). *Aspergillus*. In R. G. Labbe & S. Garcia (Eds.), *Guide to foodborne pathogens* (pp. 35–49). Wiley.

Bilgrami, K. S., & Sinha, K. K. (1992). Aflatoxins-their biological effects and ecological significance. In D. Bhatnagar, E. B. Lillehoj, & D. K. Arora (Eds.), *Handbook of applied mycology* (Vol. 5, pp. 59–86). Marcel dekker.

Cavusoglu, A., & Solusoglu, M. (2015). The effects of exogenous GA_3 on seed germination of the fruit species. *Derleme, 8*(1), 06–09.

Daskek, W. V., & Llwellyn, G. C. (1983). Mode of action of the hepatocarcinogens, aflatoxins in plant system: A review. *Mycopathologia, 81*, 83–94.

Dilip, W. S., Singh, D., Moharana, D., Rout, S., & Patra, S. S. (2017). Effect of different concentrations of GA_3 at different time intervals on seed germination and seedling growth of Rangpur lime. *Journal of Agroecology and Natural Resource Management, 4*(2), 157–165.

Dospekov, B. A. (1984). *Field experimentation, statistical procedures* (V. P. Kolykhmatov, Ed., Vol. 352). Mir Publishers.

El-Naghy, M. A., Fadl-llah, E. M., & Samhan, M. (1999). Effects of aflatoxin B_1 on germination, growth and Metabolic activities of some crop plants. *Cytobiosis, 97*, 87–93.

Hela, K., Cardwell, K., Setamou, F. M., & Phoehlig, H. M. (2000). The influence of storage practices on aflatoxin contamination in maize in four agroecological zones of Benin, West Africa. *Journal of Stored Products Research, 36*, 365–382.

Kpodo, K., Sorensen, A. K., & Jakobsen, M. (1996). The mycotoxins in fermented maize products. *Food Chemistry, 56*, 147–153.

Lentopoloys, D., Siafaka, A., & Markaki, P. (2003). Black olives as substrate for *Aspergillus parasiticus*growth and aflatoxin B_1 production. *Food Microbiology, 20*, 119–126.

Lowry, O. H., Rosenbrough, N. J., Farr, A. L., & Randall, R. J. (1951). Protein measurement with the folin-phenol reagent. *Journal of Biological Chemistry, 193*, 265–275.

Lutz, C. (1994). *The functioning of the maize market in Benin: Spatial and temporal arbitrage on the market of staple food crop*. Ph.D thesis, University of Amsterdam.

McLean, M., Watt, M. P., Berjak, P., & Dutton, M. F. (1995). Aflatoxin B_1-Its effects on an in vitro plant system. *Food Additives & Contaminants, 12*(3), 435–443.

Prasad, G., Kumari, N., Kumari, S., & Kumari, S. (2018a). Efficacy of GA_3 on amino acid contents in aflatoxin B_1 treated maize seeds. *Bioglobia, 5*(1), 26–29.

Prasad, G., Mishra, V. K., & Kumari, N. (2018b). Effects of phytohormones on reversing the inhibitory action of aflatoxin B_1 on the growth of maize seeds. *International Journal of Plant & Soil Science, 22*(5), 1–5.

Pratiwi, C., Rahayu, W. P., Lioe, H. N., Herawati, D., Broto, W., & Ambarwat, S. (2015). The effect of temperature and RH for *Aspergillus flavus* BIO 2237 growth and aflatoxin production on soyabeans. *International Food Research Journal, 22*, 82–87.

Reddy, K. R. N., Raghavender, C. R., Reddy, B. N., & Salleh, B. (2010). Biological control of *Aspergillus flavus* growth and subsequent aflatoxin B_1 production in sorghum grains. *African Journal of Biotechnology, 9*, 4247–4250.

Sakdeo, B. M. (2016). Physiological effects of seed treatment with kinetin on seedling growth under laboratory and field conditions in green gram. *International Journal of Applied Research, 2*(7), 384–387.

Sembedner, G., & Parthier, B. (1993). The biochemistry, physiological and molecular actions of jasmonates. *Annual Review of Plant Biology, 44*, 569–586.

Shah, H. U., Thomas, J., Simpson, S. A., Khanzadi, F. K., & Sajda, P. (2010). Mould incidence and mycotoxin contamination in maize kernels from Swat valley, North West West Frontier Province of Pakistan. *Food and Chemical Toxicology, 48*, 111–1116.

Sinha, K. K. (1990). Incidence of mycotoxins in maize grains in Bihar State, India. *Food Additives & Contaminants, 7*, 55–61.

Vob, U., Bishopp, A., Farcol, E., & Bennett, M. J. (2014). Modelling hormonal response and development. *Trends in Plant Science, 19*, 311–319.

Chapter 19
The Untapped Potential of Fungi in Phenol Biodegradation

Reshmi Sasi, Serin Zachariah, and T. V. Suchithra

Contents

19.1 Introduction

Water pollution caused by phenolic pollutants is due to the expulsion of wastewater from industries such as petrochemicals, oil refineries, resin manufacturing, etc. (Villegas et al., 2016). Phenolic compounds are highly toxic even at lower concentrations and have severe effects on humans and animals. Due to the lethal effects of phenol, the US environmental protection agency (EPA) has included phenol and its derivatives in the list of high-priority pollutants (EPCRA, 2014). Removal of phenolic pollutants from the wastewater can be done with the help of numerous methods, among them biological methods gained more attention as it offers complete degradation of pollutants without the use of harmful chemicals and low cost of operation (Mahiudddin & Fakhruddin, 2012). Bacteria and fungi are the major classes of microbes that are being used for phenol degradation. Bacteria are the commonly employed microbes, while fungal degradation of phenolic pollutants is a

R. Sasi · S. Zachariah · T. V. Suchithra (✉)
School of Biotechnology, National Institute of Technology, Calicut, Kerala, India
e-mail: drsuchithratv@nitc.ac.in

A. C. Shukla (ed.), *Applied Mycology*, Fungal Biology,
https://doi.org/10.1007/978-3-030-90649-8_19

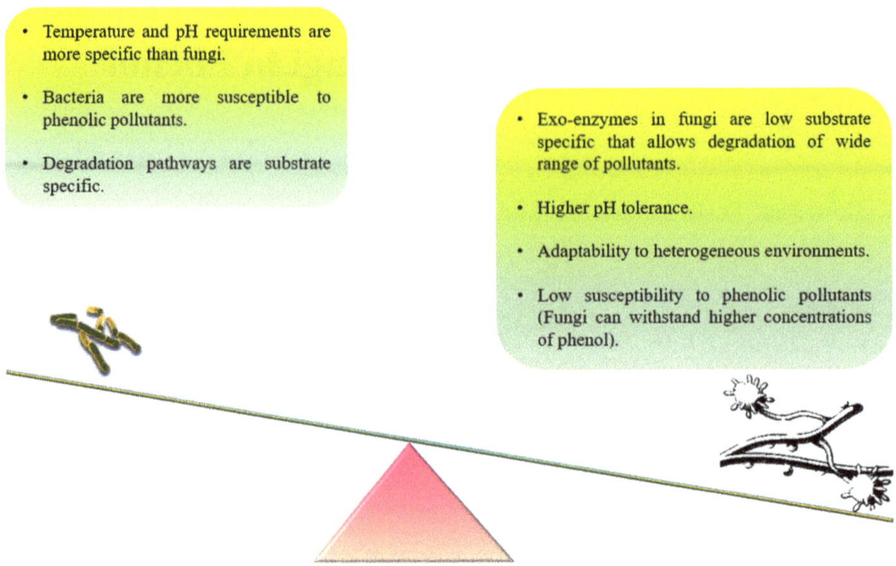

Fig. 19.1 A comparative representation of fungi v/s bacteria in phenol degradation

much less explored area. Fungi have so many advantages over bacteria such as they maintain stability in activity, especially at low pH, and are less susceptible to phenol toxicity when compared with bacteria (Fig. 19.1) (Yordanova et al., 2013). Filamentous fungi are the major groups used for phenol degradation, including white-rot fungi, *Penicillium, Fusarium,* etc. White-rot fungi are known for the degradation and mineralization of chlorophenol pollutants, and the process mainly depends on their degradation enzyme systems such as lignin degradation enzymes and cellular degradation enzymes. The lignin degradation system consists of laccase and peroxidase such as manganese peroxidase and lignin peroxidase (Valentín et al., 2013). Yeasts including candida maltose, *Oidium, Trichosporon cutaneum, Candida albicans,* etc. are good degraders of phenolic pollutants (Santos & Linardi, 2001).

19.2 Major Fungal Species for Phenol Degradation

Fungi are widely used as ideal candidates for the biodegradation of phenolic compounds. Their biochemical, metabolic, and ecological features are making them potential candidates for the biodegradation process. They have the capacity to extend their mycelial network, deeply penetrate the soil, and utilize the organic pollutants as their growth substrates. Despite other microorganisms, the unique characteristic of fungi is the presence of low substrate-specific catabolic enzymes like phenol oxidases in their extracellular matrix of the cell wall. Lignin peroxidases, manganese peroxidases, and laccases are the major ligninolytic enzymes secreted by certain white-rot fungi. Besides this, some other intracellular enzyme systems consisting of cytochrome P450 monooxygenases (CYP) and glutathione transferases are also present in some fungal species. Adaptability to an immense range of habitats makes them more suitable for large-scale bioremediation of polluted environments (Arisoy, 1998; Pointing, 2001).

Different species of basidiomycetes are one of the major fungal groups involved in the degradation of organic pollutants. The low substrate-specific exoenzymes present in saprotrophic basidiomycetes and white-rot fungi make them best suited for mycoremediation (Lang et al., 1995). Because of the low substrate specificity, ligninolytic enzymes can also target some persistent organic pollutants like polychlorinated biphenyls (PCB), chlordane, and lindane, which have structures similar to lignin (Arisoy, 1998; Pointing, 2001; Mansur et al., 2017). *Phanerochaete chrysosporium* was known as the model white-rot fungus involved in the degradation of many pesticides such as dichlorophenol and pentachlorophenol by utilizing ligninolytic enzymes exuded from the extracellular matrix (Bhalerao & Puranik, 2007; León-Santiesteban et al., 2016). *Anthracophyllum discolor, Ceriporiopsis subvermispora, Lentinus subnudus, Phlebia acanthocystis, Pleurotus ostreatus,* and *Lenzites betulina* are some other white-rot fungi effective in the biodegradation of organic pollutants such as pentachlorophenol (PCP), chlordane, and PCBs. *Phanerochaete sordida,* another saprotrophic basidiomycete, involves in the PCP degradation initiated by manganese peroxidase. Laccases are oxidases with copper as a cofactor and capable of oxidizing most phenolic and nonphenolic compounds. Laccase activity has been significantly observed in *Tinea versicolor* (Margot et al., 2013).

Mycorrhizal basidiomycetes such as *Gomphidius viscidus, Suillus variegatus*, and some edible basidiomycetes like *Agaricus bisporus and Lactarius piperatus* utilize the mechanism of biosorption to remove heavy metal contamination from phenolic compounds such as anthracene, TNT (2,4,6-trinitrotoluene), and crude oil. There are no enzyme systems or mediators involved in this degradation process (Margot et al., 2013). *Fusarium solani, Penicillium chrysogenum,* and *Scedosporium apiospermum* are some ascomycetes involved in nonligninolytic enzyme-mediated degradation. Cellulose and hemicellulose are the substrates to initiate the degradation process of some organopollutants (Tigini et al., 2009). *Fusarium incarnatum* is another ascomycete species involved in bisphenol degradation by using the laccase enzyme. Lipolytic enzymes and biosurfactants secreted by *Aspergillus niger, A. terreus, Cladosporium oxysporum,* and *Fusarium ventricosum* are used for the degradation of several pesticides (Bhalerao & Puranik, 2007). Yeasts such as *Oospora, Saccharomyces, Candida tropicalis, Debaryomyces,* and *Trichosporon cutaneum* were reported for active phenol degradation (Table 19.1).

Table 19.1 Different fungal species used for the biodegradation phenol and its derivatives

Type of fungi	Fungi	Compound	References
White-rot fungi	*Phanerochaete chrysosporium*	Phenol and Chlorophenols	León-Santiesteban et al. (2016), Reddy and Mathew (2001) and Anastasi et al. (2013)
	Phanerochaete sordida *Ceriporiopsis subvermispora* *Lenzites betulina*	PCP	Lamar et al. (1994), Lamar and Dietrich (1990, 1992) and Tortella et al. (2008)
	Anthracophyllum discolor	Chlorophenols	Acevedo et al. (2011)
Brown-rot fungi	*Laetiporus sulphureus*	Phenanthrene	Sack et al. (1997)
Tropical basidiomycetes	*Trametes villosa,* *Agrocybe perfecta* *Trichaptum bisogenum* *Peniophora cinerea*	PCP	Machado et al. (2005)
Mycorrhizal basidiomycetes	*Gomphidius viscidus*	Anthracene	Huang et al. (2010)
	Suillus variegatus	TNT	Meharg et al. (1997)
Ascomycetes	*Trichoderma harzianum*	Phenanthrene	Singh and Singh (2011)
	Doratomyces nanus *D. purpureofuscus,* *D. verrucisporus,* *Myceliophthora Thermophila* *Phomaeupyrena* *Thermoascus crustaceus* *Fusarium solani,* *Penicillium chrysogenum,* *Scedosporium apiospermum*	PCBs	Mouhamadou et al. (2013) and Tigini et al. (2009)
	Aspergillus fumigatus	Phenol	Fernandez-Fernandez et al. (2013)
	Fusarium incarnatum	Bisphenol	Chhaya and Gupte (2013)
Yeasts	*Oospora* *Saccharomyces* *Candida tropicalis* *Debaryomyces* *Trichosporon cutaneum*	Phenol	Harris and Ricketts (1962), Henderson (1961) and Neujahr and Varga (1970)

19.3 Mycoremediation Pathways

Microbes including fungi degrade phenolic pollutants either aerobically or anaerobically (Fig. 19.2). The initial step in aerobic degradation is oxygenation, in which a monooxygenase enzyme phenol hydroxylase hydroxylates the aromatic ring at the ortho position of the pre-existing hydroxyl group to form 1,2-dihydroxybenzene (Catechol). This reaction requires a reduced pyridine nucleotide ($NADH_2$). Depending on the microorganism, the resulting catechol is further metabolized via two alternative pathways: β-ketoadipate or ortho degradation pathway and metadegradation pathway. In the ortho degradation pathway, the enzyme catechol 1,2-dioxygenase cleaves the aromatic ring between the catechol hydroxyls to produce cis, cis muconate. Cis, cis muconate is further cleaved via β-ketoadipate to

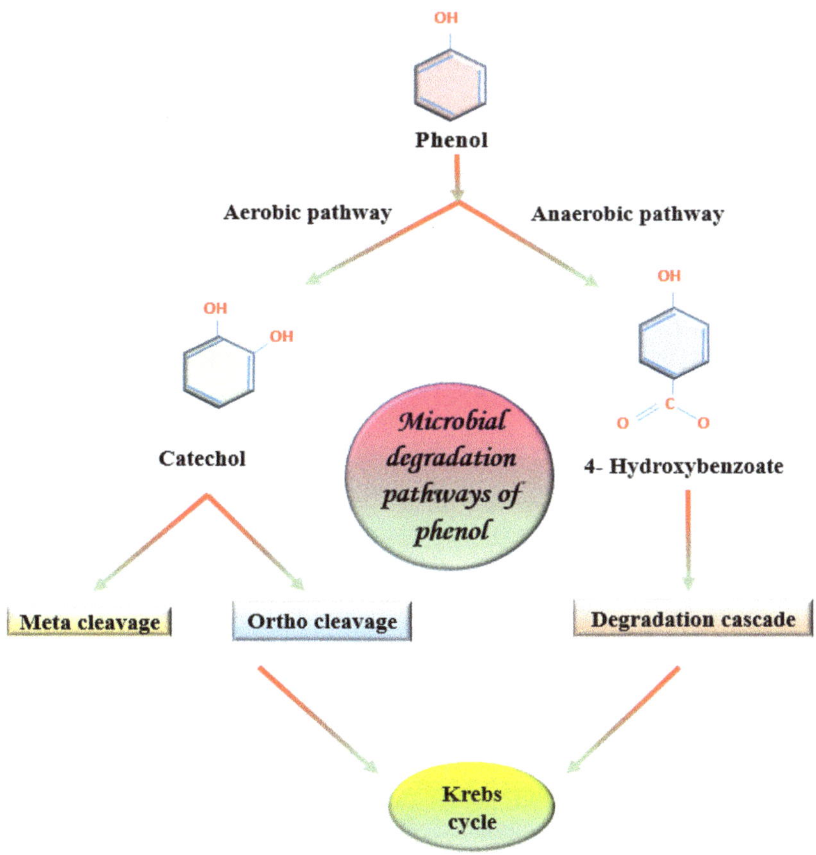

Fig. 19.2 A summary of microbial phenol degradation pathways

Krebs cycle intermediates (Evans, 1947; Kilby, 1948). In the metadegradation pathway, the estradiol fission or ring fusion happens near the two hydroxyl groups of catechol. The catechol 2,3-dioxygenases enzyme converts catechol into 2-hydroxymuconicsemialdehyde, which is further cleaved to form the Krebs cycle intermediates (Harwood & Parales, 1996; Stainer & Ornston, 1973).

Anaerobic phenol degradation is less explored than the aerobic process. This pathway shares similarities with the anaerobic benzoate pathway of *Paracoccus denitrificans*. The initial stage of the anaerobic pathway is the carboxylation of phenol at the para position by the enzyme 4-hydroxy benzoate carboxylase, which produces an intermediate called 4-hydroxybenzoate. The 4-hydroxy benzoate is further metabolized into protocatechuate by the enzyme protocatechuate 3,4-dioxygenase. It undergoes a cascade of metabolic cleavages and finally forms the Krebs cycle intermediates (Basha et al., 2010).

19.4 Fungal Immobilization Techniques for Phenol Degradation

Cell immobilization involves encapsulating cells in polymeric materials to keep them viable for a long-term biochemical process. The polymeric substance restricts the movement of cells by entrapping them inside. During phenol degradation, fungi can be immobilized using different immobilization techniques including encapsulation, entrapment, adsorption, etc. (Fig. 19.3). Several polymeric materials can be used for immobilization, including alginate, agar, collagen, polystyrene, and cellulosic materials. Among all the immobilization techniques, cell entrapment is the most practiced method for fungal immobilization during phenol degradation (Zaushitsyna et al., 2014; Karel et al., 1985).

19.4.1 Cell Entrapment

Cell entrapment is the safest and easiest method of immobilization. This method requires synthetic (polyethylene glycol conjugated PEG–lipid) or natural polymers (agar, alginate, carrageenan, agarose, etc.) to entrap cells and a cross-linking agent to strengthen the polymer network (Teramura et al., 2008). In the cell entrapment method, the cells are not in direct contact with the supporting material instead; they

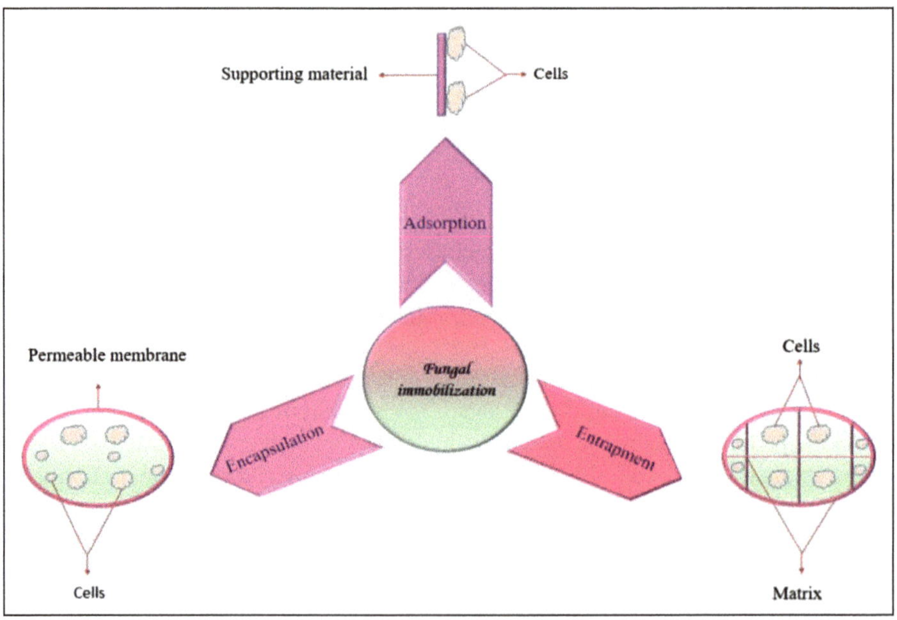

Fig. 19.3 Fungal immobilization techniques

are trapped inside the polymer matrix. The entrapment process involves mixing the cells with the polymeric solution and passing it through a needle and dropping it to a hardening solution (Gray et al., 2013).

19.4.2 Adsorption

As the name suggests, the cells are adsorbed or adhered to the surface of inert material. The adsorption mainly depends on the age of the cells and the chemical nature of the supporting material. The attachment is stabilized and maintained by various forces of attraction such as electrostatic, van der Waals, hydrophobic, and hydration forces. Organic and inorganic supporting materials are available for immobilization. Organic materials include DEAE-cellulose, DEAE-sephadex, and starch, and inorganic materials include silica gel, glass, alumina, etc. (Gray et al., 2013; Cassimjee et al., 2014).

19.4.3 Encapsulation

The encapsulation method is an altered entrapment method where a permeable membrane is used to encapsulate cell suspension. The membrane protects and supports cells during immobilization. It acts as a capsule and the liquid with active substances forms the core component and the membrane forms the outer covering. Different types of encapsulation procedures include emulsification, interfacial polymerization, coacervation, etc. (Karel et al., 1985).

19.5 Fungal Reactor Models for Phenol Removal

Large-scale biodegradation is carried out in large setups called bioreactors. Bioreactors provide optimum growth and degradation conditions for the microbes. Many fungal aerobic bioreactor models have been proposed and widely employed for removing phenol and its derivatives from wastewater. Fluidized bed reactors, batch reactors, bubble column reactors, stirred tank reactors, etc. (Fig. 19.4) are the commonly used models. Holladay et al. in 1978 reported the biodegradation of phenol using an anaerobic packed-bed reactor. They compared phenol degradation efficiency in different bioreactors, including stirred tank, packed bed, and fluidized bed reactor. They could find that the degradation efficiency is higher in stirred tank reactors when compared to packed bed and fluidized-bed bioreactor. The degradation rate inside the bioreactor majorly depends on the biomass development, flow rate, and feed concentration (Holladay et al., 1978).

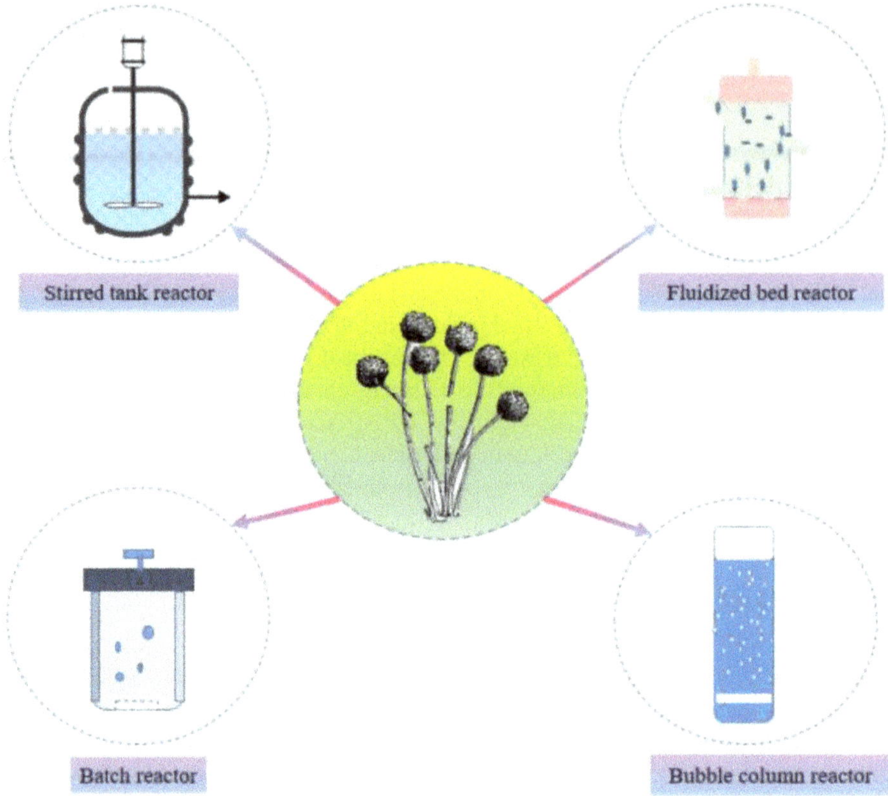

Fig. 19.4 Different bioreactor models for mycoremediation

19.6 Entrepreneurial Aspects of Fungal Bioremediation

Fungal bioremediation is the most suitable green route method for treating contaminated sites with recalcitrant pollutants. It gains more industrial attention among other microbes due to its higher pollutant removal efficiency and less substrate specificity. The environmental factors will significantly influence the industrial aspects of fungal bioremediation. Delayed lag phase, increased sludge generation, difficulty in bioprocess control, and monitoring are the major challenges faced during the scale-up process. The use of enzymes directly into the reactors over fungal biomass will reduce the bioremediation time with no lag phase, minimal sludge generation, and easy process control. Most of the enzymes used for biodegradation are highly expensive and have lesser stability with low shelf life. To tackle this situation, developments in whole-cell and enzyme immobilization techniques can be used to extend their stability and shelf life. These methods are preferably used in large-scale bioreactors like fluidized beds and cost-effective rotating biological contactors with the reuse of enzymes and by immobilization of fungus (Jebapriya & Gnanadoss, 2013).

Another novel strategy, based on biopurification systems in promoting bioremediation of organopesticides through highly active biological mixtures, particularly white-rot fungi, was significantly highlighted in the advanced studies (Rodríguez-Rodríguez et al., 2013). Mixed filamentous inoculum in a continuous large-scale bioreactor can be used for more sustainable and ecofriendly bioremediation of sewage sludge (Rahman et al., 2014). One of the recent advancements in this area is the introduction of novel permeable reactive biobarriers of *Trichoderma longibrachiatum* on nylon sponge to treat phenolic pollutants (Cobas et al., 2013).

To many entrepreneurs, fungal bioremediation opens a wide window of opportunities. So many start-up companies in the bioremediation sector enjoy the advantages of mycoremediation. Earth Fax engineering groups Pvt. Ltd. is one of the leading companies in the mycoremediation sector. They mainly focus on fungal-based soil remediation which employs the pollutant-degrading abilities of a group of white-rot fungi. They have performed the degradation of many phenolic pollutants including PCP under full-scale conditions with the use of white-rot fungi. Several online consultation companies such as Bloom, Mushroom Mountain, etc. also help and support people in their mycoremediation ventures.

19.7 Conclusion

Fungi are humans' age-old companions. In an entrepreneurial view, its uses are countless, starting from fermentation to bioremediation. The current chapter mainly embraces the role of fungi in phenol biodegradation. In detail, it discusses the essential pathways of fungal phenol degradation, major species used in phenol bioremediation, immobilization techniques for improved degradation, bioreactor models for industrial applications, and the entrepreneurial aspects of fungal remediation of phenol.

References

Acevedo, F., Pizzul, L., Castillo, M. D., Cuevas, R., & Diez, M. C. (2011). Degradation of polycyclic aromatic hydrocarbons by the Chilean white-rot fungus *Anthracophyllum discolor*. *Journal of Hazardous Materials, 185*, 212–219.

Anastasi, A., Tigini, V., & Varese, & G.C. (2013). The bioremediation potential of different ecophysiological groups of fungi. In *Fungi as bioremediators* (Soil biology) (Vol. 32, pp. 29–49). Springer.

Arisoy, M. (1998). Biodegradation of chlorinated organic compounds by white-rot fungi. *Bulletin of Environmental Contamination and Toxicology, 60*, 872–876.

Basha, K. M., Rajendran, A., & Thangavelu, V. (2010). Recent advances in the biodegradation of phenol: A review. *Asian Journal of Experimental Biological Sciences, 1*(2), 219–234.

Bhalerao, T. S., & Puranik, P. (2007). Biodegradation of organochlorine pesticide, endosulfan, by a fungal soil isolate, *Aspergillus niger*. *International Biodeterioration and Biodegradation, 59*, 315–321.

Cassimjee, K. E., Kadow, M., Wikmark, Y., Humble, M. S., Rothstein, M. L., Rothstein, D. M., & Bäckvall, J. E. (2014). A general protein purification and immobilization method on controlled porosity glass: Biocatalytic applications. *Chemical Communications, 50*(65), 9134–9137.

Chhaya, U., & Gupte, A. (2013). Possible role of laccase from *Fusarium incarnatum* UC-14 In bioremediation of Bisphenol A using reverse micelles system. *Journal of Hazardous Materials, 254–255*, 149–156. https://doi.org/10.1016/j.jhazmat.2013.03.054

Cobas, M., Ferreira, L., Tavares, T., Sanroman, M. A., & Pazos, M. (2013). Development of permeable reactive biobarrier for the removal of PAHs by *Trichoderma longibrachiatum*. *Chemosphere, 91*, 711–716. https://doi.org/10.1016/j.chemosphere.2013.01.028

Emergency Planning and Community Right-to-Know Act (EPCRA) Section 313 Chemical list for reporting year 2014. Available from: http://www.epa.gov/toxics-release-inventory-tri-program/tri-chemical-list-ry-2014-including-toxic-chemical-categories

Evans, W. C. (1947). Oxidation of phenol and benzoic acid by some soil bacteria. *The Biochemical Journal, 41*(3), 373.

Fernandez-Fernandez, M., Sanromán, M. Á., & Moldes, D. (2013). Recent developments and applications of immobilized laccase. *Biotechnology Advances, 31*(8), 1808–1825.

Gray, C. J., Weissenborn, M. J., Eyers, C. E., & Flitsch, S. L. (2013). Enzymatic reactions on immobilised substrates. *Chemical Society Reviews, 42*(15), 6378–6405.

Harris, G., & Ricketts, R. (1962). Metaboilsm of phenolic compounds by yeasts. *Nature, 195*, 473–474.

Harwood, C. S., & Parales, R. E. (1996). The β-ketoadipate pathway and the biology of self-identity. *Annual Review of Microbiology, 50*(1), 553–590.

Henderson, M. E. K. (1961). The metabolism of aromatic compounds related to lignin by some Hyphomycetes and yeast-like fungi of soil. *Journal of General Microbiology, 26*, 155–165.

Holladay, D. W., Hancher, C. W., Scott, C. D., & Chilcote, D. D. (1978). Biodegradation of phenolic waste liquors in stirred-tank, packed-bed, and fluidized-bed bioreactors. *Journal – Water Pollution Control Federation*, 2573–2589.

Huang, Y., Zhang, S. Y., Lv, M. J., & Xie, S. G. (2010). Biosorption characteristics of ectomycorrhizal fungal mycelium for anthracene. *Biomedical and Environmental Sciences, 23*, 378–383.

Jebapriya, G. R., & Gnanadoss, J. J. (2013). Bioremediation of textile dye using white-rot fungi: A review. *International Journal of Current Research and Review, 5*, 1–13.

Karel, S. F., Libicki, S. B., & Robertson, C. R. (1985). The immobilization of whole cells: Engineering principles. *Chemical Engineering Science, 40*(8), 1321–1354.

Kilby, B. A. (1948). The bacterial oxidation of phenol to beta-ketoadipic acid. *The Biochemical Journal, 43*, 1.

Lamar, R. T., & Dietrich, D. M. (1990). In situ depletion of pentachlorophenol from contaminated soil by *Phanerochaete* spp. *Applied and Environmental Microbiology, 56*(10), 3093–3100.

Lamar, R. T., & Dietrich, D. M. (1992). Use of lignin-degrading fungi in the disposal of pentachlorophenol-treated wood. *Journal of Industrial Microbiology, 9*, 181–191.

Lamar, R. T., Davis, M. W., Dietrich, D. M., & Glaser, J. A. (1994). Treatment of a pentachlorophenol- and creosote-contaminated soil using the lignin-degrading fungus *Phanerochaetesordida*: A field demonstration. *Soil Biology and Biochemistry, 26*(12), 1603–1616.

Lang, E., E'Uer, G., Kleeberg, I., Martens, R., & Zadrazil, F. (1995). Interaction of white rot fungi and soil microorganisms leading to biodegradation of soil pollutants. In *Contaminated Soil'95* (Soil and environment) (Vol. 5, pp. 1277–1278). Springer.

León-Santiesteban, H. H., Wrobel, K., Revah, S., & Tomasini, A. (2016). Pentachlorophenol removal by *Rhizopus oryzae* CDBB-H-1877 using sorption and degradation mechanisms. *Journal of Chemical Technology & Biotechnology, 91*, 65–71.

Machado, K. M. G., Matheus, D. R., Monteiro, R. T. R., & Bononi, V. L. R. (2005). Biodegradation of pentachorophenol by tropical basidiomycetes in journal of environmental science and health, part B 153 soils contaminated with industrial residues. *World Journal of Microbiology and Biotechnology, 21*, 297–301.

Mahiudddin, M., & Fakhruddin, A. N. M. (2012). Degradation of phenol via meta cleavage pathway by *Pseudomonas fluorescens PU1. ISRN Microbiology, 2012*, 741820.

Mansur, M., Arias, M. E., Copa-Patino, J. L., Flardh, M., & Gonzalez, A. E. (2017). The white-rot fungus Pleurotusostreatus secretes laccase isozymes with different substrate specificities. *Mycologia, 95*, 1013–1020.

Margot, J., Bennati-Granier, C., Maillard, J., Blanquez, P., Barry, D. A., & Holliger, C. (2013). Bacterial versus fungal laccase: Potential for micropollutant degradation. *AMB Express, 3*, 63. https://doi.org/10.1186/2191-0855-3-63

Meharg, A. A., Cairney, J. W. G., & Maguire, N. (1997). Mineralization of 2,4-dichlorophenol by ectomycorrhizal fungi in axenic culture and in symbiosis with pine. *Chemosphere, 34*(12), 2495–2504.

Mouhamadou, B., Faure, M., Sage, L., Marçais, J., Souard, F., & Geremia, R. A. (2013). Potential of autochthonous fungal strains isolated from contaminated soils for degradation of polychlorinated biphenyls. *Fungal Biology, 117*, 268–274. https://doi.org/10.1016/j.funbio.2013.02.004

Neujahr, H. Y., & Varga, J. M. (1970). Degradation of phenols by intake cells and cell-free preparations of *Trichosporoncutaneum. European Journal of Biochemistry, 13*, 37–44.

Pointing, S. B. (2001). Feasibility of bioremediation by white-rot fungi. *The Applied Microbiology and Biotechnology, 51*, 20–33.

Rahman, R. A., Molla, A. H., & Fakhru'l-Razi, A. (2014). Assessment of sewage sludge bioremediation at different hydraulic retention times using mixed fungal inoculation by liquid-state bioconversion. *Environmental Science and Pollution Research, 21*, 1178–1187. https://doi.org/10.1007/s11356-013-1974-5

Reddy, C. A., & Mathew, Z. (2001). Bioremediation potential of white rot fungi. In *British mycological society symposium series* (Vol. 23, pp. 52–78).

Rodríguez-Rodríguez, C. E., Castro-Gutiérrez, V., Chin-Pampillo, J. S., & Ruiz-Hidalgo, K. (2013). On-farm biopurificationsystems: Role of white-rot fungi in depuration of pesticide-containing wastewaters. *FEMS Microbiology Letters, 345*, 1–12. https://doi.org/10.1111/1574-6968.12161

Sack, U., Heinze, T. M., Deck, J., Cerniglia, C. E., Martens, R., Zadrazil, F., & Fritsche, W. (1997). Comparison of phenanthrene and pyrene degradation by different wood-decaying fungi. *Applied and Environmental Biology, 63*(10), 3919–3925.

Santos, V. L., & Linardi, V. R. (2001). Phenol degradation by yeasts isolated from industrial effluents. *Journal of General and Applied Microbiology, 47*(4), 213–221.

Singh, L., & Singh, V. P. (2011). Microbial decolourization of textile dyes by the fungus *Trichoderma harzianum. Journal of Pure and Applied Microbiology, 6*, 1829–1833.

Stainer, R. Y., & Ornston, L. N. (1973). The β-ketoadipate pathway. In *Advances in microbial physiology* (Vol. 9, pp. 89–151). Academic.

Teramura, Y., Kaneda, Y., Totani, T., & Iwata, H. (2008). Behavior of synthetic polymers immobilized on a cell membrane. *Biomaterials, 29*(10), 1345–1355.

Tigini, V., Prigione, V., Di, T., Fava, F., & Varese, G. C. (2009). Isolation and characterisation of polychlorinated biphenyl (PCB) degrading fungi from a historically contaminated soil. *Microbial Cell Factories, 8*(1), 1–14.

Tortella, G. R., Rubilar, O., Gianfreda, L., Valenzuela, E., & Diez, M. C. (2008). Enzymatic characterization of Chilean native wood-rotting fungi for potential use in the bioremediation of polluted environments with chlorophenols. *World Journal of Microbiology and Biotechnology, 24*, 2805–2818.

Valentín, L., Oesch-Kuisma, H., Steffen, K. T., Kähkönen, M. A., Hatakka, A., & Tuomela, M. (2013). Mycoremediation of wood and soil from an old sawmill area contaminated for decades. *Journal of Hazardous Materials, 260*, 668–675.

Villegas, L. G. C., Mashhadi, N., Chen, M., Mukherjee, D., Taylor, K. E., & Biswas, N. (2016). A short review of techniques for phenol removal from wastewater. *Current Pollution Reports, 2*(3), 157–167.

Yordanova, G., Godjevargova, T., Nenkova, R., & Ivanova, D. (2013). Biodegradation of phenol and phenolic derivatives by a mixture of immobilized cells of *Aspergillus awamori* and *Trichosporoncutaneum*. *Biotechnology & Biotechnological Equipment, 27*(2), 3681–3688.

Zaushitsyna, O., Berillo, D., Kirsebom, H., & Mattiasson, B. (2014). Cryostructured and cross-linked viable cells forming monoliths suitable for bioreactor applications. *Topics in Catalysis, 57*(5), 339–348.

Chapter 20
Mycobased Biorefinery for Gold Nanoparticles Production

Mehavesh K. Hameed, Javad B. M. Parambath, and Ahmed A. Mohamed

Contents

20.1 Introduction

Green synthesis has gained immense recognition as a reliable, sustainable, and eco-friendly strategy for the synthesis of a wide range of nanoparticles (Nasaruddin et al., 2021; Rana et al., 2020; Pei et al., 2017; Molnár et al., 2018). Accordingly, green production approaches are considered important strategies for the mass production of nanomaterials by eliminating the ruinous effects related to the traditional chemical and physical approaches. Gold nanoparticles are interesting because of their unique physicochemical properties along with their oxidation/corrosion resistance, biocompatibility, and stability, which offer a multitude of potential applications in the fields of sensors, electronics, biomedicine, and catalysis (Siddiqi & Husen, 2016).

The production of gold nanoparticles for potential entrepreneurship has progressed dramatically. Both the facile and large-scale productions have aided the reliability of this approach. Further, the wide applications of the products are growing in biomedicine among industrial interests. Fungi are widely studied in the fabrication of metal nanoparticles than bacteria because of their ability to tolerate high metal concentrations (Siddiqi & Husen, 2016; Hyde et al., 2019). Moreover, they

M. K. Hameed · J. B. M. Parambath · A. A. Mohamed (✉)
Department of Chemistry, University of Sharjah, Sharjah, UAE
e-mail: ah.mohamed@sharjah.ac.ae

A. C. Shukla (ed.), *Applied Mycology*, Fungal Biology,
https://doi.org/10.1007/978-3-030-90649-8_20

can release high concentrations of redox agents which ultimately can aid the fabrication of the metal nanoparticles. Hence, fungal protein extracts can produce gold nanoparticles at a large scale and relatively fast for potential entrepreneurship. Several studies supported the formation of small sizes of gold nanoparticles by fungi (Kumar et al., 2018).

There are a few advantages in utilizing fungi such as the thick coating of organic shells because of the redox reactions by the extracellular redox enzymes which are capable to reduce the common gold (III) precursors under static conditions (Pei et al., 2017; Molnár et al., 2018; Siddiqi & Husen, 2016). Overall, the use of microorganisms in gold nanoparticles synthesis is gaining speed because of the negligible toxicity in medical applications. There are a few merits for using fungi for the synthesis of different morphologies (Pei et al., 2017; Molnár et al., 2018). Gold nanoparticles have been fabricated as trigonal, hexagonal, spherical, and other shapes using fungi. The isotropic and anisotropic morphologies permitted wide applications in agriculture, industrial, and medical fields. Bimetallic alloy nanoparticles such as gold-silver were also synthesized using fungi for medical applications (Kumar et al., 2018).

The entrepreneurship based on fungi, apart from the basic laboratory applications, is extremely diverse such as traditional Chinese medicine, control of insects and weeds, environment-friendly biofertilizers, growth-promoting hormones, and many more. Recently, gold nanoparticles fabricated using fungi have been used in food and cosmetic industries (Kumar et al., 2018). In this chapter, we will address the routes for the synthesis of gold nanoparticles using fungi and their applications for entrepreneurship. Highlights on the fundamental mechanisms of the reduction of gold (III) precursors by redox enzymes from fungi will be discussed.

20.2 Mycobased Synthesis of Gold Nanoparticles

Fungi play a pivotal role in tackling enormous global issues for sustainable development and large-scale production. The exploration of the significance of fungi in nanotechnology is considered tremendously important. In this regard, a plethora of fungi have been investigated as they were found to be potential players in the biogenesis of gold nanoparticles. Filamentous fungi have some exceptional advantages over other biosystems like bacteria and algae, because of their high metal tolerance and bioaccumulation capability (Sastry et al., 2003). Since fungi secrete extracellular enzymes effectively, they paved the way for the industrial production of enzymes and proteins (Castro-Longoria, 2012). Moreover, high wall-binding and intracellular metal uptake are the separate privileges of exploiting them in nanoparticle synthesis (Volesky & Holan, 1995). On top of all these, fungi could produce metal nanoparticles by utilizing reducing enzymes intracellularly as well as extracellularly (Ahmad et al., 2003). It is worth mentioning that the effortless propagation and the rapid growth rate are other advantages of fungi for the green choice for the large-scale production of metal nanoparticles (Castro-Longoria, 2012).

Marsili et al., reported that maturation of *Rhizopus oryzae* in the presence of $HAuCl_4$ solution resulted in an incremental color change of the mycelia over time from off-white to purple. After 48 h, the UV-VIS spectrum of the purple mycelia exhibited a peak at 535 nm, indicating the formation of AuNPs (Das et al., 2012a, b). Sastry et al., confirmed that fungus *Fusarium oxysporum* releases reducing agents into a solution that are responsible for the formation of gold nanoparticles. Moreover, the release of proteins from the biomasses has been confirmed by gel electrophoresis (Mukherjee et al., 2002). Sastry et al., observed that an absorption peak at 270 nm is vivid and is assigned for the aromatic amino acids. Notably, the absorption peak at 270 nm occurs due to tryptophan and tyrosine amino acids (Mukherjee et al., 2002). Fungi capable of producing gold nanoparticles in a wide variety of shapes and sizes are summarized in Table 20.1.

The morphology of gold nanoparticles has an enormous impact on their characteristics and applications. Even though spherical-shaped nanoparticles have been formed most frequently, metallic ion solution, incubation circumstances, and active biomolecules produced by the fungal organisms are the major factors to accomplish a wide variety of morphologies (Moghaddam et al., 2015). Interaction of different metal ions with *F. oxysporum* resulted in the fabrication of nanoparticles in a range of sizes and morphologies. Furthermore, nanoparticles synthesized by *F. oxysporum* had asymmetrical forms displaying a total quasi-spherical morphology in the size range 20–50 nm (Bharde et al., 2006). The utilization of fungal extract could regulate the morphology of gold nanoparticles at room temperature. Consequently, nanoparticles have been fabricated with the alteration of prime growth factors like gold ion concentration, pH, and response time (Gericke & Pinches, 2006; Das et al., 2010; Sun, 2002). The exploitation of an entophytic fungus *Colletotrichum* species for the synthesis of gold nanoparticles has resulted in the formation of different morphologies like rods, flat sheets, and triangles (Shankar et al., 2003) (Fig. 20.1).

20.2.1 Mechanism of Mycobased Synthesis of Gold Nanoparticles

As far as the mechanism of the extracellular and intracellular mycosynthesis is concerned, it is presumed that metabolites such as enzymes, proteins, polysaccharides, flavonoids, alkaloids, phenolic, and acids are secreted as an outcome of the fungus resistance to metal ions when exposed to different environmental stresses and are mostly responsible for the reduction of metal ions to metallic nanoparticles (Srivastava et al., 2019). It is quite noticeable that Au (III) ions penetrated or internalized into the cell membrane and reduced there during intracellular gold nanoparticles fabrication. Das et al. observed that Au (III) has a significant role in the expression of an oxidative stress response gene, which upregulated oxidoreductase enzymes. Also, the same research group discovered that *R. oryzae* reacted with

Table 20.1 Mycosynthesis of Gold nanoparticles

	Fungi	Size (nm)	Morphology	Mechanism	References
1	*Helminthosporium solani*	15–20	Sphere, rod, triangle, star	Extracellular	Kumar et al. (2008)
2	*Epicoccum nigrum*	5–50	Spherical, rod	Extracellular	Sheikhloo et al. (2011)
3	*Neurospora crassa*	>100	Quasi-spherical	Intra, extracellular	Castro-Longoria et al. (2011)
4	*Aspergillus clavatus*	48–72	Triangle, sphere, hexagonal	Extracellular	Verma et al. (2011)
5	*Aspergillus niger*	10–30	Various shapes	Extracellular	Prakash and Soni (2012)
6	*Candida albicans*	60–80	Spherical	–	Owais et al. (2011)
7	*Fusarium oxysporum*	8–40	Spherical, triangular	Extracellular	Mukherjee et al. (2002)
8	*Rhizopus oryzae*	16–25	Spherical	Extracellular	Das et al. (2012)
9	*Fusarium solani*	20–50	Spherical	Extracellular	Gopinath and Arumugam (2013)
10	*Sclerotium rolfsii*	25	Triangle, hexagonal, decahedral, rod, isotropic spherical	Extracellular	Narayanan Sakthivel (2011)
11	*Phanerochaete chrysosporium*	10–100	Spherical	Extracellular	Sanghi et al. (2011)
12	*Hormoconis resinae*	3–20	Spherical	Extracellular	Mishra et al. (2010)
13	*Yarrowia lipolytica*	–	Various shapes depending on Au (III) concentration	Intracellular	Pimprikar et al. (2009)
14	*Fusarium semitectum*	10–80	Spherical	Extracellular	Sawle et al. (2008)
15	*Saccharomyces cerevisiae*	15–20	Spherical	Cell wall, cytoplasm	Sen et al. (2011)

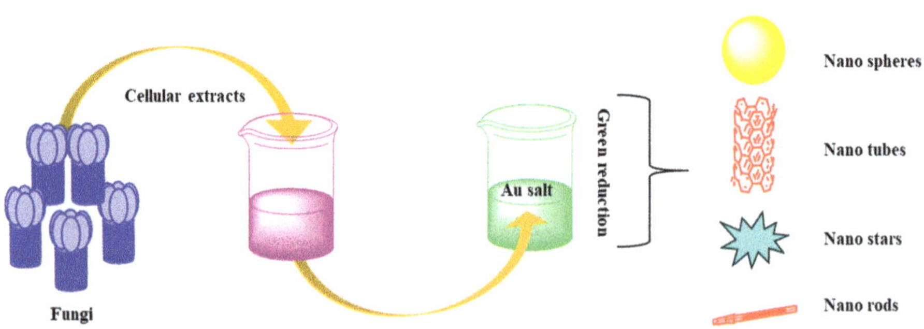

Fig. 20.1 Schematic representation of the synthesis of gold nanoparticles from fungi

concentration-dependent active biochemical detoxification when interacting with Au (III) (Das et al., 2012a, b).

AuNPs synthesis could take place in either the intracellular or extracellular space. Das et al. observed two reduction methods: (a) electrostatic interaction of Au (III) on the cell wall and (b) internalization of Au (III) into cytoplasm resulted in the synthesis of AuNPs with the aid of proteins or enzymes. Furthermore, Das et al. demonstrated the formation of AuNPs in both cell walls and the cytoplasmic region of *R. oryzae* by TEM micrograph (Das et al., 2012a, b). Even though the exact mechanism of the synthesis of AuNPs has not been elucidated, studies suggested that Au ions require NADH-reliant nitrate reductase enzyme for their reduction, which was secreted by the fungus in the extracellular environment (Rai et al., 2021; Dorcheh & Vahabi, 2017). Another study investigated that cofactors such as nicotinamide adenine dinucleotide (NADH) and reduced form nicotinamide adenine dinucleotide phosphate (NADPH)-dependent enzymes have a crucial role as reducing agents by the transfer of electrons from NADH by NADH-dependent enzymes (Kitching et al., 2014, Fig. 20.2). Observations on AuNPs biosynthesis in the presence of fungi *Fusarium oxysporum* exhibited that NADH-dependent reductase is considerably important in the biomineralization process (Mukherjee et al., 2002). Moreover, Au (III) ions are reduced by NADH or NADPH oxidoreductase either on the cell surface or in the cytoplasm. Narayanan et al. reported the participation of thermostable NADPH-dependent enzymes in the AuNPs biomineralization process (Narayanan & Sakthivel, 2011).

Another recent study on fungal AuNPs biosynthesis investigated the fungal pathogen *candida albicans* as a potential candidate for producing phytochelatins, which are composed of chain links of glucosecysteine and glycine ((c-Glu-cys)

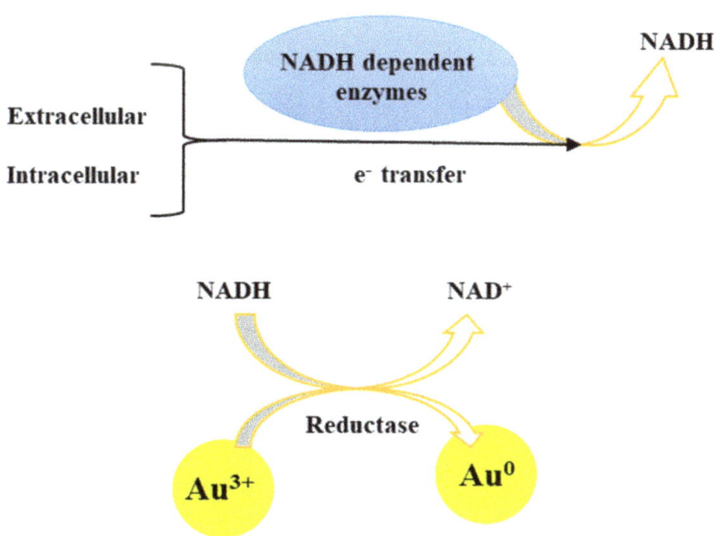

Fig. 20.2 Synthesis mechanism of AuNPs by extracellular or intracellular enzymatic reduction

n-Gly), by the transpeptidation reaction of a c-Glu-Cys dipeptide from a succession of glutathione molecules. Besides, glutathione-capped AuNPs have been formed by the reduction of Au (III) ions in the presence of glutathione (Owais et al., 2011).

Although melanin-mutant and melanin-complemented strains have not been studied, the association of biosynthesis redox mediators in the mycosynthesis of AuNPs has been reported in *Yarrowia lipolytica*. Interestingly, reduction of Au (III) ions to AuNPs occurred in the presence of melanin which was discharged by this fungus (Apte et al., 2013). Even though reduction by melanin has not been proved, further studies from the same research group revealed that melanin could induce the reduction of metal salts to their elemental form as nanomaterials. Remarkably, the phenol form of melanin oxidized to quinone, while Au (III) reduced to AuNPs (Apte et al., 2013).

Szczyyiel et al. reported that the formation of AuNPs by *M. plumbeus* is a non-enzymatic process. However, the development of AuNPs occurred by the dead cells of *M. plumbeus* is an implicit indication of the function of polysaccharides in this phenomenon. It is well-known that high temperature results in exposure of polysaccharides, and these molecules become more susceptible to the gold ions for conjugation (Maliszewska et al., 2017) (Fig. 20.3).

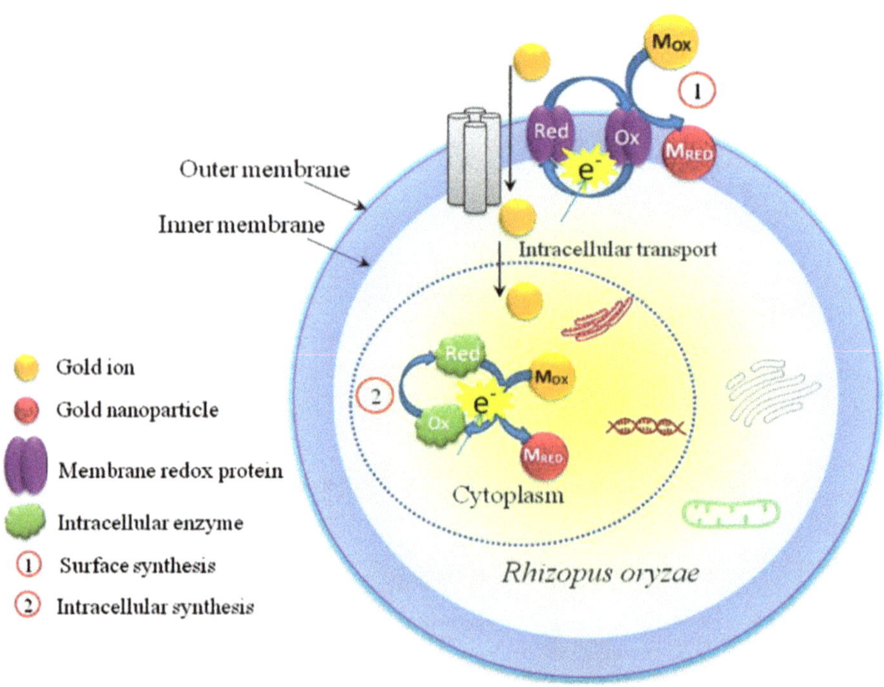

Fig. 20.3 Gold biomineralization mechanism in *R. oryzae*. (Das et al., 2012a, b)

20.3 Challenges and Scale-Up for Entrepreneurship

Mycosynthesis of gold nanoparticles has shown tremendous developments in synthesis and their applications over the last decade. However, there is a large gap between utilizing small-scale synthetic procedures and their commercial utilization. Much work is needed to improve the efficiency of synthesis, decreased toxicity, controlled size, and morphology. For example, mycosynthesis is a relatively slow process compared to other physical and chemical methods. A few challenges in mycosynthesis are addressed below:

1. Pathogenesis from fungi and fungal extracts is a challenge while scaling up for commercialization. Specifically, some extracts might contain pathogenic spores, which transmit to cause hazards to health and the ecosystem (Pantidos & Horsfall, 2014).
2. Particle size and monodispersity are important characteristics that must be controlled in the evaluation of gold nanoparticles for medicine and health care applications. The ability to control the shape, size, and size distribution of gold nanoparticles is more challenging with mycosynthesis since we are still lacking mechanistic insights (Akbari et al., 2011).
3. The stability of mycosynthesized gold nanoparticles is also a concern since the nanoparticles formed by fungi may be decomposed after a certain period.
4. Enzymes involved in the reduction process are not specifically categorized or addressed since the genes involved are hard to identify in fungi as a eukaryote. Thus, the mechanism of the reaction and enzyme roles in this process is hardly understood and is a major challenge in further recognition of mycosynthesis (Thakkar et al., 2010).

Besides the above-mentioned major challenges, gold nanoparticles synthesized using mycological route must be rigorously purified since the major applications are in biomedical fields. Downstreaming of fungal extracts is hard since it is time-consuming and requires several techniques. Even though, mycosynthetic route of gold nanoparticle generation is safer and economical. This is reported to be a non-toxic and eco-friendly "green-route" strategy. Exploiting this mycobased synthetic approach to gold nanoparticles has exciting aspects for addressing the current manufacturing challenges because this encompasses most of the 12 principles of green synthetic route (Patwardhan et al., 2018). Metabolic, zymology, and molecular mechanisms that help in synthesis should be studied in detail to overcome these challenges, increase synthetic rate, and obtain the required size and morphology of gold nanoparticles. Owing to the rich diversity of fungi, their potential as synthetic agents is still to be explored on large scale. Table 20.2 summarizes a few reported mycosynthetic gold nanoparticles with their applications.

Consequently, developments in this field of mycosynthesized gold nanoparticles are important in the industrial and pharmaceutical fields as antimicrobial, antifouling agents, catalysts, drug carriers, anticancerous agents, etc. Considering the immense need for fungi nanotechnology entrepreneurship in medical sciences, gold

Table 20.2 Synthesized gold nanoparticles using various fungus species and their applications

	Fungus species	Applications	References
1	*Rhizopus oryzae*	Pesticides removal	Das et al. (2009)
2	*Fusarium semitectum*	Optoelectronics	Sawle et al. (2008)
3	*Helminthosporium solani*	Anticancer	Kumar et al. (2008)
4	*Penicillium brevicompactum*	Target cancer cells	Mishra et al. (2011)
5	*Volvariella volvacea*	Therapeutic	Philip (2009)
6	*Candida albicans*	Liver cancer detection	Owais et al. (2011)
7	*Aspergillus japonicus* AJP01	Catalysis	Bhargava et al. (2015)
8	*Aspergillus Niger*	Kill mosquito larva	Prakash and Soni (2012)
9	*Mariannaea sp. HJ*	Nitrophenol reduction	Pei et al. (2017)
10	*Penicillium aculeatum*	Against hydatid cyst	Barabadi et al. (2017)
11	*Cylindrocladium floridanum*	Nitrophenol reduction	Narayanan and Sakthivel (2011)
12	*Pycnoporus sanguineus*	Catalysis	Shi et al. (2015)
13	*Trichoderma harzianum*	Hg^{2+} detection	Tripathi et al. (2014)
14	*Aspergillus terreus*	Bactericidal against	Priyadarshini et al. (2014)

nanoparticles have their major role in drug delivery, cancer treatment, gene editing, DNA analysis, antibacterial, biosensors, and MRI, which leads us to expand entrepreneurship of mycosynthetic approach in the synthesis of gold nanoparticles. In the energy sector, the role of economically synthesized gold nanomaterials promises to capture light, convert energy, which results in the fabrication of photovoltaics, etc. Promising research and developments will take time to scale up and bring costs down. Hopefully, the proper addressing of these challenges can lead to large-scale and commercial utilization of mycosynthesized gold nanoparticles, especially in medicine and health care system, in the coming years.

20.4 Conclusion

Fungi "nano factories" can produce large-scale materials for gram-scale industrial applications. However, the fabrication speed is still relatively slow compared to the chemical and physical approaches. This area of research is still in need to study the various factors that control the formation of the different morphologies and their effect on the potential toxicity in medical applications. Moreover, the research must expand to include easily reducible metal ions rather than the commonly used high oxidation states which are predicted to oxidize some of the components in the live cells.

References

Ahmad, A., Senapati, S., Khan, M. I., Kumar, R., Ramani, R., Srinivas, V., & Sastry, M. (2003). Intracellular synthesis of gold nanoparticles by a novel alkalotolerant actinomycete, *Rhodococcus* species. *Nanotechnology, 14*, 824–828.

Akbari, B., Tavandashti, M. P., & Zandrahimi, M. (2011). Particle size characterization of nanoparticles–a practicalapproach. *Iranian Journal of Materials Science and Engineering, 8*, 48–56.

Apte, M., Girme, G., Bankar, A., Ravikumar, A., & Zinjarde, S. (2013). 3, 4-dihydroxy-L-phenylalanine-derived melanin from *Yarrowia lipolytica* mediates the synthesis of silver and gold nanostructures. *Journal of Nanobiotechnology, 11*, 2.

Barabadi, H., Honary, S., Mohammadi, M. A., Ahmadpour, E., Rahimi, M. T., Alizadeh, A., Naghibi, F., & Saravanan, M. (2017). Green chemical synthesis of gold nanoparticles by using *Penicillium aculeatum* and their scolicidal activity against hydatid cyst protoscolices of *Echinococcus granulosus*. *Environmental Science and Pollution Research, 24*, 5800–5810.

Bharde, A., Rautaray, D., Bansal, V., Ahmad, A., Sarkar, I., Yusuf, S. M., Sanyal, M., & Sastry, M. (2006). Extracellular biosynthesis of magnetite using fungi. *Small, 2*, 135–141.

Bhargava, A., Jain, N., Gangopadhyay, S., & Panwar, J. (2015). Development of gold nanoparticle-fungal hybrid based heterogeneous interface for catalytic applications. *Process Biochemistry, 50*, 1293–1300.

Castro-Longoria, E. (2012). Production of platinum nanoparticles and nanoaggregates using *Neurospora crassa*. *Journal of Microbiology and Biotechnology, 22*, 1000–1004.

Castro-Longoria, E., Vilchis-Nestor, A. R., & Avalos-Borja, M. (2011). Biosynthesis of silver, gold and bimetallic nanoparticles using the filamentous fungus Neurospora crassa. *Colloids and Surfaces B: Biointerfaces, 83*, 42–48.

Das, S. K., Das, A. R., & Guha, A. K. (2009). Gold nanoparticles: Microbial synthesis and application in water hygiene management. *Langmuir, 25*, 8192–8199.

Das, S. K., Das, A. R., & Guha, A. K. (2010). Microbial synthesis of multishaped gold nanostructures. *Small, 6*, 1012–1021.

Das, S. K., Dickinson, C., Lafir, F., Brougham, D. F., & Marsili, E. (2012a). Synthesis, characterization and catalytic activity of gold nanoparticles biosynthesized with Rhizopus oryzae protein extract. *Green Chemistry, 14*, 1322.

Das, S. K., Liang, J., Schmidt, M., Laffir, F., & Marsili, E. (2012b). Biomineralization mechanism of gold by zygomycete fungi Rhizopousoryzae. *ACS Nano, 6*, 6165–6173.

Dorcheh, S. K., & Vahabi, K. (2017). Biosynthesis of nanoparticles by fungi: Large-scale production. In *Fungal metabolites* (pp. 395–414). Springer.

Gericke, M., & Pinches, A. (2006). Biological synthesis of metal nanoparticles. *Hydrometallurgy, 83*, 132–140.

Gopinath, K., & Arumugam, A. (2013). Extracellular mycosynthesis of gold nanoparticles using fusarium solani. *Applied Nanoscience, 4*, 57–662.

Hyde, K. D., Xu, J., Rapior, S., Jeewon, R., Lumyong, S., Niego, A. G. T., Abeywickrama, P. D., Aluthmuhandiram, J. V., Brahamanage, R. S., Brooks, S., & Chaiyasen, A. (2019). The amazing potential of fungi: 50 ways we can exploit fungi industrially. *Fungal Diversity, 97*, 1–136.

Kitching, M., Ramani, M., & Marsili, E. (2014). Fungal biosynthesis of gold nanoparticles: Mechanism and scale up. *Microbial Biotechnology, 8*, 904–917.

Kumar, S. A., Peter, Y.-A., & Nadeau, J. L. (2008). Facile biosynthesis, separation and conjugation of gold nanoparticles to doxorubicin. *Nanotechnology, 19*, 495101.

Kumar, M., Kumar, V., Prasad, R., & Wang, S. (2018). *Fungal Nanobionics: Principles and applications*. Springer Singapore.

Maliszewska, I., Tylus, W., Chęcmanowski, J., Szczygieł, B., Pawlaczyk-Graja, I., Pusz, W., & Baturo-Cieśniewska, A. (2017). Biomineralization of gold by Mucorplumbeus: The progress in understanding the mechanism of nanoparticles' formation. *Biotechnology Progress, 33*, 1381–1392.

Mishra, A. N., Bhadauria, S., Gaur, M. S., & Pasricha, R. (2010). Extracellular microbial synthesis of gold nanoparticles using fungus Hormoconisresinae. *Journal of Metals, 62*, 45–48.

Mishra, A., Tripathy, S. K., Wahab, R., Jeong, S.-H., Hwang, I., Yang, Y.-B., Kim, Y.-S., Shin, H.-S., & Yun, S.-I. (2011). Microbial synthesis of gold nanoparticles using the fungus *Penicillium brevicompactum* and their cytotoxic effects against mouse mayo blast cancer C2C12 cells. *Applied Microbiology and Biotechnology, 92*, 617–630.

Moghaddam, A. B., Namvar, F., Moniri, M., Tahir, P. M., Azizi, S., & Mohamad, R. (2015). Nanoparticles biosynthesized by fungi and yeast: A review of their preparation, properties, and medical applications. *Molecules, 20*, 16540–16565.

Molnár, Z., Bódai, V., Szakacs, G., Erdélyi, B., Fogarassy, Z., Sáfrán, G., Varga, T., Kónya, Z., Tóth-Szeles, E., Szűcs, R., & Lagzi, I. (2018). Green synthesis of gold nanoparticles by thermophilic filamentous fungi. *Scientific Reports, 8*, 3943.

Mukherjee, P., Senapati, S., Mandal, D., Ahmad, A., Khan, M. I., Kumar, R., & Sastry, M. (2002). Extracellular synthesis of gold nanoparticles by the fungus *Fusarium oxysporum*. *Chembiochem, 3*, 461.

Narayanan, K. B., & Sakthivel, N. (2011). Synthesis and characterization of nano-gold composite using *Cylindrocladium floridanum* and its heterogeneous catalysis in the degradation of 4-nitrophenol. *Journal of Hazardous Materials, 189*, 519–525.

Nasaruddin, R. R., Chen, T., Yao, Q., Zang, S., & Xie, J. (2021). Toward greener synthesis of gold nanomaterials: From biological to biomimetic synthesis. *Coordination Chemistry Reviews, 426*, 213540.

Owais, M., Chauhan, A., Tufail, S., Sajid, M., Suri, C. R., & Owais, M. (2011). Fungus-mediated biological synthesis of gold nanoparticles: Potential in detection of liver cancer. *International Journal of Nanomedicine, 2011*, 2305.

Pantidos, N., & Horsfall, L. E. (2014). Biological synthesis of metallic nanoparticles by bacteria, fungi and plants. *Journal of Nanomedicine & Nanotechnology, 5*, 5.

Patwardhan, S. V., Manning, J. R., & Chiacchia, M. (2018). Bioinspired synthesis as a potential green method for the preparation of nanomaterials: Opportunities and challenges. *Current Opinion in Green and Sustainable Chemistry, 12*, 110–116.

Pei, X., Qu, Y., Shen, W., Li, H., Zhang, X., Li, S., Zhang, Z., & Li, X. (2017). Green synthesis of gold nanoparticles using fungus Mariannaea sp. HJ and their catalysis in reduction of 4-nitrophenol. *Environmental Science and Pollution Research, 24*, 21649–21659.

Philip, D. (2009). Biosynthesis of Au, Ag and Au–Ag nanoparticles using edible mushroom extract. *Spectrochimica Acta Part A: Molecular and Biomolecular Spectroscopy, 73*, 374–381.

Pimprikar, P., Joshi, S., Kumar, A., Zinjarde, S., & Kulkarni, S. (2009). Influence of biomass and gold salt concentration on nanoparticle synthesis by the tropical marine yeast *Yarrowia lipolytica* NCIM 3589. *Colloids and Surfaces B: Biointerfaces, 74*, 309–316.

Prakash, S., & Soni, N. (2012). Synthesis of gold nanoparticles by the fungus *Aspergillus niger* and its efficacy against mosquito larvae. *Reports in Parasitology, 2*, 1–7.

Priyadarshini, E., Pradhan, N., Sukla, L. B., & Panda, P. K. (2014). Controlled synthesis of gold nanoparticles using aspergillus terreus IF0 and its antibacterial potential against gram negative pathogenic bacteria. *Journal of Nanotechnology, 2014*, 1–9.

Rai, M., Bonde, S., Golinska, P., Trzcińska-Wencel, J., Gade, A., Abd-Elsalam, K. A., Shende, S., Gaikwad, S., & Ingle, A. P. (2021). *Fusarium* as a novel fungus for the synthesis of nanoparticles: Mechanism and applications. *Journal of Fungi, 7*, 139.

Rana, A., Yadav, K., & Jagadevan, S. (2020). A comprehensive review on green synthesis of nature-inspired metal nanoparticles: Mechanism, application and toxicity. *Journal of Cleaner Production, 272*, 122880.

Sanghi, R., Verma, P., & Puri, S. (2011). Enzymatic formation of gold nanoparticles using *Phanerochaete chrysosporium*. *Advances in Chemical Engineering and Science, 01*, 154–162.

Sastry, M., Ahmad, A., Khan, M. I., & Kumar, R. (2003). Biosynthesis of metal nanoparticles using fungi and actinomycete. *Current Science, 85*, 162–170.

Sawle, B. D., Salimath, B., Deshpande, R., Bedre, M. D., Prabhakar, B. K., & Venkataraman, A. (2008). Biosynthesis and stabilization of Au and Au–Ag alloy nanoparticles by fungus, *Fusarium semitectum. Science and Technology of Advanced Materials, 9*, 035012.

Sen, K., Sinha, P., & Lahiri, S. (2011). Time dependent formation of gold nanoparticles in yeast cells: A comparative study. *Biochemical Engineering Journal, 55*, 1–6.

Shankar, S. S., Ahmad, A., Pasricha, R., & Sastry, M. (2003). Bioreduction of chloroaurate ions by geranium leaves and its endophytic fungus yields gold nanoparticles of different shapes. *Journal of Materials Chemistry, 13*, 1822.

Sheikhloo, Z., Salouti, M., & Katiraee, F. (2011). Biological synthesis of gold nanoparticles by fungus Epicoccum nigrum. *Journal of Cluster Science, 22*, 661–665.

Shi, C., Zhu, N., Cao, Y., & Wu, P. (2015). Biosynthesis of gold nanoparticles assisted by the intracellular protein extract of *Pycnoporus sanguineus* and its catalysis in degradation of 4-nitroaniline. *Nanoscale Research Letters, 10*, 47.

Siddiqi, K. S., & Husen, A. (2016). Fabrication of metal nanoparticles from fungi and metal salts: Scope and application. *Nanoscale Research Letters, 11*, 98.

Srivastava, S., Bhargava, A., Pathak, N., & Srivastava, P. (2019). Production, characterization and antibacterial activity of silver nanoparticles produced by *Fusarium oxysporum* and monitoring of protein-ligand interaction through in-silico approaches. *Microbial Pathogenesis, 129*, 136–145.

Sun, Y. (2002). Shape-controlled synthesis of gold and silver nanoparticles. *Science, 298*, 2176–2179.

Thakkar, K. N., Mhatre, S. S., & Parikh, R. Y. (2010). Biological synthesis of metallic nanoparticles. *Nanomedicine: Nanotechnology, Biology and Medicine, 6*, 257–262.

Tripathi, R., Gupta, R. K., Singh, P., Bhadwal, A. S., Shrivastav, A., Kumar, N., & Shrivastav, B. (2014). Ultra-sensitive detection of mercury (II) ions in water sample using gold nanoparticles synthesized by *Trichoderma harzianum* and their mechanistic approach. *Sensors and Actuators B: Chemical, 204*, 637–646.

Verma, V. C., Singh, S. K., Solanki, R., & Prakash, S. (2011). Biofabrication of anisotropic gold nanotriangles using extract of endophytic *Aspergillus clavatus* as a dual functional reductant and stabilizer. *Nanoscale Research Letters, 6*, 16.

Volesky, B., & Holan, Z. R. (1995). Biosorption of heavy metals. *Biotechnology Progress, 11*, 235–250.

Index

A. C. Shukla (ed.), *Applied Mycology*, Fungal Biology,
https://doi.org/10.1007/978-3-030-90649-8

T

U

V

W

GPSR Compliance
The European Union's (EU) General Product Safety Regulation (GPSR) is a set of rules that requires consumer products to be safe and our obligations to ensure this.

If you have any concerns about our products, you can contact us on

ProductSafety@springernature.com

In case Publisher is established outside the EU, the EU authorized representative is:

Springer Nature Customer Service Center GmbH
Europaplatz 3
69115 Heidelberg, Germany

www.ingramcontent.com/pod-product-compliance
Ingram Content Group UK Ltd.
Pitfield, Milton Keynes, MK11 3LW, UK
UKHW022216051026
15624UKWH00001B/193
9783030906481